UNIVERSITY OF DAYTON ROESCH LIBRARY

NOV 06 1980

D1716253

S. Yariv · H. Cross

Geochemistry of Colloid Systems

For Earth Scientists

With 86 Figures

DISCARDED
FROM
UNIVERSITY OF DAYTON
ROESCH LIBRARY

Springer-Verlag
Berlin Heidelberg New York 1979

UNIVERSITY OF DAYTON ROESCH LIBRARY

Dr. Shmuel Yariv
The Hebrew University of Jerusalem, Israel

Dr. Harold Cross
Israel Patent Office, Jerusalem, Israel

ISBN 3-540-08980-2 Springer-Verlag Berlin Heidelberg New York
ISBN 0-387-08989-2 Springer-Verlag New York Heidelberg Berlin

Library of Congress Cataloging in Publication Data. Yariv, Shmuel, 1934—. Geochemistry of colloid systems for earth scientists. Bibliography: p. Includes index. 1. Colloids. 2. Geochemistry. I. Cross, Harold, 1930—, joint author. II. Title. QE515.Y37. 551.9. 78-11407

This work is subject to copyright. All rights are reserved, whether the whole or part of the material is concerned, specifically those of translation, reprinting, re-use of illustrations, broadcasting, reproduction by photocopying machine or similar means, and storage in data banks. Under § 54 of the German Copyright Law, where copies are made for other than private use, a fee is payable to the publisher, the amount of the fee to be determined by agreement with the publisher.

© by Springer-Verlag Berlin Heidelberg 1979.
Printed in Germany.

The use of registered names, trademarks, etc. in this publication does not imply, even in the absence of a specific statement, that such names are exempt from the relevant protective laws and regulations and therefore free for general use.

Typesetting, offsetprinting and bookbinding: Brühlsche Universitätsdruckerei, Lahn-Gießen. 2132/3130-543210

QE
515
Y31

Preface

81 03123

Colloid science has been applied by soil chemists and clay mineralogists for many years, and some of the most important studies on the behavior of colloids have been contributed by them. Barring a few notable exceptions, only in the last decade have geochemists applied colloid science in their research and in this period much work has been published.

It seemed to the authors that it would be useful at this stage to attempt to summarize the progress made and to try to examine what colloid science has contributed and can further contribute to geochemistry.

This book is based partly on a course of the same title given to graduate students by one of the authors (S. Y.) between 1972 and 1977 at the Department of Geology at the Hebrew University of Jerusalem. Consequently many fundamental concepts of the subject are included that will be of use to graduate students in geology, geochemistry, soil science, and oceanography.

So that specialists interested in certain sections may find their subjects comprehensively covered, a few topics are dealt with in more than one chapter so that readers may ignore sections not especially of interest to them. However the chapters more fully treating certain topics are cross-referenced. In such cases the subjects are treated from different viewpoints and the citations used represent these differing viewpoints.

Since this branch of geochemistry is in its infancy, a few of the conclusions are still speculative and some of these are even highly speculative. In those cases where conclusions were regarded by us as speculative, allusion was made to this fact. However, in our five years of accumulating the literature it has been seen that conclusions once regarded as highly speculative have been shown to be solidly based, and it is probable that the conclusions now regarded as speculative will also in time be shown to be correct.

We would like to acknowledge the invaluable advice and encouragement of Professor V. G. Gabriel of New York given during the entire gestation period of the manuscript and for his careful review of the work in progress.

The various chapters were critically reviewed by experts in the fields involved, many of whom supplied us with unpublished data, pointed out further sources in the literature, and corrected errors and misconceptions. We are grateful to the following, in alphabetical order, for this generous and invaluable contribution: Miss. N. Agron, Geological Survey of Israel, Jerusalem; Z. Aizenshtat, Hebrew University of Jerusalem; F. Bartoli, Centre National de la Recherche Scientifique, Vandoeuvre, France; Mrs. L. Ben-Dor, Hebrew University of Jerusalem; A. Ben-Naim, Hebrew University of Jerusalem; Miss M. Bielsky, Hebrew University of Jerusalem; G. P. Briner, Division of Agricultural Chemistry at the Department of Agriculture, Victoria, Melbourne; A. E. Foscolos, Geological Survey of Canada, Calgary; Z. Garfunkel, Hebrew University of Jerusalem; Mrs. L. Heller-Kallai, Hebrew University of Jerusalem; W. H. Huang, Texas A & M University, College Station; W. D. Keller, University of Missouri, Columbia; B. Kohn, Ben-Gurion University of the Negev, Beer-Sheva; N. Lahav, Hebrew University, Rehovot; A. A. Levinson, The University of Calgary; B. Luz, Hebrew University of Jerusalem; E. Mendelovici, Instituto Venezolano de Investigaciones Cientificas, Caracas; U. Mingelgrin, Agricultural Research Organization – The Volcani Center, Bet-Dagan; B. D. Mitchell, The Macaulay Institute for Soil Research, Aberdeen; M. M. Mortland, Michigan State University, East Lansing; A. M. Posner, The University of Western Australia, Nedlands; I. Rozenson, Hebrew University of Jerusalem; Mrs. S. Saltzman, Agricultural Research Organization–The Volcani Center, Bet-Dagan; E. Sass, Hebrew University of Jerusalem; A. Singer, Hebrew University, Rehovot; B. Spiro, Hebrew University of Jerusalem; E. Vansant, University of Antwerp; D. H. Yaalon, Hebrew University of Jerusalem.

We are grateful to the authors and publishers who granted us permission to reproduce illustrations from their books, articles, and journals. These sources are noted in the legends of the figures.

Thanks are also due to Mr. Peter Grossman for the preparation of some of the illustrations, to Messrs. Shlomo Shoval, Eli Yariv, and Shalom Yariv for their help in preparing the typescript, to the librarians of the Department of Geology, Hebrew University, Miss Yona Metzger and Mrs. Zelda Kolodner for their help in collecting the literature and lastly to our wives for their forbearance during the long period of the book's preparation.

During the final preparation of the book, one of us (S. Y.) spent a sabbatical at the Instituto Venezolano de Investigaciones Cientificas (I. V. I. C.). He would like to acknowledge the contribution of discussions he had with colleagues from the Institute, and the secretarial help of Mrs. Barbara Kamal, Mrs. Paula Morales and Mrs. Monica Sanhueza.

October, 1978

S. Yariv
H. Cross

Contents

Introduction

1. What is Colloid Science?

Colloid systems may be defined as systems containing at least two components: 1) a continuous dispersing medium and 2) a disperse phase. For many years the science of colloids was concerned mainly with the description of the behavior of very small particles. The classical definition of a colloid system described the disperse phase as being comprised of particles or macromolecules smaller than 1000 nm in diameter, but larger than 1 nm. Particles smaller than 1 nm do not exist as a discrete phase, and any system containing them cannot be considered as heterogeneous. This definition is to some extent dependent on the resolving power of the optical microscope and on the sizes of the pores of conventional filter paper. Objects are seen by means of ordinary light of wavelengths between 400 and 750 nm. The optical microscope is limited in its resolving power by the wavelength of the light used for illumination. Consequently particles smaller than about a few hundred nm in diameter are submicroscopic. Colloid systems in which a liquid is the dispersing medium appear to form true solutions. These solutions pass unchanged through ordinary filter paper. The classical science of colloids dealt with particles that are so small that they behave in some respects like molecules and with molecules so large that they behave in some respects like particles. The term colloid was used for very finely divided matter or for very large molecules.

When a solid is broken into smaller and smaller parts, the total surface area increases as the particle size decreases. Natural laws governing the behavior of matter in the molecular state or in bulk also apply to the colloidal state. The two important phenomena that characterize the behavior of small particles as distinct from non-colloid material are 1) dispersion of the material and 2) chemical reactions at the surfaces of the colloids. Colloid chemistry deals with these two phenomena.

Colloid chemistry is also concerned with describing the methods of formation and stability of those substances that form colloid systems.

One should be aware of the fact that no system of classification can impose rigid limits on its subdivisions. The upper limit of the size of colloid material as given by the preceding classical definition is somewhat ambiguous. The stated size of 1000 nm is related to spherical shapes. However, when the envelope of the sphere is defective or the shape of the particle is not spherical, which leads to a ratio of total surface area to total mass, O/m, greater than that of a sphere having the same diameter, or if the chemical activity of the surface of the particle is very high, then particles with a diameter greater than 1000 nm can behave as colloids. Soil scientists generally regard a diameter of 2000 nm as the upper limit of particle size of colloids.

Colloid systems characteristically undergo fast reactions such as sorption, ion exchange, and dispersion. These and other similar reactions can be readily followed under normal experimental conditions in the laboratory. Similar reactions can occur with large particles, but because of the slow reaction rate, or their relatively small extent, they cannot be detected experimentally. In geologic systems and laboratory experiments, "time" and "dimensions of the reaction vessel" have different meanings. Many reactions, which in the laboratory are characteristic of colloid systems only, may in nature control the geochemical behavior both of particles having diameters much greater than 1000 nm and of solutions that are in contact with these large particles. In this treatment particles whose size greatly exceeds that of particles classically defined as "colloids" will also be considered.

2. Classification of Colloid Systems

2.1 Based on States of Matter

A colloid system requires the simultaneous presence of at least two phases. There are nine possible combinations of two phases. A mixture of gases, however, always forms one single gaseous phase, and therefore, one of the nine combinations does not provide a heterogeneous system and cannot be considered as a colloid disperse system. The various combinations are listed in Table 1.

Table 1. Types of heterogeneous dispersions in two-phase systems

Disperse Phase	Dispersing Medium	Names of Colloid System	Examples of Colloid Systems
1. Gas	Liquid	Foam	Bubbles of volatiles in magma; bubbles of air in seawater and rivers
2. Gas	Solid		Pumice, vesicular, charcoal; air in soil
3. Liquid	Gas	Aerosol	Mist, fog; glowing avalanche, ash cloud
4. Liquid	Liquid	Emulsion	Droplets of sulfidic melt in magma
5. Liquid	Solid		Water in soil and sediments, fluid inclusions in minerals
6. Solid	Gas	Aerosol	Volcanic smoke, glowing avalanche, dust
7. Solid	Liquid	Colloid solutions and suspensions	Mud; solid particles in seawater and rivers
8. Solid	Solid		Amygdaloid, colored glasses, precious stones

In some cases it is difficult to distinguish between the dispersing and the disperse phase. When one continuous phase is solid, the other phase may also be continuous, or both materials may be composed of very small crystals mixed together, as seen in many rocks. Highly porous solids, such as silica gel, exemplify colloids consisting of solid and gaseous phases, both of which are continuous. When the gas in the silica gel system is replaced by water the system consists of continuous solid and liquid phases.

2.2 Based on Chemical Properties of the Components

Comparison of the stability and particle structure of colloids shows that these substances may be divided into two groups, namely, lyophilic (stable) colloids and lyophobic (metastable) colloids. For simplicity, let us demonstrate the differences between these two types of colloids using liquid as the dispersing medium. Dispersions of lyophilic colloids are formed when macromolecules dissolve in suitable solvents or when dissolved small molecules are spontaneously associated to form micelles. The free energy decreases in the dissolution process and the disperse system formed is thermodynamically stable.

Lyophobic colloids have a characteristic property, a true interface with a defined surface tension σ, which exists between the disperse particles, drops, or bubbles and the dispersing medium. Every particle consists of many molecules or groups of atoms and exists as a discrete phase. As will be shown in the next chapter, the free energy of the system is dependent on the specific particle surface. The lyophobic colloids are characterized by having excess free energy. The ratio of total surface area to total mass should spontaneously decrease with a decrease in the free energy, unless an energy barrier can be generated between the particles, preventing them from approaching one another. Hence, solutions of lyophobic colloids are metastable.

When water is the dispersing phase, the colloids are called *hydrophilic* or *hydrophobic,* respectively. When an organic solvent is the dispersing medium, the colloids are called *organophilic* or *organophobic,* respectively. The same terms are also used to describe the wetting affinity of surfaces. Hydrophobic surfaces are those surfaces that remain unwetted when they are in contact with water. On the other hand, hydrophilic surfaces become wet when they are in contact with water. Most of the minerals, although belonging to the group of hydrophobic colloids, have hydrophilic surfaces and they sorb water. Only a few minerals have hydrophobic surfaces. These minerals include graphite, sulfur, talc, pyrophyllite, and sulfide minerals such as molybdenite and stibnite (Feurstenau and Healy, 1972).

Let us consider the changes in energy that take place during the dispersion of material A in a solvent B. In a dispersed system, surfaces of an A particle may either interact with the surfaces of other A particles to form A–A aggregates, or they may be solvated to form A–B interfaces. The solvent B–B must dissociate to single B molecules before it can be sorbed on the surfaces of the dispersed particle. ΔG_{AA} is the free energy of association of A particles. ΔG_{BB} is the free energy of association of B molecules (work of *cohesion*[1]). ΔG_{AB} is the free energy of sorption of molecules B onto the surfaces of A with the formation of an interface A–B (work of *adhesion*), all given per one mole of B. In a stable dispersed system, A will be a lyophilic colloid and

$$2\,\Delta G_{AB} < \Delta G_{AA} + \Delta G_{BB}. \tag{1}$$

1 The term adhesion refers to the attractive forces exerted between a solid or liquid surface and a second phase (either liquid or solid) whereas cohesion refers to the association forces of molecules or particles belonging to the same phase.

On the other hand in a metastable dispersion, A will be a lyophobic colloid and

$$2\,\Delta G_{AB} > \Delta G_{AA} + \Delta G_{BB}. \tag{2}$$

From these two equations it follows that the formation of a lyophobic or a lyophilic colloid depends on the ratio of the sum of the association energies of the two phases to the energy of the interaction occurring at the interface between the two components. Hence, this classification of colloids depends on the chemical properties of both components of the system. If A is polar, electrostatic interaction between the particles predominates and minus ΔG_{AA}[2] will be very high. It will not be high if A is a nonpolar material and van der Waals interaction between the particles predominates. Van der Waals interaction between particles increases with increasing contact area and increasing size of particles. The same is true for B and for minus ΔG_{BB}.

If both components are polar, solvation occurs via electrostatic attraction and minus ΔG_{AB} should also be high. In this case A can be a lyophilic colloid. This is the case found, for example, in the protein-water system. Proteins are amphoteric substances derived essentially from the combination of amino acids [R – CH(NH_2) – COOH] and characterized by the presence of the polypeptide chains made up of repetitions of the peptide unit:

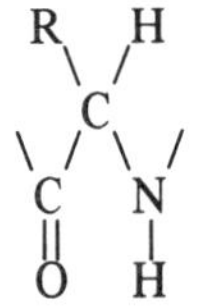

where R is a side chain. Proteins contain paraffinic, alcoholic, phenolic, SH, S–S, and/or S–CH_3 groups. The interaction of water is closely related to hydration of specific polar groups such as NH and CO. The electrostatic interaction of water with these polar groups leads to a high minus ΔG_{AB}. On the other hand minus ΔG_{AA} is lower than expected from polar interaction because of the branching of the protein chains.

The highly polarized oxides and hydroxides of highly charged cations, such as silicon, iron, and aluminum, when crystallized, will form lyophobic colloids in polar solvents because of their very high minus ΔG_{AA}. On the other hand, if the atomic structures of networks of the polymeric groups that form the crystals are not fully developed, the same compounds will form hydrophilic colloids. This will be further discussed in Chapter 3. Also clay minerals have very high minus ΔG_{AA}, thus giving rise to lyophobic colloids in polar solvents.

If both components A and B are nonpolar, then minus ΔG_{AB} will not be high, since A–B bonds are of the van der Waals type. However, because minus ΔG_{AA} and minus ΔG_{BB} are also low, it is likely that the disperse component will become a lyophylic colloid. This is the case for solutions of organic nonpolar macromolecules in organic solvents.

2 Since ΔG of cohesion and adhesion have negative values it is convenient to use minus ΔG values to show the sequence of stabilities of these reactions.

If one of the components A or B is nonpolar, bonds between A and B are weak and minus ΔG_{AB} is low. The probability of A becoming a lyophilic colloid is very low.

3. Crystal Chemistry of Silicates

The most common oxide of silicon is silicon dioxide, SiO_2. This combines with basic oxides to form silicates varying widely in composition. Silicon dioxide displays a marked tendency to form gels with water. Silicon dioxide, unlike carbon dioxide, seems never to exist as discrete SiO_2 molecules, but rather in a crystalline form based on a three-dimensional lattice of alternating silicon and oxygen atoms. Each Si atom is surrounded by four oxygen atoms in an almost regular tetrahedral arrangement. The bonds are mainly covalent in character, and the partial charges for silicon and oxygen are +0.45 and −0.23, respectively, indicating the low polarity of the bond.

The electronic configuration of a Si atom is designated by $1s^2\ 2s^2\ 2p^6\ 3s^2\ 3p^2$, where 1, 2, and 3 are the numbers of the electronic shells, s and p are the subshells, and the superscripts indicate the numbers of electrons in each subshell. The outer electrons of silicon are two in one s orbital and one in each of two p orbitals: $3s^2\ 3p_x^1\ 3p_y^1$. The third p orbital remains empty. The above configuration, as it stands, enables the formation of two single covalent bonds, which may be formed with other atoms having orbitals with single electrons. If "pure p" bonds, involving only the p orbitals of the Si atom, are formed, the bonds will be directed in space at an angle of 90° from one another. A higher stability is obtained if the p orbitals are hybridized with the s orbital, resulting in four orbitals directed to the corners of a regular tetrahedron, each orbital having a single electron. The hybridization increases the stability of silicon compounds for two reasons: 1) four covalent bonds are formed instead of two and 2) the distance between two adjacent atoms that are bonded to one Si atom increases, since the tetrahedral angle (109°28′) is greater than 90°. Such hybridization is designated as sp^3.

The electronic configuration of an O atom is designated by $1s^2\ 2s^2\ 2p_x^2\ 2p_y^1\ 2p_z^1$. An atom of oxygen possesses the requisites for the formation of two covalent bonds with other atoms. Here again, to achieve a higher stability, the p orbitals are hybridized to some extent with the s orbital. As a result, the two single bonds are not oriented at a right angle, but are closer to the regular tetrahedral angle.

The differences in the crystal chemistry of silicon and carbon are explained by the fact that the Si atom has empty $3d$ orbitals in the outer shell, which, after hybridization, can participate in the chemical bonds. When combined with fluorine, an ion $[SiF_6]^{-2}$ is obtained in which the Si atom has the coordination number 6. Also silicon–oxygen compounds formed in the mantle, under high pressure, such as stishovite, contain Si atoms with the coordination number 6. In these compounds electron pairs are partly donated to the Si atom by the F or O atoms. Such a hybridization is designated by sp^3d^2. The six orbitals thus obtained are directed to the corners of a regular octahedron.

The empty orbitals of a Si atom can also contribute to the stability of compounds in which this atom is tetrahedrally coordinated. In this case they can form π orbitals. Thus, paired electrons of O atoms can also contribute to the stability of the Si–O bond by taking part in the formation of this Π interaction. It should be noted that this

π interaction occurs to a very small extent only, and the character of the Si–O bond is predominantly that of a single bond.

The double bond character is revealed by comparing the experimental and theoretical Si↔O distances. The covalent radii of O and Si atoms are 0.73 and 1.11 Å, respectively. That means that if the bond is completely covalent, the distance Si ↔ O should be 1.84 Å. The ionic radii of O^{-2} and Si^{+4} are 1.40 and 0.41 Å, respectively. An ionic bond should result in a distance of 1.81 Å. X-ray diffraction indicated that the distance Si ↔ O is 1.55 Å in the quartz crystal and 1.54 Å in crystals of cristobalite and tridymite. This distance becomes shorter in those silicates in which the SiO complex acquires a negative charge. A decrease in this distance indicates an increase in the double bond character of the Si–O bond.

The electronic configuration of a C atom is designated by $1s^2\, 2s^2\, 2p_x^1\, 2p_y^1$. This atom has no empty d orbitals in its outer shell and thus the sp^3 type hybridization cannot be stabilized by the formation of π orbitals. There are two other possibilities leading to hybridization, namely sp and sp^2. The first type of hybridization leads to a coordination number 2, the two orbitals obtained lying on a straight line. The second type of hybridization leads to a coordination number 3, the three orbitals being directed to the apices of a regular triangle, with the angle between the bonds being 120°. All four electrons of the outer shell of the C atom participate in bond formation with oxygen in any of the coordination numbers, either in σ orbitals or in π orbitals. On the other hand, distances between neighboring O atoms are very short in the sp^3 type hybridization. Electrostatic repulsion between two neighboring O atoms results in species such as CO_2 and CO_3^{-2}. When carbon interacts with hydrogen no π orbitals can be formed, and 4 is the only coordination number that enables all four electrons of the outer shell to take part in the formation of the bonds.

Crystal phenomena in silicates are based on a common structural principle, namely, the formation of tetrahedra of O atoms surrounding each Si atom. If, for stoichiometric reasons, there are fewer than four O atoms for each Si atom this principle is still maintained through the possibility of O atoms being shared by neighboring tetrahedra. From the preceding discussion of the hybridization of the electrons of O atoms, it follows that the nonlinear orientation of two Si atoms to the common oxygen atoms is extremely important, significantly affecting the structure of the silicates, as will be shown. Silicates have O and Si atoms in a ratio higher than 2. The SiO networks are therefore to be expected to be negatively charged. The presence of ions in the silicate crystal is shown both by their electric conductivity and by ionic migration. The SiO networks can be regarded as anions. The metallic cations involved in the structure of the crystal are situated between the negatively charged SiO networks in such a manner that they lie as far apart as possible from each other, and, as is general in ionic structures, have a maximum number of negative ions around them. Because the Si–O bond is polarized and the O atom is therefore negative, the metal cation is coordinated by the O atom and not by the Si atom.

Another feature of the silicates is the common substitution of Al atoms for Si atoms. The ease of this substitution is due to the fact that the covalent radii of both elements are similar. They are 1.18 and 1.11 Å for Al and Si respectively. Such a substitution is associated with a change in the charge of the Si–O complex anion. Since the Al atom has three electrons in its outer shell, any tetrahedron that has an Al atom at its center must gain an additional electron from some source in order to fill all four σ orbitals, which are derived from the sp^3 type of hybridization. The substitution of Al for Si increases the negative charge of the SiO complex by one negative unit per one Al atom.

Atoms larger than Si, such as Mg and Fe, occupy a coordination number six in oxygen compounds. The O atoms are directed to the apices of a regular octahedron. In the silicates such an octahedron can be combined with a tetrahedron by sharing an O atom. The Al atom in silicates occupies coordination numbers four and six.

According to their fine structure, silicates are divided into those with island structures (neso-, soro-, and cyclosilicates), with chain or band structures (inosilicates), with sheet structures (phyllosilicates), and with three-dimensional network structures (tectosilicates).

3.1 Island Structure
(Fig. 1)

This term is applied when *island* groups, made up of a finite number of Si and O atoms, are present. Typical of this is the mineral olivine, a nesosilicate, $(Mg, Fe)_2SiO_4$, which contains both magnesium and iron in varying proportions and is intermediate between

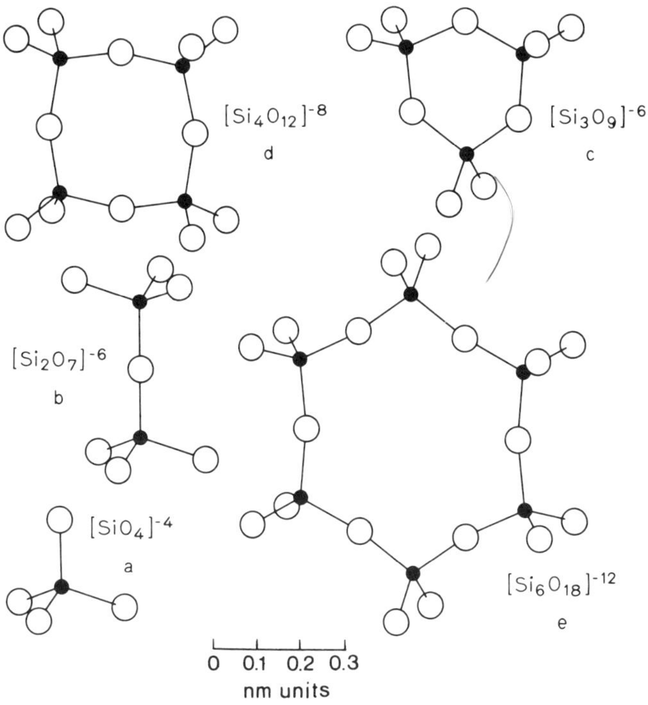

Fig. 1a–e. Types of linkage of silicon–oxygen tetrahedra, in island structures, **a** independent tetrahedra. **b** double tetrahedra. **c–e** ring structures. (From Mason and Berry, W. H. Freeman and Company. Copyright 1968)

the comparatively rare forsterite Mg_2SiO_4 and fayalite Fe_2SiO_4. X-ray study of the atomic structure shows the SiO_4 groups are independent of each other.

3.2 Chain and Ribbon Structures
(Fig. 2)

The SiO_4 tetrahedra can be linked together to form chains by the sharing of one O atom between two adjacent tetrahedra. In each tetrahedron two O atoms are shared by the tetrahedra on either side. Since the chain is of indefinite length, each Si atom is associated stoichiometrically with $2 + (2 \times 1/2) = 3$ O atoms. The polymeric anion is therefore designated as SiO_3^{-2} (or sometimes $Si_2O_6^{-4}$). These chains lie parallel to the vertical crystal axis and are bound together laterally by metallic cations. This structure is shown in minerals of the pyroxene group. Examples of the last group are enstatite, $MgSiO_3$, diopside, $(Ca,Mg)SiO_3$, and spodumene, $LiAl(SiO_3)_2$.

In building a "ribbon" from two chains, one additional O atom is eliminated from each group of four tetrahedra. Thus a ribbon $[Si_4O_{11}]^{-6}$ of indefinite length is formed. Here again, the ribbons lie parallel to each other and are bound together by metallic cations. This structure is shown in minerals of the amphibole group. In addition to the

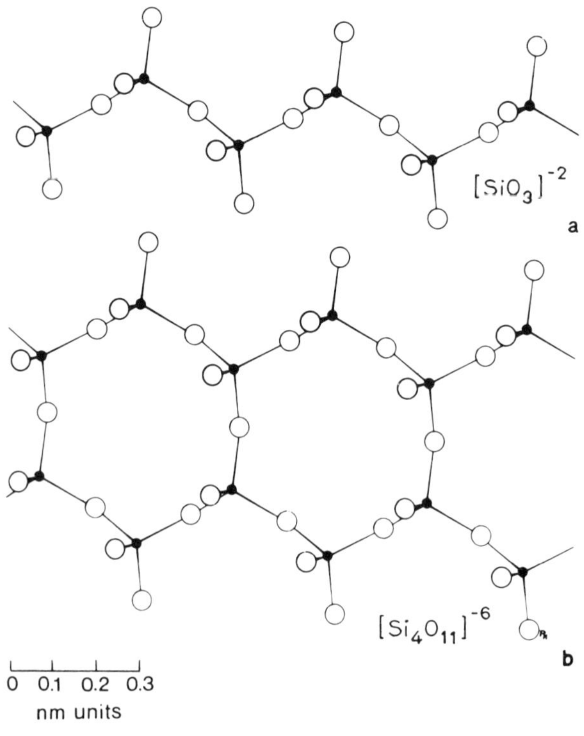

Fig. 2a and b. Types of linkage of silicon–oxygen tetrahedra, in chain and ribbon structures. a single chains, b double chains. (From Mason and Berry, W. H. Freeman and Company. Copyright 1968)

ribbon anion, other negative ions of small size may participate in the structure of such lattices, e.g., F^-, OH^-, and O^{-2}. Tremolite, $Ca_2Mg_5[Si_8O_{22}](OH)_2$, riebeckite, $Na_2Fe_3''Fe_2'''[Si_8O_{22}](OH)_2$, and hornblende, $NaCa_2(Mg,Fe,Al)_5[(Si,Al)_8O_{22}](OH)_2$ are examples of this group.

3.3 Layer Structure
(Fig. 3)

The continuous broadening of the $Si_4O_{11}^{-6}$ ribbon produces a planar network or sheet. In such a sheet, the silica tetrahedral groups are arranged to form a hexagonal network, which is repeated indefinitely to form a layer of composition $[Si_4O_{10}]^{-4}$. The tetrahedra are arranged so that all their apices point in the same direction, and all their bases are in the same plane. Such a sheet can be considered to be composed of three planes: O atoms, Si atoms, and again O atoms. It is common to call this sheet "the tetrahedral sheet" and to designate it by T.

Micas and most clay minerals are examples of minerals having a layer structure. Each layer here is composed of two types of sheets. One type is the tetrahedral sheet and the other type is the octahedral sheet, designated by the letter O. This sheet is composed of three planes, two of which are of closely packed O atoms or OH groups, and between them a plane of Al, Fe, or Mg atoms. Each of these atoms is coordinated

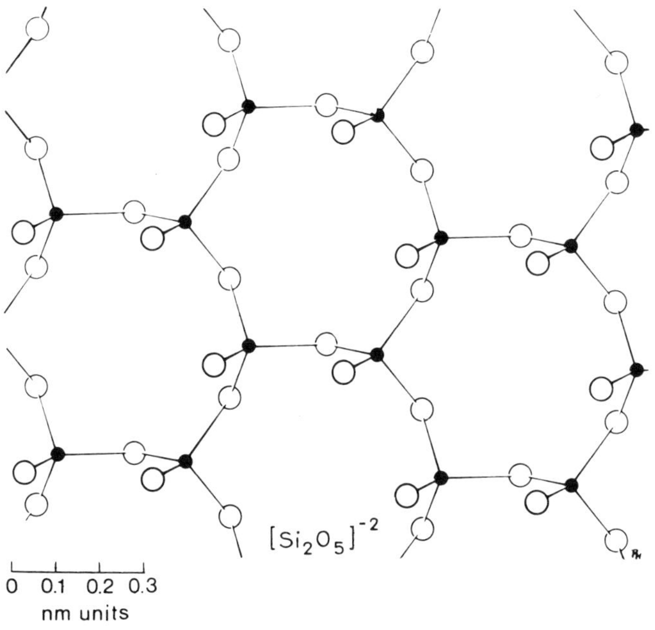

Fig. 3. Types of linkage of silicon–oxygen tetrahedra: sheet structure. (From Mason and Berry, W. H. Freeman and Company. Copyright 1968)

by six O atoms. Such a sheet is obtained by the condensation of single octahedra. Each O atom is shared by 3 octahedra, but two octahedra can share only two neighboring O atoms. The minerals brucite $Mg(OH)_2$ and gibbsite $Al(OH)_3$ have such a sheet structure. The charge of the sheet is balanced by protons, and the two outer planes are actually made up of hydroxyls and not of oxygens. All the octahedra are filled with Mg atoms in brucite, but only 2/3 of the octahedra are filled with Al atoms in gibbsite. Derivatives of brucite and gibbsite are called tri- and di-octahedral minerals respectively. Serpentine and talc represent trioctahedral minerals whereas kaolinite and pyrophyllite (Fig. 4) are dioctahedral.

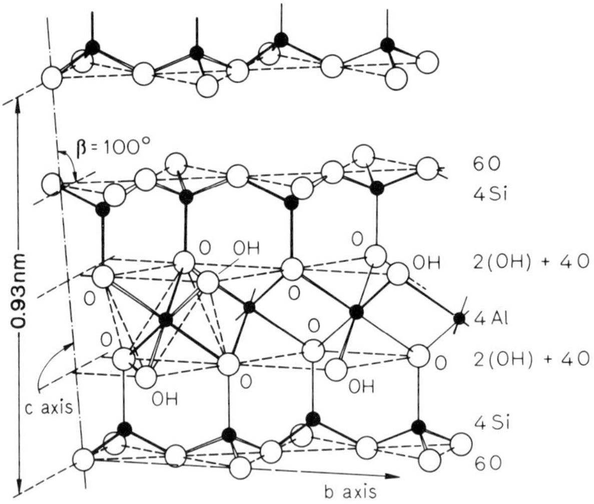

Fig. 4. Structure of pyrophyllite, $Al_4(Si_4O_{10})_2(OH)_4$. Octahedral $Al_4O_8(OH)_8$ linked to two tetrahedral Si_4O_{10} sheets. (From Mason and Berry, W. H. Freeman and Company. Copyright 1968)

Serpentine is composed of a single tetrahedral sheet condensed with a single magnesium octahedral sheet into one unit layer designated by 1 : 1 or by TO. In this unit layer the apices of the silica tetrahedra and one of the OH planes together form a single plane that becomes common to both the octahedral and the tetrahedral sheets. Two-thirds of the O atoms in this common plane are shared by the Si and Mg atoms and the remainder are shared by Mg and H atoms. Such units are stacked one above the other in the *c* direction, the forces between the units being mainly of the van der Waals type. The charges within the structural unit are balanced. The structural formula is $Mg_6Si_4O_{10}(OH)_8$. In chrysotile, which is one of the serpentine minerals, the layers are curved into a cylindrical shape.

Talc is composed of layers made up of two tetrahedral sheets with a central Mg octahedral sheet. This unit is designated by 2 : 1 or by TOT. The layers are combined so that the apices of the tetrahedra of each Si sheet and one of the hydroxyl planes of the octahedral sheet form a plane common to both sheets. Two common planes are obtained on both sides of the octahedral sheet. These common planes are the same as the plane obtained for serpentine. Two-thirds of the O atoms in these common planes are shared by the Si and Mg atoms and the remainder are shared by Mg and H atoms. The layers are stacked one above the other in the *c* direction, the forces between them being mainly of the van der Waals type. There is a perfect cleavage between the layers. The structural formula is $Mg_6Si_8O_{20}(OH)_4$.

The basic structural unit of biotite mica is similar to that of talc, except that approximately one-fourth of the Si atoms are replaced by Al atoms so that the charge deficiency is about 2 per unit cell. The resultant charge deficiency is balanced by potassium ions that are situated between unit layers. The mineral can be considered as a *salt-like* compound consisting of macroanions continuous in the *a* and *b* directions and of K cations. A crystal is built having a high electrostatic attraction between cations and anions. The structural formula of the biotite mica is $K_2(Mg,Fe)_6(Si_6Al_2)O_{20}(OH)_4$ and that of phlogopite is $K_2Mg_6(Si_6Al_2)O_{20}(OH)_4$.

Muscovite mica is dioctahedral. Only two-thirds of the possible octahedral positions are filled, and the octahedral layer contains Al atoms. As in biotite, approximately one-fourth of the Si atoms in the tetrahedral sheet are replaced by Al atoms and the charge deficiency is balanced by K ions located between unit layers. The structural formula is $K_2Al_4(Si_6Al_2)O_{20}(OH)_4$ (Fig. 5).

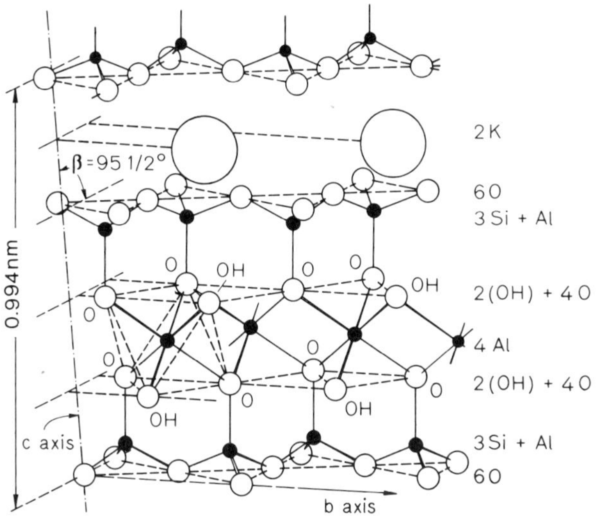

Fig. 5. Structure of muscovite. Pyrophyllite layers with one Al substituted for one out of four Si in each tetrahedral layer, linked toghether by K atoms in twelvefold coordination with O. (From Mason and Berry, W. H. Freeman and Company. Copyright 1968)

3.4 Three-dimensional Network
(Fig. 6)

If each SiO_4 tetrahedron shares all four of its O atoms with neighboring tetrahedra, a three-dimensional skeleton is built up. Since the structure is of indefinite size, each Si atom is associated with $4 \times 1/2 = 2$ O atoms. The resulting compound is silicon dioxide, SiO_2.

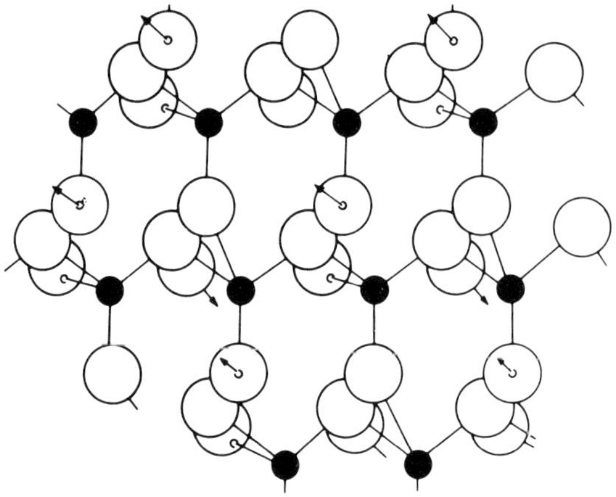

Fig. 6. Types of linkages of silicon–oxygen tetrahedra in quartz, a three-dimensional network

Si atoms can be replaced by Al atoms in some of the tetrahedra. Positive ions must be incorporated simultaneously in the structure to compensate for the resulting charge. Typical minerals are those of the feldspar group, orthoclase, $KAlSi_3O_8$, and the plagioclase feldspar series, albite, $NaAlSi_3O_8$, and anorthite, $CaAl_2Si_2O_8$, which are completely miscible and together form an isomorphous series ranging from pure soda feldspar at one end to pure lime feldspar at the other end.

4. Water

Water is an important constituent of geologic colloid systems. It is also all-important in the structure of animal and vegetable organisms. It is found in the lithosphere, hydrosphere, biosphere, and atmosphere and plays an especially important role in geologic processes. It occurs either as a dispersed phase or as a dispersing medium. A few important examples are listed below.

1. The greater part of the earth's surface is covered by an aqueous blanket called the hydrosphere. The oceans form by far the greatest and most important part of the hydrosphere. These bodies, as well as rivers, lakes, and inland seas, contain hydrophilic and hydrophobic suspended materials. In these colloid systems liquid water acts as a dispersing medium. The amount of water in oceans and rivers is 1.41×10^{18} and 5.1×10^{14} tons, respectively.

2. Glaciers, icebergs, and snow are solid–solid dispersion systems in which solid H_2O serves as a dispersing medium. They acquire rock debris of various sizes as a disperse phase. The average moraine content of antarctic glaciers is about 1.6% by volume (Evteev, 1964). It is calculated that 10.1% of the total land area and about 25% of the ocean are now covered by ice and snow (Rankama and Sahama, 1950).

3. Water vapor is an important component of the dispersing gaseous phase in pyroclastic flows occurring during volcanic eruptions.

4. Magmas contain water. Depending on composition, temperature, and pressure, water can be dissolved in the magma but can also be dispersed in the form of gas bubbles or salt solution droplets in the magma melt.

5. Liquid water is dispersed in sediments and soils. The amount of water in sediments is approximately 3×10^{17} tons (Wedepohl, 1971). *Ground water* fills pores, cracks, and larger openings in the soils and rocks. Water in the capillaries of rocks is known as *hygroscopic* and *capillary water.* Water can also be chemically sorbed by various minerals or organic materials dispersed in soils and sediments.

6. It occurs as water vapor and also as dispersed water droplets and ice particles in the atmosphere. Snow and rain droplets contain many dispersed solids. These solid particles can enter the snow crystals or the rain droplets in the following ways: (a) as a nucleation seed of the primary ice crystals; (b) as the condensation seed of the cloud droplets; (c) by attachment of solid particles from the atmosphere; and (d) by precipitation of insoluble salts from interaction of substances dissolved in raindrops (Ishizaka, 1972, 1973).

The heat of formation of the O–H radical is 99.4 kcal/mol whereas the heat of formation of H_2O is 219 kcal/mol and differs by only about 10% from twice the heat of formation of a single O–H radical (Coulson, 1961).

4.1 Fine Structure of Water

In the H_2O molecule two covalent single bonds are formed between one oxygen and two hydrogen atoms. The two hydrogen nuclei and the midpoint of the oxygen atom together form an angle of 104.5°, indicating that the electron orbitals of the oxygen are hybridized.

The physical and chemical properties of H_2O are governed by the following two characteristics.

1. Being nonlinear, the H_2O molecule is strongly bipolar in character. Polar molecules are those in which the centers of gravity of the positive and negative electric charges do not coincide. The oxygen is more electronegative than the hydrogens and attracts the electrons to a greater extent than do the hydrogens. The negative charge is located in the vicinity of the oxygen nucleus while the positive charge is centered midway between the two hydrogen nuclei. This is illustrated in Figure 7, in which the water molecule is considered as a rigid tetrahedron. The oxygen atom is located at the center of the tetrahedron, and four partial charges are located at the apices, two of which are positive partial charges and have the mass of a hydrogen atom and two of which are negative partial charges and represent the two electron pairs on the oxygen atom. The partial charge magnitude is 0.912×10^{-10} esu. This gives a definite dipole moment to the water molecule.

2. Having a partial positive charge, the hydrogen atom of a single water molecule has a very small electronic sphere, permitting very close approach of its nucleus to the electrons of an atom of another molecule. If the other atom bears a partial negative charge, these electrons may then be attracted by the hydrogen nucleus with sufficient energy to form a weak bond. The attraction is predominantly due to electrostatic forces acting between centers of opposite charges. There is a possibility of transitory

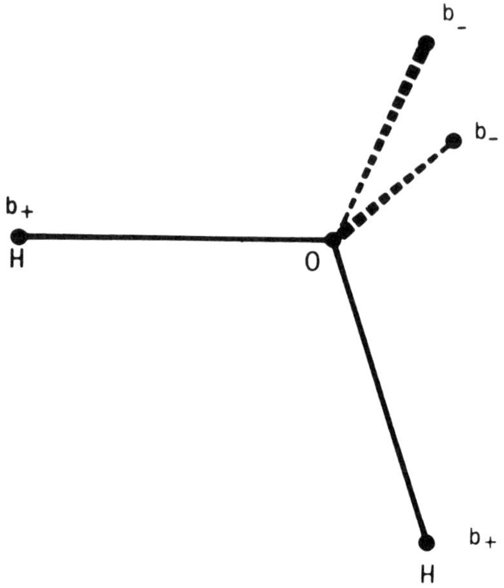

Fig. 7. Schematic representation of the water molecule. The molecule is a tetrahedron with the oxygen at the center. The two hydrogens and the two negative partial charges are located 0.1 nm from the oxygen. (From Briant and Burton, 1975; used with permission of the American Institute of Physics and of the authors)

occupancy of the hydrogen orbitals by a second pair of electrons donated by the newly bound, negatively charged atom, thus contributing a weak covalent character to the bond. This type of bonding, where a *hydrogen bridge* is obtained, is termed a *hydrogen bond.* In typical hydrogen bridges, the hydrogen is not equidistant between the two atoms but closer to the one to which it is covalently attached. The bond to the more distant atom is usually of about 5 to 7 kcal/mol. The three nuclei that take part in the hydrogen bridge formation all lie on a straight line. It is unlikely that hydrogen bond bending occurs to any considerable extent.

During the formation of hydrogen bonds water can act both as a proton donor and a proton acceptor. Consequently one water molecule may be involved in 1 to 4 hydrogen bridges.

The association of two water molecules results in the formation of a dimer, and two different orientations of the two water molecules in a dimer can be obtained (Fig. 8). According to Briant and Burton (1975) those with large energies appear to have the linear arrangement seen in Figure 8a, characteristic of the formation of a hydrogen bond. A high energy bond is not necessarily linear, but of the dipole–dipole interaction type seen in Figure 8b. This is a less favorable configuration.

Ordinary ice has the structure of tridymite, in which each oxygen atom is surrounded tetrahedrally by four hydrogen atoms and each hydrogen by two oxygen atoms. An eightfold aggregate can be regarded as a unit of this crystal and is called "ice molecule". From Figure 9 it is obvious that ice is very open in structure.

Water is an atypical fluid. For example, the molar volume of liquid water is smaller than that of solid common ice at 0 °C. Further, the liquid contracts on warming up to 4 °C and then expands. The high melting and boiling points of water have long been taken as evidence for the importance of intermolecular hydrogen bondings. From this basic premise, many theories on the fine structure of liquid water have been evolved (Ben-Naim, 1974).

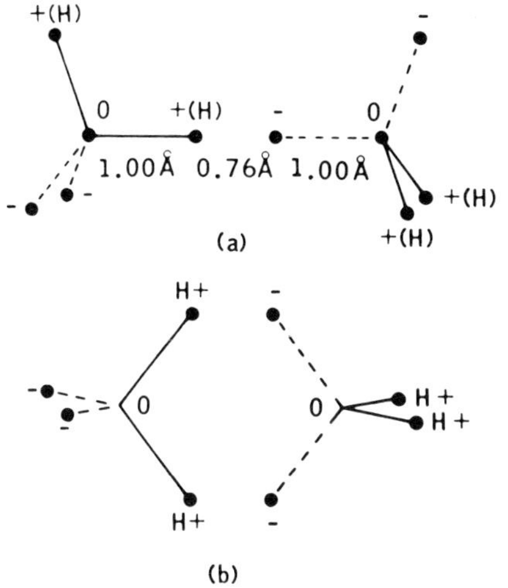

Fig. 8a and b. Two different orientations of two water molecules: **a** two molecules at the minimum potential energy configuration, and **b** a less favorable configuration. (From Briant and Burton, 1975; data taken from Rahman and Stillinger, 1971; used with permission of the publishers and of the authors)

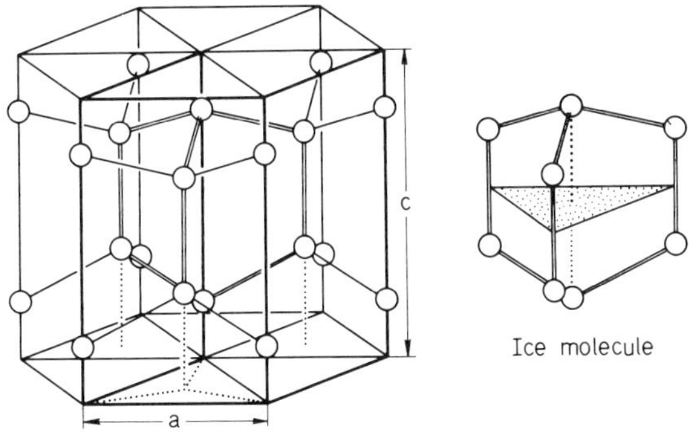

Fig. 9. Crystal lattice of ice and structure of an $(H_2O)_8$ molecule (*ice molecule*), a = 0.45 nm, c = 0.73 nm. (From Remy, 1956; used with permission of the publisher)

Water molecules exist in different molecular environments in the liquid phase. For example, there may be water molecules involved in 4, 3, 2, 1, or 0 hydrogen bridges. Hydrogen bonding between water molecules is a cooperative process. Association of two water molecules by hydrogen bonding promotes association of these molecules with other water molecules, resulting in a cluster of water molecules having an open structure similar to that of the ice molecule. At room temperature an average cluster contains 40 molecules. The lifetime of a cluster of water molecules is ca. 10^{-11} s. In water at low temperatures there are low density networks of hydrogen-bonded water molecules. With increase in temperature, the extent of intermolecular hydrogen bonding decreases and the cluster structure breaks down. This reaction is reversible and with decrease in temperature the extent of intermolecular hydrogen bondings increases. Heating and cooling result in "structure breaking" and "structure making", respectively.

Water vapor is an ideal gas at a pressure less than about 90% of saturation at about room temperature. Water vapor deviates considerably from ideality, and aggregates exist in significant concentrations even at temperatures above 100 °C, near saturation.

The distance between water molecules in these aggregates is comparable to the intermolecular distance of water molecules in ice. Measurements of irradiated water vapor have led to the determination of relative concentrations of the hydrates $H^+(H_2O)_n$ and thermodynamic constants for the reactions $H_2O + H^+(H_2O)_{n-1} = H^+(H_2O)_n$, in which $1 \leqslant n \leqslant 8$ (Christian et al., 1970).

4.2 Fine Structure of Aqueous Inorganic Salt Solutions

The structure of water can be affected by adding a solute (Blandamer, 1970). The term hydration can apply either to the interaction between an ion and the proximate water molecules (which is the primary hydration shell, also known as the *self-atmosphere* of the ion) or to the interaction between an ion and the solvent over the range $0 < x < \infty$. The structure of water is affected at some distance from an ion, beyond the nearest neighbour water molecules. The solvent in aqueous solutions cannot be treated as a bulk continuum. A simple model for a salt solution assumes that the ions are randomly distributed. There is a close link between the properties of salt solutions and corresponding solid salts. In very dilute solutions the ions are separated by many water molecules, and the properties of these solutions are influenced by the structure of water. The interaction between an ion and water is usually strong. For simple inorganic ions in aqueous solution it is common to use a model that shows three zones of water structure around each ion (Fig. 10). Zone A contains all those water molecules for which an electrostatic ion–water dipole interaction dominates and is known as the zone of *electrostricted* water molecules. Zone C extends to infinity and includes those water molecules that have essentially the same arrangement as in pure water. The water between the two spheres, zone B, is subjected to the competing demands of water structures associated with zone C and with zone A. These two influences counteract each other and the water structure in this fault zone is broken down. Zone B increases with increase in ion size, and consequently the extent of water structure breaking increases. Conversely with decrease in ion size, zone B contracts and may disappear, as with lithium and fluoride. In these systems the hydrated ions are

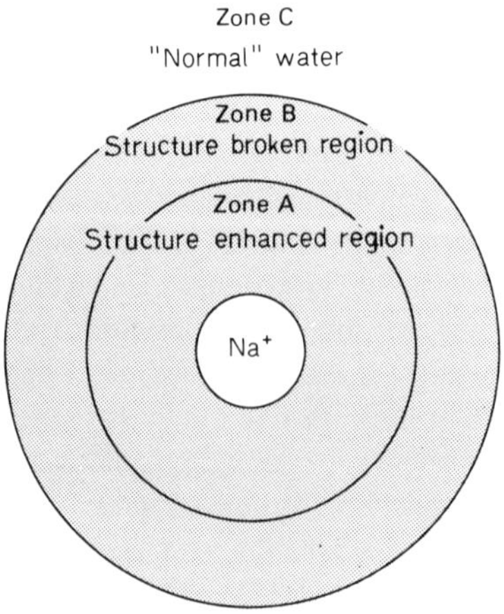

Fig. 10. Multizone hydration atmosphere of ions in aqueous solution. (From Horne and Courant, 1973)

accommodated within the structure of water. Furthermore, as a result of the high polarizing power of the small ion on the water molecules, the number of the intermolecular hydrogen bondings increase. Such ions are *structure makers.*

The dipolar axis of each solvent molecule in zone A passes through the ion center. Zone A decreases with increasing ionic size and increases with increasing ionic charge. Zone A does not exist around iodide so that zone B extends from the surface of the ion and the extent of intermolecular hydrogen bondings of water adjacent to the ion is reduced. Such ions are *structure breakers.*

The concept of structure breaking by ions has also been used in relation to *negative hydration.* This approach to ionic solvation takes into consideration the molecular mobility of solvent molecules in the vicinity of the ion. If, in comparison to those in pure water, the mobility of these solvents molecules is more rapid, the ion is negatively hydrated whereas, if less rapid, the ion is positively hydrated.

4.3 Fine Structure of Aqueous Solutions of Small Nonpolar Molecules and Radicals

The presence of nonpolar solutes, such as nonpolar gases and small hydrocarbon molecules, gives rise to an increase in the degree of structuring of water. This enhancement of water–water interactions results in a polyhedron made up of water molecules, which has the structure of a clathrate hydrate containing a nonpolar "guest". These structured water regions are often denoted as "icebergs".

Addition of small amounts of alcohols to water also enhances water–water interactions. Quaternary ammonium salts probably enhance water structuring for reasons similar to those described for nonpolar solutes. Van der Waals interactions between the water network and the organic chain stabilize those water molecules involved in a hydrogen-bonded lattice-type structure, such water structures being able to form sufficiently large cavities to accommodate the alkyl chain. The preceding model, derived for alkylammonium ions, has been successfully extended to account for properties of aqueous solutions of various organic quaternary salts not having functional groups for specific localized ion–solvent interaction. A water structure of this character is called a *hydrophobic hydration* structure.

4.4 Liquid Boundary Layer

When water is in contact with a nonmiscible phase, the structure of the liquid boundary layer depends on the polarity of the phase. When the phase is polar a hydrophilic hydration structure is obtained, and the liquid is exothermally adsorbed by the phase; on the other hand, the water is not adsorbed by a nonpolar phase and a hydrophobic hydration structure is obtained. The hydrophilic hydration structure and the properties of the boundary layer result from the mutual interaction of both phases, whereas the structure of the hydrophobic hydration zone results from the thermal motion of molecules in the water phase and from the attraction forces between surface water molecules and the bulk water phase. For both cases it is common to use a model that specifies three zones of water structure in every system. Zone E is the *vicinal water boundary layer.* Here the molecular mobility is the lowest. Zone D is a *transition zone* and zone C is the *bulk water solution.* Analogous to the broken structure region surrounding ions, D is a disordered zone that separates the ordered vicinal water boundary layer of zone E and "normal" bulk water at some greater distance from the surface.

4.4.1 Hydrophilic Hydration Boundary Layer

The water molecules at the surface are preferentially oriented with the dipolar axis perpendicular to the surface. Whether the negative or positive pole of the water molecule points outwardly depends on the net surface charge of the polar phase. In some cases, where the net surface charge of the polar phase is negative and the distribution of the charge in the surface is such that a very high charge is localized at a few sites, the hydration occurs through one hydrogen of water forming a hydrogen bond with specific atoms at the boundary of the polar phase in such a way that the second hydrogen can still form a hydrogen bond with another water molecule outside the primary hydration layer. Hydrogen bonds, with water acting as the proton acceptor, may be formed if the polar-phase boundary is populated by protons.

Water-soluble ions having a charge opposite to the net charge of the polar phase (named *counterions*), can penetrate into zones D and E due to electrostatic attraction. Ions having a charge of the same sign as the polar phase (*co-ions*) also penetrate into these zones due to their thermal motion, but to a much smaller extent, and an electric *double layer* is obtained (see Ch. 2). As a result of the enhancement of water structuring of zone E, ions in the immediate vicinity of the interface are adsorbed without their hydration shells, as "bare" ions. Since the primary hydration layer (zone A) is more stable in cations than in anions, the latter are preferentially adsorbed. Fully hydrated ions can be found in zone D. In the bulk of the electrolyte solution the self atmosphere is symmetric about the given ion center, but in zone D the degree of symmetry is reduced. The thickness of zones D and E depends chiefly on the surface charge density and on the concentration of soluble salts in the water medium. This will be further discussed in later chapters.

4.4.2 Hydrophobic Hydration Boundary Layer

When in contact with nonpolar phases the water molecules at the surface are oriented such that the negative oxygens point out from, and the positive hydrogens into, the bulk solution. The boundary layer thus formed, which has an enhanced structuring of water, extends only a few molecules in depth (Fig. 11). The exact structure of zone E is not known. It is unlike any of the structures previously mentioned. There is some evidence that water of zone E is denser than water of zone C, at least near solid surfaces. Transport processes such as diffusion of molecules through this layer, viscous flow, and electric conductivity may be slower (Ben-Naim, 1974).

In aqueous solutions of inorganic salts the concentration of the ionic species in the boundary is significantly different from that in the bulk phase and is much smaller. This is because many of the requirements of ionic hydration cannot be fulfilled, and the hydration shell of electrostricted water molecules cannot be formed when the ion is situated in zone E. When the hydrated ion is situated in zone D, the symmetry of its hydration atmosphere is perturbed. However, as a result of a lower hydration energy, the difference in concentration between negative and positive monovalent ions

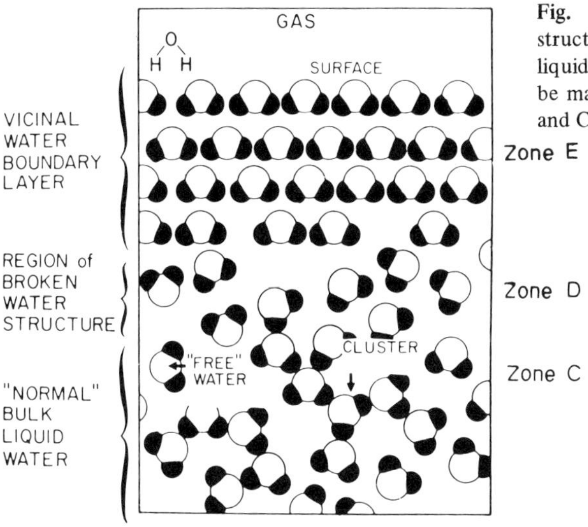

Fig. 11. Schematic diagram of the structure of pure liquid water at the air/liquid interface (the boundary layer may be many molecules thick). (From Horne and Courant, 1970)

increases in the direction of the liquid boundary layer and consequently a negative electric potential at the liquid surface is developed. It is obvious that the absolute value of the negative electric potential increases with increasing size or decreasing charge of anion and decreasing size or increasing charge of cation.

A negative surface potential is developed when the liquid water is in contact with nonpolar phases, such as the nonpolar atmospheric gases oxygen and nitrogen or aliphatic hydrocarbons. In a similar manner the proximate hydration atmosphere of a nonpolar solute in aqueous solution becomes negatively charged.

The orientation of the water molecules at the liquid boundary with the positive pole toward the bulk solution is due to the net negative charge gained by zone D as a result of the selective *desorption* of highly hydrated cations.

4.5 Aqueous Solutions of Hydrophilic Colloids

Comparison of the stability and particle structure of hydrophilic colloids shows that these substances may be divided into two groups: (1) macromolecular solutions and (2) association or micelle colloids.

4.5.1 Macromolecular Solutions

Substances classed as macromolecules have molecular weights ranging from about 10,000 up to several million. Each molecule is in itself a colloid particle and is sometimes called a *high polymer.* The natural high polymers include all the common organic structural materials such as wood, cotton, silk, horn, hair, wool, proteins and starch, lignin, and humic substances. Many of these high polymers are composed of one or more types of small molecules, called *monomers.* The simplest type of high polymer is made by the repeated addition of the same monomer unit to a parent molecule. The solubility mechanism of protein in water has been discussed previously. Water, because of its ability to form hydrogen bonds, is a good solvent for substances having many groups that can form hydrogen bonds. Examples of these groups in order of decreasing affinity for water are $-COOH$, $-CONH_2$, $-OH$, $-NH_2$, $=NH$, $\equiv N$, and R–O–R. The natural macromolecules mentioned above contain such polar groups. During the dissolution of such molecules a hydrophilic hydration boundary layer is formed.

Inorganic polymeric ions also occur in nature, and under certain conditions they may form hydrophilic colloid solutions (see Ch. 3, 6).

4.5.2 Association Colloids and Micelles Formation (Fig. 12)

Micelle formation is an extreme example of "hydrophobic bonding", which is of foremost importance in organic geochemistry. Micelles, which are of colloid dimensions, are obtained by the reversible association of a large number of amphipathic molecules, 20 to 100 in many systems, under the effect of the solvent. Amphipathic monomers are compounds of low molecular weights, 100 to 500, which have a peculiar structural feature, namely, that they contain a nonpolar hydrophobic part, usually a straight or branched hydrocarbon chain, which is attached to a polar hydrophilic group, which can be nonionic, ionic, or zwitterionic. The properties of dilute solutions of amphipathic molecules differ from those of inorganic salts or from those of smaller or nonamphipathic molecules. Water in the vicinity of the noncharged moiety of the amphipathic molecule will have the "hydrophobic hydration" structure whereas the polar group will be hydrophilically hydrated, either through electrostatic interactions or through hydrogen bonding. At higher concentrations, the reversible association of the

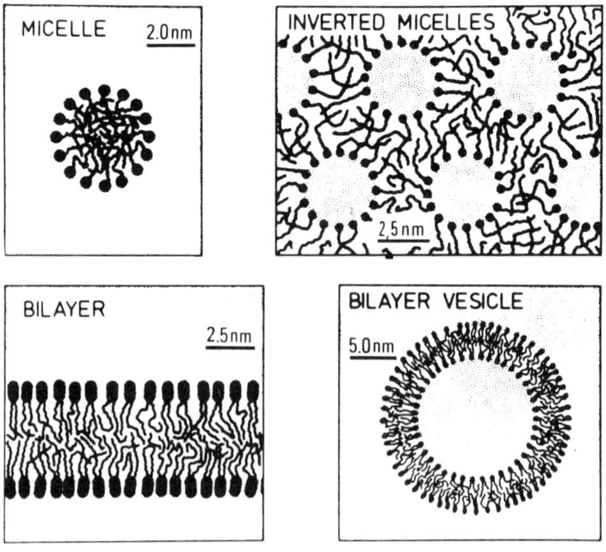

Fig. 12. Micelle structures. (From Israelachvili and Ninham, 1977. Used with permission of Academic Press, Inc. and of the authors)

monomers and the formation of the micelles occur. The concentration at which this sharp change occurs is the *critical micelle concentration* (c.m.c.). The characteristic feature that distinguishes association colloids from other hydrophilic colloids is that the colloidal units of the former, the micelles, are in association–dissociation equilibrium with the monomers. The c.m.c. depends on the size of the hydrophobic chain and becomes very low with increasing chain length. For high chain lengths, the c.m.c may be as low as 10^{-6} mol/l for nonionic association colloids and 10^{-4}–10^{-3} for ionic ones. The c.m.c. decreases with added salt concentration, since more water molecules become electrostricted, and for many systems the data show a linear dependence of log c.m.c. on the salt concentration.

During the formation of the micelle a hydrophilic hydration boundary layer is obtained and the exposure of the hydrocarbon surface to water is minimal. In many systems the micelle appears to be a compact spheroid particle, 1.2–3.0 nm in radius, in which the hydrophilic groups are at the surface. In the case of ionic compounds, some of the counterions are fully dissociated and move from the micelle surface toward the bulk solution, whereas the rest of the counterions remain attached to the surface. The hydrocarbon core of the micelle is liquid-like, resulting in the ability to dissolve water-insoluble and oil-soluble materials in micelles. Their heat capacities and compressibilities are similar to those of liquid hydrocarbons. The primary driving force for micelle formation is the tendency of water in the vicinity of the hydrophobic moiety of the monomer to associate with itself rather than to remain in close proximity with the hydrocarbon chain. Chemical interactions in which the formation of stable water–water linkages is the driving force for the breaking of the solute–water interface are called *hydrophobic interactions* and they give rise to the formation of *hydrophobic associations.*

At concentrations lower than the c.m.c. amphipathic molecules can be either monomeric or dimeric. Since replacement of the hydrocarbon–water interface in the dimer is likely to be the main driving force for dimerization, as for micelle formation, the structure of the dimer can be regarded as one in which the hydrophilic parts are far apart, thus minimizing electrostatic repulsion, and the flexible hydrophobic moieties are intertwined.

Amphipathic molecules are specifically adsorbed by a water–air interface, the hydrophilic groups are in contact with the boundary water layer, and the hydrophobic moieties are oriented away from the polar water. The surface-adsorbed molecules can be compacted into films in which hydrophobic bonding between the nonpolar moieties and hydrophilic hydration at the water boundary layer occur, and the exposure of hydrocarbon surfaces to water is minimal. Here again, the primary driving force for this specific surface adsorption is the breaking of the hydrocarbon–water interface and the formation of more stable water–water linkages. Such amphipathic molecules are known as *surface active.*

References

Ben-Naim, A.: Water and aqueous solutions. New York: Plenum Press 1974

Blandamer, M. J.: Structure and properties of aqueous salt solutions. Quart. Rev. *24*, 169–184 (1970)

Briant, C. L., Burton, J. J.: Molecular dynamics study of water microclusters. J. Chem. Phys. *63*, 3327–3333 (1975)

Christian, S. D., Taha, A. A., Gash, B. W.: Molecular complexes of water in organic solvents and in vapor phase. Quart. Rev. *24*, 20–36 (1970)

Coulson, C. A.: Valence, 2nd ed. London: Oxford Univ. Press 1961

Evteev, S. A.: Geological activity of the east antarctic ice sheet: Results of investigation of the IGY Program (in Russian). Glaciology, No. *12*, Moscow Publ. House Akad. Nauk. SSSR. (1964)

Feurstenau, D. W., Healy, T. W.: Principles of mineral flotation. In: Adsorptive bubble separation techniques. Lemlich, R. (Ed.). New York: Academic Press, 1972, pp. 91–131

Horne, R. A., Courant, R. A.: The structure of sea-water at the air/sea interface. Proc. Symp. Hydrogeochem. Biogeochem., Tokyo, 1970, *1*, 558–566 Washington D. C.: The Clarke Company, 1973

Ishizaka, Y.: On materials of solid particles contained in snow and rain water: Part 1. J. Meteorol. Soc. Jpn *50*, 362–375 (1972)

Ishizaka, Y.: On materials of solid particles contained in snow and rain water: Part 2. J. Meteorol. Soc. Jpn *51*, 325–336 (1973)

Israelachvili, J. N., Ninham, B. W.: Intermolecular forces–the long and short of it. J. Colloid Interface Sci. *58*, 14–25 (1977)

Mason, B., Berry, L. G.: Elements of mineralogy. San Francisco: W. H. Freeman and Co., 1968

Rahman, A., Stillinger, F. H.: Molecular dynamics of liquid water. J. Chem. Phys. *55*, 3336–3359 (1971)

Rankama, K., Sahama, Th. G.: Geochemistry. Chicago: Univ. Chicago Press 1950

Remy, H.: Treatise on inorganic chemistry. Amsterdam: Elsevier, 1956, Vol. I

Wedepohl, K. H.: Geochemistry. Althaus, E. (translator). New York: Holt, Rinehart and Winston, Inc., 1971

Chapter 1
Some Geologic Colloid Systems

1. Mineralogy of Colloids in the Sedimentary Cycle

The sedimentary cycle includes the hydrosphere, the atmosphere, the pedosphere, and sedimentary rocks in the lithosphere. Clay minerals, zeolites, hydrated oxides and hydroxides, mainly of Si, Fe, Mn, and Al, and some organic macromolecules are responsible for most colloid properties of geologic systems that constitute the sedimentary cycle. The clay minerals, which generally comprise the greater part of the colloid fraction, are so-called from the term used by sedimentologists and soil scientists for the fraction of particles having a very small size, the "clay fraction". Although certain clay deposits contain well-defined crystalline particles, most clay minerals occur as particles too small to be resolved by the ordinary microscope. Furthermore, a wide distribution of particle sizes is frequently present.

Using X-ray diffraction analysis it can be shown that most clay minerals, even in their finest size fraction, are composed of crystalline particles and that the number of different crystalline minerals likely to be found is limited. The clay minerals are essentially hydrous layer aluminosilicates (phyllosilicates), with magnesium and iron acting as proxy wholly or in part for the aluminum in some minerals and with alkali metals and alkaline earth metals present as essential constituents in some of them. Allophane and imogolite are amorphous aluminosilicates and do not show X-ray diffraction. These will be discussed in Chapter 6.

1.1 Structure of the Common Clay Minerals

The present section is based on Grim (1968), van Olphen (1963), and Weaver and Pollard (1973). The terminology used here is that of Bailey et al. (1971). There are three types of phyllosilicates, namely, (1) the 1:1 type unit layer silicates (also known as the TO type), in which there is one tetrahedral sheet for every octahedral sheet, (2) the 2:1-type unit layer silicates (TOT type), in which there are two tetrahedral sheets per one octahedral sheet, and (3) the 2:1:1-type unit layer, in which brucite-type sheets lie between the parallel 2:1-type silicate layers. The structures of both unit layers, the 1:1 and the 2:1 types, were described in the Introduction.

The phyllosilicates are divided into groups according to the net charge of the unit layer. A further subdivision is based on the number of available octahedral sites occupied: Those minerals containing divalent cations are called *trioctahedral* since all the available cation sites are filled, whereas minerals containing trivalent cations are called *dioctahedral,* having only two-third of the octahedral sites filled. Another subdivision is based on whether or not the stacked 1:1- or 2:1-type unit layers can be expanded and on the ease of swelling of the interlayer space either with water or with organic liquids. Each subgroup contains the different clay mineral species, which are distinguished by their chemical compositions.

The common phylloclay minerals occurring in soils and sediments are listed in Table 1.1, divided into groups. An additional group includes those clay particles with more than one type of layer present, termed mixed-layered minerals. Apart from the phyllosilicates, chain-structure types of clay minerals also exist. These minerals belong to the sepiolite-palygorskite-attapulgite group. Modified amphibole double chains are linked together by octahedral groups of oxygens and hydroxyls containing Al and Mg atoms. Minerals of this group are virtually restricted to playa lake environments and arid soils and also appear in a few isolated hydrothermal localities.

Table 1.1. Classification scheme and chemical composition of clay minerals common in soils and sediments

Type	Charge per formula unit	Group	Subgroup and minerals	Chemical composition
TO (or 1:1)	0	Kaolin-serpentine	Nonexpanding. Dioctahedral series	
			Kaolin subgroup	
			Kaolinite	$Al_2Si_2O_5(OH)_4$
			Dickite	$Al_2Si_2O_5(OH)_4$
			Nacrite	$Al_2Si_2O_5(OH)_4$
			Nonexpanding. Trioctahedral series	
			Serpentine subgroup	
			Chrysotile	$Mg_3Si_2O_5(OH)_4$
			Antigorite	$Mg_3Si_2O_5(OH)_4$
			Lizardite	$Mg_3Si_2O_5(OH)_4$
			Amesite	$(Mg_2Al)(SiAl)O_5(OH)_4$
			Expanding. Dioctahedral series	
			Kaolin subgroup	
			Halloysite	$Al_2Si_2O_5(OH)_4 . 2H_2O$
TOT (or 2:1)	0	Talc-pyrophyllite	Nonexpanding. Dioctahedral series	
			Pyrophyllite	$Al_2Si_4O_{10}(OH)_2$
			Nonexpanding. Trioctahedral series	
			Talc	$Mg_3Si_4O_{10}(OH)_2$
TOT (or 2:1)	0.25–0.6	Smectite	Expanding. Dioctahedral series	
			Beidellite	$[(Al_{1.98}Fe'''_{0.02}Mg_{0.01})(Si_{3.48}Al_{0.52})O_{10}(OH)_2]Na_{0.50}$ [a]
			Montmorillonite	$[Al_{1.48}Fe'''_{0.05}Mg_{0.52})(Si_{4.0})O_{10}(OH)_2]Ca_{0.20}Na_{0.02}$ [b]
			Nontronite	$[(Al_{0.51}Fe'''_{1.58}Mg_{0.01})(Si_{3.37}Al_{0.63})O_{10}(OH)_2]Na_{0.34}$ [c]

Table 1.1 (continued)

Type	Charge per formula unit	Group	Subgroup and minerals	Chemical composition
			Expanding. Trioctahedral series	
			Hectorite	$[(Al_{0.02}Mg_{2.65}Li_{0.33})(Si_{4.0})O_{10}(OH)_2]Ca_{0.01}Na_{0.28}K_{0.01}$ [d]
			Saponite	$[(Al_{0.04}Mg_{2.85}Fe'''_{0.01})(Si_{3.70}Al_{0.30})O_{10}(OH)_2]Na_{0.45}$ [e]
TOT (or 2:1)	0.6–0.9	Vermiculite	Expanding. Dioctahedral series	
			Vermiculite	$[(Al_{1.44}Fe'''_{0.16}Ti_{0.14}Mg_{0.27}Mn_{0.05})(Si_{2.90}Al_{1.10})O_{10}(OH)_2]Mg_{0.18}Na_{0.43}H_{0.30}$ [f]
			Expanding. Trioctahedral series Vermiculite	
		Illite	Nonexpanding. Dioctahedral series	
			Illite	$[(Al_{1.51}Fe'''_{0.23}Fe''_{0.10}Mg_{0.29})(Si_{3.45}Al_{0.55})O_{10}(OH)_2]Ca_{0.02}Na_{0.02}K_{0.52}$ [g]
			Glauconite. Dioctahedral illite rich in iron	
			Nonexpanding. Trioctahedral series Illites rich in magnesium and ferrous iron	
[TOT]O[TOT] (or 2:1:1)	variable	Chlorite	Nonexpanding. Dioctahedral series. There is some evidence for the existence of such a mineral, e.g., donbassite	
			Nonexpanding. Trioctahedral series. Chlorites with structures related to $[(Mg, Fe)_{3-x}(Al,Fe)_x(Si_{4-x}Al_x)O_{10}(OH)_2][(Mg, Fe, Al)_3(OH)_6]$ e.g., clinochlore	
			Nonexpanding. Di-trioctahedral series Sudoite	

[a] Beidellite from Black Jack Mine, Beidell, Colo., USA.
[b] Montmorillonite from Santa Rita, N. Mex., USA.
[c] Nontronite from an alteration zone in gneiss, Spruce Pine, N. C., USA.
[d] Hectorite from hot springs alteration of zeolite in alkaline lake, Hector, Calif., USA.
[e] Saponite, hydrothermal alteration of dolomitic limestone, Milford, Utah, USA.
[f] Vermiculite from lateritic red earth (alteration of mica), Taiwan.
[g] Illite, Pennsylvania underclay, near Fithian, Ill., USA.

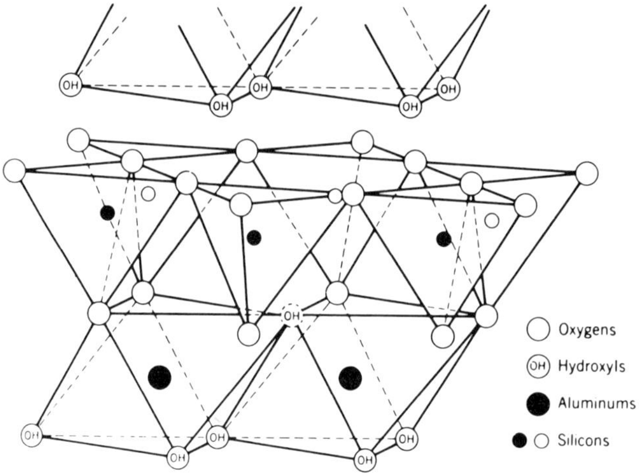

Fig. 1.1. Diagrammatic sketch of the structure of kaolinite. (From Grim, Copyright, 1968, by McGraw Hill, Inc. Used with permission of the publisher)

1.1.1 1-Type Unit Layer Minerals (Fig. 1.1)

Unit layers of the 1:1-type are stacked one above the other in the *c* direction. The serpentine minerals, which were described in the Introduction, belong to the trioctahedral variety of the kaolin–serpentine group while the kaolin minerals belong to the dioctahedral variety. The unit layer is called serpentine- or kaolin-like layer, depending on the subgroup. The charge distribution in the layers of minerals having the ideal composition is shown in Scheme 1.1. As a result of the polarization of the Si–O and O–H groups the surface oxygen and proton planes become negatively and positively charged, respectively. These charges contribute greatly to the stacking of the unit layers one above the other by electrostatic-type attraction forces (Giese, 1973) in addition to van der Waals interaction (Cruz et al., 1972). Members of the kaolin group differ according to the way in which the unit layers are stacked above one another (Frank-Kamenetskii et al., 1974).

Analyses of many kaolinites have shown that isomorphous substitutions are rare (see, e.g., Ferris and Jepson, 1975). Only small amounts of iron can be truly integrated within the kaolinite structure. This type of iron is called "kaolinite structural iron" (Herbillon et al., 1976). Serpentines, on the other hand, show very large amounts of isomorphous substitutions of Fe, Al, and many other metallic cations for Mg in the octahedral sheet and for Si in the tetrahedral sheet (Faust and Fahey, 1962). Substitution of trivalent ions in the tetrahedral and octahedral sheets increases the polarization of the oxygen and proton planes and the delocalization of some hydrogen atoms. In addition to the electrostatic and van der Waals types of interactions between the unit layers, the superposition of oxygen and hydroxyl planes of successive layers within a single serpentine crystal gives rise to pairing of O and OH groups belonging to substituted

Scheme 1.1. Schematic representation of layer structure, charge distribution in the layers, intersheet distances in the unit layer, and interlayer space of minerals of the 1:1 type

Interlayer spacing	Intersheet spacing	Plane (Serpentine)	Plane (Kaolinite)	Interlayer spacing	Plane (Halloysite)	Charge
		O_6	O_6		O_6	−12
	0.22 nm	Si_4	Si_4		Si_4	+16
		$O_4(OH)_2$	$O_4(OH)_2$		$O_4(OH)_2$	−10
	0.22 nm	Mg_6	Al_4		Al_4	+12
0.715 nm		O_6	O_6	1.01 nm	O_6	−12
		H_6	H_6		H_6	+ 6
					$(H_2O)_4$	
		O_6	O_6		O_6	
		Si_4	Si_4		Si_4	
		$O_4(OH)_2$	$O_4(OH)_2$		$O_4(OH)_2$	
		Serpentine	Kaolinite		Halloysite	

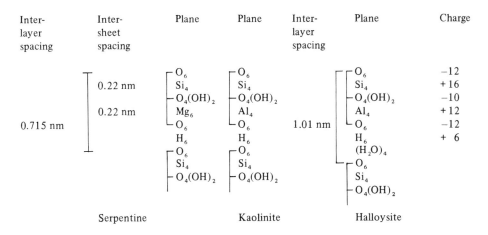

tetrahedra and octahedra sheets, respectively, which results in interlayer hydrogen bond formation (Heller-Kallai et al., 1975). A similar hydrogen bond is formed between two sheets of amesite, a serpentine with the highest substitution of Al for Si (Serna et al., 1977).

As indicated by the chemical formulas in Table 1.1, the kaolin- and serpentine-like layers are electrically neutral, but in reality they carry small negative charges due to small amounts of isomorphous substitutions. These "permanent" negative charges are not pH dependent and are responsible for the small exchange capacities of these minerals under acid conditions (Schofield and Samson, 1953; Range et al., 1969; Mashali and Greenland, 1975; Bolland et al., 1976).

Halloysite is composed of kaolin-like layers between which a single monolayer of water molecules is interposed. The basal spacing increases from 0.72 nm (that for kaolinite) and may reach 1.01 nm (Churchman et al., 1972). Greater expansions are possible with the adsorption of several polar organic compounds (Carr and Chih, 1971). The interlayer water molecules are linked to each other by hydrogen bonds, but the interactions between the water sheet and the kaolin-like layers are primarily due to dipole attraction (Yariv and Shoval, 1975). Halloysite occurs mainly as tubular-shaped particles, but spheroidal particles are also known (Askenasy et al., 1973; Dixon and McKee, 1974). Expanding minerals with serpentine-like layers have not been detected.

1.1.2 2:1-Type Unit Layer Minerals
(Fig. 1.2, Scheme 1.2)

The joining of 2:1-type unit layers results in a great variety of minerals, the properties of which depend on the charge density of the unit layer. Talc and pyrophyllite are tri-

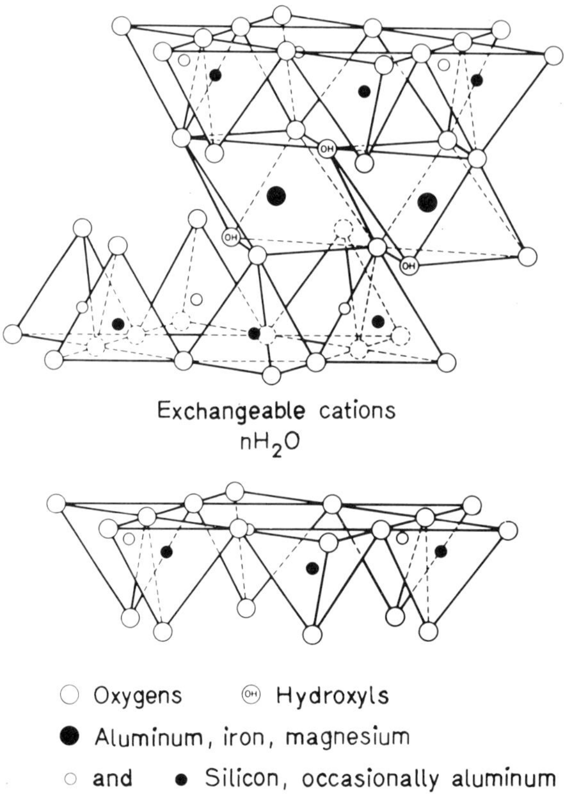

Fig. 1.2. Diagrammatic sketch of the structure of smectite. (From Grim, Copyright, 1968 by McGraw-Hill, Inc. Used with permission of the publisher).

and dioctahedral 2:1-type minerals with no tetrahedral or octahedral substitution and no electric charge. In the absence of any charge there is no intercrystalline swelling.

Minerals of the smectite group (known also as the montmorillonite group), are analogous to pyrophyllite and talc but differ from them in that a small fraction of the tetrahedral Si atoms are substituted by Al atoms, and/or octahedral atoms (Al or Mg) are substituted by atoms of a lower oxidation number. The resulting charge deficiency is balanced by exchangeable cations, mainly Na, Ca, and Mg, which are located between parallel layers.

Water and other polar molecules may penetrate between the layers causing the expansion of the structure in a direction perpendicular to the layers. Under ordinary conditions a smectite with exchangeable Na ions contains water as a monomolecular sheet resulting in a c axis spacing of about 1.2 nm; Ca and Mg smectites usually contain bimolecular water sheets and the c axis spacing ranges from 1.45 to 1.55 nm. With increasing moisture the c spacing may increase further, indicating the additional swelling of the clay mineral. The expansion properties are reversible. The water sheet comprising the exchangeable cations is positively charged. The resulting structure of

Scheme 1.2. A schematic representation of layer structure, charge distribution in the unit layers, intersheet distances in the unit layer, and interlayer space of minerals of the 2:1 type

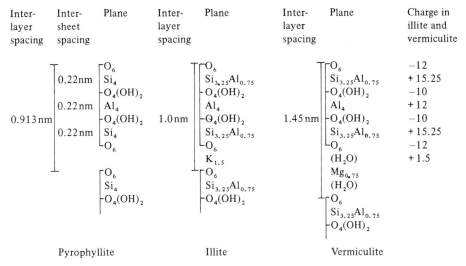

Inter-layer spacing	Inter-sheet spacing	Plane	Inter-layer spacing	Plane	Inter-layer spacing	Plane	Charge in illite and vermiculite
		O_6		O_6		O_6	−12
	0.22nm	Si_4		$Si_{3.25}Al_{0.75}$		$Si_{3.25}Al_{0.75}$	+15.25
		$O_4(OH)_2$		$O_4(OH)_2$		$O_4(OH)_2$	−10
	0.22nm	Al_4		Al_4		Al_4	+12
0.913nm		$O_4(OH)_2$	1.0nm	$O_4(OH)_2$	1.45nm	$O_4(OH)_2$	−10
	0.22nm	Si_4		$Si_{3.25}Al_{0.75}$		$Si_{3.25}Al_{0.75}$	+15.25
		O_6		O_6		O_6	−12
				$K_{1.5}$		(H_2O)	+1.5
		O_6		O_6		$Mg_{0.75}$	
		Si_4		$Si_{3.25}Al_{0.75}$		(H_2O)	
		$O_4(OH)_2$		$O_4(OH)_2$		O_6	
						$Si_{3.25}Al_{0.75}$	
						$O_4(OH)_2$	
		Pyrophyllite		Illite		Vermiculite	

the smectites is comprised of parallel negatively charged 2:1-type unit layers between which positively charged water sheets are interposed. The force acting between the water sheets and the silicate layers is weakly electrostatic.

Beidelite and saponite are smectites with mainly tetrahedral substitution. The charges of montmorrillonite and hectorite are located predominantly in the octahedral sheet. Nontronite is an iron-rich smectite. Montmorillonite is the most common mineral of this group.

Vermiculite is an expanding 2:1-type mineral, the unit layers of which are separated by sheets of water molecules occupying a definite space, 0.498 nm, which is about the thickness of a bimolecular water sheet. Megascopic vermiculite is invariably trioctahedral; dioctahedral varieties are also found in clay fractions. Vermiculites usually have greater layer charge densities than smectites. In the natural mineral the balancing exchangeable cation is Mg, sometimes with a small contribution from Ca or Na.

Vermiculites differ from smectites in that the expansion with water is limited to about 0.498 nm (Walker, 1961). Both minerals sorb certain organic molecules into the interlayers but a thinner organic sheet is formed in the interlayer of vermiculite. These characteristics may be the result of the relatively larger particle size of the vermiculite layers and also due to the fact that the charge of the vermiculite is located mainly in the tetrahedral sheet. This, togehter with the higher layer charge density, results in the stacks of the unit layers being held together by forces that are comparatively greater than the forces that hold together the layers in smectites.

Illite is a nonexpanding 2:1-type phyllosilicate. It differs from talc and pyrophillite in that a substitution occurs, predominantly in the tetrahedral sheets. The cations that compensate the negative layer charge are K ions, which have a very low hydration energy. Those ions that are located in the interlayers are not available for ion-exchange reactions. The negative charge of illite layers is usually much greater than that of smectite layers, but not much greater than that of vermiculites. The layers in a stack are held together by electrostatic forces of the same order of magnitude as those in highly charged vermiculites, but since the hydration energy of K ions is low, illites do not expand.

The basal spacing of the illites as well as that of the micas is about 1.0 nm. Illites differ from true micas in several ways:

1. Al replacement of Si in the tetrahedral sheets is much less than that which occurs in micas.
2. Small amounts of Ca and Mg ions are present in the interlayers of illites but not in those of micas.
3. The stacking of layers in illite is more random than in micas.
4. Illites appear as very fine particles whereas micas form relatively coarse particles.

1.1.3 2:1:1-Type Minerals
(Fig. 1.3)

Chlorite (Fig. 1.3) consists of 2:1-type layers, separated from each other by quasi-brucite sheets that are octahedral Mg hydroxides with replacements of Mg by Al and Fe, and consequently the sheets are positively charged. The 2:1-type silicate layer is negative as a result of substitution of Al for Si, and the deficiency of charge is balanced by an excess charge in the quasi-brucite sheet. The entire structure is neutral. The two parts of the structure are held together by electrostatic forces and by hydrogen bonds formed between the hydroxyls of the quasi-brucite sheet and oxygens of the 2:1-type unit layers (Hayashi and Oinuma, 1967). Hydrogen bonds are formed when oxygens and hydroxyl groups belonging to substituted tetrahedra and octahedra, respectively, are paired.

Various members of the chlorite group differ from each other in the kind of metal and amount of substitution within the brucite sheet and within the tetrahedral and octahedral sheets of the 2:1-type layer. Well-crystallized chlorite has been found as a trioctahedral species. Fine-grained chlorites are found in some clay-size materials. Clayey material chlorites differ from well-crystallized chlorites in having somewhat random stacking of unit layers and perhaps in being somewhat hydrated. The *c* spacing of chlorite is equal to 1.4 nm.

1.2 Hydrous Oxides of Aluminum and Iron

The most common oxides and hydrated oxides of aluminum in nature are corundum, Al_2O_3, gibbsite, $Al(OH)_3$, and the monohydrates boehmite and diaspore, AlO(OH). Corundum is characteristic of undersaturated syenites, ultrabasic rocks, and contact-metamorphosed limestones. Diaspore has been reported in sedimentary fireclays but has not been observed as a weathered product in soils. This mineral should form only at elevated temperatures. Gibbsite and boehmite form at ordinary pressures and temperatures and both are relatively common in highly leached soils in tropical areas (Loughnan, 1969).

The structure of gibbsite was described in the Introduction, in connection with the structure of the phyllosilicates. Boehmite has a more complex structure comprised

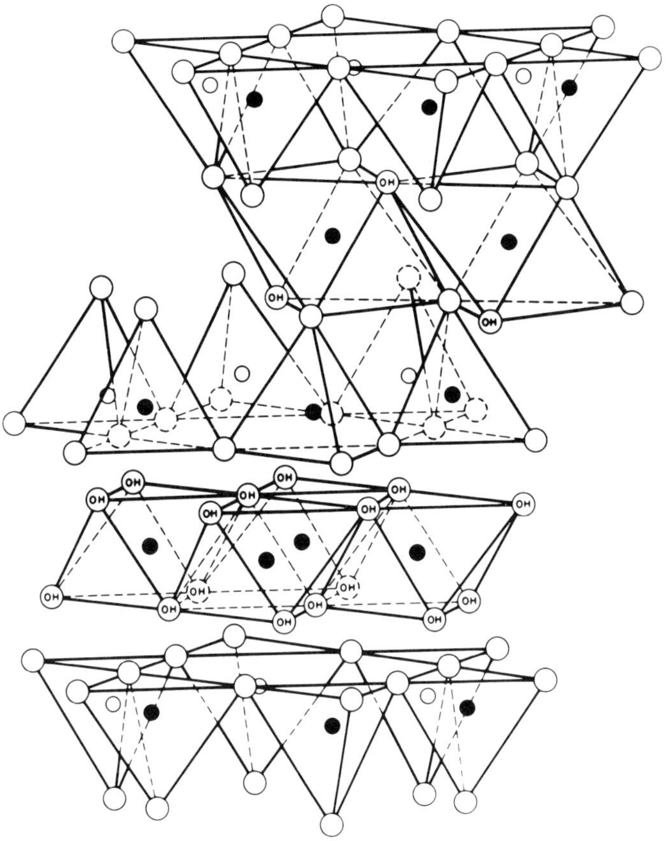

Fig. 1.3. Diagrammatic sketch of the structure of chlorite, for explanation of symbols, see Fig. 1.2. (From Grim, Copyright, 1968 by McGraw-Hill, Inc. Used with permission of the publisher)

of a double sheet of octahedra with Al ions at their centers. There are two types of oxygens in this structure: (1) those in the center of the structure are shared by four octahedra and (2) those on the outside are linked to only two. The sheets are held together by hydrogen bonds.

The difference in structure between boehmite and gibbsite results in differences in stability. Boehmite is almost inert to alkalis and acids whereas gibbsite is amphoteric and dissolves in alkaline solutions at pH values above 10 or in acid solutions at pH values below 4.

The most common oxides of iron in nature are the two polymorphic forms of Fe_2O_3, hematite and maghemite, and magnetite, Fe_3O_4, a member of the spinel group, which contains di- and trivalent iron. Iron does not form a stable trihydrate, like gibbsite, but rather forms two monohydrates, both having the formula FeO(OH), goethite and lepidocrocite, which are isostructural with diaspore and boehmite, respectively.

Magnetite is characteristic of igneous and metamorphic rocks and may persist as a detrital grain in sediments, where it is converted to oxides having higher oxidation states. Maghemite also has a spinel-like structure. It occurs in laterites, where it generally appears together with hematite.

In most soils and sedimentary rocks much of the iron is present as an oxide amorphous to X-ray diffraction. This poorly organized material contains indefinite amounts of water. It is called limonite because of its yellow color.

2. Soils and Sediments as Colloid Systems

The constituents of sediments and of soils exist in solid, liquid, and gas states forming a disperse system. The solid phase is represented mainly by minerals and by some amorphous inorganic and organic particles with size variation from submicroscopic colloidal to visually discrete particles. The solid phase also contains water molecules sorbed on the particles. The liquid phase is water that contains soluble salts and soluble organic compounds, monomers, micelles, and hydrophilic macromolecules. In some sediments petroleum hydrocarbons also form a liquid phase. The gaseous phase in soils and in the upper horizons of sediments approximates the composition of air but with high content of water vapor and CO_2. In deep-seated horizons volatiles released from magma form the gaseous phase. In some sediments organic compounds of low molecular weights (mainly methane) also comprise the gaseous phase. The behavior and properties of such a system depend on the characteristics of each separate phase and on the nature of the manifold mutual relationships existing between them. The solid phase is considered to be the continuous dispersing medium, with voids or pore space between the particles. A soil system is very dynamic and the equilibrium between the three phases changes continuously. For example, during a rainfall or during irrigation the liquid phase is increased and some of the gases are removed. Furthermore, the soil can be fluidized, i.e., the water becomes the continuous dispersing medium and the solid particles become the disperse phase. Deep-seated sediments can also become fluidized by the action of volatiles released from magma (see Sect. 3 and 4).

Soil is the finely divided material that lies on a consolidated rock bed, at or near the earth's surface. Soils may be said to be the products of the mechanical processes and chemical interactions in the four geologic spheres–the lithosphere, the hydrosphere, the atmosphere, and the biosphere. Soil formation is confined to the surface of the earth because it is only here that the lithosphere, the hydrosphere, and the atmosphere are in contact (Mattson, 1938). Soils are formed by the weathering or disintegration of rock material and decomposition of organic matter. We can differentiate between two types of soils:

1. The loose material resulting from rock weathering is eroded from its place of origin by wind, rainfall, rivers, and glaciers and is deposited elsewhere. These are "transported soils".

2. Products of weathering remain at their place of origin forming "residual soils" (Fig. 1.4).

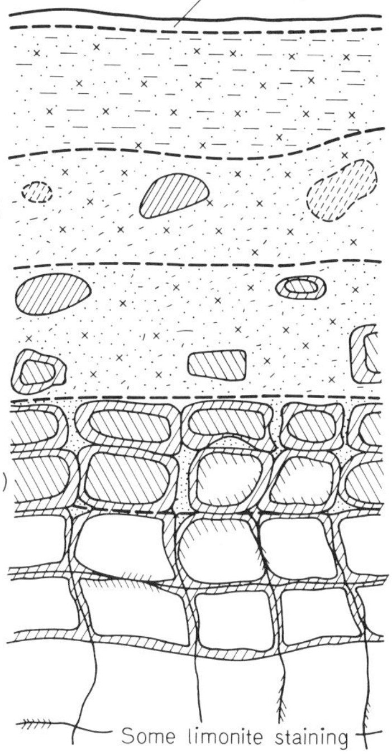

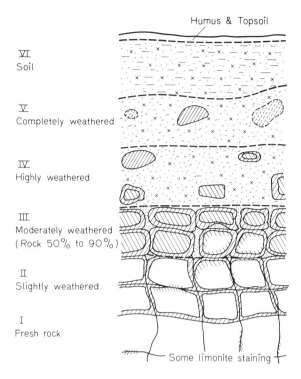

Fig. 1.4. Soil formation by rock decomposition and morphological definition of the degree of rock decomposition. (From Gidigasu, 1974. Used with permission of the publisher)

Another important factor in soil formation is the displacement or migration of soluble or very fine-grained suspended particles under the influence of percolating solutions. These migrations lead to the development of impoverished horizons, called *A horizon* and accumulative horizons, called *B horizon*. The deepest horizon, which shows the greatest similarity to the parent rock, is called *C horizon*.

About 90% of the earth's continental area is covered with a veneer of sedimentary rocks. Sedimentary rocks are formed at the surface of the earth by the following sedimentation processes:

1. Mineral fragments produced by chemical and physical weathering processes are eroded from their place of origin to a basin of sedimentation. Erosion may take place either in the hydrosphere or in the atmosphere, the mineral fragments being dispersed. The erosion process is accompanied by sorting and mineralogic differentiation. Depending upon the environment of deposition, sediments are divided into *terrestrial* and *marine sediments*.

2. Biogenous products and chemical precipitates accumulate from seawater and other solutions.

After the sedimentation process the deposit is subjected to compaction due to the rise of the lithostatic pressure with depth, to mineralogic diagenesis due to the rise in temperature with depth, and to cementation, which results from chemical reactions between the accumulated sediments and the pore solutions.

More than 99% of the total volume of sedimentary rocks is made up of only three types: shales (including clay sediments and siltstone), sandstones (including graywacke), and limestone (including dolomites).

2.1 Solid Fraction in Soils and Sediments

The grains may vary in size within very wide limits, from colloid clay particles to large boulders. For the purpose of classification it is necessary to determine the size of the grains that constitute a particular system and the percentage of the total weight represented by the various grain fractions. Common classifications of soils and sediments are based on the size of the major particles, and divide the systems into sandy, silt, and clay. In all natural systems there is more than one fraction present. The very coarse fraction, called *gravel,* consists of rock fragments that are composed of one or more minerals. The coarse fraction, or sand, is a mixture consisting mainly of quartz grains, but also of feldspar, calcite, and mica. The proportion of the flat, plate-like particles to the bulky ones increases as grain size decreases. Sand minerals form grains that have nonactive surfaces and that do not form strong interparticle linkages either in the presence or in the absence of water (Ch. 8) and are therefore regarded as *cohesionless* particles.

The size of clay particles is of the order of microns. The clay minerals are the major component of this fraction. Various surface forces, chemical and electric in nature, act at the surfaces of clay grains; these are responsible for the peculiarities of clay behavior. Clay minerals have a sheeted structure, thin plate-shaped or elongated, and needle-shaped particles are also very common. The ratio of the diameters of plates to their thickness may vary between limits as wide as 10 to 1 and 250 to 1, depending upon the character of the clay mineral and the exchangeable ions. Clay particles form stable aggregates and are regarded as *cohesive particles.*

The bulk volume of a mass unit of soil or sediment is made up of the three parts:

1. The total volume of the solid material in the particle
2. The volume of the spaces between the solid particles
3. The pore volume of the particles

It is a complex function of particle shape and size, surface polarity, density, and compaction. The cohesion of mineral particles will be favored when they are sufficiently close together for their surfaces to touch each other and their attraction fields to overlap (Ch. 8). The voidage of the system is highest when cohesion between the particles is at a maximum. Cohesion of the particles with each other makes it difficult for the particles to move into positions of minimum potential energy. In practical terms this will result in a maximum in the bulk volume and hence in voidage. Clay minerals, being cohesive, have a bulk volume in the dry state greater than that of the cohesionless quartz (Fig. 15, a_1 and b).

The effect of gravitation on the bulk volume is of prime importance when sedimentation occurs under wet conditions. The adsorption of water molecules on the solid surface causes a reduction of the surface forces per unit area of the solid. This allows greater freedom of movement of solid particles, which are then able to achieve a state of lower potential energy. In practice, this will result in a minimum in the bulk volume and hence in voidage (Fig. 1.5, a_1 and a_2).

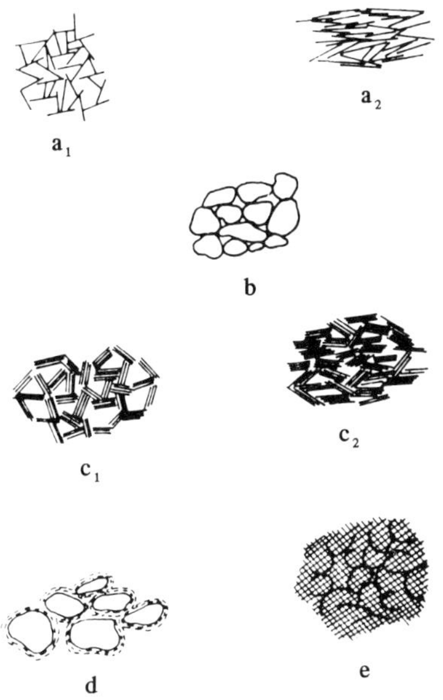

Fig. 1.5a–e. Schematic representations of elementary particle arrangements: a individual primary clay platelet interaction, b individual silt or sand particle interaction, c clay tactoid interaction, d clothed silt or sand particle interaction, and e partly discernible particle interaction. (From Collins and McGowan, 1974. Used with permission of the Institution of Civil Engineering, London)

Porosity is the ratio of the volume of voids to the total volume of the system. It is generally expressed as a percentage and is defined by the formula

$$n\,(\%) = 100\,\frac{V_m + V_g}{V} \tag{1.1}$$

where V, V_m, and V_g are the total volume, volume of water, and of air, respectively.

The porosity of soils and sediments varies over a wide range. It depends on the origin of the system, on the uniformity of its grain-size distribution, and to a great extent on the shape of the grains. Thus, for example, sands and gravels laid down by the swift and varying current of flooding rivers are loose, whereas sands deposited from still water or slow-flowing rivers are densely packed. In general, the more uniform the grain-size distribution the greater is the porosity. A well-graded soil is composed of particles varying in size such that the finer particles can readily fill the voids formed by the larger grains. Spherical sand particles form deposits with a greater porosity than plate-like clay particles. Porosity of sediments decreases with burial as a result of compaction, which causes particles to be in maximal contact, even if rough particles disaggregate or their edges are broken. Cementation that occurs during burial also results in decreasing porosity (Fig. 15, c, d and e).

Porosity of clay sediments that contain smectite minerals depends on the water content of the sediment. These minerals expand when they adsorb water, and thereby the total volume of the solid material increases whereas that of the space between solid particles decreases. Clay horizons are therefore impermeable to water migration.

2.2 Water in Soils and Sediments

Due to the effect of gravity, water draining in soil voids is downward, reaching the ground water at a particular depth. The level where the draining water comes to rest is called the *water table.* Below this level all the pores in the system are filled with water. However, voids may also be saturated by water above the water table, due to capillary action. In the vicinity of the ground surface the pores are occupied by both air and water. The configuration of the two different substances, water and air, in the voids is determined by their respective attraction to the solid phase. Surfaces of mineral particles are polar, and the sorption of water occurs from the electrostatic interaction between dipoles of water molecules and charged sites on the surfaces of the minerals. Air molecules are nonpolar, and they interact with surfaces of minerals predominantly by van der Waals interactions. These bonds are much weaker than the polar bonds that are formed with water. Surfaces of most minerals are hydrophilic and aerophobic.

The extent to which the solid fraction is hydrophobic or aerophobic depends on the types of minerals present. Clay minerals and amorphous hydrous oxides increase hydrophilicity of soils whereas the presence of quartz or lime decreases hydrophilicity and increases aerophilicity of soils. The presence of organic compounds that have a low proportion of polar groups may also increase aerophilicity of soils. On the other hand, organic compounds that have a high proportion of polar groups will increase hydrophilicity of soils.

Under stable equilibrium conditions, the water tends to assume a configuration such as to wet the grains over the largest possible surface area. The pore–air, on the other hand, has a tendency to assume a position in which the air–liquid interface area is at a minimum. As a result the liquid phase always occupies the smaller voids, whereas the air accumulates in the larger voids, forming spherical bubbles therein.

Soil and sediment water can be classified as follows:

1. *Ground water* is the subsurface water that fills the voids below the water table. This water is subjected only to gravitational force. This pore water has the normal fine structure of liquid water. It exhibits the physical and chemical properties of ordinary liquid water. This type of water is capable of moving under hydrodynamic force unless restricted in its free movement, e.g., when entrapped between air bubbles, or by retention due to capillary forces that in fine pores may overcome the hydrodynamic force. This pore water can dissolve salts and can take part in the chemical weathering and erosion of the minerals.

2. *Capillary water* is the water lifted by surface tension so as to be brought into contact with minerals above the water table. To a certain height above the water table, the capillary water fills all the pores in the system. Above this level a zone of open capillaries is located, in which the capillary water held in a network of fine pores is in contact with the ground water. On the other hand water is displaced by air in the larger pores above the water table.

3. *Adsorbed water* is held on the surface of mineral particles or as interlayer water by minerals with an expanding lattice structure. The fine structure of this water differs from that of ordinary water. This is actually the thin film that forms the solid–liquid interface and its fine structure depends on the nature of the ions sorbed by the mineral.

The adsorption forces are extremely great, and this kind of water cannot migrate by means of normal hydrodynamic forces.

Evaporation of water from sediments and soil drying occur in three stages. The first stage is characterized by a relatively high evaporation rate controlled by atmospheric conditions such as humidity, temperature, and winds. When the soil water can no longer be transported to the surface fast enough to meet the evaporative demands, the second stage begins. Here the evaporation rate declines, the soil surface experiences rapid drying, with the mode of transfer shifting primarily from liquid to vapor movement. The third stage is characterized by a low, relatively constant evaporation rate controlled by desorption reaction acting over molecular distances at the liquid–solid interfaces in the soil.

Pore water is expelled from sediments by overburden load, the water migrating upward from deep horizons, where the hydrostatic pressure is high, to shallower horizons with a lower hydrostatic pressure. At great depths, where the volume of the spaces between the solid particles is extremely low, capillary and adsorbed water migrate and the permeability is extremely low. Under these circumstances the drainage of water is very poor, and very high hydrostatic pressures are obtained at depths.

Pore solutions are present in practically all sediments that are not completely consolidated, and they often contain more dissolved electrolytes than seawater. If the aqueous phase was seawater when the sediment was formed, its composition was in most cases changed by physical, chemical, mineralogic, and biologic processes. Pore solutions provide the chemical components for diagenetic alterations.

2.3 Gases in Soils and Sediments

The gaseous phase contained in the voids of soil may be of two types:

1. Gas in Contact with the Atmosphere. This can communicate freely between the pores and the atmosphere. Its pressure is equal to the atmospheric pressure. Should the latter change, air will move in or out of the pores, until the pressure reaches equilibrium.

2. Gas Separated from the Atmosphere. Entrapped air bubbles are formed when voids full of air are filled by water from above. Under special morphologic conditions of the system, the free communication of the pore air with the atmospheric air is blocked. This air is in continuous interaction with the surrounding pore water.

Entrapped air is also found in sediments. Deep-seated gas may reach sediments by igneous processes. Organic gases originate in organic diagenetic processes. The water and the entrapped gas bubbles form a disperse system. This mixture, known as a foam, is formed in voids, with the gas as disperse phase and the liquid as the dispersing medium. Air bubbles contained in the pore water tend to reduce the permeability of the soils and sediments to water.

Air is important for maintaining biogenous activity in soils. It is a natural concomitant of aeolian deposits whose grains, having settled from air, are covered with only a thin molecular water layer. Water-laid sediments are found under normal conditions in a state of nearly 100% saturation by water. Favorable conditions for air invasion exist only when the surface of such a deposit is exposed to drying.

2.4 Organic Matter in Soils and Sediments

The nonliving organic fraction of soils and sediments consists of a complex system of *biopolymers,* substances that are organic residues of plant, animal, insect, and microorganism origin and products of their transformation under the action of biological, chemical, and physical factors. The transformation products contain compounds of low molecular weights as well as compounds of very high molecular weights, the latter known as *geopolymers.* Sediments are divided according to the organic content of the sediment into concentrated and diluted. The first group includes petroleum, asphalts, and coals. The diluted sediments include hydrocarbons, bitumens, and kerogens. Petroleum is a complex mixture of hydrocarbons. Those that have high molecular weigths are solids and are soluble in the liquid hydrocarbon phase. Petroleum contains small amounts of other organic compounds, such as porphyrines, which are soluble in the organic liquid. Asphalts are solid or semisolid hydrocarbons having high melting points (> 65 °C) with a high content of cycloparaffin hydrocarbons. Coals are geopolymers with extremely high molecular weights and with a high degree of aromaticity. Kerogens are also geopolymers but with a much lower degree of aromaticity. Bitumen is the organic fraction of the sediment characterized by being extractable in CS_2. It is of a much lower molecular weight than kerogen and is composed mainly of a mixture of hydrocarbons. The geoplymers found in soils are humic and fulvic acids. They have lower molecular weights than kerogens and under certain conditions they may form hydrophilic colloid solutions in aqueous systems. Humic and fulvic acids are sometimes transferred into the hydrosphere whereas kerogen and coal are not.

2.4.1 Soil Organic Matter

The organic matter of soils is divided into two groups: (a) nonhumic substances and (b) humic substances (humus) (Kononova, 1961). Nonhumic substances consist of various nitrogenous and non-nitrogenous compounds having still recognizable chemical constituents, e.g., proteins and their decomposition products, peptides and amino acids, carbohydrates, fats, waxes, resins, pigments, and other low-molecular-weight substances. These compounds form 10–15% of the total amount of soil organic matter. Carbohydrates constitute the largest organic fraction of defined composition and are present as polysaccharides (Cheshire et al., 1975).

Humic substances are amorphous, brown or black, polydisperse substances of molecular weights ranging from several hundred to tens of thousands (Schnitzer and Khan, 1972). Based on their solubility in alkali and acid, humic substances are usually divided into three main fractions: (1) *humic acid*, which is soluble in dilute alkaline solution but is precipitated by acidification of the alkaline extract; (2) *fulvic acid,* which is that humic fraction which remains in the aqueous acidified solution, and (3) *humin*, that soil fraction which cannot be extracted by dilute base and acid. Structurally the three humic fractions are similar to one another. They differ in molecular weight, ultimate analysis, and functional group content. The fulvic acid fraction has a lower molecular weight and a higher content of oxygen-containing functional groups per unit weight than humic acid and the humin fraction. The chemical structure and properties of the humin fraction are similar to those of humic acid. Its insolubility arises from the strength of the bonds of its combination with inorganic soil constituents.

Humic fractions are resistant to microbial degradation. In developed soils this group forms up to 85–90% of the total amount of soil organics. These compounds, which act as ion exchangers, form stable water-soluble and water-insoluble salts and complexes with metal ions and they interact with clay minerals and hydrous oxides. These reactions and the properties of the products so formed are of considerable importance to colloid properties of soils.

The predominant elements in humic substances are carbon and oxygen. Carbon content in humic acid and humin ranges from 50–60%, oxygen content from about 30–35%, hydrogen 4–6%, nitrogen 2–4%, and the sulfur content may vary from about 0–2%. Fulvic acids contain less carbon and nitrogen but more oxygen than do humic acids and humins. The carbon content of fulvic acid ranges from 40–50%, the oxygen content 44–50%, nitrogen 1–3%, and sulfur from about 0–2%.

About 20–55% of nitrogen in humic substances consists of amino acid nitrogen and 1–10% is amino sugar nitrogen. Small amounts of nitrogen compounds, such as purine and pyrimidine bases, have been found in humic substances. Most of the combined amino acid nitrogen occurs in amino acid bound by peptide linkages. Amino sugar nitrogen is present in the form of glucosamine and galactosamine.

Humic materials originating from different soils do not differ markedly in amino acid composition. The amino acid compositions of humic and fulvic acids and humin extracted from different soils were found to be fairly similar.

The major oxygen-containing functional groups in humic substances are carboxyls (2–9%), phenolic and alcoholic hydroxyls (2–6 and 2–4% respectively), carbonyls (1–5%), and methoxyls (0.3–1.7%). Due mainly to the presence of dissociable hydrogen in aromatic and aliphatic COOH and phenolic OH groups, these substances reveal acidic properties and behave like ion exchangers. Dialyzed solutions of humic acids have a pH of approximately 3.5. They are weakly dissociated acids with an equivalence point at a pH of about 8–9. The fact that humic acids are weakly dissociated determines to a large degree an important property of the soils, namely the buffering capacity.

Humic substances have a complex structure, built of several main components (structural units), each component being a polymer. The size and molecular weight of humic substances from one soil sample vary from a few hundred (fulvic acid) to a few thousand, and one should consider only the average molecular weight. The humic acid molecules are much larger than fulvic acid molecules, and the reported molecular weights generally are in the 5000–50,000 range (Schnitzer and Khan, 1972).

Humic substances do not have the same integrity of structure and rigid chemical configuration as other natural macromolecules. They may be regarded as polycondensates of random collections of the structural units.

Haworth (1971) concludes that humic acid contains a complex aromatic core to which are attached, chemically or physically, (1) polysaccharides, (2) proteins, (3) simple phenols, and (4) metals. The means of attachment of the groups is uncertain. The protein core attachment is stable toward chemical and biologic attack. Humic substances are linked to polypeptides by hydrogen bonds. Some parts of the carbohydrates and nitrogenous molecules are attached very loosely to the humic materials.

According to Schnitzer and Khan (1972) fulvic acid is made up of phenolic and benzenecarboxylic acids joined by hydrogen bonds to form a polymeric structure of considerable stability. The structure can adsorb organic and possibly also inorganic compounds. Any weakening of the hydrogen bonds will cause the structure to break and permit the extraction of the "building blocks" and also of the compounds that are adsorbed.

One of the characteristics of the structure is that it is perforated by voids or holes of different dimensions which can trap or fix organic molecules such as alkanes, fatty acids, dialkyl phthalates and possibly carbohydrates and peptides as well as inorganic compounds such as metal ions and oxides, provided that these compounds have suitable molecular sizes. This mechanism does not exclude interactions between peripheral or functional groups of the acids with some of the organic and especially the inorganic compounds. The proposed structure is quite loose or open.

Organic matter of river water contains fulvic acid isolated from soils. Elemental and functional analyses are similar for soil and water fulvic acid samples. Wilson and Weber (1977) studied fulvic acid isolated from soil and water at one site in New-Hampshire. The soil and aquatic fulvic acid average molecular weights are 644 and 626 respectively. They concluded that humic materials in soil and water degrade on the average to similar sized fulvic acid molecules.

2.4.2 Coals in Sediments

Coal is a heterogeneous mixture composed primarily of aromatic organic compounds, which comprise mainly carbon (above 65%), hydrogen (3–6%), oxygen (15–20%), nitrogen (1–2%), and sulfur (less than 1%). Inorganic constituents such as pyrite rarely fall short of 0.2% or exceed 5% of bituminous and higher rank coals. Coal ash consists mainly of mineral matter. The minerals commonly occurring in coals are clay minerals, such as illite, kaolinite and mixed-layer clay minerals, quartz, calcite, and pyrite. Iron sulfate minerals are common in extremely fresh coal samples, resulting from the oxidation of pyrite and marcasite (Rao and Gluskoter, 1973). Most trace elements in coal are present in quantities that approximate the composition of the earth's continental crust. Elements normally related to one another in other geologic environments are also related in coals (Ruch et al., 1973).

Coalification occurs when plant material becomes buried and gradual changes in its chemical composition and physical properties take place. The changes may lead to gradual maturation of the coal and enrichment in carbon content and the formation of lignite, brown coal, bituminous coal, and anthracite, respectively. Pure carbon (graphite) should be the final product of metamorphism. According to Tan Li-Ping (1965) the time required to attain an apparent equilibrium in the metamorphism of the Miocene coals of Taiwan was 60 million years or longer. The subbituminous coals have been covered by 3500–4500 m of overburden whereas the more mature, highly volatile bituminous coals have been buried to depths of 5000–7000 m. Artificial coalification under laboratory conditions requires higher temperatures than natural coalifications to achieve the same results.

Coals are highly aromatic and contain few short alkyl substituents. The coal network is generally assumed to be formed from small aromatic units held together by short cross-linkers, mainly methylene, but also including ethylene, propylene, and ethers. The di- and triaromatic compounds represent the simplest *monomeric* moieties in the coal structure. Oxygen functional groups remaining in the di- and triaromatic fractions are aromatic ethers. The furan systems have been identified but other ether structures are known to be present. A coal particle appears to act like a sponge. It is permeated by large and small fissures (from a few tenths to 10 nm) that are more or less accessible to small molecules such as ammonia and hydrocarbons. Most of these low-boiling materials are dispersed in the pores of coal and are not actually chemically bonded. When the pores are small, adsorbed molecules are able to enter them at high temperatures and pressure but are trapped therein upon cooling, resulting in *tarry asphaltenes.* When the pores are large they permit unrestricted entrance and exit of the adsorbed molecules. Organic solvents elute all the material naturally entrained therein. However, chemically bonded entities cannnot be eluted from the coal particles. Traces of these low-boiling materials may have been obtained by thermal decomposition of the periphery of the coal particle. Oxygen is the most abundant heteroatom present in coal. Almost half of the oxygen is present as ethers, and originally most of the carboxylic acids were probably esters. Structures such as:

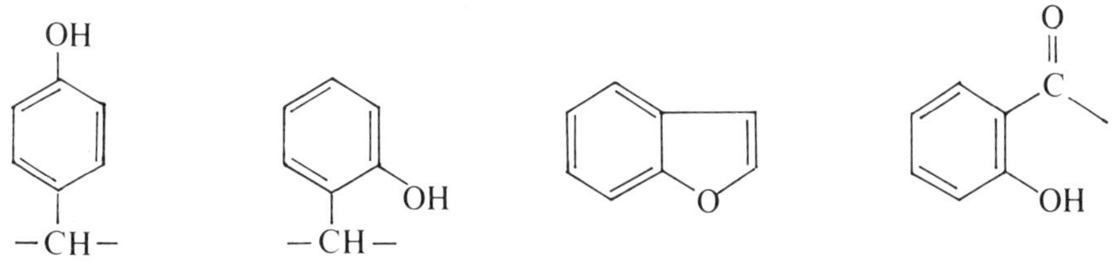

as well as more complex molecules are to be found in coal. Free and hydrogen-bonded OH and NH groups and traces of carbonyl groups have also been identified (Ruberto et al., 1977).

3. Fluidized Beds

A mixture of solid particles suspended in a turbulent gas or liquid behaves in many respects like a fluid. This state of suspension is called "fluidization". In the fluidized state, the mechanical behavior of granular material is radically altered and the resulting fluid-particle system flows readily under the influence of a hydrostatic head. The overall bed flow behavior and the upward injection are obviously of importance in problems of solid transport in plutonic and volcanic igneous activity and in postdepositional events in sediments. For example, volatiles, after separation from the silicate melt, are under a high pressure and can serve as fluidizing agents for beds of solid particles. Examples of fluidization that occur in nature are revealed by some dikes and sills, miniature sand or mud volcanoes, polygonal patterns or pseudomud cracks, and other minor surface features (Dzulynski and Walton, 1965).

When a gas percolates slowly through a layer of sediment, it does not agitate the individual particles. As the flow rate of the gas increases, the bed expands. The expansion of the bed is directly related to the porosity of the sediment. With further increase in the rate of gas flow, the bed continues to expand until a critical stage is reached at which gas bubbles form. These bubbles may burst at the top of the bed, or they may collapse, and the gas dissipates into the sediment. At this stage the particles in the expanded bed are violently agitated and the bed is said to be fluidized. The minimum velocity of gas needed to support the weight of individual particles is called *minimum fluidization velocity*. With a further increase in the rate of gas flow, more and more bubbles are formed until the solid particles become entirely entrained and are transported by the gas. Without the existence of bubbles there is no particle movement. The expanded bed has a density analogous to that of a liquid, so that objects of relatively low specific gravity float on its upper surface.

Bubbles in a fluidized bed bear a strong resemblance to gas bubbles in a liquid or to immiscible liquid globules in liquid. A bubble in a liquid is characterized by a continuous envelope at the interface while the bubble in the fluidized bed has a visible but porous envelope, allowing gas to flow through it while the particle cloud flows around it. The streamlines of gas flow converge toward the gas bubble, thus acquiring in the process a high gas velocity in the cloud around the bubble. The velocity decreases inside the bubble, giving rise to a sustained higher static pressure than at the bubble's exterior.

Bubbles may grow by accumulation of gas or by coalescence with other bubbles. The coalescence of bubbles in a fluidized bed is analogous to that of gas bubbles in a liquid. The bubbles, which have a density less than that of the surrounding medium, travel upward. There is a critical size of bubbles at which they can rise through the bed. Below this size the bubbles are stationary. This critical size depends on the rate of the gas flow and on the cohesion and internal friction between the particles that comprise the bed.

The phenomena observed with solid–gas fluidization are different from those with solid–liquid fluidization. Solid–liquid fluidization is generally described as being homogeneous, the bed being relatively uniform and the liquid a continuous phase. Solid–gas fluidization is regarded as nonhomogeneous since the gas is present as discontinuous bubbles (De Jong and Nomden, 1974).

Gas-fluidized beds have been observed to be susceptible to bubble formation because of bed inhomogeneity. However, a nonuniform flow field in the liquid fluidized bed produces fluid channeling and bulk circulation of solids. The two most important features of a gas-fluidized system that distinguish it from a liquid system are the electric charge of the particles induced by mutual collisions and the large difference in the density of materials constituting the various phases.

Three-phase fluidization occurs when a bed of solid particles is fluidized by two fluids, a liquid and a gas. The liquid forms a continuous phase whereas the gas is present as a discontinuous bubble phase. Two basic parameters affect the performance of this type of fluidization: (1) the diameter of the bubbles and (2) the degree of expansion of the bed (Bruce and Revel-Chion, 1974).

The diameter of the bubbles and their rising velocity increase with both increasing gas and liquid velocities. At the same time, the diameter of the bubbles and their rising velocity decrease with increasing particle size. At very low liquid velocities the bubble size is reduced by an increase in the gas flow rate, but when the liquid velocity is very high, the bubble size becomes independent of the gas flow rate.

With regard to bed expansion, the three-phase fluidization is unique in that when there is an increase in gas velocity, a reduction of bed expansion may occur. Contraction of the bed in this case is due to the formation of gas-liquid bubbles that reduce the residence time of the liquid. On the other hand, an increase of liquid velocity produces an increase of bed expansion.

There are great behavioral differences with different rock materials. With beds of minerals of very fine particle-size, interparticle forces tend to be larger than the gravitational forces acting upon them and the bed will not be well fluidized. With particles of a density less than 1.4 g/cm^3 and of mean diameter between 20 and 100 μm, the bed is likely to continue to expand uniformly beyond the point at which it becomes fluidized. The minimum fluidization velocity of the gas may often be exceeded by a factor of three before the system becomes unstable and bubbles form. With minerals of a density within the range 1.4–4.0 g/cm^3 and in the size range of 40–500 μm, bubbling begins at or only slightly in excess of the minimum fluidization velocity. Large and dense particles are characterized by a somewhat different behavior altogether. The rising bubbles induce particle circulation within the bed and have a marked effect on the overall bed flow properties of fluidized solids (Botterill and Bessant, 1973).

Reynolds (1954) described a number of nonvolcanic rocks that she considered to have been formed under fluidization and stressed the importance of this as a petrogenetic process with a bearing on the problem of intrusive granites. She distinguished two main fluidization systems and their corresponding products that depend on the gas velocity, (1) expanded systems and (2) entrained systems. In the first there is no bulk transport of solids and no flow banding. The second system contains material of alien derivation and there may be a flow banding because of a decreased gas flow rate near the margins of the fluidized system.

Sacchi (1971) described fluidization system products in volcanic vents in the Southern Alps. An uprushing gas fluidized a rock material, which consisted of a slurry of crystals in a liquid phase. The liquid reduced attrition rounding of solid particles. The resultant phase, which had the characteristics both of a solid–gas and a liquid–gas aerosol, rushed upwardly through cracks in the host rocks and tore fragments loose from the walls. Sintering completed the process and resulted in a rock with the appearance of an andesite.

Dionne (1976) described upward injection features such as miniature mud volcanoes[1], dikes forming polygonal patterns, and isolated patches of clay that were formed in tidal flats of cold regions. Although not directly related to the action of ice, it plays a role in their formation, because when ice lenses melt the clayey sediments arc fluidized and tend to flow upward when overloaded. Overloading by ice floes alone may create conditions appropriate to induce upward movement of fluidized clay. Neither seismic shock nor slumping is required to induce upward injection features.

---

1 The expression mud or sand volcano is used in sedimentary geology to designate various volcano-like features (Dionne, 1973).

Kimberlites are formed from highly mobile fluidized systems composed of solid xenoliths, a liquid phase (kimberlite magma), and a gas phase continually exsolved from the liquid. The gas, whose chief components are water and CO_2, can be generated in quantity at depths below the Moho discontinuity. The kimberlite magma is generated at a very great depth and its explosive ascent takes place at depths below 100 km. The solid phase contains mantle debris, including abundant fragments of garnet–peridotite and of eclogite that crystallize at high pressures. The upward flow of the fluidized system from the mantle to the crust occurs with violent turbulence. The resulting kimberlites are serpentinized porphyritic phlogopite peridotites that normally form minor intrusions (Harris and Middlemost, 1970).

4. Magma as a Colloid System

Magma is a molten rock material of a rather complex composition that contains volatile materials and is located within the earth. All igneous rocks are formed by the congealing of magma. If magma is erupted to the surface as lava, or exploded to the surface as ash, it forms *volcanic* rocks but if it crystallizes beneath the earth's surface, it forms *plutonic* rocks.

The chemical composition of a melt depends on its origin and history. Together with the main constituents, which are silica and metal oxides, a melt contains small amounts of sulfides, chlorides, carbonates, and volatiles, of which water is the most abundant. *Basaltic* magmas are rich in magnesium and iron and low in silicon and are therefore also called *mafic* magmas. *Granitic* magmas consist of sodium, potassium, aluminum, and silicon with large amounts of water. Silica is present in excess of the stoichiometric amount necessary to form feldspar, so that crystallization of the fluid results in feldspar and quartz. Because of its composition this magma is sometimes called *felsic*.

The "acidity" of magmas is defined on the basis of their silica content. A magma poor in silica, such as basalt (48–51%), is defined as "basic". Ultrabasic magmas contain even less silica than basalts. Granite with a high silica content (~73%, Wedepohl, 1971) solidifies from an "acidic" magma.

At atmospheric pressure pure silicon dioxide melts at 1723 ± 5 °C (JANAF, 1971). Mg olivines, which have high lattice energies, melt at temperatures above 1800 °C while the melting temperatures of most other rock-forming minerals fluctuate within the limits of 1000–1500 °C, which is lower than the melting point of pure SiO_2. Under these conditions basalts melt at 985–1260 °C and granites, at 1215–1260 °C. Temperature measurements of lavas from Stromboli, Etna, and Vesuvius show values of 1000–1100 °C and those from Kilauea values of 1000–1200 °C (Mcdonald, 1972). Owing to higher pressure at depth, the magma contains a considerable quantity of volatile components that lower its melting temperature. Therefore, it may be assumed that basaltic magmas only rarely attain a temperature of 1250 °C, often less than 1000 °C, and sometimes even less than 850 °C. Acid magma has a temperature of 850 °C or even somewhat less.

Some processes of differentiation of magmas that lead to inhomogeneity in the chemical and mineralogic composition can be attributed to colloid behavior of the magma. A magmatic melt should be considered as a system with a certain energy reserve (Zavaritskii and Sobolev, 1964). This energy can be transmitted to the rocks surrounding the magma chamber by heat transfer, accompanied by a decrease in temperature of the magma. The energy can also be exchanged during upward movement of the magma. As the liquid rises toward the surface, the pressure resulting from the weight of the overlying rocks decreases. If temperature and pressure are sufficiently high, the magma consists of a single liquid phase, containing soluble polymeric species of high molecular weights, defined as *magmaphilic colloids.* On cooling, crystallization of solid particles of a new silicate phase begins. These solid particles are dispersed in the system and are therefore considered as *magmaphobic colloids.* Given the appropriate chemical composition, the appearance of globules of another liquid phase, such as a sulfide phase, is possible. As crystallization progresses, the volume of remaining liquid becomes smaller and smaller leading to the concentration of volatiles, until the solution becomes supersaturated and bubbles of the volatiles, which are dispersed in the residue of the molten magma, appear, giving rise to foam. These processes, which may be summarized as phenomena of polymerization, differentiation, crystallization, and degasification, occur at the expense of the energy reserve and lead to the formation of colloid systems. The occurrence of interface boundaries in magma is responsible for many chemical reactions that are either catalyzed by the interface or in which the interface participates. These reactions determine many of the properties of the resulting rock, such as the distribution of certain elements, the oxidation state of certain elements, the crystal growth of certain minerals, and other properties that will be now described.

As long as any of the condensed silicate species in magma contain a small number of atoms, it may be regarded as comprising one phase with the melt. However, when it contains a few hundred atoms or more, and the structural order of atoms in the species has a longer range than in liquid, it must be regarded as a separate phase and the system will be colloidal. Colloids dispersed in the magma can be either magmaphilic or magmaphobic, depending on whether the energy obtained by their solvation by the magma liquid is higher or lower, respectively, than the sum of their aggregation energy and the dissociation energy of the liquid silicate (see pages 3–5).

4.1 Magmaphilic Polymeric Structures

It is difficult to obtain an accurate picture of the fine structure of discrete molecules and ions among the species that comprise a molten magma. The structures to be described are therefore mainly speculative. They are inferred from the viscosity of melted silicates and from data for liquids in general, from silicates in the solid state as well as from oxides of phosphorus that have low molecular weights even at room temperature and from the chemical properties of the molten magma. This difficulty arises chiefly from the fact that conventional methods for the study of the structure of liquids are applicable at temperatures that are much lower than those of molten magmas. X-ray and other studies indicate that there is some similarity of structure between a liquid and a solid. In the liquid, "crystallinity" extends over only a few molecular diameters. The arrangement of nearest neighbors around any particular molecule in a liquid just above the melting point exhibits "short-range order" and is very similar to that in the crystal just below the melting point.

4.1.1 Polymers in Molten Silica

The species that comprise the bulk of the molten SiO_2 should contain "monomers" of only a small number of atoms and "polymers" that result from the clustering of the monomers. The concentrations of the various polymeric and monomeric species should be in equilibrium, depending on the composition, pressure, and temperature of the melt. When equilibrium is disturbed by a decrease in temperature some of the monomeric species polymerize and the small polymers become larger. When equilibrium is disturbed by an increase in temperature the opposite occurs.

As was explained in the Introduction, monomeric SiO_2 is unstable, in contrast with monomeric CO_2, which is stable. A tetramer such as Si_4O_8 (Scheme 1.3) can be

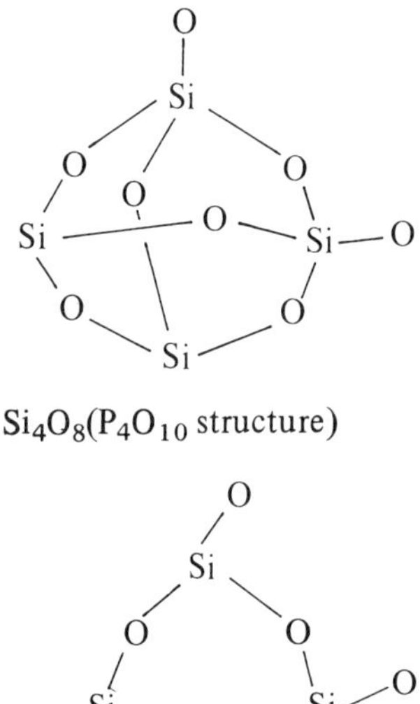

Si_4O_8(P_4O_{10} structure)

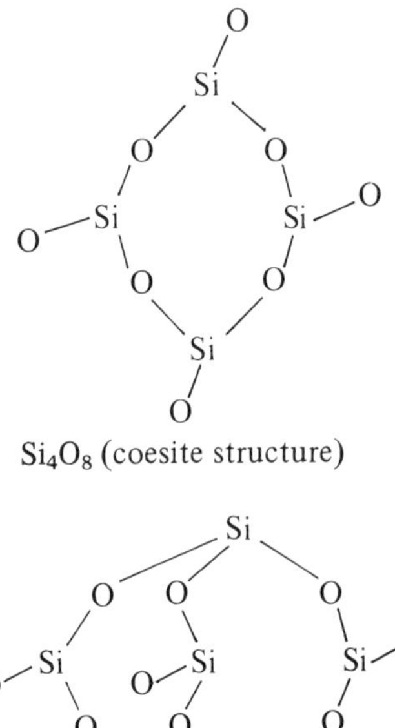

Si_4O_8 (coesite structure)

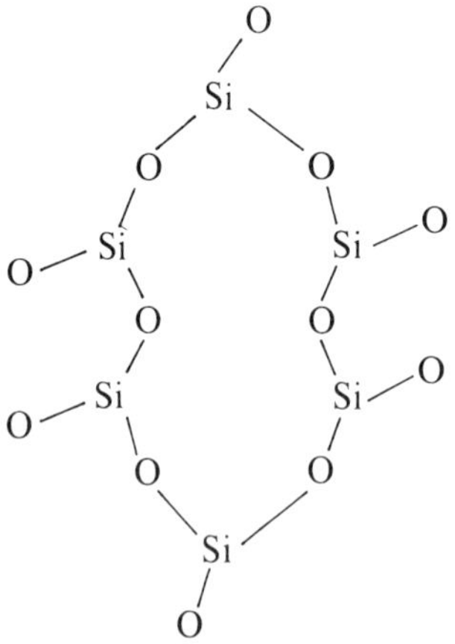

Si_6O_{12} (tridymite structure)

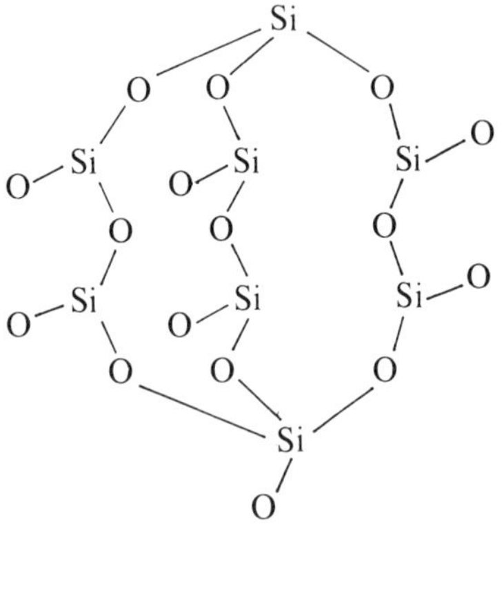

Si_8O_{16} (ice molecule, tridymite structure)

The Si ↔ O distances for individual tetrahedra in tridymite structure range from 0.158–0.162 nm. The Si ↔ Si nearest-neighbor distances range from 0.303–0.318 nm, the Si–O–Si angles are averaging 148.3° and range from 139.7–173.2°. The O–Si–O angles range from 106.2–114.1°.

Scheme 1.3. Possible basic units of Si_nO_{2n} in polymers of molten silica considered as a simple basic unit (monomer) in which every Si atom occupies the apex of a tetrahedron and which satisfies as nearly as possible the chemical demands of silicon and oxygen atoms discussed previously in connection with the crystal chemistry of silicates. This molecule has a structure similar to that of gaseous P_4O_{10}, and is labile because two Si and two O atoms participating in the molecule have one unsatisfied valency. This molecule tends to polymerize and to form a cluster. The tetramer, Si_4O_8, with a higher degree of undersaturation, has a ring structure and is the basic unit of coesite (Naka et al., 1976). Other forms for small molecules are Si_6O_{12}, a six-member ring, or Si_8O_{16}, which has a structure similar to that of the "ice molecule". Both species can be considered as basic units of tridymite. Half of their atoms also have unsatisfied valencies and they tend to form clusters.

Scherer et al., (1970) claim that the low values of entropy of fusion for the quartz and cristobalite polymorphs of silica indicate a similarity in the coordination polyhedra and their arrangement in the liquid and crystal modifications. The polymers obtained from the melting of these polymorphs are suggested to be composed mainly of the Si_6O_{12} ring units and the Si_8O_{16} "ice molecule" units. Both units have an unoccupied central space thus giving them a large *atomic volume* (i.e., volume per atom). The result of increased pressure is to form a more dense liquid that may feature silicon atoms in Si_4O_8 rings, similar to that existing in coesite. At still higher pressures silicon atoms achieve partial octahedral coordination such as in stishovite. The melting points of SiO_2 rise with the applied pressure (Jackson, 1976).

Oxygen atoms that form covalent bonds with two silicon atoms are called "bridging" oxygens whereas those that form covalent bonds with only one silicon atom are called "nonbridging" oxygens. The polymerization process involves both the interaction between two nonbridging oxygens, which gives rise to one bridging oxygen and one ionic O^{-2} as follows.

$$\geq Si-O^- + {}^-O-Si\leq \rightleftharpoons -Si-O-Si\leq + O^{2-}, \qquad (1.I)$$

and/or the linking between a nonbridging oxygen and a silicon having unsatisfied valencies as follows

$$\geq Si-O^- + {}^+Si\leq \rightleftharpoons \geq Si-O-Si\leq. \qquad (1.II)$$

The average size of the clusters depends on the heat content of the system and becomes smaller with a rise of temperature. In a liquid, bonding occurs at only one site between neighboring monomeric species so that the species are free to rotate about one another, and the short-range order and the long-range disorder are maintained.

The formation of small species during the melting of SiO_2- occurs to a very limited extent because the cohesive forces between the small species are very high in the fraction that first depolymerizes at the melting point. This is apparent from the molar heat of fusion, ΔQ_m, which is equal to 1.95 kcal/mol and is much smaller than the molar heat of vaporization, ΔQ_{vap}, which is equal to 123.5 kcal/mol (small species must be formed during vaporization). Furthermore, the specific heat of silica is affected only slightly by fusion. At the melting point the specific heat is 18.47 and 21.66 cal/mol per one degree for the solid and the liquid phases respectively (Carmichael et al., 1974).

4.1.2 Polymers in Molten Silicates

The probability of species smaller than Si_6O_{12} existing at the melting point of silicon dioxide is extremely low, and it increases with increasing temperature. In the presence of basic oxides, such as MgO and FeO, the stability of the small species increases due to the interaction of the molten silica with the oxide. Two types of interactions, between the silica species and the basic oxide can be distinguished:

1. The interaction can be regarded as a solvation process in which the metallic cation with a coordination number other than four, forms bonds with the nonbridging oxygen atoms of silicate species. The electrostatic field induced by the charged metallic ion favors a high density of oxygen atoms in its vicinity. Thus, a rearrangement of the "low-density" monomeric six-member ring or "ice molecule" units occurs. Polymeric chains and double chains, similar to those of pyroxenes and amphiboles, but much shorter, are formed. The length of the chains depends on the temperature and increases with decreasing temperature. Metals that participate in rearrangement of the structure of molten silica and lead to the formation of shorter polymers are called "structure breakers". The Al atom can take part in both 4- and 6-fold coordinations and is not necessarily a structure breaker. Waff (1975) showed that pressure-induced coordination changes of Al from tetrahedral into octahedral sites are likely to occur at pressures lower than 35 kbar in tholeitic and andesitic melts.

2. The basic-oxide donates O^{2-} to the silica species, the oxygen forms a bond with an unsaturated Si atom, giving rise to negatively charged silicate ions such as SiO_4^{-4}, $Si_2O_6^{-4}$, and $Si_4O_{11}^{-6}$. The process can be expressed by

$$\geq Si^{+} + O^{-2} \rightleftharpoons \geq Si{-}O^{-}. \tag{1.III}$$

From (1.I), (1.II), and (1.III) it is obvious that the degree of polymerization in a magma can be expressed by means of O^{2-} activity. A melt with no polymers (and no bridging oxygens) has by definition an O^{2-} activity of 1.

The concentration of the small species is very low at the melting point of silicates, as can be concluded from the fact that silicate minerals fuse with only a small increase in entropy and in volume (5–15%). This suggests that the silicates are compounds in which strong structural similarities exist between the crystalline phases and their melts, as occurs in silica systems. From X-ray studies Urness (1966) showed that the melting process of silicates involves the destruction of long-range order and that short-range order is preserved.

Under high pressure water reacts chemically with silicate melts, serving as a structure breaker. It enters the melt by reacting with bridging oxygens, forming –Si–OH groups at the edges of the polymers as follows:

$$\geq Si{-}O{-}Si + HOH \rightleftharpoons 2 \geq Si{-}OH \tag{1.IV}$$

and thereby reduces the melting temperature of the silicates. The effect of water on the structure of polymers in the silicate melt is highly pronounced in the presence of metals having a coordination number 6, such as Mg, Fe, and Al. These cations form hydroxy complexes of the type $[O_3Si{-}O{-}Mg(OH)_x{-}O{-}SiO_3]^{-(6+x)}$, which, under high pressure, are stable at relatively high temperatures. These complexes have a tendency to polymerize, giving rise to layered structures, similar to those of talc or micas but much smaller in size.

Magmas rich in alkali metals also retain high contents of water dissolved in the molten silicate. In such compositions the final crystallization is delayed to a much lower temperature than in less alkaline and less volatile-rich magmas (Sood and Edgar, 1970).

The solubility of H_2O in a $NaAlSi_3O_8$ melt is nearly 10% greater than in a SiO_2 melt under a pressure of 2 kbar and at 1100 °C. According to Burnham (1975) a process in addition to depolymerization is involved in the dissolution and is responsible for this increment in the dissolution. This process is an ion exchange of Na by H according to the following equation

$$NaAlSi_3O_8 + HOH \rightleftharpoons H^+ + AlSi_3O_8^- + Na^+ + OH^-. \qquad (1.V)$$

The exchange of any cation not having a tetrahedral coordination by a proton results in very high electric mobility of the melt. This proton is electrostatically bonded to the ionic polymer and its electric mobility is very high, whereas the proton that belongs to the –Si–OH group and that is covalently bonded to a nonbridging oxygen, is only slightly mobile (Fig. 1–6a, b).

CO_2, NH_3, or HCl in the presence of water, raises the melting point of granite, whereas addition of HF or P_2O_5 has the reverse effect as does the presence of water only (Carmichael et al., 1974). The solubility of CO_2, NH_3, and HCl is predominantly physical, the magmaphobic solutes occupying interpolymer voids in the liquid phase. Bigger silicate polymers form bigger voids, which may serve as cages for the dissolved gas molecules. Such solutes are therefore "structure makers". The polymerization process is of a "magmaphobic solvation" type, and there are some analogies between this process and hydrophobic hydration formation in liquid water under the influence of hydrophobic solutes (see Introduction). The solubility of fluorine is of a chemical type. The F^- ion (0.133 nm) because of its similarity to O^{2-} (0.132 nm) will break up some of the Si–O chains, forming edge Si–F bonds instead (Buerger, 1948), or exchange OH groups in the complex $[O_3Si–O–(Mg, Al, Fe)(OH)_x–O–SiO_3]$ (Bailey, 1977).

The following chemical equations show that CO_2, NH_3, and HCl can be considered as polymerizing agents (structure makers) and HF or P_2O_5 can be considered as depolymerizing agents (structure breakers):

$$2\ R–Si–OH + CO_2 \rightarrow R–Si–O–Si–R + \underline{H_2O.CO_2} \qquad (1.VI)$$

$$2\ R–Si–OH + NH_3 \rightarrow R–Si–O–Si–R + \underline{HOH.NH_3} \qquad (1.VII)$$

$$2\ R–Si–OH + HCl \rightarrow R–Si–O–Si–R + \underline{H_2O.HCl} \qquad (1.VIII)$$

$$R–Si–O–Si–R + HF \rightarrow R–Si–OH + R–Si–F \qquad (1.IX)$$

$$R–Si–O–Si–R + 2\ H_2O + P_2O_5 \rightarrow 2\ R–Si–O–PO_3H_2 \qquad (1.X)$$

where R is a silicate radical.

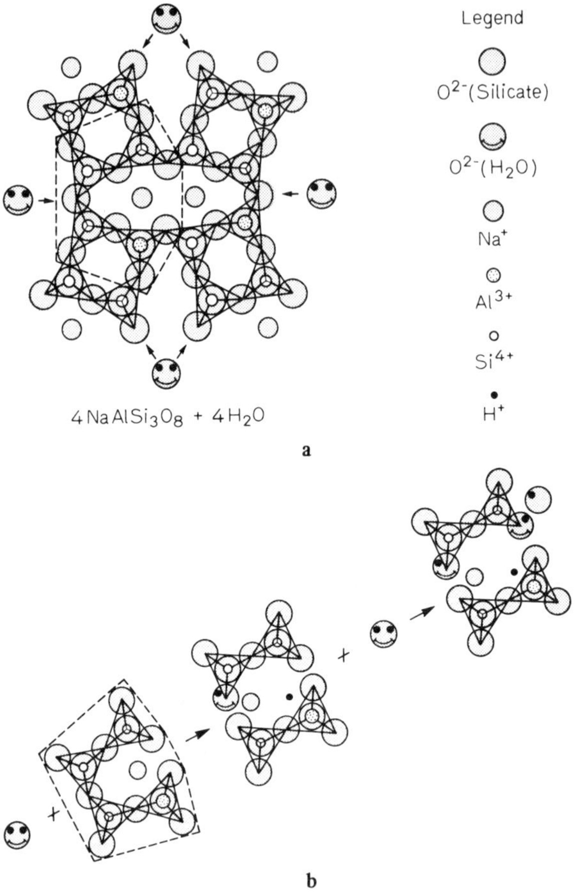

Fig. 1.6a and b. Representative schemes for the interaction of water with polymers of silicates. **a** Projection of the relative atomic position in albite, showing proposed bridging oxygen sites for the initial reaction with H_2O; **b** proposed reaction scheme for the dissolution of H_2O in $NaAlSi_3O_8$ melt. (From Burnham, 1975)

The association products of water and volatiles, underlined in the chemical equations (1.VI)–(1.VIII) comprise *quasi-inverted micelles,* located inside the cages, in which the water component of the association product acts as a bridge between the silicate liquid and the appropriate volatile (Introduction, Fig. 12). Such water associations are unstable under atmospheric conditions at high temperatures of the order of those found in magmas. In magmas at high temperatures and high pressures these associations are stabilized as the result of the magmaphobic character of the appropriate volatiles and the magmaphilicity of water molecules. Water supplies protons, which act as catalysts in the above reactions. The occurrence of association products of water and volatiles is an important factor in the generation of fluids during the evolution of magmas. According to Holloway (1976) if there are some volatiles in the magma chamber that are structure makers, H_2O must be part of the coexisting volatile phase even when the magma is undersaturated with respect to water.

Magmaphilic colloids remain dispersed in the magma until conditions arise that allow them to become magmaphobic. The magmaphilic colloids are never found as such in igneous rocks. Magmaphobic colloids will, under suitable magmatic conditions, settle out and separate from the melt. After consolidation these magmaphobic colloids (generally crystals) comprise the igneous rocks. Let us examine this phenomenon for a few cases.

4.1.3 Three-dimensional Polymerization

Three-dimensional polymerization of SiO_2 may lead to macromolecules having an atomic packing similar to that of quartz but extending over several hundred atomic diameters only. Since the bonds within the macromolecule are primarily covalent, the energy released from the solvation of the macromolecule by the magma melt is low. Cohesive forces that lead to aggregation are primarily van der Waals. These forces are very weak unless the dimensions of the polymer are huge. This quasi-quartz colloid is magmaphilic as long as the particle is small, but becomes magmaphobic when the size of the particle increases and the cohesive van der Waals forces also increase. Quartz, therefore, does not separate from the magma, unless the concentration of silica in the magma is high enough to form huge macromolecules.

If an Al atom is substituted for a Si atom, the polymer gains a negative charge. The macromolecule then has an atomic packing similar to that of feldspars. Due to the surface charge, solvation energies (adhesion) are high in the quasi-feldspar polymer, in comparison to those of the quasi-quartz. However, polymers of much smaller size than those of quasi-quartz become magmaphobic because cohesive forces leading to aggregation are primarily electrostatic and are strong. The stage at which the quasi-feldspar colloid changes from magmaphilic into magmaphobic depends on the charge of the particle and the metallic cation. The quasi-Ca plagioclase precedes the quasi-Na plagioclase because of its higher charge density. Quasi-orthoclase becomes magmaphobic at temperatures lower than quasi-Na plagioclase because the K ion has a larger radius than the Na ion, resulting in a lower electrostatic cohesive force.

4.1.4 Two-dimensional Polymerization

Two dimensional polymerization occurs in the presence of Al, Mg, or Fe. The packing of the atoms in the macromolecule is similar to that found in talc or in micas, but it extends over several hundred atomic diameters only. Quasi-talc and quasi-pyrophyllite polymers are nonpolar. Cohesive forces between the layers are of the van der Waals type. The chemical compositions of most natural magmas, which are at equilibrium with respect to temperature and pressure, do not allow for extensive growth of quasi-talc and quasi-pyrophyllite parallel to *a* and *b* axes. Hence the van der Waals forces are weak and the layers do not stack above the other in the *c* direction. These colloids are magmaphilic and do not separate from the magma.

The quasi-mica polymers are charged. The layers stack one above the other in the *c* direction even if their diameters are small. Due to the high electrostatic cohesive forces these polymers are magmaphobic and separate from the magma.

4.2 Magmaphobic Colloids

Solids in the magma originate from two sources: (1) from solid rocks by disintegration and (2) from the molten magma by crystallization.

4.2.1 Disintegration of Solid Rock Material

Initially there is a partial melting of preexisting silicate rocks. As the melt phase increases in volume in proportion to the solid, it circulates with greater ease, so dispersing the remaining solid crystals in the system.

The melt may segregate from the surrounding grains, leaving the bed of solid grains behind. A small fraction of melt segregates at depths below 60 km to form certain alkaline basalts (Green, 1971). In midoceanic ridges segregation begins at less than 30 km (Kay et al., 1970). According to Sleep (1974) favorable conditions for the segregation of melt from an upwelling fluidized bed include a high concentration of melt, a narrow conduit containing the fluidized bed, a large grain size, and a large ratio of the viscosity of the fluidized bed and the viscosity of the pure melt.

4.2.2 Formation of a New Solid Phase by Condensation

According to Bowen (1928), with decreasing temperature of an ideal mafic magma, two groups of minerals crystallize, one according to a discontinuous reaction series and the other according to a continuous reaction series. This is best illustrated by the following scheme:

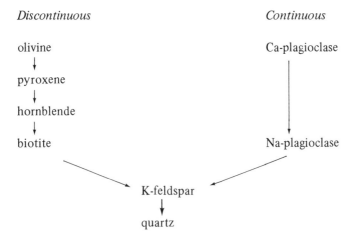

Discontinuous — *Continuous*

olivine → pyroxene → hornblende → biotite

Ca-plagioclase → Na-plagioclase

biotite → K-feldspar ← Na-plagioclase

K-feldspar → quartz

The effect of temperature on the chemical composition of the crystallized feldspar (the continuous reaction series) was discussed previously. The effect of temperature on the mineralogy of the condensed phase in the discontinuous reaction series is shown by (1) the relation between thermal dissociation and chemical equilibrium of polymerization–the average size of the silicate polymer increases with the fall in temperature–and (2) the relation between the temperature of crystallization and the lattice energy of the mineral–minerals with high lattice energies are stable at high temperatures while those with low lattice energies become stable only at lower temperatures.

For a gram-formula of a mineral having an ionic crystal, the lattice energy U is given by

$$U = N\left(-\frac{Az_1z_2e^2}{r} + \frac{B}{r^n}\right) \tag{1.2}$$

where N is the Avogadro number, A is the Madelung constant, a geometric factor that takes into account the various ions in the crystal, B and n are constants of repulsion, z_1 and z_2 are the ionic charge numbers of the cations and anions, respectively, e is the electric charge on one electron, and r is the internuclear separation between the cation and anion. In complex crystals such as those of the silicates, it is difficult to evaluate the various constants in Eq. (1.2). Nevertheless, a satisfactory correlation between lattice energies and polymerization of the silicates has been found.

The internuclear separation in the following calculations is defined as the distance between the centers of the M and Si atoms, where M is the metallic cation. This value is valid for the nesosilicates. As an approximation, the internuclear separation is assumed to be the same distance for all silicates having polymeric anions. In this case the *effective charge* of the silicate anions is obtained by dividing the ionic charge by the number of Si atoms in the ion. This effective charge is –4 for SiO_4^{-4} and –3 for $Si_2O_7^{-6}$, in the neso- and sorosilicates respectively, –2 and –1.5 in the chain and ribbon structures, respectively. From Eq. 1.2 it follows that the lattice energy, and hence temperature of crystallization, decreases from olivine to pyroxenes and from pyroxenes to amphiboles.

The newly formed crystals dispersed in the molten magma may be removed from the melt by settling out. The accumulated crystals are surrounded by intercumulus liquid (Wager et al., 1960) forming a liquid in solid system. The particles may form aggregates before separation from the melt, as was observed by Flood et al. (1977) in rhyodacite, in which hornblende and biotite nucleated in part as separate grains, but also nucleated about the pyroxenes of the plagioclase/pyroxene aggregates.

As the magma cools, the solid particles that remain dispersed in the chamber of the magma react with the melt to form the next mineral in the series. Rims of pyroxene around olivine represent failure to maintain complete equilibrium between dispersed particles and molten phase during cooling.

The formation of quartz and alkali feldspar depends not only on the cooling of the system, but also on the removal of the first-formed crystals of the Bowen reaction series from the system. The differentiation sometimes observed in thick sills and flows of mafic rocks, such as the development of an olivine-rich layer just above the basal chilled border or the development of an interstitial granophyre (a mixture of quartz and K-feldspar) near the top of the body, is due to the settling out of heavy, early formed olivine crystals, with a resulting enrichment of the remaining magma in silica. A quantitative model for the settling of olivine crystals in a Newtonian fluid in a sill was developed by Fujü (1974).

4.3 Emulsions of Immiscible Liquids

4.3.1 Emulsion of Sulfide Globules

The four important variables controlling the solubility of sulfur in magmas are (1) the bulk composition of the melt, (2) the temperature, (3) the partial pressure of oxygen, and (4) the partial pressure of sulfur above the melt. Sulfur dissolved in the silicate melt at high partial pressure of oxygen occurs in the form of the sulfate species, whereas sulfur under low partial pressure of oxygen occurs as the dissolved sulfide species. The dissolved sulfur species in mafic magmas are sulfides. The solubility of sulfur in mafic magmas increases with ferrous iron content. At lower temperatures the capacity of a magma to dissolve sulfur is lessened (see Ch. 2).

The formation of immiscible sulfide liquids in mafic magmas was discussed by Haughton et al. (1974). As the magma cools, it reaches saturation and commences separating an immiscible sulfide liquid. If the magma is subjected to continuous movement, the separating sulfide globules will remain suspended and not accumulate. If the magma chamber is large and relatively stable, the sulfide liquid separates and accumulates. The common observation that many layered intrusives have a sulfide-rich layer near their base indicates that the parent magmas already contained sulfide globules when they were intruded.

As the magma cools, competing factors will influence sulfide precipitation. The principal factors are compositional. Assuming the system to be closed to sulfur, the crystallization of any silicates or oxides will increase the relative concentration of sulfur in the melt. On the other hand, separation of plagioclase and of pyroxene, which are less iron-rich than the parent magma, will cause the magma to become enriched in FeO and may cause the magma to be undersaturated with respect to sulfide. Sulfide precipitation should, therefore, cease. When chromite, ilmenite, or magnetite start crystallizing, the magma becomes depleted in FeO and sulfide liquid may again be able to separate. The sulfide globules may then sink and accumulate with the above oxides.

4.3.2 Emulsion of Carbonate Globules

A magma rich in carbonates, under certain conditions, can be composed of two immiscible melts. The exact composition of the two melts is unknown, but one is certainly a silicate-rich melt and the other carbonate-rich. Immiscibility of the two melts occurs in soda-rich magmas but not in the lime-rich systems. Immiscibility may be an important phenomenon in the derivation of carbonatite fluids from parent carbonated silicate magmas. Carbonate-rich melts and silicate-rich melts are immiscible in natural ijolitic magmas. Other phases such as CO_2 vapor and an aqueous, saline, carbonate bearing fluid coexist with these two melts. According to Rankin and LeBas (1974) a hyperalkaline silicate parent magma can produce an emulsion of carbonatite magma in ijolite magma.

4.4 Foam of Volatiles

The residual fluid from a crystallizing magma shows a concentration of substances having low melting and boiling points, the so-called volatiles, which were originally dissolved in the molten silicate. The volatiles consist mainly of water, and also of carbon dioxide, hydrogen halides, sulfur compounds, nitrogen compounds, boron compounds, and noble gases. Their content varies from one body of magma to another. Two groups of volatiles differing greatly in solubility can be distinguished. The first group consists of those gases that do not react chemically with the silicate melt but may penetrate into the interpolymer vacancies in the melt and that have a very low solubility. The solubility of the gases increases in the order: $Ar = N_2 < CO_2 < NH_3 = HCl$. The second group consists of those gases that interact chemically with silicates, breaking down their polymers. These are highly soluble components. Their solubility increases in the order: $H_2O < SO_2 < HF$.

Separation of a gas phase from a magma can be brought about by a change in pressure as well as by increase of the concentration of the volatiles in the liquid phase during crystallization. Magma that rises from depth and/or loses heat, becomes supersaturated and the volatiles separate as bubbles. The stage at which the formation and separation of the new phase begins is called the "second boiling point". The first portion of fluids released by the "boiling" magma is rich in components with a low solubility. The most rapid release of the soluble components, H_2O, SO_2, and HF occurs in the final stages of the process.

At first the bubbles are very small. The decreasing pressure allows them to expand. Collisions between bubbles may lead to their coalescence. However, not all the collisions are effective. As more bubbles form, the frequency of collisions increases, resulting in a greater coalescence.

Water is the predominant volatile in most magmas, followed by carbon dioxide. The amount of water dissolved in a granitic liquid increases indefinitely as the pressure is raised, but cannot exceed 15% by weight at temperatures and pressures to be expected in a cooling magma. The solubility of water is only slightly affected by a change in composition of the molten magma. Granite dissolves more water than basalt and basalt dissolves more water than melts of ultrabasic composition (Scarfe, 1973). The solubility of water decreases with increasing partial pressure of CO_2 (Kadik and Lukanin, 1973). The solubility of carbon dioxide in mantle magmas is less than about 5 wt % at depths less than 80 km. Solubility increases abruptly to about 40 wt % at depths greater than 80 km (Wyllie and Huang, 1976).

Two types of water bubbles are dispersed in the magma, one a less dense vapor bubble, consisting of molecules belonging to the gaseous phase, the other consisting of molecules belonging to a dense quasi-liquid phase. In the first case the system has the properties of a foam whereas that of the bubbles of the quasi-liquid phase has the properties of an emulsion. Those droplets comprising the quasi-liquid phase are actually aqueous solutions of salts, mainly NaCl. It should be remembered that the magma is at temperatures above the critical temperature of water–374 °C for pure water. At this temperature no chemical interaction between H_2O molecules can occur and water will not liquefy under any pressure. Since salts have strong polarizing effect on water molecules, bonds between these molecules can form in the presence of salts, if a suitable pressure is applied at temperatures higher than the critical point. For example, a 2% solution of NaCl has a critical temperature of 399 °C and a 5% solution, of 424 °C. According to Ryabchikov and Hamilton (1971) the vapor holds 5 ± 2 wt % NaCl and the quasi-liquid may hold up to 70 ± 10 wt % NaCl.

Compounds much less soluble than NaCl behave differently. Silica, for example, raises the critical temperature of water only slightly, even when excess solid is present, simply because its aqueous solubility is very low at temperatures around 400 °C and also because water molecules are only slightly polarized by the silica.

The generation of separate, water-rich fluids of magmatic affiliation has been discussed by Burnham (1967). In general, saline supercritical fluids are alkali chloride-bearing aqueous fluids with variable concentrations of CO_2 and minor amounts of sulfur. Alkali chlorides are almost always the dominant constituent, and in many cases are present in concentrations of 40 or more wt %. In natural environments, fluid inclusions in crystals provide the best evidence for the existence of supercritical fluids. Fluid inclusions in minerals are exceedingly small metastable chemical systems, very effectively isolated from their surroundings (Roedder, 1971).

The appearance of the bubbles in the system drastically changes the mechanism of crystallization of the magma. Relatively large, well-formed crystals of quartz and feldspar grow as the proportion of the water-rich phase increases. Crystallization takes place chiefly from the viscous, silicate-rich phase, but the existence of a water-rich fraction provides space for the growing crystals to acquire the texture of pegmatites.

Being lighter than the surrounding liquid, the bubbles tend to rise. If the magma is stationary, all the gas bubbles in time may rise to the top and escape. However, because the liquid is very viscous, the bubbles rise slowly, and if the magma is rising toward the surface, the rate of rise of the bubbles may be barely faster than that of the surrounding liquid. The upward migration of the saline supercritical fluid bubbles leads to an upward enrichment in alkali metals within the magma body (Ewart, 1965).

5. Colloid Systems in Volcanic Eruptions

The present model is based on that of Itamar (1975) supplemented by data from Macdonald (1972) and from more recent publications. A volcano is an aperture in the earth's crust from which molten rocks or gases discharge from the earth's interior. Flow systems that erupt from the volcanoes are colloid systems in which the dispersing medium is either a liquid or a gas, and they are called *lava* and *volcanic smoke*, respectively.

As the magma rises, the lowering of the pressure allows the gases to exsolve from the liquid in which they are held. Initially the gases are dispersed in the liquid as bubbles forming a foam. As the volume of gas increases in relation to that of the liquid, the surface tension of the liquid is overcome and the colloid system changes from a foam into an aerosol of droplets dispersed in the gaseous phase. The gas released from the melt can diffuse from the magma chamber through the solid rock comprising the chamber wall, or through beds of nonconsolidated rock material and may act as a fluidizing agent. When the rate of gas release from the magma is higher than the rate of gas escape from the magma chamber through the surrounding rocks, a hydrostatic

Table 1.2. Classification of ejecta. (After Macdonald, 1972)

Size of fragments (average diameter in mm)	Shape of fragments	Condition of ejection	Names: individual fragments	Names: accumulation of fragments
> 64	Round to subangular	Plastic	Bombs	Agglomerate
	Angular	Solid	Blocks	Breccia
64.0–2.0	Round to angular	Liquid or solid	Lapilli	Lapilli agglomerate or lapilli breccia
2.0–0.062	Generally angular, but may	Liquid or solid	Coarse ash	Ash when unconsolidated Tuff when
< 0.062	be round		Fine ash	consolidated

Table 1.3. Types of colloid systems in volcanic eruptions. (After Itamar, 1975)

Colloid system	Dispersing medium	Dispersed material	Volcanic system
Solid in gas (plus small amounts of lava globules)	Mainly air	Ash, fragments, crystalline material, pumice	Ash cloud
Liquid in gas (plus vitreous solid particles)	Mainly volatiles	Lava globules, ash, fragments, pumice	Pyroclastic flow (Glowing avalanche and ash flow)
Solid in liquid	Silicate melt	Fragments, crystals	Lava and lava flow
	Water (ocean and sea)	Ash, fragments, crystals, pumice	Ash suspension and subaqueous debris flow
	Water (crater-lake and stream)	Ash, fragments, crystals, pumice	Mud flow
Liquid in liquid	Silicate melt	Sulfide globules	Lava and lava flow
	Water	Silicate melt	Pillow lava flow
Gas in liquid	Silicate melt	Volatiles	Lava and froth flow
Gas in solid	Silicate glass	Volatiles	Pumice and scoria
Solid in mixture of liquid and gas	Frothy, gas-filled glassy lava	Crystalline material	Ground surge

pressure is obtained. When this pressure becomes sufficiently high to break the overlying rocks, a volcanic explosion takes place. Volcanic explosions sometimes result from the expansion of external water coming in contact with a magma (Schminke et al., 1973).

The released gases may carry with them frothy bits of molten lava and chunks of rock torn from the fissures through which they move. Fluid lava commonly follows the original outburst of gas, welling up along fissures opened by the pressure from beneath. As the eruptions continues, fluid lava containing fewer gas bubbles continues to pour out.

Volcanic explosions range in intensity from the weak spattering that commonly accompanies the eruption of very fluid basaltic lava to cataclysmic blasts that throw debris many kilometers up into the atmosphere. Fragments thrown out by volcanic explosions, called *ejecta,* accumulate to form *pyroclastic rocks.* Ejecta may also be dispersed in the ocean and in the atmosphere. Their classification is given in Table 1.2. Ejecta may derive from the molten magma itself or from rock previously solidified from magma of the same or even earlier eruptions, or from still older unrelated rocks from the crust underlying the volcano.

Some of the common colloid systems in volcanic eruptions are described in Table 1.3.

5.1 Lava

This fluid is new rock-forming material brought up from the depths. The disperse phases include solid particles, globules of immiscible liquids, such as sulfides, bubbles of volatiles, and magmaphilic polymers. The solid materials dispersed in the lava reach to the surface together with the ejected melt, or are incorporated from older rocks or from solidification of the melt. When the lava is flowing in a channel, the upper section quickly cools to a semisolid, highly viscous state that comprises a mixture of amorphous magmaphobic and magmaphilic polymers. The crust over a flow can be broken up by the drag of the lava flowing beneath it and the solid particles then become suspended in the flowing lava.

Jostling of blocks as they are moved by the lava results in abrasion and the formation of a certain amount of fine debris with particle sizes of sand and gravel. The fragmented portion ranges from a few percent to more than 50% of the flows. The concentration of the solid material has a great effect on the viscosity of the lava and hence on the nature of the rock obtained after solidification (Christiansen and Lipman, 1966).

Volatiles, either dissolved in the melt or as gas bubbles, may reach the surface together with the ejected melt. Bubbles are also formed when lava flows move over wet ground, by formation of steam which rises as bubbles into the lava. Lava flows may be shattered by steam explosions yet still retain the general form of their flow (Fiske et al., 1963).

If the solidification of the lava is very slow, all the gas bubbles may escape, but if cooling is rapid, the liquid freezes and entrains the gas bubbles. The voids caused by the bubbles are called vesicles and the rock is said to be vesicular. The vesicles may be nearly perfect spheres, but more often they are elongated by the flow of the surrounding liquid as it solidified, forming ellipsoidal or almond-shaped openings. In some cases the openings are twisted and greatly distorted (Whitford-Stark, 1973).

Postmagmatic solutions moving through the rock after it has solidified carry dissolved mineral matter that may be deposited in the vesicles. If they are completely filled with mineral matter, the filled voids are called *amygdules* and the rock is called *amygdaloid.*

5.2 Volcanic Smokes

Aerosols obtained from volcanic eruptions are called herein "volcanic smokes". The dispersing gas medium consists of volatiles and of air. These systems may be divided into two groups:

(1) *ash cloud,* in which the dispersing gas is mainly atmospheric air, and (2) *pyroclastic flow* (which includes *glowing avalanche* and *ash flow*), in which the dispersing medium is mainly volcanic volatiles. The densities of smokes of the first group are much lower than those of the second group. The former are borne aloft by atmospheric currents, sometimes reaching heights of tens of kilometers, whereas smokes of the second group are fluidized by exsolved magmatic gases or by air trapped beneath the advancing flow and travel as a dense mass, subjected to the influence of gravity and flow, just as liquids, along topographic contours.

The temperature of the ash cloud is that of the surrounding atmosphere, whereas that of the pyroclastic flow is above 700 °C. Most of the dispersed fraction in the ash cloud belongs to the solid phase while smokes of the second group have liquid globules as the main disperse fraction during most of the flow period. Solidification of the liquid continues to the last stages of the flow, when the disperse material settles out. Settling of particles of pyroclastic flow takes place over a period of only a few hours while settling of the ash cloud is a much longer process and may take several years.

The original exsolved volatiles of a magma are to a great extent lost during smoke eruption. When the ash escapes from the base of the eruption column, it entraps some of the exsolved volatiles, to which are added gases that are continuously exsolved from the hot globules during movement of the cloud, as well as atmospheric air engulfed by the moving cloud. Much of this gas is lost in transit. The contribution of the volatiles exsolved from the hot globules to the gaseous phase is important in pyroclastic flows but is much less significant for ash clouds where the dispersed fraction is composed mainly of solid particles and not of liquid globules.

Volcanic rocks so highly vesicular as to resemble solidified froth, are called *pumice,* if the rock is granite, and *scoria*, if the rock is basaltic. Volcanic bombs of pumice and scoria, subcolloid systems in the volcanic smokes, range in size from a few millimeters to tens of centimeters. The pumice is extremely vesicular, more so than the scoria. It has a specific gravity of as low as 0.3. These systems are formed when drops of magma are rapidly cooled to temperatures below the freezing point of magma. The exsolved gas is trapped in the solidified magma drops and forms bubbles therein. In pyroclastic flows the pumice floats to the top of the flow. Scoria, far less abundant than pumice, has a lower silica content than the latter.

The characteristics of the pyroclastic flow give rise to a rock with a structure that mantles the topography around the volcanic vent with a roughly uniform layer of rock material. This rock is clearly distinguishable from an ash cloud deposit (known as a *fall deposit*), which shows the effects of its passage through the air after ejection, for example an internal stratification.

5.2.1 Ash Cloud

Most of the air in the system comes directly from the atmosphere and is incorporated after the smoke has left the crater. This atmospheric air is sucked into the system during expansion of the cloud and mixture by air currents. The resulting cloud is cooler than the original volcanic volatiles.

The solid phase consists of particles of older volcanic and nonvolcanic rocks. It may result from disruption of a crust that solidifies at the surface of a lava pool or of a dome that forms in or over the vent. Lapilli and ash are formed from drops of lava that are ejected in very fluid condition and solidify in the colder atmospheric air. Most ash is vitreous although olivine and augite, which crystallized prior to eruption, are also blown out from volcanoes.

The rate of movement of the ash cloud in the atmosphere depends on wind velocities. The rate of travel can sometimes reach 100 km/h. The ash can be carried by the wind to distances of as great as 2500 km. Ash from the cloud settles out to form a relatively well-sorted deposit.

5.2.2 Pyroclastic Flow

The subject has been treated by Schmincke (1974) and Sparks (1976). Pyroclastic flows move at a very high velocity, averaging 165 km/h. The temperature of the aerosol, as it leaves the crater, is approximately 1000 °C, but it may drop to 700 °C before the particles settle.

A pyroclastic eruption is composed of two colloid systems, (1) an aerosol of incandescent rock fragments called a glowing avalanche and (2) a cloud of dust rising above the avalanche. The percentage of air is higher in the dust cloud than in the avalanche itself. The avalanche, forming a pyroclastic flow, acts as a heavy fluid whose movement depends on gravity and on topography of the land surface and sinks into valleys whereas the dust cloud spreads laterally, sometimes to considerable distances.

The dispersed liquid phase consists of droplets of magma that are carried along by the flowing gas. This molten phase solidifies, forming vitreous silicates of which a large percentage is pumice and some of which is composed of near-spherical masses. This glassy material is only one of the components of the dispersed solid particles. Other components are debris and fragments of foreign solid rock material that are picked up from the basement beneath the volcano by the rising magma and crystals that precipitated from the magma. All these components are carried out by the flowing gas by the mechanism of segregation and fluidization. Another type of dispersed solid material is the eroded earth underlying the avalanche.

The predominant solid material is of ash size. Scattered throughout the material there usually are lumps of pumice or remnants of collapsed pumice. In some systems fine material is present in relatively minor amounts and lumps of pumice predominate.

The ratio oi large blocks to ash in the avalanche varies greatly. The foreign fragments are generally nonvesicular and considerably denser than the pumice lumps. The bulk density of the gas-particle cloud is approximately equal to the density of the pumice lumps, which is 0.2–0.8 g/cm^3.

Pyroclastic flows normally deposit little or no material on the steep flanks of the volcano. It is not until they lose velocity on the gentler slopes at the base of the volcano that they deposit their load. The deposits consists of an intimate mixture of finely pulverized rock material (ash) with angular blocks of rock ranging up to several meters in diameter, often including many fragments of pumice. The great turbulence in the avalanche prevents any significant degree of sorting of the material (Murai, 1961), although in some deposits there is a slight tendency for the larger blocks to be segregated toward the bottom. The very light-weight pumice fragments tend to be carried farther than the denser fragments. The dust that forms the cloud above the avalanche often settles to form a thin layer of fine material on the top of the avalanche deposit and extends sometimes as much as 2 km laterally beyond the edges of the main deposit.

Some pyroclastic flows reach distances of more than 100 km from the source. The fact that pyroclastic flows can extend over such broad areas with a temperature still high enough to bring about welding of the fragments after the flow comes to rest implies very rapid extrusion and spreading of the fragmental material, little cooling due to adiabatic expansion or to admixture of air, and little dissipation of the system upward into the atmosphere. The fluidized state is determined by two factors: (I) forces preventing the settling of particles by gravity and (II) properties such as a) high density of the dispersing medium and b) surface tension of the bulk fluidized system, both of which prevent dissipation of the particles into the atmosphere, as follows:

I. The aerosol is stabilized by a high degree of turbulence that commonly continues to exist until the flow becomes immobile. The high degree of turbulence results from the expansion of gas. There are two reasons for the expansion of this phase: (1) the release of gas from the molten silicate droplets is not restricted to the initial explosion, but continues for some time as they are carried along by the avalanche (Fenner, 1923) and (2) the cold air entrapped between particles in the avalanche is heated causing it to expand (MacTaggart, 1960). Because of the turbulence, fragments of all sizes settle more or less at random resulting in lack of sorting in the deposits (Sheridan, 1971; Walker, 1971).

II.a. A relatively high content of carbon dioxide makes the gas of the fluidized system denser than the atmosphere, thus lessening its tendency to expand adiabatically and to mix with the surrounding air.

II.b. In the fluidized system gas bubbles cause particles and droplets to collide and aggregate and later to disaggregate. The strength of the interparticle bonds in the aggregates depends on the area of contact between the particles (Chap. 8). Usually this area is very small and the contact area can be considered as a point-contact. The presence of small molecules of liquid silicates and of volatiles strengthens these bonds because they penetrate into the interparticle voids and act as adhesives either through electrostatic interactions (polar molecules and polar surfaces) or through van der Waals interaction (nonpolar molecules and nonpolar surfaces). The strength of the bond depends on the surface properties of the solid particles and the liquid globules. As will be shown in Chapter 4, the chemical activity of the surfaces of solids is very high if the particles are the products of explosion. However, this property is weakened with time. The *surface tension of the bulk system* is a measure of the forces that prevent the dissipation of the particles into the atmosphere. These forces are primarily the bonds between the particles and can be represented by the average time fraction that a particle is associated with other particles. The average time increases with increasing bond strength, concentration, size, and absolute surface area of particles.

The lumps of pumice contribute greatly to the surface tension of the fluidized system. This solid material has simultaneously a high surface area and a large particle size. Thus the ratio of surface area per mass is very high, similar to that found in very small rock fragments. Also the shape of the lumps permits contact between two lumps at several points. It is therefore to be expected that bonds in which lumps of pumice take part are very strong and that their aerosols are stable.

6. The Ocean as a Colloid System

The ocean will be treated as a model for colloid systems in the hydrosphere, water being the dispersing medium. The disperse phase consists either of solid particles, liquid droplets, or gas bubbles of inorganic or organic compounds, but also can be micelles of amphipathic organic molecules.

The bulk of inorganic material is contributed to the oceans by rivers as dissolved, suspended, and colloid material. Great quantities of detritus are produced by glaciers and disseminated in the oceans by icebergs (Lisitzin, 1972). Additional but minor contributors of terrigenous material are winds and sea waves. In addition, appreciable amounts of sediment are contributed locally by volcanoes. When sedimentation rates are minimal, the concentration of cosmogenic suspended material becomes appreciable.

Biogenous material originates in the ocean. The living organism extracts those elements required for the generation of living tissue from salt solutions contributed to the oceans by rivers. As long as the organism is alive, its behavior is controlled by biologic rules. After the organism dies, the inorganic residual particle behaves according to the physicochemical rules that govern the behavior of disperse systems. The organic material decomposes to various organic compounds belonging to well-known groups of organic chemistry, e.g., proteins and their decomposition products, carbohydrates, organic acids, fats, waxes, resins, etc. The alteration of the tissue of the organism results also in the formation of humic substances.

Organic compounds in the ocean originate from the following sources:

1. They are transported to the sea from the land, mainly by rivers and to a smaller extent by winds. The soluble organic content of river waters, which may be as high as 20 mg carbon/l arises mainly from the leaching of humic materials and other decomposition products of vegetable matter from the soil.

2. They are produced by the decay of dead organisms. The decay takes place by two mechanisms: bacterial action and autolysis. In autolysis, decomposition is brought about by enzymes present in the dead cells. Both mechanisms occur together, the

extent of each varying according to the conditions of death and to the availability of the necessary enzymes and bacteria.

3. Algae liberate compounds produced by photosynthesis into the surrounding water by excretion.

4. Zooplankton and marine animals contribute nitrogenous compounds such as urea, purines, trimethylamine oxide, and amino acids by excretion.

The ocean water column is divided into four regions: the surface, intermediate, deep, and bottom water masses. Each region is characterized by its own horizontal and vertical circulations (Lisitzin, 1972). In the water column vertical salinity variations are insignificant.

The surface water mass extends to a depth of 200–250 m. This mass is characterized by climatic zonal distribution of temperature and salinity. It is here that the most active physicochemical exchanges with the atmosphere take place and the basic planktonic organisms develop. The climatically controlled interactions between the atmosphere and the hydrosphere determine the formation of surface regions with production of abundant biogenous material.

Current velocities in this layer of the open ocean are 100–200 cm/s. Winds induce waves reaching 25 m in height. These waves have a great effect on suspension properties up to a depth of 100 m. In the high latitudes ice and glaciers influence water dynamics by damping waves. They also reduce the amount of living organisms in the surface layer. The pH of surface water usually averages 8.2.

Intermediate water masses extend to a depth of 1000–2000 m. Current velocities in this layer are 2 to 8 cm/s on the average. The resulting transportation is directed from the high latitudes toward the equator. The pH of intermediate water masses drops from 8.15 at the top to 7.7–8.0 at the bottom of the region.

Deep water masses extend to a depth of about 4000 m with current velocities of 0.2–0.8 cm/s. This layer plays a leading role in the interlatitudinal exchange of suspended material. The pH of deep water masses varies from 7.9 to 8.2 and the usual value at 4000 m depth is 8.0–8.1.

Bottom water masses extend from 4000 m depth down to the ocean bottom. Theoretically, current velocities average 0.1–0.2 cm/s. It was found that they are as much as ten times higher in some localities. There are strong advective flows characteristic of bottom currents and turbulent flows in these water masses. The salinity (3.47–3.49%) and temperature (0–3 °C) of bottom water masses are practically constant. Seasonal climatic changes do not reach the ocean bottom (Lisitzin, 1972).

6.1 Air–Water Interface

The chemical composition of the air–water interface differs from that of the ocean bulk. The interface contains a mixture of amphipathic organic compounds that are highly surface active. The adsorbed molecules can be compacted into films that damp capillary waves and produce light reflectance anomalies known commonly as "slicks". Natural surface films are monomolecular layers. They are observed most frequently in regions of high biologic activity or of extensive pollution by film-forming chemicals, near continents or island groups. In the open ocean, the dispersive forces of waves and surf together with a lower concentration of surface-active material reduce the probability of coherent monolayer formation. In relatively calm weather of several hours duration, a large area of the sea can become slick in appearance due to the adsorption of film-forming materials in the surface layer.

Garret (1972) determined the chemical constitution of the upper sea surface, a layer of 0.15 mm thickness. The major water-insoluble components of the sea surface consist of a complex mixture of fatty esters, free fatty acids, fatty alcohols, and varying quantities of hydrocarbons. Although proteins and carbohydrates have been identified qualitatively in surface water, their contribution to surface effects is small because of their high hydrophilicity and great water solubility. As the film ages the less soluble species having high molecular weights are strongly adsorbed at the air–water interface, while the more soluble or less surface-active compounds are forced out of the surface film by the more active molecules. Thus, aging progressively decreases the solubility of the film.

Daumas et al. (1975) suggest the occurrence of a stratification at the air–water interface. The uppermost microsurface layer contains inert particles and bacteria, while the liquid film below is distinguished by photosynthetic cells and dissolved substances. Particulate fatty acids are more abundant in the uppermost microsurface layer whereas dissolved fatty acids are more abundant in the lower liquid film. The particles gathered at the interface generally have lower concentrations of unsaturated fatty acids than the particles of the underlying layer.

There is a high accumulation of dissolved and particulate hydrocarbon in the surface microlayer with respect to the underlying water, having an enrichment factor of 50 on the average. This enrichment is generally greater for dissolved than particulate alkanes, except in very rich or polluted zones, in which the concentration of amphipathic molecules is high and the enrichment may become greater for the particulate alkanes (Marty and Saliot, 1976).

6.2 Gas Bubbles in the Ocean

The present section is based on the treatment of de Vries (1972) on the morphology of foam bubbles, supplemented by some geochemical data.

Large-scale turbulence near the surface, raindrops, and the breaking of waves result in gas bubbles in the water. Free gas bubbles in the sea may be produced by the decomposition of detrital matter and also may originate from volcanic emanations. When the gas bubbles are uniformly dispersed in the water layer a foam is obtained. As a result of the increased gas–liquid interface there is a concomitant increase of surface free energy excess. Because of this free energy excess the foam, thermodynamically, is an unstable system. The spontaneous decrease of the interfacial area, which should ultimately result in a complete separation of the gaseous and liquid phase, often progresses at a relatively low rate in many air–ocean systems.

As a consequence of the positive surface tension existing at any gas–liquid interface, the gas bubbles will tend to adopt a spherical shape spontaneously, regardless of the particular method by which the foam was obtained. The stability of the foam decreases with increasing gas concentration. Low-density foams, in which more than 74% of the total volume is occupied by nearly spherical gas bubbles, may persist at the upper layer of the sea for a considerable period of time if the bubbles are of unequal sizes.

As the result of the movement of water, the original spherical bubbles begin to be distorted after a certain period of time, the duration of which increases with the degree of polydispersity of the bubbles. This leads to changes in their surface curvatures, which lose their uniformity along the gas–liquid interface, and vary continuously as a function of the degree of distortion. When several bubbles collide, the possibility of coalescence to a more stable configuration is likely.

Pure water does not foam; the bubbles break as soon as they reach the surface. The ease with which foam is formed depends on the substances dissolved and suspended in the water mass. According to Garret (1967) the longevity of air bubbles at the air–water interface is a function of the chemical activity of surface of the water and the chemical conditions of the water interface. In the absence of a sea-surface film, the lifetime of an air bubble is determined by the surface-active material it adsorbs during its passage upward through the vertical ocean column. The presence of a compressed monomolecular layer (sea slick) at the sea surface will significantly reduce bubble stability and act as an antifoaming agent even though the underlying water may be rich in bubble-stabilizing substances. The gas bubble may sorb soluble and insoluble surface-active material and be an important agent for the transport of organic surface-active material to the sea surface (Sutcliffe et al., 1963).

The presence of salt in water has a pronounced effect on the size of air bubbles. With increasing salt concentraiton the number of air bubbles first decreases sharply but shows no further drop when the salt concentration is greater than 0.6 M. Total surface area of an air foam in pure water shows practically no temperature dependence but oxygen foam in sea water shows an increase in surface area with rising temperature. This dependence is especially pronounced between 15 and 20 °C (Zieminski et al., 1976).

6.3 Suspended Inorganic Solids in the Ocean

This subject has been treated by Lisitzin (1972) and the present section is based on his treatment supplemented by more recent publications. The classification of suspended solid matter is based on (1) grain size, (2) material generic type, and (3) chemical composition of organic or inorganic type.

Table 1.4 is a summary of the various groups of clastic sediments, a classification based on the size of the particles commonly dealt with by sedimentologists and oceanographers. The classification used by colloid scientists for systems of material suspended in water is also shown. The following five material-genous types of suspended matter have been identified in the ocean: terrigenous, volcanogenic, cosmogenic, chemogenic, and biogenous. The most important material types are terrigenous and biogenous carbonates and silica. In the open ocean and in adjacent seas the role of chemical precipitation of authigenic minerals is almost nil, except in unique localities. Particulate matter, both organic and inorganic, is present at all depths in the oceans.

In the direction from the coasts toward the pelagic regions of the oceans, the concentration of the suspended material is determined by two processes: (1) the

Table 1.4. Classification of suspended solid matter in the ocean and of sediments, based on particle size and on the properties of the colloid particles

Diameter of particles in nm	Sediment group	Sediment	Solution	Optical properties	Separation methods in laboratory
$> 2 \times 10^6$	Rudite, psophite	Gravel	Suspension	Visible	Retained by ordinary filter paper
$62 \times 10^3 - 2 \times 10^6$	Arenite, psamite	Sand	Suspension	Visible	
$4 \times 10^3 - 62 \times 10^3$	Arenite, psamite	Silt	Suspension	Visible in microscope	
$2 \times 10^3 - 4 \times 10^3$	Lutite, pelite	Shale or clay	Suspension	Visible in microscope	
$1 \times 10^3 - 2 \times 10^3$	Lutite, pelite	Shale or clay	Sol or colloid solution	Visible in microscope	
800–1000	Lutite, pelite	Shale or clay	Sol or colloid solution	Visible in microscope	Not retained by ordinary filter paper
5–800	Lutite, pelite	Shale or clay	Sol or colloid solution	Visible with ultramicroscope	
1–5	Lutite, pelite	Shale or clay	Sol or colloid solution	Not visible	
< 1	Lutite, pelite	Shale or clay	True solution	Not visible	

dispersion of the terrigenous material and (2) the biogenous activity. Concentrations of suspensions are highest in nearshore areas and in small marine basins. Concentrations increase appreciably in areas of phytoplankton bloom. The range of suspension concentrations in the oceans is much narrower than in the seas. Concentrations of 0.5–1.5 mg/l and averaging 0.5–1.0 mg/l, are commonly found at the ocean surface. In the northern part of the Pacific Ocean concentrations are often as high as 5–7 mg/l.

The predominant suspended material nearshore is of terrigenous origin, whereas the proportion of biogenous material increases seawardly. Both are subjected to seasonal variations, particularly in the colder parts of the humid zones where in winter many rivers freeze over, the sea surface is covered with ice, and phytoplankton cease developing. Maximum suspension concentrations are obtained in the surface waters during the spring phytoplankton bloom. Concentrations of suspended material decrease in summer as the terrigenous and biogenous supplies decrease. Outside the zone of seasonal freezing, the change in suspension supply from the land varies with rainfall. In the arid zones, where sedimentary supply from the land and biogenous supply are insignificant the year round, seasonal changes are infinitesimal.

The seasonal changes of plankton biomass depend on latitude. In the upper latitudes of the Arctic, a single annual maximum occurs during the summer. In warmer regions, such as those of the Norwegian and British coasts, two annual maxima of plankton development occur, in spring and autumn. In the tropical zones the annual distribution of plankton shows one maximum of long duration during the winter months. Seasonal production of zooplankton such as foraminifers and radiolarians generally follow those of phytoplankton.

The Pacific Ocean is the largest of the earth's basins, encompassing all the climatic zones. This ocean exemplifies the relationship between the quantitative distribution of suspended material and climatic zonality. High suspension concentrations are found in three latitudinal zones: a northern, a southern, and an equatorial belt. These belts are clearly related to the major climatic zones and humid types of sedimentary genesis and are separated by the latitudinal arid zones of minimum suspension concentrations. In other words, the terrigenous and biogenous factors combine to form latitudinally elongated, high concentration belts in the ocean corresponding to the humid climatic zones. The largest areas of high suspension concentrations are found near the continental areas of the humid zones, in the western part of the Pacific near Japan and Kamchatka and in the Bering and Okhotsk seas. The smallest areas exist in the arid zones off the Australian and South American coasts.

The distribution of particles in the ocean depends on the rates of the following processes:

1. Migration of particles due to horizontal and vertical ocean currents
2. Spreading by vertical turbulent diffusion
3. Settling, the rate of which increases with particle size and with the differences between the density of the suspended material and that of water (Stokes' law)–consequently, liquid density boundaries are layers of high suspension concentrations
4. Dissolution of particles
5. Submarine rising of particles from bottom sediments due to drag forces caused by tidal currents and seismic sea waves

As well as the permanent currents, which are especially strong in the surface water masses and sometimes in the bottom masses, tidal currents also influence the distribution of suspended material. Seismic sea waves are another important dynamic factor. These waves are 100–300 km long and are propagated at velocities of 500–700 km/h. Major seismic sea waves occur once in several years and lesser ones several times a year. Their effect on the concentration of suspended material is appreciable. After the occurrence of seismic sea waves the turbidity of bottom water masses increases sharply. The turbidity decreases to more normal values only after several years.

Suspended material is sorted by currents and waves giving rise to a distribution from coarse nearshore to fine in deep-sea basins. Volcanogenic material is introduced into suspensions azonally. Since this material is only slightly scattered, its effect on grain size and mineral distribution is only local in character.

At the surface of the open ocean the proportion of particles coarser than 0.05 mm does not exceed 2.5% of the total solid fraction. The proportion of fine silty particles can sometimes be as high as 20–30%. The bulk of suspensates are of clayey size. The latter usually comprise more than 90%, but in some localities they comprise more than 99% of the suspensates. Material finer than 1000 nm comprises 30–90% of suspensates. A major portion of the finest material suspended at the surface does not reach the ocean bottom because of its dissolution in the water column. As a result the suspended material is much finer than the sediments deposited on the ocean bottom.

Suspensions in the active, 200 m thick surface water masses are irregularly distributed. At greater depths concentrations decrease and become more uniform. Three types of vertical distributions can be distinguished, each related to basin type and size: (1) that occurring in shallow basins and bays; (2) that characteristic of shelves and shelf seas, and (3) a deep-sea and oceanic type.

1. Bottom sediments in shallow bays and basins are periodically disturbed. Thus, such areas are characterized by two concentration maxima, the one in the upper active layer due to plankton, and the other in the near-bottom layer due to roiling.

2. On the oceanic shelves and in the nearshore areas the vertical suspension distribution is similar to that of shallow bays. In coastal areas sediments are roiled by storms and tides. In deeper localities, as many as three suspension maxima can usually be observed in profile. The upper maximum is caused by blooming phytoplankton. The second maximum is associated with the *Thermocline,* the zone of water in which the temperature decreases rapidly with depth, i.e., at a depth between ca. 200 and 1000 m in temperate and subtropical waters. The third maximum is observed near the bottom and is attributable to tidal currents and to the movement toward deeper water of fine sedimentary material from the shelf, where these particles cannot be deposited because of hydrodynamic conditions.

3. The deep seas and oceans are characterized by an irregular distribution of suspension. The upper turbid layer associated with photosynthesis and the second maximal layer at the pycnocline are well distinguished. Suspensate concentrations diminish with depth due to dissolution. In the peripheral parts of the basins secondary maxima are formed by suspensions that extend from the shelf. Similar maxima are formed above submarine ridges.

The average amount of suspended sediment for the entire water column is usually higher over the shelf and continental slope than in the open ocean. In the pelagic ocean areas the average suspension concentration increases in the proximity of submarine rises and ridges.

During the course of settling the large particles break into smaller fragments and small particles coagulate. An organic "skin" may preserve the small particles from dissolution (Suess, 1970).

Suspended particulate material in sea water causes light scattering and turbidity that may be roughly related to the concentration of the suspended material. Regions of turbid water having very intense scattering properties distinguishing them from other oceanic regions have been located. Below these regions are the floors of the great oceanic basins where vast quantities of sediments have accumulated. Nephelometric profiles, relating the strength of light scattering as a function of depth, normally show strong scattering in the upper part of the water column, which is an indication of the presence of abundant organic particulate matter. The nephelometric profiles of some localities in the ocean show a strong light-scattering bottom layer characteristic of a region that is called the *nepheloid layer.* Studies of North American and Brazilian basins indicate higher concentrations of particulate matter in nepheloid layers (0.5 ± 0.1 mg/l) compared to clear bottom water (0.1 ± 0.1 mg/l). The concentration of material suspended in turbid water usually ranges from 0.44–0.56 mg/l compared to 0.03–0.24 mg/l for normal clear water (Jacobs et al., 1973).

Density currents are defined as underflow currents produced by fine-grained sediments in suspension. They are driven by both suspended sediment and differences in the temperature and salinity of the inflowing water (Normark and Dickson, 1976).

A *turbidity current* is defined as a downslope density flow of turbid water. It plays an important role in the transport of sediment from the continental slopes to basin floors (Heezen and Ewing, 1952). Turbidity currents are of short duration, episodic and randomly located. A plausible cause of turbidity currents might be winter storms that increase turbulent mixing and introduce large quantities of river sediments onto a basin shelf within a relatively short time (Fleischer, 1972). The chemical composition of water in a basin floor can be altered by near-bottom transport processes such as turbidity currents. The highly concentrated suspensions are very dense, and they barely mix with the water through which they flow. The chemical composition of the interparticle solution does not change during the transport. Sedimentologic evidence presented by Sholkovitz and Soutar (1975) indicates that turbidity currents transport sediment, benthic organisms, and sea water from depths near the sill (440–480 m) to the center of the basin (590 m) in the Santa Barbara Basin. Water sampled at less than 2 m from the basin floor in December, 1969, had the chemical composition of water originating from the sill depth with notably higher oxygen and nitrate content than usual.

6.3.1 Terrigenous Material

This type contains material supplied from the land, including clastic rock debris or mineral grains as well as clayey particles, the weathering products of continental rocks.

Two major types of weathering products can be distinguished. The first is the product of physical disintegration of the rocks. As will be shown in Chapter 3, this type of weathering does not change the chemical composition and the mineralogic constitution of the rock material. However, the surface properties of the disintegrated material differ from those of the original rock and depend on the type of disintegration. Physical weathering is characteristic of cold as well as of dry climates, but the resulting surfaces of the weathered products do not have similar properties.

The second type of weathering is of a predominantly chemical character. The resulting solid material is usually composed of clay minerals. Chemical weathering is typical for the humid intertropical zone and occurs to a smaller extent in semitropical regions.

Terrigenous material is very abundant in humid zones. In arid zones the supply of terrigenous material is minor in relation to the formation of carbonates. A special type of suspended material is due to iceberg melting. Here the particles are composed of moraine material including very fine pelitic varieties. In parallel with the geographic distribution of land and ocean, terrigenous material is more abundant in the northern hemisphere.

Most coarse-clastic materials composed of boulders, pebbles, and gravels are found in the nearshore regions of high water energy and their concentrations generally decrease with distance from the shore. Coarse clastic material is also dispersed over vast areas in the ice and cold regions of the humid zones, by means of sea ice, river ice, or icebergs, the particles being incorporated into the ice mass. The maximum distances to which rock fragments are transported by sea ice are usually about 500 km and seldom are as much as 1000 km.

In the humid zones beyond the ice distribution region, large quantities of rock material are transported by gigantic algae, tens and sometimes hundreds of meters long. These algae have cavities that provide excellent buoyancy, enabling them to transport rock particles several kilograms in weight.

The mineralogy of particles in the size range of sand in the nearshore parts of the oceans is closely related to local conditions. The very existence of many heavy and light minerals depends not only on their occurrence in the bed rock of a drainage basin but also on the nature and extent of weathering, which ultimately transforms a considerable part of the bed rock minerals into clays. Nonclayey grains are best preserved in high latitudes where physical weathering predominates over chemical weathering.

Garnet, hornblende, and black iron ore minerals are to be found mostly in the silty and finer particles suspended in the oceans. Garnet is widespread in suspensions in the Pacific sectors of the Antarctic Ocean. This can be accounted for by its extremely common occurrence in the ancient rocks of Antarctica. Garnet does not occur on the sub-Antarctic islands and thus is likely to be the most widely distributed and clearly definable mineral indicator of the Antarctic province both in suspensions and in sediments.

The regions of high concentrations of suspended garnet ore are confined to definite localities. The garnet content of the heavy fraction of suspensates is as high as 30–40% in some localities in the vicinity of Antarctica but decreases with increasing distance from the coast. The zone containing 2–10% garnet extends as far as 2500–3000 km from Antarctic coasts in the west and 1000 km in the east.

Common hornblende in suspension forms two maxima in the vicinity of Antarctica, where this mineral is supplied to the ocean. The maximum quantity of this mineral in the suspensions is 15–20% of the heavy fraction of suspended material. Over considerable areas within these two regions the amount of hornblende makes up 5–10%. The two regions of high hornblende content are separated by a zone where hornblende is not supplied from the land and the quantities do not exceed 1–5%. Similar low hornblende concentrations occur beyond the distribution range of the Antarctic minerals.

A different distribution is found for monoclinic pyroxenes, which are widely distributed in the Antarctic source province. In some localities suspensions contain 5–10% of monoclinic pyroxene of the total solid dispersed material. At the same time pyroxenes are also present in the volcanogenic provinces of the sub-Antarctic islands. Therefore the distribution spectra of Antarctic pyroxenes sometimes overlap those of volcanogenic pyroxenes, causing deviations from the close correspondences to the source provinces observed for garnet and hornblende. Off Antarctica the coincidence is more noticeable and is close to the patterns of garnet and hornblende occurrence. The dual source of monoclinic pyroxenes found in the Antarctic Ocean is apparent. Terrigenous pyroxenes are distributed according to the laws of scattering of terrigenous heavy minerals, whereas volcanogenic pyroxenes form local zones with high concentrations decreasing rapidly with distance from the volcanic centers. Not all volcanic centers are revealed by the nature of the suspension. The erosion products of volcanoes rising above the ocean surface are encountered most often, whereas pyroclastic materials from the numerous underwater volcanoes as a rule do not contribute to the mineralogy of surface suspensions.

Black iron ore minerals contribute 10–20% of the heavy suspended fraction over large areas, with maximum values of 30–47%. They are indicators of large land masses of metamorphic and silicic rocks. The dispersal zone of suspended black ore minerals resembles the quartz dispersal zone. In the western part of the Indian Ocean the region rich in iron minerals stretches as far north as South Africa and becomes narrower eastward. The maxima regions of these minerals also coincide in the southern part of the ocean. As in the case of quartz, the black ore minerals are supplied from the African source province. In the western region the two source provinces merge.

In the central Indian Ocean the amounts of black ore minerals in suspensions are negligible. Only toward the southern extremity of India and in the Red Sea, near regions of ancient strata, do the amounts of black ore minerals in suspension increase again.

Of the light minerals the most frequently occurring in suspensions are quartz and plagioclase. Their contents in suspension are sometimes as high as 30–35%, with the rich localities mainly near Antarctica. The distribution range of suspended quartz is close to that of garnet. Unlike garnet, additional quantities of suspended quartz are contributed by southern Africa, and a zone with 15–20% quartz contents occurs at the southern extremity of the continent.

Most clay minerals that are distributed in the ocean are typical terrigenous minerals formed during weathering on land and subsequently are distributed in the manner of similarly sized clastic terrigenous material. A very small portion of the suspended clay minerals is *authigenic,* that is, they form in the oceans through the interaction between silica and alumina compounds.

The most frequently occurring clay mineral in the oceans is illite, usually constituting more than 50% and sometimes more than 80% of the suspended clay minerals. Only in the tropics does its proportion drop to 20–40% or less. The two other widespread minerals, kaolinite and montmorillonite, usually do not constitute more than 40–60%, and chlorite is not more than 30%. Compared to these four major clay minerals, the amount of other clay minerals suspended in the ocean is insignificant.

The relative concentration of the various clay minerals in suspension depends mainly on two factors: (1) the minerals furnished by streams and (2) the transformation of one clay mineral into another clay mineral.

6.3.2 Inorganic Biogenous Material

Carbonates and silica comprise the most important biogenous-type suspended matter. The major carbonate accumulators in the nearshore marine areas and on shelves are benthonic organisms such as mollusks, bryozoa, and coral biocoenoses, and in the pelagic areas–planktonic foraminiferes, cocolithophorids, and pteropodes. The tests of these organisms are composed of calcium carbonate. Their magnesium carbonate content is insignificant.

The most abundant siliceous organisms in suspension are diatomaceous algae, the main producers of organic matter in the oceans. They are also the main producers of suspended amorphous silica. Radiolarians are the second and silicoflagellates are the third most important silicate formers. Most of the organisms that accumulate skeletal silica are located in the upper 200-m layer of the ocean. At greater depths only dead frustrules of these organisms and rare living radiolarians are found.

Silica concentration of surface waters ranges between 0 and 35% of the total dry dispersed solid phase. Opal content of the surface water is 0.13–1000 μg/l. The amount of suspended biogenous silica generally is 1/5–1/50 of that of dissolved silica. Thus the organisms succeed in converting only an insignificant part of dissolved silica into particulate silica. Amorphous silica concentrations are distributed very regularly in three belts: a southern belt of siliceous suspension encompassing the entire globe and much less pronounced equatorial and northern belts, the latter found in the Pacific. The vast spaces between these belts are occupied by water containing less than 0.5% suspended opal.

The global surface water belts that are characterized by large amounts of suspended silica in the photosynthetic zone are also silica-rich throughout the entire water column and in the underlying sediments.

The surface ocean layer is not as rich in carbonate organisms as it is in most siliceous organisms. The maximum foraminiferal population is in the upper 10–50 m layer and decreases sharply below 200 m. The greatest carbonate content falls within 50 °N to 50 °S latitudes with a maximum in the equatorial and tropical zones. Suspensions in the cold Antarctic waters contain less than 1% calcium carbonate. In the Pacific ocean maximum values of up to 3% occur near the equator. Surface suspensions contain 5–10% calcium carbonate even in such areas of recent accumulation as the Mediterranean and Red Seas.

The amount of terrigenous material, biogenous silica, and organic carbon is usually much greater than that of calcium carbonate in the suspended fraction in the surface water masses. During settling to the ocean bottom, the unstable components of the suspension, such as amorphous silica and particularly organic carbon, are removed from the suspension, causing suspensions in the carbonate accumulation zones to become progressively more enriched in calcium carbonate.

The solublility of calcium carbonate increases with decreasing temperature and pH of solution or with increasing hydrostatic pressure, i.e., with depth in the ocean. Maximum absolute amounts of carbonates in suspension occur in the surface water mass and decrease with depth primarily due to the slow dissolution of the finest particles. Carbonates universally begin to dissolve rapidly at depths of about 3100–3500 m (Bramlette, 1961; Turekian, 1965). This is known as the *lysocline* (Berger, 1968). The depth at which the rate of carbonate supply equals the dissolution rate and the carbonate content drops to a few percent of the total suspended material is defined as the *carbonate compensation depth.* From thermodynamic data it is obvious that only the warm upper layer of marine waters is saturated with respect to calcium carbonate, and the saturation horizon is everywhere shallower than the compensation horizon (Pytkowicz and Fowler, 1967; Edmond and Gieskes, 1970; Edmond, 1972). Calcite should therefore disappear from regions in the water column much higher than the compensation depth. Peterson (1966) showed experimentally that the dissolution of small spheres of calcites placed in the upper part of the oceanic water column is negligible within a period of several months. At 3700-m depth the dissolution increases sharply and becomes more severe at still greater depths. The solubility of aragonite is much higher than that of calcite, resulting in a compensation depth for aragonite at 2000–3000 m. According to Edmond (1974) the dissolution of calcium carbonate in undersaturated solutions is a very slow process, but the presence of turbulent conditions and of active currents in the bottom waters (below the compensation depth), exerts a marked acceleration on the dissolution rate. The dissolution of calcite is inhibited by various ions that are capable of forming surface coatings on the crystals of the mineral (Terjesen et al., 1961). It has been suggested by Chave and Suess (1970) that in seawater adsorbed ions and organic molecules control the dissolution rate of calcite. The dissolution rate should become rapid when conditions are such that the surface coatings become unstable. Berner and Morse (1974) studied the effect of phosphate on the mechanism of dissolution of calcite. They showed that the dissolution rate in seawater exhibits a sudden acceleration at a critical degree of undersaturation that increases with increasing phosphate concentration in solution. With this degree of undersaturation the diffusion of the Ca^{2+} and CO_3^{2-} ions through the adsorbed phosphate coating layer becomes high enough for the dissolution process to reach high rates.

Unlike with calcium carbonate, no critical depth for silica exists, and siliceous residues occur in suspension down to maximum oceanic depths. All other things being equal, an increase in depth results in a decrease of the total amount of opal due to dissolution, as well as a decrease in median particle sizes and in the variety of species. The near-bottom layers of the ocean are reached by only a tenth to a hundredth of the silica bonded by organisms, as frustules at the surface water masses.

Diatom frustules are destroyed most rapidly in the upper 100-m water layer. Radolarians and to a smaller extent silicoflagellates are preserved in deep-water suspensions and reach the bottom without appreciable loss. Due to the dissolution of the smaller particles in the water column, deep-water suspensions appear enriched in large particles.

6.4 Organic Compounds in the Ocean

This subject has been reviewed by Riley and Chester (1971) and the present section is based on their treatment supplemented by more recent publications. The organic matter in the ocean is usually classified into two categories, dissolved and particulate. The dissolved organic matter, together with the hydrophilic colloid material, can be separated from the particulate material by passing the solution through a 500-nm membrane filter. The amount of the dissolved organic matter in the sea usually exceeds the suspended solid organic fraction by a factor of 10 or 20. According to Sharp (1973) a particulate cut-off with a membrane filter of about 1000 nm permits almost the entire range of colloid matter to be classified as dissolved. In oceanic waters about 98% of the total organic carbon is thus considered as dissolved.

6.4.1 Dissolved and Colloid Organic Matter in the Ocean

The total soluble organic carbon content of ocean waters ranges between 0.3 and 3.0 mg C/l. In coastal waters values as high as 20 mg C/l may be found. The total soluble organic nitrogen content of ocean waters lies in the range of 0.005–0.300 mg N/l.

The molecular weight of the dissolved organic matter is distributed over wide ranges. The major fraction of the colloid organic carbon in the sea consists of a complex material of a very high molecular weight that has been named *Gelbstoff* (yellow material) because of the yellow color that it imparts to the water. This material is resistant to bacterial attack. According to Kalle (1966), *Gelbstoff* is not a single compound but a complex mixture formed in the sea and may be regarded as the equivalent of the humic substances formed in soils. It seems to be formed by the condensation of polyphenol with carbohydrates and proteins, all of which are of algal origin. Despite the considerable amount of *Gelbstoff* liberated into the sea by the brown algae, its concentration level in the oceans is relatively low (ca. 1 mg/l).

According to Akiyama (1972) high-molecular-weight proteinaceous substances produced by bacterial degradation may in part be altered to particulate form utilized by organisms and in part altered to *Gelbstoff*, the more stable and more complex substances that are resistant to biologic attack. Sieburth (1969) pointed out the importance of phenolic compounds in the formation of *Gelbstoff*. According to Khailov et al. (1969), the high-molecular-weight (> 50,000) organic substances in seawater contain only small amounts of protein (1.6–7.4%) and carbohydrates (1.6–23.0%). There are significant amounts of carbohydrates and proteinaceous substances from biogenous origin with molecular weights of 1000 to 10,000 in ocean water (Maurer, 1976).

Some humic material derived from the leaching of soils and having a composition somewhat different from that of *Gelbstoff* enters the ocean in river water. According to Nissenbaum and Kaplan (1972) terrigenous and marine humic acids appear to be similar. However, marine humic acids contain more nitrogen and sulfur and the $^{13}C:^{12}C$ ratios in normal marine humic acids are enriched in ^{13}C to a greater degree than in normal continental humic acids. According to Sieburth and Jensen (1968) humic materials from continental sources rapidly precipitate on reaching the sea. Humic substances occurring in the oceans at distances far from the continents are assumed to be authigenic in the marine environment. According to Nissenbaum and Kaplan (1972) the mechanism of formation of marine humic substances is not the same as that of terrigenous humic substances. Marine humates are formed by condensation of carbohydrates, amino acids, and possibly other simple molecules. The condensation is accompanied by cyclization of sugars to hydroaromatic and hydroxyaromatic acids. Terrigenous humic acids, on the other hand, are formed by degradation of lignins into quinonoid and phenolic compounds that react with amino acids and other molecules, polymerizing to give humic acids.

Typical concentrations of some dissolved organic compounds of molecular weights less than those of the polymeric fractions are given in Table 1.5. The principal amino acids found in the oceans are α-alanine, serine, threonine, glycine, and valine. Most of these acids enter the sea as a result of protein decomposition during the decay of organic tissue and excreta (Degens, 1970). Dissolved nitrogen is more abundant in the form of peptides and amino acids combined with carboxylic and phenolic compounds than in the form of a free monomeric amino acid (Daumas, 1976).

Several carboxylic acids have been detected in seawater, including acetic, glycollic, citric, and malic acids. These acids are excreted by organisms and are common extracellular metabolites. It seems that they are major components of the low-molecular-weight organic matter, but owing to their high solubility it is difficult to separate them from the water system, and not much is known about their distribution in the oceans.

Table 1.5. Representative concentrations of some dissolved organic compounds in marine surface waters. (After Riley and Chester, 1971)[a]

Compounds or class of compounds	Concentration in $\mu g/l$	Locality
Methane	1	
Paraffinic hydrocarbons[b]	400	South Pacific
Pristane (2,6,10,14-tetramethylpentadecane)	trace	Cape Cod Bay
Pentoses	0.5	Pacific off California
Hexoses	14–36	Pacific off California
Glucose		Saragaso Sea
Malic acid	300	Atlantic Coast
Citric acid	140	Atlantic Coast
Triglycerides and fatty acids[c]	200	South Pacific
Amino acids[d]	10–25	Irish Sea
Peptides[d]	10–100	Irish Sea
Vitamin B_{12}	0.01	Irish Sea
Thiamine (Vitamin B_1)	0.021	Long Island Sound
Biotin	0.01	Gulf of Mexico
Urea	80	English Channel
Adenine	100–1000	Gulf of Mexico
Uracil	300	Gulf of Mexico
p-hydroxy-benzoic acid	1–3	Pacific off California
Vanillic acid	1–3	Pacific off California
Syringic acid	1–3	Pacific off California

[a] For an adequate list of references see Riley and Chester, 1971, page 189.
[b] C_{12}, C_{14}, C_{16}, C_{18}, C_{20}, and C_{24} paraffins.
[c] A large number of saturated and unsaturated fatty acids were present, with oleic and palmitic predominating.
[d] Mainly alanine, serine, glycine, threonine, and valine with lesser amounts of glutamic and aspartic acids.

Because of the high salinity of ocean water, carboxylic acids and other amphipathic molecules are present as micelles.

Several factors are responsible for the presence of saturated and unsaturated hydrocarbons in the oceans (Swinnerton and Lamontagne, 1974). These are organic decay, the result of anaerobic conditions, gas and oil seepage from oil well rigs, and man-made sources, such as shipping activities. The surface waters of the open ocean have relatively low concentrations of saturated hydrocarbons. Methane in the surface waters of the open ocean is nearly in equilibrium with the atmosphere, while in near-shore areas the methane concentration may be over 100-fold higher. The average concentration of methane in open ocean surface water is 49.5 nl/l. This methane concentration is about the value one would expect on the basis of its partial pressure in the atmosphere. Concentrations of ethane and propane in the open ocean average 0.50 and 0.34 nl/l, respectively. Contaminated nearshore samples usually contain 100–200 times these concentrations. Swinnerton and Lamontagne report on high levels of ethane and propane in the Norwegian Sea. This may be due in part to an outflow of the North Sea into the Norwegian Sea. There are considerable oil and gas reserves under the North Sea, many of which are being exploited.

In oxygenated waters, concentrations of saturated hydrocarbons generally tend to decrease with increasing depth, but in anoxic areas the situation is reversed. Methane is produced by anaerobic decomposition of organic matter. Higher-molecular-weight hydrocarbons may also be formed in the same manner. Unsaturated compounds, however, appear only as traces in truly anoxic environments, and ethylene and propylene, for example, the two simple unsaturated hydrocarbons, are produced mainly by biologic activity. Laboratory experiments involving marine organisms have shown substantial production of unsaturated hydrocarbons and only minimal production of saturated hydrocarbons. In the upper oxygenated layers of the oceans to a depth of 150 m, the unsaturated hydrocarbons generally are higher in concentration than their saturated homologs. They are found in relatively high concentrations in both the open ocean and nearshore contaminated water. Their concentrations in the surface layers of the open ocean average 4.8 and 1.4 nl/l for ethylene and propylene, respectively. The concentrations of the unsaturated hydrocarbons decrease with increasing depth in the water column, reaching trace levels at 150–200 m. The production of unsaturated hydrocarbons in nearshore areas is equal to or higher than that in the open sea, because there can be considerably higher biologic activity in the nearshore areas. In a truly anoxic environment the unsaturated hydrocarbon concentrations decrease to trace levels.

The aqueous solubility of hydrocarbons decreases with increasing chain length and hence their concentrations in the oceans also decrease accordingly. Long-chain aliphatic saturated hydrocarbons are usually found in trace amounts in the open ocean. Their concentrations increase in nearshore contaminated areas and in truly anoxic environments.

6.4.2 Particulate Organic Matter in the Ocean

Organic particulate matter in the sea comprises both living organisms and detritus. The near-surface waters contain about 10 to more than 50% organic matter in the form of living organisms. Usually less than 10% of the organic matter in deep waters is living. Different size groupings of particulate organic matter show increasing percentages of nonliving matter within the smaller size groups (Beers and Stewart, 1969). Average values for particulate organic carbon in the deep ocean are about 10 μg/l (Sharp, 1973).

Dead organisms, such as plants and zooplankton, constitute one of the most important sources of detritus. This material is subjected to slow decay. In the nearshore areas waters contain appreciable amounts of wind- or river-transported terrestrial and industrial debris, such as wood, spores, and carbonaceous combustion products. These terrestrial substances may be carried by currents into deeper waters. Particulate matter is also obtained by the aggregation of soluble or hydrophilic colloids.

The composition of the detritus material in the near-surface waters consists to a large extent of carbohydrates, proteins, and plant pigments and their degradation products. Most of these compounds are not resistant to bacterial activity, and only the resistant parts of the organisms, such as the cell walls of phytoplankton and the exoskeletons of zooplankton and precipitated humic substances reach depths greater than 100–200 m. The majority of the material present in deep water is resistant to bacterial activity and may be related to the *Gelbstoff* fraction of the hydrophilic-colloid organic material. Sheldon et al. (1967) suggest that little of the particulate matter present below 50 m originated in the surface waters and that it must have been formed in situ.

The major inorganic constituent of marine organic particulate matter is Na and to a lesser extent Mg, Ca, and K, which are present in proportions related to their normal concentrations in seawater. The total amount of inorganic constituents increases with depth. Fe and Mn, which form complexes with organic compounds, are scavenged from the water column to some extent, Fe more effectively than Mn, and are thus considerably enriched in the organic particulate matter. Mn is scavenged more effectively by the flocculation of hydrophilic organic colloids than by the normal bio-geo-processes of particle formation, and is therefore more enriched in the in situ-created aggregates than in the normal particles originating from organisms. Organic particulate matter is responsible for the migration and sedimentation of many inorganic constituents. The utilization of organic material in the upper part of the water column by various organisms results in the fragmentation of the larger particles and in a decrease in sinking rate with depth (Hirsbrunner and Wangersky, 1976).

7. The Atmosphere as a Colloid System

The gaseous phase is the dispersing medium of the atmosphere. It is composed of a number of gases with nitrogen, oxygen, argon, carbon dioxide, and water vapor making up 99.997% of it by volume below 90 km. The atmosphere is divided into five rather well-distinguished horizontal layers, mainly on the basis of temperature. The pattern consists of three relatively warm layers, the maximum temperatures of which occur near the surface, between 50 and 60 km and above about 120 km, separated by two relatively cold layers. The minimum temperatures of the cold layers occur between 10 and 30 km and at about 80 km. They are designated from bottom to top as troposphere, mesosphere, thermosphere, and ionosphere (Fig. 1.7). The transition layer between the atmosphere and outer space is the exosphere. The base of the exosphere varies from 500 to 750 km. Here neutral atomic oxygen, ionized oxygen, and hydrogen atoms form the tenuous atmosphere and the gas laws cease to be valid. The troposphere contains 75% of the total molecular or gaseous mass of the atmosphere and virtually all the water vapor and the aerosols. It is in this zone that weather phenomena and atmospheric turbulence are most marked.

A certain amount of matter, other than the gaseous constituents, is always present in the atmosphere. This matter consists of inorganic and organic constituents and organisms. Three chief classes of aerosols have been identified in the atmosphere: (1) those whose particles are nonvolatile and nonsoluble, either solids or liquids, such as dusts, fumes, smokes, and volcanic smokes; (2) those whose particles are nonvolatile but soluble in water, such as sea salts, and (3) those whose particles are volatile and usually aqueous, such as mists, clouds, and fogs.

Aerosols are common trace constituents of the lower 30 km of the atmosphere. These atmospheric aerosols are primarily of terrestrial origin, either transported upward from the earth's surface or formed in situ by various processes of accretion. Particulate aerosols observed in the mesosphere and lower thermosphere are generally considered to be extraterrestrial, with the exception of a small overall contribution

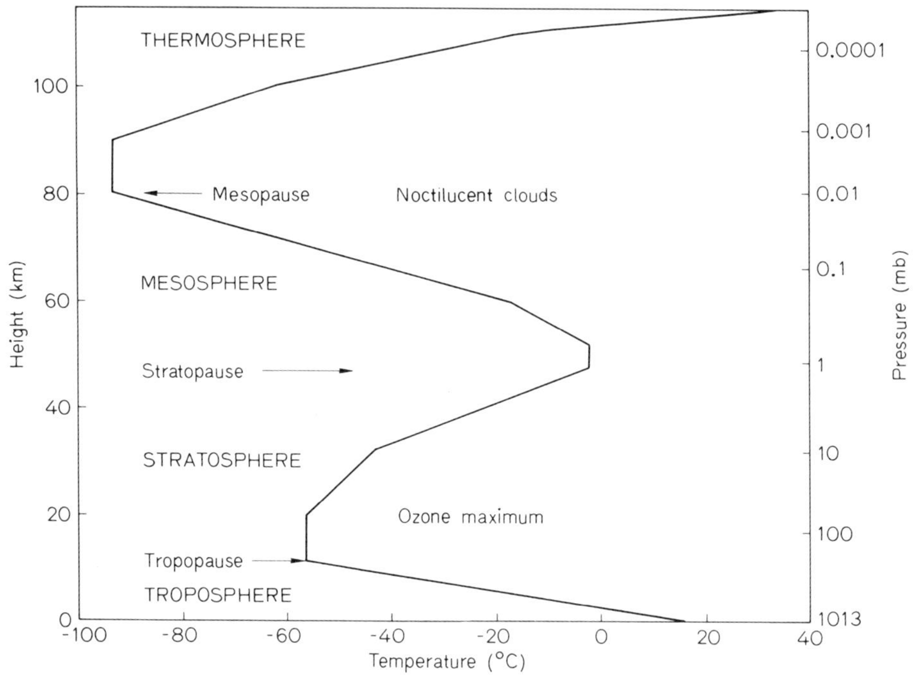

Fig. 1.7. Vertical division of the atmosphere into horizontal layers, on the basis of temperature. (From Barry and Chorley, 1970; used with permission of the authors)

from man-related phenomena, such as the by-product of rocket and satellite programs and high-altitude nuclear testing. Aerosols are important as effective scatterers of light. The lower atmospheric aerosols play an important role as condensation nuclei for cloud droplets and ice crystals formation (Bach, 1976). Upper atmospheric aerosols are involved in the photochemistry of the upper region by providing active surfaces for reactions (Olivero, 1974).

The two following air convections are responsible for the transportation of aerosols in the atmosphere:

1. Free convection occurs in fluids that are able to circulate internally and distribute heated parts of the mass. The low viscosity of air serves to make this characteristic the chief method of atmospheric heat transfer. The constantly maintained difference in temperature existing between the poles and the equator provides the energy necessary to drive the planetary atmospheric circulation by the conversion of thermal energy into kinetic energy. However, vertical movements are generally much less in evidence than horizontal movements, which may cover vast areas and persist for a few days and up to several months. Also the average horizontal wind speeds are of the order of one hundred times greater than average vertical movements, though individual exceptions occur, particularly in convection storms.

2. Forced convection is due to the development of eddies as air flows over uneven surfaces, even when there is no surface heating to set up thermal free-convection.

Forced convection is sometimes called mechanical turbulence. At a certain speed the air motion may become turbulent and consequently particles are dragged and dispersed into aerosol systems.

7.1 Tropospheric Aerosols

Aerosols are obtained either by the disintegration of bigger particles or by the condensation of small units present in the gaseous state. Meteors of varying sizes are constantly piercing the atmosphere. Meteoric dust is formed from the friction against the air blanket and reaches the troposphere because of the earth's gravitational field. Volcanic eruptions and explosions may throw out considerable quantities of finely divided material high into the atmosphere. Winds pick up and transport much inorganic and organic material such as rock debris, sand, volcanic ash, dry soil, and industrial dust, especially from semiarid and arid regions. Other major contributors are forest fires and other vegetation fires.

Winds also raise water spray from oceans, seas, lakes, and rivers. The water drops may contain inorganic and organic, dissolved, or suspended contaminations. As the water drop evaporates, the solid suspended material may stay dispersed in the atmosphere. Aerosols of nonvolatile solutes, such as sea salts, are obtained by a different mechanism. The droplet becomes supersaturated and a sudden condensation of the solute molecules or ions results in the formation of very fine particles. In clouds, water droplets or snow particles are formed by condensation of water vapor.

Condensation occurs under varying conditions, which in one way or another are associated with a change in one of the linked parameters of air volume, temperature, pressure, or humidity. For instance, condensation takes place when the temperature of the air is reduced but its volume remains constant or when the volume of air is increased without addition of heat (adiabatic expansion), or when a joint change of temperature and volume reduces the moisture-holding capacity of the air to below its existing moisture content. The formation of particles and liquid droplets is controlled by the formation and presence of nuclei; that is, moisture must generally find a suitable surface upon which it can condense. If air free of particulate matter is cooled, water droplets, which may serve as nuclei, are obtained only at extreme supersaturation. Usually condensation occurs on surfaces of hydrophilic particles, the so-called *hygroscopic nuclei.* These particles can be dust, smoke, sulfur dioxide, salts such as NaCl, and so on.

The aerosol present in the middle and upper tropospheres is thought to be the background tropospheric aerosol on which other aerosols, such as freshly generated land dust, pollution-produced particles, and sea spray, are locally superimposed. When particles are raised from ground level into the atmosphere, at a height of a few kilometers the size distribution exhibits a characteristic form deficient in both large and small particles. According to Blifford (1970), during the aging of the aerosol small particles coagulate into larger ones and large particles are removed by sedimentation. The middle and upper tropospheres contain a uniform aerosol composed of 90–95% continental material and 5–10% marine material. In the continental profiles a continuous regime extending from ground level up to 9 km exists, whereas in the oceanic profiles two regimes are clearly delineated. The lower or *marine regime* extends upward only 1 or 2 km, whereas above the marine regime the aerosol composition is the same as that found for high tropospheric *continental* regimes (Delamy et al., 1973).

The settling and removal of particles can be classified (in inverse order of importance) as (1) dry gravitational sedimentation or *fallout*; (2) impaction against the ground; (3) *rainout* (or *snowout*)–where particles are incorporated into cloud droplets in the cloud, some of which are hygroscopic nuclei, and (4) *washout*–in which falling rain or snow scavenges the particles below cloud level. Statistically, smaller particles remain longer in the atmosphere and travel longer distances than do larger particles. The average tropospheric residence time of a small particle with a radius of about 100 nm is of the order of 1 month, enough time for such a particle to make a trip around the world (Schneider, 1972). The removal processes depend upon local meterologic conditions.

Snow scavenging is 28–50 times more efficient than rain per equivalent water content. The dynamic electric properties of the atmosphere are crucial to the scavenging process, since electrostatic attraction between charged aerosols and raindrops or snow particles is a more important scavenging mechanism than is diffusive or impactive scavenging (Graedel and Franey, 1975).

7.1.1 Continental Aerosols

The possiblity that a grain will be winnowed from the rock or soil increases with the degree of turbulence and speed of wind and decreases with increasing size and specific weight of the grain (Ch. 5). It also depends on the water content of the parent rock or soil and decreases with increasing humidity up to a certain degree (Ch. 9). The average wind speeds range from 1 to 30 m/s. However, the speeds that are effective are 15–20 m/s. In the troposphere silty and clayey grains are held by ascending components that are five times slower than the average speed of wind. Strong winds increase the concentration of the aerosol and at speeds exceeding 6 m/s sand grains are also present in the suspended load. In eolian transportation the size of 0.05 mm forms a natural boundary: Below it, grains are transported predominantly in suspension whereas above it they move by traction and saltation, although it is found that wind can support grains of a maximum diameter of 3 mm in suspension (Kukal, 1971).

Initiation of movement of a particle is most frequently by saltation, less often by traction or a direct lifting into an aerosol system. The formation of aerosols from large soil particles was studied by Rosinski and Langer (1974). The surfaces of large soil particles were examined with a scanning electron microscope and were found to consist mainly of small particles attached to their basic surfaces (Yaalon and Ganor, 1973). When such soil particles are lifted by winds, the concentration of aerosol particles increases by the absolute number of particles shed from the surfaces of large particles that collide with obstacles, provided the shed particles become airborne. The mechanism of particle shedding is not a breakage of loose agregates or more compact agglomerates. In this case particles present in large numbers in the interstices of a parent particle are released into air and form an independent aerosol colloid system leaving the parent particles practically unchanged.

Gillette et al. (1972) measured aerosol size distributions and vertical fluxes of aerosols on rural land subjected to wind erosion. They concluded that the size distribution of the aerosol bears a strong resemblance to the size distribution of the soil itself. For a radius size range of 300–1000 nm, the number of soil particles and of soil-derived aerosol particles increased slightly with decreasing radius. With larger particles, the effect of the radius on the size distribution of soil and aerosol particles was even more drastic. In more populated regions of a continent where many sources of aerosol contribute to the total mixture of aerosol, the aerosol size distribution is influenced by the addition of submicron particles from sources other than soil. Physical settling also acts upon the aerosol to remove large particles in preference to small particles. According to these authors, independent submicron particles in the rural soil were only slightly more numerous than particles larger than 10,000 nm. Clay coating was present on all particles. The failure of wind erosion mechanisms to disintegrate totally the aggregated soil particles results in the relative lack of particles having radii smaller than 1000 nm compared to the concentration of the clay minerals in the soil. Thus, an aerosol derived from a rural soil has a size distribution that seemingly follows the size distribution of the soil except for the very small particles, where soil aggregation takes place, and for particles with radii larger than 10,000 nm, where gravitational sedimentation effectively removes particles from the air.

Johansson et al. (1976) established a baseline for nine trace element constituents of nonurban aerosol particles as a function of particle size at ground-level stations in northern Florida. Elements contained in the largest particles display the greatest degree of average concentration differences between sites, a result suggesting short atmospheric residence times and pointing out the importance of local dispersion sources and atmospheric processes in regulating the particle concentrations in air. Elements contained in particles of radii smaller than 1000 nm show little average concentration differences between sites unless they are influenced by local sources of pollution. This suggests that the concentration of these small suspended particles is regulated by large-scale sources and transport processes. K, Ca, Ti, Mn, Fe, and Zn appear to be regulated in the main by terrestrial source processes, Cl by marine source processes, but Br and Pb appear to be accounted for adequately as originating mainly from automotive fuel combustion. These two elements were found mainly in the smallest particles in urban areas whereas Cl was found in the largest particle size fraction. No evidence was found for the dependence of particulate sulfur concentrations on local pollution sources.

The grain-size distribution and the amount of material in dusts together with their mineralogic compositions have been studied at many places on the earth's surface and have been summarized by Kukal (1971). The mineral composition of terrigenous, nonurban dust, where no obvious industrial components are detected, is uniform on the whole. It contains about 70% quartz, feldspar (the amount of which decreases with the decreasing grain size), micas, calcite, heavy minerals, and aggregates of clay minerals. High carbonate dusts, in which the nonclayey fraction contains 40–60% calcite and 15–30% dolomite, are also known to occur (Ganor and Yaalon, 1974). The biogenous components may be divided into four categories: (1) fungal spores; (2) pollen grains; (3) diatoms, and (4) opal phytoliths. Darby et al. (1971) examined arctic airborne dust from northern Alaska containing no industrial components. They found that most of the material was less than 5 μm in diameter. Organic material accounted for the largest diameter grains. The largest grain measured, 73 x 22 μm, appeared to be a plant fragment. The largest mineral grain measured in most sites was 40 μm. Somewhat larger grains were found at a site 1.6 km off-shore with a maximum grain diameter of 100 μm. Larger amounts of plant debris–up to 80% were noted in those samples.

The *man-made contribution* to aerosol systems is 5–50% of the total mass of suspended particles. The exact ratio depends on site location. In the midlatitudes of the northern hemisphere, where most of the human activity producing these emissions takes place, it is probable that the man-made particle production is comparable to that of nature (Schneider, 1972). In addition to the increase in the mass of the total particulate load, Schneider points out that there is another important consequence of the injection of man-made particles into the atmosphere. The average size of man-made particles is smaller than the average size of naturally produced particles. When the man-made contribution is combined with the natural or *background* distribution of particles, the average size of the suspended particles is decreased. Increases in particle concentrations can have significant effects on climate. The chemical composition and the particle size distribution of the atmospheric aerosols are major factors in determining the toxic effects of airborne particulate matter (Natusch and Wallace, 1974).

The chemical compositions of various man-made contributions to aerosol systems together with the chemical composition of the natural background aerosol systems were summarized by Miller et al. (1972). Some of the more important are:

Automobile Emissions. Aerosol emissions from auto exhaust are mostly lead-bearing material and nonvolatile hydrocarbons.

Fuel Oil Emissions. The fuels most commonly used are natural gas and fuel oil. Fuel oil produces much larger quantities of aerosols than natural gas. The composition of fly ash arising from fuel oil combustion is highly variable. It may contain 18–58% non-carbonate carbon, 17.5–25% SO_4^{2-}, and oxides of various metals such as aluminum, iron, sodium, and vanadium at concentrations of up to 10% and oxides of other metals at concentrations of less than 1%. Vanadium concentrations in atmospheric particulate matter have been used to monitor fuel oil consumption (Martens et al., 1973a).

Based on the preceding data Miller et al. (1972) developed a method for calculating, on a chemical element-by-element basis, the contribution of various sources to an aerosol from a polluted atmosphere. They reached the following conclusions on man-made contributions to aerosol systems in Pasadena: Mg, Ca, and Ba show the presence of anthropogenic sources in addition to rock dispersion. Ba is from additives to diesel fuel as smoke suppressants. Cu, Fe, and Zn can be ascribed to anthropogenic sources, particularly metal working facilities. ZnO, a component of rubber tire treads, may have contributed to the zinc concentration. Mn is unique among the heavy metals since soil dust appears to be the major source of manganese. Iodine is introduced into the atmosphere mainly by man through the use of organic iodides in industrial processes. Automobile emissions contribute most of the bromine to the aerosol. Primary sources of carbon-containing compounds include tarry material present in automobile exhaust emissions, tire dust, diesel exhaust, and soot from jet aircraft. Soot comprises about 96% of the particulate matter emitted by jet turbine engines.

Several aerosol components, such as sulfate, nitrate, and ammonium salts as well as hydrocarbon-derived materials, are produced by condensation reactions in the atmosphere. Atmospheric sulfate and nitrate arise from the oxidation of SO_2 (Arrowsmith et al., 1973) and nitrogen oxides, respectively. Ammonia is adsorbed from the gas phase. Animal and plant metabolic processes are natural sources of ammonia. Man increases the ammonium ion concentration through the use of fertilizers and petrochemicals.

7.1.2 Marine Aerosols

According to Chesselet et al. (1972) bubbles bursting at the sea surface inject into the atmosphere quantities of ionic fractionated aerosols large enough to affect the geochemistry of atmospheric particles. At the very beginning of their existence in the atmosphere, some of the aerosols produced by bubbles bursting at the surface of the sea have higher elemental K/Na weight ratios than those existing in the bulk seawater. This ratio shows large variations according to time and place for particles sampled in open-sea marine air. This situation reflects the presence of a complex population of atmospheric particles whose composition is largely controlled by settling processes. The particles in the marine atmosphere include: (1) sea spray particles, having the seawater ionic ratios; (2) sea spray particles having ionic ratios other than those in seawater (ionic fractionated marine aerosols) and (3) terrigenous dust of small sizes, probably enriched with cations of primary continental origin in a water-soluble phase. Data from the atmosphere over the open sea demonstrate that the smaller size categories of marine aerosols are the more fractionated. In upper layers of the marine atmosphere, their concentrations are sufficient to compete with the terrigenous dust load and to determine the cationic ratios.

The aerosol Cl/Na ratios remain close to the seawater value. Thus, when compared with the sodium concentrations, potassium and calcium concentrations in particulate matter appear to be essentially controlled by a fractionation process at the sea–air interface. The concentrations of chlorine and sodium in marine aerosols increase with increasing wind speed. This relation indicates the marine origin of the dispersed salt. There is a trend to higher values for the Na/K ratios with increasing wind speed. From the previous discussion on the liquid boundary layer (Introduction), one would expect the depletion of both Na and K ions in the air–liquid interfaces. Since the hydration shell of Na is more stable than that of K, the latter is depleted to a smaller extent. The Na/K ratio in the atmosphere is determined by two factors: (1) the ratio in the ocean–air boundary layer and (2) the ratio in the bulk solution. At low wind speeds the total number of droplets formed is small. They are predominantly made up of water molecules and salts from the boundary layer. With increasing wind speed the rate of formation of droplets also increases and the second factor becomes more important. There is a higher production rate of particles less enriched in potassium. Wada and Kokubu (1973), in a study of marine aerosols near Japan, found that Ca to a small extent and F to a very high extent are augmented in the aerosol system in comparison with their partial concentration in the sea solution. They concluded that on one hand sea spray and on the other hand continental materials including blown-off soil dust, gaseous and smoky industrial material waste as well as volcanic ejecta and exhalation constitute the major sources of Ca and F salts.

Mészáros and Vissy (1974) studied the compositions of aerosols sampled over the oceans of the southern hemisphere, where the contribution of continental material is minimal. The total aerosol concentrations in this hemisphere are about 2/3 of the values measured over the north Atlantic Ocean. The atmoshperic aerosol particles with radii larger than 30 nm are composed of sodium chloride (larger particles) and of ammonium sulfate (smaller particles). These salts often coagulate to form particles of mixed type. The size distribution of these particles shows a maximum radius of about 70–100 nm. In some cases, droplets of sulfuric acid are also present. These authors believe that if the sea salt component is not taken into consideration, the aerosol particles in the global background atmosphere consist of ammonium sulfate or sulfuric acid, depending on the available ammonia gas.

Gaseous chlorine is released from the marine aerosol into the atmosphere (Junge, 1963). Martens et al. (1973b) studied chlorine loss from the slightly polluted Puerto Rican and the highly polluted San Francisco Bay area marine aerosols and found losses averaging 13% and 54%, respectively, of the particulate marine Cl originally present. These authors suggest that the Cl loss is controlled by the interaction of H_2SO_4 and HNO_3 vapors with the NaCl droplets, which lower the pH and cause the release of HCl to the atmosphere. Background levels of gaseous NO_2 and SO_2 may account for the narrow range of Cl loss observed in nonpolluted marine atmosphere.

Studies on organic carbon content of marine aerosols have been reviewed by Hoffman and Duce (1974). Bubbles formed in the oceans are responsible for transporting both dissolved and particulate organic matter to the sea surface. When the bubbles burst they eject some fractions of this organic material into the atmosphere along with sea salt. The organic material may be associated with certain inorganic constituents in seawater and may be responsible for transporting these inorganic species to the atmosphere in relative concentrations higher than those found in bulk seawater. The relative excesses of P, I, Cu, and K observed in marine aerosols compared with those observed in seawater are the result of the association of these elements with the organic material transported to the atmosphere by bursting bubbles. Compressed films of surface-active material are found on sea-salt particles. The surface-active film/salt mass ratio varies from 0.3 to 0.7 in these particles. The percentage of organic material varies with sea-salt particle size, the smaller particles in marine air containing greater proportions of organic material than the larger particles. Concentrations of marine aerosols collected by Hoffman et al. (1974) in Bermuda ranged from 0.15 to 0.47 $\mu g/m^3$. The mass of organic carbon was from 1–19% of the mass of the sea salt, and this percentage decreased with increasing atmospheric sea salt concentrations.

Wind transport is an important means for the introduction of eolian land-derived solids to marine aerosols. The solids are then settled out onto the oceans and subsequently become part of deepsea sediments. Dust loadings in the southeast trade winds of the Atlantic are considerably lower than those in the northeast trade winds. The southern African desert (the Namib and the Kalahari) have a much smaller effect on southeast trade midocean dust loadings than the Sahara has on northeast trade midocean dust loadings. Dust loadings in the northeast trades range between 10 and

100 $\mu g/m^3$. In the southeast trades they are one or two orders of magnitude smaller than those of the northeast trades. Eolian dust makes a significant contribution to the inorganic land-derived material in the upper few meters of seawater in the area of the northeast trades but only a very minor contribution to the inorganic suspended material in the upper layer of the ocean (Chester et al., 1972).

7.2 Stratospheric Aerosols

The existence of a semipermanent stratospheric aerosol layer was first suggested by Gruner and Kleinert in 1927 on the basis of twilight observations. This was experimentally established in 1961 by Junge and co-workers using balloon-borne impactors. The *Junge layer* has since been verified and extensively studied by many investigators employing both in situ and remote sensing techniques. The layer concentration peak is at 20 km plus or minus a few kilometers. Before the Agung volcanic eruption in 1963 this layer was observed to be relatively stable and independent of latitude. However, since that time, considerable variability in the altitude and concentration of the layer as a function of latitude and time has been noted. Observations by Cunnold et al. (1973) and others have been interpreted as evidence of an additional stratospheric aerosol layer at around 50 km. This layer may possess a spatial structure.

The 20-km stratospheric aerosol layer has been directly sampled on many occasions, and sulfuric acid, ammonium sulfate and sulfite, and nitric acid along with several other materials, such as salts of Na, Ca, Mg, F, and Cl in minor amounts, have been found. According to Toon and Pollack (1973) nitric acid cannot be present as an aerosol particle in the lower stratosphere. Sulfuric acid aerosol particles are composed of H_2SO_4 (75 wt %) and water. These particles are solid at stratospheric temperatures and pressures. Ammonium sulfate aerosol particles are also solid. Polar molecules may form surface coatings on these aerosol and may serve as effective catalysts for further chemical reactions.

A commonly used term for the lower stratospheric aerosol layer is *sulfate layer.* The three most important possibilities for the entry of sulfur into the stratosphere are:

1. Volcanic injections in the form of sulfur-bearing gases or sulfate compounds
2. Upward convection of tropospheric sulfur-bearing gases, for gases in tropical regions
3. The influx of tropospheric sulfur-bearing particles having diameters smaller than 100 nm

The first two possibilities require gas phase reactions to form particles. The mechanisms proposed for such conversions are based on either the absorption of SO_2 by water droplets and the aqueous phase oxidation of the sulfite enhanced by catalytic activity, or gas phase photochemical reactions. A further, nonphotochemical mechanism for the reaction between NH_3 and SO_2 in the absence of free water is heteromolecular nucleation, which converts atmospheric gas to particulate matter (Arrowsmith et al., 1973; Scargill, 1971).

Hofmann et al. (1973) pointed out that the distribution of the particles in the stratosphere cannot be correlated with the preceding three sources. Similar concentrations of sulfur aerosols are found at the North and the South Poles, although the northern hemisphere is the dominant sulfur producer. The average particle size below 9 km was observed to be smaller than in the stratosphere, probably indicating a different particulate source for the two regions. The differences in concentrations between the hemispheres would be reduced considerably in air entering the stratosphere since the residence time of sulfur varies from 3 h near sources of pollution to about 4 days, the time required for SO_2 or other tropospheric sulfur gases to reach the stratosphere. The final distribution of stratospheric sulfate is controlled principally by dynamic stratospheric processes. Higher sulfate concentrations than those found at 20 km or lower altitudes are observed at altitudes as high as 37 km. The residence time for SO_2 in the stratosphere is 24 years. Under such circumstances it is conceivable that SO_2 diffuses to more severe oxidizing conditions at higher altitudes. It is also possible that sulfate particles are transported to altitudes as high as 37 km by dynamic processes (Lazarus and Gandrud, 1974).

References

Akiyama, T.: Chemical composition and molecular weight distribution of dissolved organic matter produced by bacterial degradation of green algae. Geochem. J. *6,* 93–104 (1972)

Arrowsmith, A., Hedle, A. B., Beer, J. M.: Particle formation from NH_3–SO_2–H_2O air gas phase reactions. Nature Phys. Sci. (London) *244*, 104–105 (1973)

Askenasy, P. E., Dixon, J. B., McKee, T. R.: Spheroidal halloysite in a Guatemalan soil. Soil Sci. Soc. Am. Proc. *37*, 799–802 (1973)

Bach, W.: Global air pollution and climatic changes. Rev. Geophys. Space Phys. *14*, 429–474 (1976)

Bailey, J. C.: Fluorine in granitic rocks and melts, a review. Chem. Geol. *19*, 1–42 (1977)

Bailey, S. W., Brindley, G. W., Johns, W. D., Martin, R. T., Ross, M.: Summary of national and international recommendations on clay mineral nomenclature. Clays Clay Miner. *19*, 129–132 (1971)

Barry, R. G., Chorley, R. J.: Atmosphere, weather and climate. New York: Holt, Rinehart and Winston, Inc. 1970

Beers, J. R., Stewart, G. L.: Microzooplankton and its abundance relative to larger zooplankton and other seston components. Mar. Biol. *4*, 182–189 (1969)

Berger, W. H.: Planktonic foraminifera: selective solution and paleoclimatic interpretation. Deep-Sea Res. *15*, 31–44 (1968)

Berner, R. A., Morse, J. W.: Dissolution kinetics of calcium carbonate in sea water. IV. Theory of calcite dissolution. Am. J. Sci. *274*, 108–134 (1974)

Blifford, I. H., Jr.: Tropospheric aerosols. J. Geophys. Res. *75*, 3099–3103 (1970)

Bolland, M. D. A., Posner, A. M., Quirk, J. P.: Surface charge on kaolinites in aqueous suspension. Aust. J. Soil Res. *14*, 197–216 (1976)

Botterill, J. S. M., Bessant, D. J.: The flow properties of fluidized solids. Powder Technol. *8*, 213–222 (1973).

Bowen, N. L.: The evolution of the igneous rocks. Princeton, N. J.: Princeton Univ. Press 1928

Bramlette, M. N.: Pelagic sediments. In: Oceanography. Sears, M. (ed.). Publ. Am. Assoc. Adv. Sci. No. *67*, 1961, pp. 345–366

Bruce, P. N., Revel-Chion, L.: Bed porosity in three phase fluidization. Powder Technol. *10*, 243–249 (1974)

Buerger, M. J.: The structural nature of the mineraliser action of fluorine and hydroxyl. Am. Mineral. *33*, 744–747 (1948)

Burnham, C. W.: Hydrothermal fluids and the magmatic stage. In: Geochemistry of hydrothermal ore deposits. Barnes, H. L. (ed.). New York: Holt, Rinehart and Winston, Inc., 1967, pp. 34–76

Burnham, C. W.: Water and magmas, a mixing model. Geochim. Cosmochim. Acta *39*, 1077–1084 (1975)
Carmichael, I. S. E., Turner, F. J., Verhoogen, J.: Igneous petrology, New York: McGraw-Hill (1974)
Carr, R. M., Chih, H.: Complexes of halloysite with organic compounds. Clay Miner. *9*, 153–166 (1971)
Chave, K. E., Suess, E.: Calcium carbonate saturation in sea water. Effect of organic matter. Limnol Oceanogr. *15*, 633–637 (1970)
Cheshire, M. V., Goodman, B. A., Mundie, C. M.: The composition of soil humus. Welsh Soil Diss. Group. Rep. *16*, 73–90 (1975)
Chesselet, R., Morelli, J., Buat Menard, P.: Some aspects of the geochemistry of marine aerosol. Nobel Symposium 20, The changing chemistry of the oceans. Dyrssen, D., Jagner, D. (eds.), Stockholm: Almquist and Wilksell, 1971, pp. 93–114
Chesselet, R., Morelli, J., Buat-Menard, P.: Variations in ionic ratios between reference sea water and marine aerosols. J. Geophys. Res. *77*, 5116–5131 (1972)
Chester, R., Elderfield, H., Griffin, J. J., Johnson, L. R., Padgham, R. C.: Eolian dust along the margins of the Atlantic Ocean. Mar. Geol., *13*, 91–105 (1972)
Christiansen, R. L., Lipman, P. W.: Emplacement and thermal history of a rhyolite lava flow near Fortymile Canyon, Southern Nevada. Geol. Soc. Am. Bull. *77*, 671–684 (1966)
Churchman, G. J., Aldridge, L. P., Carr, R. M.: The relationship between the hydrated and dehydrated states of an halloysite. Clays Clay Miner. *20*, 241–246 (1972)
Collins, K., McGowan, A.: The form and function of microfabric features in a variety of natural soils. Geotechnique *24*, 223–254 (1974)
Cruz, M., Jacobs, H., Fripiat, J. J.: The nature of the cohesion energy in kaolin minerals. Proc. Int. Clay Conf., Madrid, Serratosa, S. W. (ed.) Madrid: Div. Ciencias C. S. I. C., 1973, pp. 35–46
Cunnold, D. M., Gray, C. R., Merrutt, D. C.: Stratospheric aerosol layer detection. J. Geophys. Res. *78*, 920–931 (1973)
Darby, D. A., Burckle, L. H., Clark, D. L.: Airborne dust on the Arctic pack ice, its composition and fallout rate. Earth Planet. Sci. Lett. *24*, 166–172 (1974)
Daumas, R. A.: Variations of particulate proteins and dissolved amino acids in coastal seawater. Mar. Chem. *4*, 225–242 (1976)
Daumas, R. A., Laborde, P. L., Marty, J. C., Saliot, A.: Influence of sampling method on the chemical composition of water surface film. Limnol. Oceanogr. *21*, 319–326 (1976)
Degens, E. T.: Molecular nature of nitrogenous compounds in sea water and marine sediments. In: Proc. Symp. Organic Matter in Natural Waters, 1968. Wood, D. H. (ed.). Univ. Alaska: Occ. Pub. No. *1*, Inst. Mar. Sci., 1970, 77–106
De Jong, J. A. H., Nomden, J. F.: Homogeneous gas-solid fluidization. Powder Technol. *9*, 91–97 (1974)
Delamy, A. C., Pollock, W. H., Shedlovsky, J. P.: Tropospheric aerosol, the relative contribution of marine and continental components. J. Geophys. Res. *78*, 6249–6265 (1973)
de Vries, A. J.: Morphology, coalescence and size distribution of foam bubbles. In: Adsorptive bubble separation techniques. Lemlich, R. (ed.). New York: Academic Press, 1972, pp. 7–31
Dionne, J. C.: Monroes, a type of so-called mud volcanoes in tidal flats. J. Sediment. Petrol. *43*, 848–856 (1973)
Dionne, J. C.: Miniature mud volcanoes and other injection features in tidal flats. James Bay, Québec. Can. J. Earth Sci. *13*, 422–428 (1976)
Dixon, J. B., McKee, T. R.: Internal and external morphology of tubular and spheroidal halloysite particles. Clays Clay Miner. *22*, 127–137 (1974)
Dzulynski, S., Walton, E. K.: Sedimentary features of flysch and greywackes. Amsterdam: Elsevier 1965
Edmond, J. M.: The thermodynamic description of the CO_2 system in seawater, development and current status. Proc. R. Soc. Edinburgh, Sect. B *72*, 371–380 (1972)
Edmond, J. M.: On the dissolution of carbonate and silicate in the deep ocean. Deep Sea Res. *21*, 455–480 (1974)

Edmond, J. M., Gieskes, J. M. T. M.: On the calculation of the degree of saturation of seawater in respect with calcium carbonate under in situ conditions. Geochim. Cosmochim. Acta *34*, 1261–1291 (1970)
Ewart, A: Mineralogy and petrogenesis of the Whakamaru ignimbrite in the Maraetai area of the Taupo volcanic zone, New Zealand, N. Z. J. Geol. Geophys. *8*, 611–677 (1965)
Faust, T. G., Fahey, J. J.: The serpentine group minerals. U. S. Geol. Survey Prof. Paper, *384–A* (1962)
Fenner, C. N.: The origin and mode of emplacement of the great tuff deposits in the Valley of the Ten Thousand Smokes. Techn. Pap. Nat. Geogr. Contr. *1*, 74 (1923) (Quot. by Sparks, 1976)
Ferris, A. P., Jepson, W. B.: The exchange capacities of kaolinite and the preparation of homoionic clays. J. Colloid Interface Sci. *51*, 245–259 (1975)
Fiske, R. S., Hopson, C. A., Waters, A. C.: Geology of Mount Rainier National Park, Washington. U. S. Geol. Survey Prof. Paper, *444* (1963)
Fleischer, P.: Mineralogy and sedimentation history of Santa Barbara Basin, California. J. Sediment Petrol. *42*, 49–58 (1972)
Flood, R. H.,Vernon, R. H., Shaw. S. E., Chappell, B. W.: Origin of pyroxene-plagioclase aggregates in rhyodacite, Contr. Miner. Petrol. *60*, 299–309 (1977)
Frank-Kamenetskii, V. A., Kotov, N. V., Goilo, E. A., Tomashenko, A. N.: Polytypism and transformation of the minerals of kaolinite group under hydrothermal conditions. Clay Sci. *4*, 199–204 (1974)
Fujii, T.: Crystal settling in a sill. Lithos, *7*, 133–137 (1974)
Ganor, E., Yaalon, D. H.: Dust in the environment. I. The composition of dust in Israel (abs.). Israel Ecol. Soc. 5th Cong., Tel. Aviv (1974)
Garrett, W. D.: Stabilization of air bubbles at the air-sea interface by surface active material. Deep-Sea Res. *14*, 661–672 (1967)
Garrett, W. D.: Impact of natural and man-made surface films on the properties of the air-sea interface. In: The changing chemistry of the oceans. Stockholm: Almqvist and Wiksell, 1972, pp. 75–90
Gidigasu, M. D.: Degree of weathering in the identification of laterite materials for engineering purpose–a review. Engineer. Geol. *8*, 213–266 (1974)
Giese, R. F., Jr.: Interlayer bonding in kaolinite, dickite and nacrite. Clays Clay Miner. *21*, 145–149 (1973)
Gillette, D. A., Blifford, I. H., Jr., Fenster, C. R.: Measurements of aerosol size distributions and vertical fluxes of aerosols on land subjected to wind erosion. J. Appl. Meteorol. *11*, 977–988 (1972)
Graedel, T. E., Franey, J. P.: Field measurements of submicron aerosol washout by snow. Geophys. Res. Lett. *2*, 325–328 (1975)
Green, D.: Composition of basaltic magmas as indicators of conditions of origin. Application to oceanic volcanism. R. Soc. London Phil. Trans., Ser. A *268*, 707–725 (1971)
Grim, R. E.: Clay mineralogy, 2nd ed. New York: McGraw-Hill: 1968
Gruner, P., Kleinert, H.: Die Dämmerungserscheinungen. In: Probleme der kosmischen Physik, Hamburg: Henry Grand, 1927, Vol. *10*, pp. 1–113
Harris, P. G., Middlemost, E. A. K.: The evolution of kimberlite. Lithos *3*, 77–88 (1970)
Haughton, D. R., Roeder, P. L., Skinner, B. J.: Solubility of sulfur in mafic magmas. Econ. Geol. *69*, 451–467 (1974)
Haworth, R. D.: The chemical nature of humic acid. Soil Sci. *111*, 71–79 (1971)
Hayashi, H., Oinuma, K.: Si–O absorption band near 1000 cm^{-1} and OH absorption bands of chlorite. Am. Mineral. *52*, 1206–1210 (1967)
Heezen, B. C., Ewing, M.: Turbidity currents and submarine slumps and the 1919 Grand Banks earthquake. Am. J. Sci. *250*, 849–873 (1952)
Heller-Kallai, L., Yariv, S., Gross, S.: Hydroxyl-stretching frequencies of serpentine minerals. Mineral. Mag. *40*, 197–200 (1975)
Herbillon, A. J., Mestdagh, M. M., Vielvoye, L., Derouane, E. G.: Iron in kaolinite with special reference to kaolinite from tropical soils. Clay Miner. *11*, 201–220 (1976)

Hirsbrunner, W. R., Wangersky, P. J.: Composition of the inorganic fraction of the particulate organic matter in seawater. Mar. Chem. *4*, 43–49 (1976)
Hoffman, E. J., Duce, R. A.: The organic carbon content of marine aerosols collected on Bermuda. J. Geophys. Res. *79*, 4474–4477 (1974)
Hoffmann, D. J., Rosen, J. M., Pepin, T. J., Pinnick, R. G.: Particles in the polar stratosphere. Nature (London) *245*, 369–371 (1973)
Holloway, J. R.: Fluids in the evolution of granitic magmas. Consequences of finite CO_2 solubility. Geol. Soc. Am. Bull. *87*, 1513–1518 (1976)
Itamar, A.: Colloid systems in volcanic eruptions, (in Hebrew). Jerusalem: The Department of Geology, The Hebrew University 1975
Jackson, I.: Melting of the silica isotypes SiO_2, BeF_2 and GeO_2 at elevated pressures. Phys. Earth Planet. Inter. *13*, 218–231 (1976)
Jacobs, M. B., Thorndike, E. M., Ewing, M.: A comparison of suspended particulate matter from nepheloid and clear water. Mar. Geol. *14*, 117–128 (1973)
JANAF Thermochemical Tables. Washington, D. C.: NSRDS–NBS *37*, 1971
Johansson, T. B., van Grieken, R. E., Winchester, J. W.: Elemental abundance variation with particle size in North Florida aerosols. J. Geophys. Res. *81*, 1039–1046 (1976)
Junge, C. E.: Vertical profiles of condensation nuclei in the stratosphere. J. Meteorol. *18*, 501–509 (1961)
Junge, C. E.: Air chemistry and radioactivity. New York: Academic Press 1963
Junge, C. E., Chagnon, C. W., Manson, J. E.: Stratospheric aerosols. J. Meteorol. *18*, 81–107 (1961b)
Junge, C. E., Manson, J. E.: Stratospheric aerosol studies. J. Geophys. Res. *66*, 2163–2182 (1961a)
Kadik, A. A., Lukanin, O. A.: The solubility-dependent behavior of water and carbon dioxide in magmatic processes. Geochem. Intern. (Eng. translation, published, January, 1974, pp. 115–129). Geokhimiya, No. *2*, 163–179 (1973)
Kalle, K.: The problem of Gelbstoff in the sea. Ocean. Mar. Bio. Ann. Rev. *4*, 91–104 (1966)
Kay, R., Hubbard, N., Gast, P.: Chemical characteristics and the origin of oceanic ridge volcanic rocks. J. Geophys. Res. *75*, 1585–1613 (1970)
Khailov, K. M., Semenov, A. D., Burlakova, Z. P., Semonova, I. M.: Some information on chemical nature and properties of macromolecular organic substances dissolved in seawater and involved in bubble. Gidrokhim. Mater. *52*, 82–91 (1969)
Kononova, M. M.: Soil organic matter, its nature, its role in soil formation and in soil fertility. Oxford: Pergamon Press 1961
Kukal, Z.: Geology of recent sediments. Prague: Acad. Pub. Czechoslovak Acad. Sci. 1971
Lazrus, A. L., Gandrud, B. W.: Stratospheric sulfate aerosol. J. Geophys. Res. *79*, 3424–3431 (1974)
Lisitzin, A. P.: Sedimentation in the world ocean. Soc. Econ. Paleontol. Mineral. Spec. Publ. No. *17*, (1972)
Loughnan, F. C.: Chemical weathering of the silicate minerals. New York: American Elsevier 1969
Macdonald, G. A.: Volcanoes. Englewood Cliffs, New Jersey: Prentice-Hall, Inc. 1972
MacTaggart, K. C.: The mobility of nuées ardents. Am. J. Sci. *258*, 369–382 (1960)
Martens, C. S., Wesolowski, J. J., Harris, R. C., Kaifer, R.: Chlorine loss from Puerto Rican and San Francisco Bay area marine aerosols. J. Geophys. Res. *78*, 8778–8792 (1973b)
Martens, C. S., Wesolowski, J. J., Kaifer, R., John, W., Harris, R. C.: Sources of vanadium in Puerto Rican and San Francisco Bay area aerosols. Environ. Sci. Technol. *7*, 817–820 (1973a)
Marty, J. C., Saliot, A.: Hydrocarbon (normal alkanes) in the surface microlayer of seawater. Deep-Sea Res. *23*, 863–873 (1976)
Mashali, A., Greenland, D. J.: Dependence of charge characteristics of kaolinites on pH and electrolyte concentration. Proc. Int. Clay Conf., Mexico. Bailey, S. W. (ed.) Wilnette, I. L.: Applied Publishing Ltd., 1975, pp. 240–241
Mattson, S.: The constitution of the pedosphere. Ann. Agric. College Sweden *5*, 261–263 (1938)
Maurer, L. G.: Organic polymers in seawater: Changes with depth in the Gulf of Mexico. Deep-Sea Res. *23*, 1059–1064 (1976)

Meszaros, A., Vissy, K.: Concentration, size distribution and chemical nature of atmospheric aerosol particles in remote oceanic areas, Aerosol Sci. *5*, 101–109 (1974)

Miller, M. S., Friedlander, S. K., Hidy, G. M.: A chemical element balance for the Pasadena aerosol. J. Colloid Interface Sci. *39*, 165–176 (1972)

Murai, I.: A study of textural characteristics of pyroclastic flow deposits in Japan. Bull. Earth Res. Inst. Tokyo Univ. *39*, 133–248 (1961)

Naka, S., Ito, S., Kameyama, T., Inigaki, M.: Crystallization of coesite. Mem. Fac. Eng. Nagoya Univ. *28*, 266–316 (1976)

Natusch, D. F. S., Wallace, J. R.: Urban aerosol toxicity: the influence of particle size. Science *186*, 695–699 (1974)

Nissenbaum, A., Kaplan, I. R.: Chemical and isotopic evidence for the in situ origin of marine substances. Limnol. Oceanogr. *17*, 570–582 (1972)

Normark, W. R., Dickson, F. H.: Man-made turbidity currents in Lake Superior. Sedimentology, *23*, 815–831 (1976)

Olivero, J. J.: Surface catalytic reactions on upper atmospheric aerosols, J. Geophys. Res. *79*, 476–478 (1974)

Peterson, M. N.: Calcite rate of dissolution in a vertical profile in the central Pacific. Science *154*, 3756 (1966)

Pytkowicz, R. M., Fowler, G. A.: Solubility of foraminifera in seawater at high pressures, Geochem. J. *1*, 169–182 (1967)

Range, K. J., Range, A., Weiss, A.: Fire-clay type kaolinite? Experimental classification of kaolinite-halloysite minerals. Proc. Int. Clay Conf., Tokyo, Heller, L. (ed.) Jerusalem: Israel Univ. Press, 1969, pp. 3–13

Rankin, A. H., LeBas, M. J.: Liquid immiscibility between silicate and carbonate melts in naturally occurring ijolite magma. Nature (London) *250*, 206–209 (1974)

Rao, C. P., Gluskoter, H. J.: Occurrence and distribution of minerals in Illinois coals. Ill. State Geol. Surv., Circular *476*, (1973)

Reynolds, D. L.: Fluidization as a geological process and its bearing on the problem of intrusive granites. Am. J. Sci. *252*, 577–614 (1954)

Riley, J. F.. Chester, R.: Introduction to marine chemistry. London: Academic Press, 1971

Roedder, E.: Metastability in fluid inclusions, Soc. Mining Geol., Japan, Spec. Issue *3*, Proc. IMA–IAGOD, Meetings 1970, 1971 pp. 327–334

Rosinski, J., Langer, G.: Extraneous particles shed from large soil particles. Aerosol Sci. *5*, 373–378 (1974)

Ruberto, R. G., Cronauer, D. C., Jewell, D. M., Seshadri, K. S.: Structural aspects of sub-bituminous coal deduced from solvation studies. I. Anthracene-oil solvents. Fuel *56*, 17–24 (1977)

Ruch, R. R., Gluskoter, H. J., Shimp, N. F.: Occurrence and distribution of potentially volatile trace elements in coal (an interim report). Ill. State Geol. Surv., Environ. Geol., Notes, *61* (1973)

Ryabchikov, I. D., Hamilton, D. L.: Possible separation of concentrated chloride solutions during crystallization of felsic magma. Dokl. Akad. Nauk. SSSR, Earth Sci. Sec. *197*, 219–220 (1971) (Americ. Geol. Inst. Trans.)

Sacchi, R.: Fluidization phenomena in the Southern Alps basement. Boll. Soc. Geol. Ital. *90*, 271–281 (1971)

Scarfe, C. M.: Water solubility in basic magmas, Nature (London) Phys. Sci. *246*, 9–10 (1973)

Scargill, D.: Dissociation constants of anhydrous ammonium sulfite and ammonium pyrosulfate prepared by gas phase reactions. J. Chem. Soc. A, 2461–2466 (1971)

Scherer, G., Vergano, P. J., Uhlman, D. R.: A study of quartz melting. Phys. Chem. Glasses *11*, 53–58 (1970)

Schmincke, H. U.: Volcanological aspects of peralkaline silicic welded ash-flow tuffs. Bull. Volcanol. *38*, 594–636 (1974)

Schmincke, H. U., Fisher, R. V., Waters, A. C.: Antidune and chute and pool structures in Base Surge deposits from the Laacher See area (Germany). Sedimentology *20*, 1–24 (1973)

Schneider, S. H.: Atmospheric particles and climate: Can we evaluate the impact of man's activities? Quat. Res. (N. Y.) *2*, 425–435 (1972)

Schnitzer, M., Kahn, S. U.: Humic substances in the environment. New York: Marcel Dekker, Inc. 1972
Schofield, R. K., Samson, H. R.: The deflocculation of kaolinite suspensions and the accompanying change-over from positive to negative chloride adsorption. Clay Miner. Bull. *2*, 45–51 (1953)
Serna, C. J., Velde, B. D., White, J. L.: Infrared evidence of order-disorder in amesites. Am. Mineral. *62*, 296–303 (1977)
Sharp, J. H.: Size classes of organic carbon in seawater, Limnol. Oceanogr. *18*, 441–447 (1973)
Sheldon, R. W., Evelyn, T. P. T., Parsons, T. R.: On the occurrence and formation of small particles in seawater. Limnol. Oceanogr. *12*, 367–375 (1967)
Sheridan, M. F.: Particle size characteristics of pyroclastic tuffs, J. Geophys. Res. *76*, 5627–5634 (1971)
Sholkovitz, E., Soutar, A.: Changes in the composition of the bottom water of the Santa Barbara Basin: Effect of turbidity currents. Deep-Sea Res. *22*, 13–21 (1975)
Sieburth, J. M.: Studies on algal substances in the sea. III. The production of extracellular organic matter by littoral marine algae. J. Exp. Mar. Biol. Ecol. *3*, 290–309 (1969)
Sieburth, J. M., Jensen, A.: Studies on algal substances in the sea, Part I., J. Exp. Mar. Biol. Ecol. *2*, 174–189 (1968)
Sleep, N. H.: Segregation of magma from a mostly crystalline mush. Geol. Soc. Am. Bull. *85*, 1225–1232 (1974)
Sood, M. K., Edgar, A. D.: Melting relations of undersaturated alkaline rocks from the ilimaussaq intrusion and Gronnedal-Ika Complex, South Greenland, under water vapor and controlled partial oxygen pressure. Medd. om Grønland *181*, 12 (1970)
Sparks, R. S. J.: Grain size variations in ignimbrites and implications for the transport of pyroclastic flows. Sedimentology *23*, 147–188 (1976)
Suess, E.: Interaction of organic compounds with calcium carbonate: Associated phenomena and geochemical implications. Geochim. Cosmochim. Acta *34*, 157–158 (1970)
Sutcliffe, W. H., Baylar, E. R., Menzel, D. W.: Sea surface chemistry and Langmuir circulation. Deep-Sea Res. *10*, 233–243 (1963)
Swinnerton, J. W., Lamontagne, R. A.: Oceanic distribution of low-molecular-weight hydrocarbons baseline measurements. Environ. Sci. Technol. *8*, 657–663 (1974)
Tan, Li-Ping: The metamorphism of Taiwan Miocene coals. Taiwan Geol. Surv. Bull. *16*, 1–44 (1965)
Terjesen, S. G., Erga, O., Thorsen, G., Ve, A.: Phase boundary processes as rate determining steps in reactions between solids and liquids. Chem. Eng. Sci. *74*, 277–288 (1961)
Toon, O. B., Pollack, J. B.: Physical properties of the stratospheric aerosols. J. Geophys. Res. *78*, 7051–7056 (1973)
Turekian, K.K.: Some aspects of the geochemistry of marine sediments. In: Chemical oceanography. Riley, S. P., Skirrow, G. (eds.) New York: Academic Press 1965, pp. 81–126
Urnes, S.: X-ray diffractions of glasses and methods of interpretation. In: Selected topics in high temperature chemistry. Oslo: Universitetsforlaget 1966, pp. 97–124
van Olphen, H.: Introduction to clay colloid chemistry. New York: Interscience Publ. 1963
Wada, S., Kokubu, N.: Chemical composition of maritime aerosols. Geochem. J. *6*, 131–139 (1973)
Waff, H. S.: Pressure-induced coordination changes in magmatic liquids. Geophys. Res. Lett. *2*, 193–196 (1975)
Wager, L. R., Brown, G. M., Wadsworth, W. J.: Types of igneous cumulates. J. Petrol. *1*, 73–85 (1960)
Walker, G. F.: Vermiculite minerals. In: The X-ray identification and structures of clay minerals. In: The X-ray identification and structures of clay minerals. Brown, G. (ed.) Great Britain: Mineralogical Society, Monograph, 1961, Chap. VII, pp. 199–223.
Walker, G. P. L.: Grain-size characteristics of pyroclastic deposits. J. Geol. *79*, 696–714 (1971)
Weaver, C. E., Pollard, L. D.: The chemistry of clay minerals. Amsterdam: Elsevier 1973
Wedepohl, K. H.: Geochemistry. Althous, E. (translator), New York: Holt, Rinehart and Winston, Inc. 1971
Whitford-Stark, J. L.: Vesicles and related structures in lava. J. Geol. *8*, 317–332 (1973)

Wilson, S. A., Weber, J. H.: A comparative study of number-average dissociation-correlated molecular weights of fulvic acids isolated from water and soil. Chem. Geol. *19*, 285–293 (1977)

Wyllie, P. J., Huang, W. L.: Carbonation and melting reactions in the system $CaO-MgO-SiO_2-CO_2$ at mantle pressures with geophysical and petrological applications. Contribut. Mineral. Petrol. *54*, 79–107 (1976)

Yaalon, D. H., Ganor, E.: The influence of dust on soils during the quaternary. Soil Sci. *116*, 146–155 (1973)

Yariv, S., Shoval, S.: The nature of the interaction between water molecules and kaolin-like layers in hydrated halloysite. Clays Clay Miner. *23*, 473–474 (1975)

Zavaritskii, A. N., Sobolev, V. S.: The physicochemical principles of igneous petrology. Kolodny, J., Amoils, R. (translators). Jerusalem: Israel Program for Scientific Translations 1964

Zieminski, S. A., Hume III, R. M., Durham, R.: Rates of oxygen transfer from air bubbles to aqueous NaCl solutions at various temperatures. Mar. Chem. *4*, 333–346 (1976)

Chapter 2
Physical Chemistry of Surfaces

The behavior of colloid systems is governed primarily by their large *interfacial area.* The interface is defined as the boundary region between the adjoining bulk phases that comprise the colloid system. When one of the phases is a gas or a vapor, the term *surface* is commonly used for the boundary region. The foundation for thermodynamic treatment of surfaces was established in the last century by Gibbs, who published a considerable contribution to the field in 1878. His work still forms the basis of modern thermodynamics (Gibbs, 1928).

Molecules and ions at an interface have different properties and energy characteristics from those in the bulk. Within the phases (Fig. 2.1, phases 1 and 2), whether solid, liquid, or gaseous, the long- and short-range chemical and physical forces acting on a molecule or an ion from all directions are balanced. In the immediate vicinity of the surface the molecules interact with adjacent identical molecules and simultaneously with molecules that form the other phase. This is of fundamental importance and will be described in more detail in the following chapters. At this stage let us examine this relationship for molecular compounds.

The magnitude of molecular forces depends both on the types of molecules and on the density of molecules. Forces of attraction between the molecules of one phase differ from those between molecules of the other phase. They are greatest between molecules in the solid phase, decrease in the liquid phase, and are least in the gaseous state. Molecules in the interface are attracted toward one phase to an extent unequal to that with which they are attracted toward the second phase. There is a finite distance across an interface in which the properties (e.g. concentration) gradually change from those of one adjacent bulk to those of the other (Fig. 2.1, phase 3). The

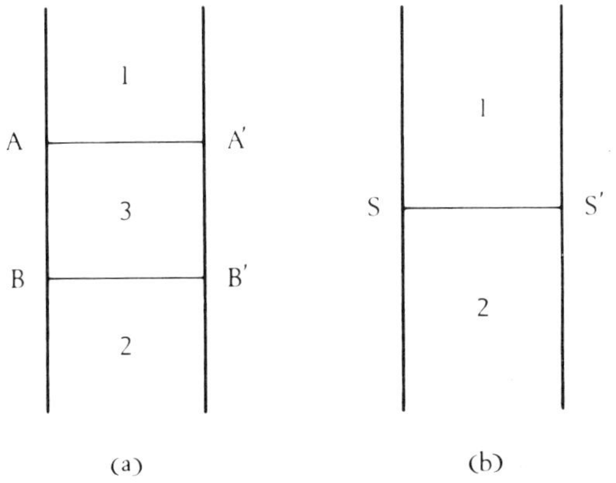

Fig. 2.1a and b. Representative scheme for interface boundary: **a** real system, **b** idealized system

extra free energy in the surface stems from the imbalance of intermolecular or interatomic forces in the surface.

We have said that near the interface the state of the molecules or ions differs from that within the phases, but more precise information regarding the distances from the boundary was not given. Precise information is difficult to obtain since the state of the molecules changes gradually as the distance from the boundary decreases. However, owing to the fact that the intermolecular forces decrease rapidly with distance, over a distance of a few molecular diameters they become infinitesimally small. It is therefore clear that the characteristics of the phases are generally established at a very short distance from the boundary.

1. Thermodynamics of Heterogeneous Systems

As was explained in Section 3 of the Introduction, it follows that lyophobic disperse systems carry a great excess of surface free energy, owing to their highly developed interface area and are therefore thermodynamically unstable. This is not true for the lyophilic disperse systems because of the great strength of the chemical bonds formed between the disperse and dispersing phases. One can legitimately apply thermodynamic ideas of phase equilibrium to the equilibrium lyophilic disperse systems, but it is questionable to what extent they may be applied to the nonequilibrium lyophobic systems. As a result of an energy barrier between the lyophobic particles, which prevents them from approaching one another, the lyophobic disperse system has a certain stability that enables it to exist in the form of the disperse system for a long period of time. If this time period is sufficient for the components to move by molecular diffusion from the bulk of both phases to the boundaries and vice versa, according to the requirements of the conditions of equilibrium, then it is obvious that the state so obtained can be successfully described using the appropriate thermodynamic concepts. Such systems are common in geologic systems. With fluids as dispersing media (e.g., magma, atmospheric air, or hydrospheric water) turbulence is a source of excess kinetic energy that prevents the trapping of the disperse particles in the potential well[1] and dispersal lasts for long periods. With solid materials such as soils and sediments being dispersing media, due to the huge dimensions of the continuous phase, the disperse phase (e.g., water) stays for very long periods in contact with solid surface. Confining ourselves to these arguments, we shall first describe the thermodynamic functions of an interface and then we shall give examples showing how they can be evaluated to explain some geochemical phenomena (see also Babcock, 1963).

1 A curve describing the interaction energy between disperse particles versus the interparticle separation has a minimum, which is defined as the *potential well*. In a disperse system the interparticle separation must be greater than that which corresponds to the potential well (Ch. 8).

1.1 Thermodynamic Properties and Quantities of the Interface

There are two approaches for expressing the thermodynamic functions of the interface. In one approach the interface is considered as a distinct phase having a finite thickness. It must be of at least one molecular diameter in thickness, but in fact it extends over several molecular thicknesses. The *extensive thermodynamic functions* and numbers of moles that appear in thermodynamic expressions relating to the interface phase are total quantities.

In the second approach, which will be used here, an idealized system is described in which the interface is regarded as a mathematical dividing plane (sometimes called the *Gibbs surface*), which divides the phases. The properties are identical throughout the entire volume up to this mathematical plane, and here they change abruptly into the properties of another phase. This is illustrated in Fig. 2.1b, where phases 1 and 2 are homogeneous up to the planes AA′ and BB′ respectively. The dividing surface is designated by SS′ and has zero thickness and volume and is placed between AA′ and BB′. The adsorption of the *i*th component is measured by its *surface excess*, defined as the amount of the *i*th component per unit area of the region between AA′ and BB′, less the amount that would be in the same region if phases 1 and 2 extended unchanged to SS′. In other words, the surface excess is the extra amount of the *i*th component between AA′ and BB′ by virtue of the presence of the interface. Parameters defined in terms of the Gibbs surface will be denoted by a subscript g.

A *primary property* is a property that has a numerical magnitude directly determinable by experimental observations. The numerical value of a primary property is independent of the history of the system.

A *fundamental thermodynamic quantity* is a simple quantity that cannot be defined in terms of other quantities regarded as simpler. *Absolute temperature, T, entropy, S, pressure, P,* and *volume, V,* can be regarded as fundamental quantities.

The *mechanical work, dW,* done during the expansion of bulk phases is the product of the increase of this volume, dV, and the pressure and is given by

$$dW = P\,dV. \tag{2.1}$$

By analogy, the mechanical work dW_g, required to increase the area O of the interface by dO is proportional to the increase of this area and is given by

$$dW_g = -\sigma\,dO. \tag{2.2}$$

Here, if we assume that dO corresponds to dV, then the proportionality factor σ corresponds to $-P$. According to Eq. (2.2), σ is the work required to increase the interface area by one unit area. This factor is known as *interface* (or *surface*) *tension.*

It is possible to extend the area of a surface in two ways. One way is to create fresh surface having the same properties as the original, and the other is to stretch the surface already present. This can easily be understood by referring to simple processes involving liquid systems. Under usual experimental conditions the extension of the surface of a pure liquid will result in a surface having the same surface tension as the original. There is a reservoir of molecules of the liquid to maintain the surface composition constant. The extension of an isolated segment of thin film would be an example of the second type of surface formation. Here there is effectively no reservoir of material to maintain the surface at constant composition, so the stress in the surface changes with the extension. This is also the case with most solids in which the number of broken bonds increases with the extension. It takes a considerable time until these atoms reach a new stable state. With anisotropic solids[2] several values for surface tension may be defined, depending on which crystallographic plane is exposed.

Entropy is a quantity that expresses the degree of randomness of molecules or atoms in the bulk phase or on the surface, due to thermal energy. Randomness denotes the number of different ways in which the molecules or atoms can arrange themselves with respect to their geometric locations and their energy content. When a substance absorbs heat from its surroundings in any reversible process at constant temperature, T, its increase in entropy is given by the amount of heat absorbed, dQ, divided by the absulute temperature,

$$dS = dQ/T. \tag{2.3}$$

The entropy of every substance in complete internal equilibrium is zero at 0° K; however, it has positive values at all other temperatures. While S is independent of process path, Q and W in general are not. Therefore, Q and W are not properties of a system.

The system to be considered will consist of two bulk phases 1 and 2 and their mutual interface, and it will be supposed throughout that phases 1 and 2 and the interface are in equilibrium with each other. V_1 and V_2 are the corresponding volumes of phases 1 and 2, respectively; S_1 and S_2 are the respective entropies. They are determined by the products $s_1 V_1$ and $s_2 V_2$ where s_1 and s_2 are entropies per unit volume (specific entropies).

Properties such as entropy, volume, and mass are characterized by the fact that their values for the entire system equal the sums of the values for the individual phases that make up the system. Such properties are called *extensive properties.* Properties such as entropy per unit volume, density, pressure, and temperature are constant throughout the entire phase as long as the phase is considered *homogenenous* by definition, and do not depend on magnitude of the phase, which is in a definite state. Such properties are called *intensive properties.*

Neglecting surface effects for the moment, the total entropy of this system is given by

$$S_{\text{ID}} = s_1 V_1 + s_2 V_2. \tag{2.4}$$

Since this system contains an interface, a correction to Eq. (2.4) is introduced. This correction should be proportional to the area of the interface O, and can be expressed

2 Isotropic solids are crystals of the isometric crystallographic system with three equal rectangular axes, whereas anisotropic solids are crystals in which at least one of the crystallographic axes is unequal to the rest.

by $S_g = s_g O$, where s_g is the specific surface entropy (entropy per unit area of surface) and is determined by the properties of the substances that form phases 1 and 2. The correct expression for the total entropy of the real system is

$$S = s_1 V_1 + s_2 V_2 + s_g O. \tag{2.5}$$

From Eq. 2.5 it follows that S_g is the entropy of the mathematical dividing plane (Gibbs surface). It is a property of the interface, and therefore for any change in state is independent of the process path.

As a result of various reactions that take place in the interface, s_g can change. A negative Δs_g indicates a decrease in the degree of randomness of molecules in this phase, while a positive Δs_g indicates an increase in the degree of randomness of molecules in the interface.

This method of comparing functions of ideal systems, in which surface effects are neglected, to real systems, is similarly used for determination of all other extensive thermodynamic properties that characterize the interface.

If the heterogeneous system is a multicomponent one, the composition of the interface differs from those of both phases in the bulk. Let us denote by n_{1i} and n_{2i} the numbers of moles of the ith component in phases 1 and 2, respectively. Neglecting surface effects on concentration for the moment, the total number of moles of the ith component in the system is given by

$$n_{\mathrm{ID}i} = n_{1i} + n_{2i}. \tag{2.6}$$

n_{gi} is the number of moles that would have to be added to $N_{\mathrm{ID}i}$ to obtain the correct number of moles of the ith component contained in the real system. The correct number is

$$n_i = n_{1i} + n_{2i} + n_{gi}. \tag{2.7}$$

We can introduce a quantity sometimes called *surface concentration,* but more often known as *surface excess,*

$$\Gamma_i = n_{gi}/O. \tag{2.8}$$

When this magnitude is positive we say that we have a *positive sorption* and when it is negative we have a *negative sorption.* A negative sorption means that the concentration of the ith component is smaller in the interface than in the bulk phase.

In dealing with variable composition, n_i, it is convenient to define a number of quantities such that if X is any macroscopic extensive property of a system then

$$\overline{X_i'} = \left(\frac{\delta X'}{\delta n_i}\right)_{T,\,P,\,n_j} \quad \text{and} \quad \overline{X_i''} = \left(\frac{\delta X''}{\delta n_i}\right)_{T,\,V,\,n_j},$$

where subscript j denotes that all mole numbers other than that of the ith component are held constant; $\overline{X_i'}$ and $\overline{X_i''}$ are partial molar quantities. A coefficient μ_i, called the *chemical potential,* is a partial molar quantity that expresses the tendency of component i to change its concentration in the corresponding phase, either by reacting chemically with another component in the same phase, or by migrating to another phase. In a sufficiently dilute solution of a concentration C_i, we have $d\mu = RT\, dC_i/C_i$. The chemical potential is an expression of the chemical energy of one mole of material. The total chemical energy of the ith component in the system is obtained by multiplying this coefficient by n_i. The *nonmechanical work* due to change of composition is given by the sum $\Sigma\mu_i dn_i$, taken over all the i components. In a system at equilibrium the chemical potentials of the ith components in all phases are equal to one another.

In heterogeneous systems the number of moles of the ith component that are transferred between the bulk and the interface, i.e., dn_{gi}, is determined by differences between the chemical potentials of both the bulk phase and the interface. As soon as the chemical potential of each component is constant throughout and

$$\mu_{1i} = \mu_{2i} = \mu_{gi}, \tag{2.9}$$

the system has reached an equilibrium.

Thermodynamic functions that characterize the interface make it possible to introduce the concept of a *surface phase.* Although this phase is not a classical bulk phase because it has zero volume, its energy, entropy, and the amount of substances sorbed in it are proportional to the surface area in a manner exactly analogous to the three-dimensional case where the energy, entropy, and mass are proportional to the volume. The heterogeneous system can be described by the parameters given in Table 2.1. Let us illustrate now how thermodynamic relations and functions can be established by means of the parameters listed in the Table.

Table 2.1. Parameters describing heterogeneous systems. (After Sheludko, 1966)

For the bulk phase	For surface phase	For heterogeneous system
P	σ	P, σ
V	O	V, O
T	T	T
n_i	n_{gi}	n_i, n_{gi}
S	S_g	S, S_g

1.1.1 Internal Energy

Internal energy, U, is a property of the system, and thus, for any change in state, ΔU is independent of process path. For a constant temperature throughout the system, the variation of the total internal energy is given by the first law of thermodynamics

$$dU = dQ - dW + \Sigma\mu_i dn_i. \tag{2.10}$$

If the work done is associated entirely with volume and surface area changes, this mechanical work is expressed as

$$dW = P_1 dV_1 + P_2 dV_2 - \sigma dO. \tag{2.11}$$

The heat brought into the system is

$$dQ = TdS_1 + TdS_2 + TdS_g. \tag{2.12}$$

The nonmechanical work due to changes of composition is

$$\Sigma\mu_i dn_i = \Sigma_1 \mu_i dn_i + \Sigma_2 \mu_i dn_i + \Sigma_g \mu_i dn_i. \tag{2.13}$$

Equation (2.10) becomes

$$dU = [TdS - PdV + \Sigma\mu_i dn_i]_1 + [TdS - PdV + \Sigma\mu_i dn_i]_2 + [TdS + \sigma dO + \Sigma\mu_i dn_i]_g. \tag{2.14}$$

At constant composition ($dn_i = 0; dn_{gi} = 0$), constant volume and surface area ($dV = 0$; $dO = 0$), we obtain:

$$dU = dQ = TdS_1 + TdS_2 + TdS_g, \tag{2.15}$$

and for a single phase, if dQ is the heat of reaction, the internal energy expresses the cohesion.

1.1.2 Free Energy

Free energy is an extensive property that is used to predict the direction of the chemical reaction and chemical equilibrium. Both the Gibbs and Helmholtz free energy functions may be applied for surfaces. Consider first the Helmholtz function $A = U - TS$ by which we have

$$A = (A_1 + A_2 + A_g) = (U_1 + U_2 + U_g) - T(S_1 + S_2 + S_g), \tag{2.16}$$

where $U_1 + U_2 + U_g$ is the total internal energy and $S_1 + S_2 + S_g$ is the total entropy. Taking into account that $d(TS) = TdS + SdT$ and applying Eq. (2.14) we obtain:

$$dA = [-SdT - PdV + \Sigma\mu_i dn_i]_{1,2} + [-SdT + \sigma dO + \Sigma\mu_i dn_i]_g. \tag{2.17}$$

At constant temperature ($dT = 0$), volume ($dV = 0$), and composition ($dn_i = 0; dn_{gi} = 0$) we have

$$dA = \sigma\, dO. \tag{2.18}$$

From Eq. (2.18) a new definition of surface tension becomes apparent, namely,

$$\sigma = (\delta A/\delta O)_{T, V, n_i} = a_g, \tag{2.19}$$

that is, the surface free energy per unit surface area, is equal to σ when the composition in the interface is independent of the interfacial area. This equation is therefore useful for a one-component liquid, if the Gibbs surface is placed such that $\Gamma = 0$.

For solids, however, the two quantities are in general not equal. When a fresh solid surface is formed, the atoms at the surface may take a considerable time to reach their equilibrium state, and the specific surface free energy of the metastable state differs from that of the stable state. Let us consider the work $dW = -d(Oa_g)$, done in the extension of the surface of a one-component isotropic solid by an amount dO. Suppose that the work is done against a force per unit length in the surface, i.e., surface tension σ. Then, according to Eq. (2.2) dW_g is given by $-\sigma dO$. Thus, $\sigma dO = d(Oa_g)$ so that $\sigma = (d/dO)\,(Oa_g)$ and hence

$$\sigma = a_g + O(da_g/dO). \tag{2.20}$$

Consider now the Gibbs free energy function $G = A + PV$, whereby, for the heterogeneous system we have

$$G = (G_1 + G_2 + G_g) = (U_1 + U_2 + U_g) - T(S_1 + S_2 + S_g) + P(V_1 + V_2 + V_g). \tag{2.21}$$

Taking into account that $d(PV) = PdV + VdP$ and that V_g and dV_g are equal to zero, we obtain

$$dG = [-SdT + VdP + \Sigma\mu_i dn_i]_{1,2} + [-SdT + \sigma dO + \Sigma\mu_i dn_i]_g. \tag{2.22}$$

Given a constant pressure, temperature, and composition, a state that is always possible in the absence of adsorption, we have

$$dG = \sigma dO \tag{2.23}$$

and

$$(\delta G/\delta O)_{P,T,n_i} = \sigma. \tag{2.24}$$

1.1.3 Exact and Inexact Differentials

The thermodynamic functions of internal energy and Gibbs and Helmholtz free energies, as well as the fundamental thermodynamic quantities entropy, temperature, volume, and pressure, are functions of state, i.e., for any change in the system the state is independent of process path. We can therefore always integrate form k to h states as follows

$$\int_k^h dF = F_k - F_h$$

where F is the thermodynamic function. An integration from k to h state and then returning to k state, i.e., an integration in a closed system, results in zero. Mathematically dF is an *exact differential.* Differentials of quantities that are not properties of a system, such as dQ and dW, are *inexact differentials* since one cannot merely obtain them by the differentiation of the functions of state. Also one cannot just integrate dQ or dW to obtain Q or W. For example, if $\int_k^h dW = \int_k^h \sigma dO$, it is impossible to integrate σdO just by knowing k and h states, since, as will be shown later, the surface tension depends on the temperature, which may change along the integration.

1.2 Pressure Dependence of Chemical Potential and of Water Migration in Compacting Sediments

Equations (2.14), (2.17), and (2.22) allow the thermodynamic parameters to be expressed by the derivatives of the characteristic functions. Let us recall this using Eq. (2.22) as example of a single bulk phase. Let us define Gibbs free energy as a function of temperature, pressure, and composition, $G = G(T, P, n_i)$, so that

$$dG = (\delta G/\delta T)_{P,n_i} dT + (\delta G/\delta P)_{T,n_i} dP + \Sigma(\delta G/\delta n_i)_{T,P,n_j} dn_i.$$

For a single bulk phase Eq. (2.22) becomes $dG = -SdT + VdP + \Sigma\mu_i dn_i$. Since dT, dP, and dn_i are independent variables, it follows from the comparison of the two last expressions that

$$(\delta G/\delta T)_{P,n_i} = -S,\ (\delta G/\delta P)_{T,n_i} = V, \text{ and } (\delta G/\delta n_i)_{P,T,n_j} = \mu_i.$$

The last expression enables us to have an alternative definition of the chemical potential. It also supplies quantitative means for describing the chemical potential. Thus, the partial molal Gibbs free energy and the chemical potential are equal to one another at fixed pressure and temperature. In similar ways two other expressions for the chemical potential,

$$\mu_i = (\delta U/\delta n_i)_{S,V,n_j} \text{ and } \mu_i = (\delta A/\delta n_i)_{V,T,n_j}$$

are obtained.

Making use of the rule of independence of the order of differentiation, i.e., that for $f = f(x, y)$ $\frac{\delta}{\delta y}\frac{\delta f}{\delta x} = \frac{\delta}{\delta x}\frac{\delta f}{\delta y}$ we obtain

$$\Big[\frac{\delta}{\delta P}\underbrace{\Big(\frac{\delta G}{\delta n_i}\Big)_{P,T,n_j}}_{=\mu_i}\Big]_{T,n_i,n_j} = \Big[\frac{\delta}{\delta n_i}\underbrace{\Big(\frac{\delta G}{\delta P}\Big)_{T,n_i,n_j}}_{=V}\Big]_{P,T,n_j}$$

or

$$\Big(\frac{\delta \mu_i}{\delta P}\Big)_{T,n_i,n_j} = \Big(\frac{\delta V}{\delta n_i}\Big)_{P,T,n_j} \equiv \overline{v_i}, \tag{2.25}$$

where $\overline{v_i}$ is the molar volume of the ith component. For a liquid of which the volume does not change with the pressure (an incompressible liquid), integration gives $\Delta\mu_i = \overline{v_i}\Delta P$.

If the system consists of only one component, we have

$$\Delta\mu = \bar{v}\Delta P. \tag{2.26}$$

When applied to an ideal gas, where volume changes with pressure according to $v = RT/P$, from (2.25) we obtain

$$(\delta\mu/\delta P)_T = \bar{v} = RT/P.$$

Integration yields the following expression for the change of its chemical potential when the pressure is change from p_1 to p_2 at a constant temperature

$$\Delta\mu = RT \ln p_2/p_1. \tag{2.27}$$

One of the geologic processes in which a chemical potential gradient plays a role is the upward migration of water during the compaction of sediments from a point with hydrostatic liquid pressure p_1 to a point with a pressure p_2. Equation (2.26) describes the change in the chemical potential of water as it passes through the rock. The rate of flow of water across a porous medium is given by $\bar{U} = \Delta\mu/\bar{r} = \bar{v}_{\mathrm{HOH}}\Delta P/\bar{r}$, where $\bar{r}$ is the resistance to flow of water in the medium (Clark, 1962).

1.3 Surface Tension, the Strength of Intermolecular Forces and Cleavage of Mineral Crystals

By analogy with Eq. (2.5), one can write (2.16) in the following way:

$$A = a_1 V_1 + a_2 V_2 + a_g O, \tag{2.28}$$

where a_g is the free surface energy per unit surface area. It has been shown that at constant temperature, volume, and composition this magnitude is equal to the surface tension (see Eq. 2.18). Let us consider a slight change dT in a system at temperature T, at constant volume, composition, and surface area. By differentiating Eq. (2.17) with respect to T, we obtain $(\delta A/\delta T)_{V,O} = -S$. Let us introduce Eq. (2.5) into this expression and compare it to the differential of Eq. (2.28) with respect to T as follows:

$$\left(\frac{\delta A}{\delta T}\right)_{V,O} = \left(\frac{\delta a_1}{\delta T}\right)_{V_1,O} V_1 + \left(\frac{\delta a_2}{\delta T}\right)_{V_2,O} V_2 + \left(\frac{\delta a_g}{\delta T}\right)_{V_1,V_2,O} O \tag{2.17*}$$

and

$$\left(\frac{\delta A}{\delta T}\right)_{V,O} = s_1 V_1 + s_2 V_2 + s_g O. \tag{2.5*}$$

From these equations and from Eq. (2.18) for the interface phase we obtain

$$(\delta a_g/\delta T)_{V,O} \equiv \delta\sigma/\delta T = -s_g.$$

According to Eq. (2.16) and (2.19) $\sigma = a_g = u_g - Ts_g$, hence

$$\sigma = u_g + T(\delta\sigma/\delta T). \tag{2.29}$$

This expression indicates that the surface tension is composed of two terms, the entropy term, i.e., the term of randomness in the interface, and the internal energy term, which at constant volume and composition represents cohesion, i.e., the force of attraction existing between the molecules. Experimental investigations of the dependence of the surface tension on temperature show two phenomena: (1) For temperatures not very close to the critical temperature, the surface tension decreases linearly with increasing temperature. According to Eq. (2.29), this means that in the given temperature interval, $\delta\sigma/\delta T$ is constant, and specific energy u_g is independent of temperature. This is in agreement with the fact that the intermolecular forces that determine u_g depend only slightly on temperature. (2) The slope of the curve that describes surface tension as a function of temperature, at points not too close to the critical temperature, is very small. According to Eq. (2.29) this means that $(\delta\sigma/\delta T) \to 0$, and that in the first approximation, $T(\delta\sigma/\delta T)$ can be disregarded as compared with u_g. This makes it possible to evaluate σ on the strength of intermolecular forces acting in the liquid.

Cleavage is the tendency of a crystallized mineral to break in certain definite directions, yielding more or less smooth surfaces. Actually this process is an extension of the surface area of the solid mineral, and according to Eq. (2.2) the work dW done during the cleavage is given by $-\sigma dO$. Although the evaluation of σ on the basis of the strength of intermolecular forces can be made only in those systems in which $\sigma = a_g$ (liquids), it seems plausible to correlate cleavage with interatomic forces, because this geologic process may proceed slowly enough to enable the newly formed surface to reach a stable state. The value (da_g/dO) of Eq. (2.20) becomes almost zero and as a first approximation $O(da_g/dO)$ can be disregarded as compared to a_g. The term "cleavage" is used here in a wider sense than commonly used by mineralogists, as a tool to characterize minerals. Processes similar to "cleavage" widely occur in geology, e.g., during crystal growth, and they play an important role in the formation of surfaces.

From the preceding discussion it follows that cleavage indicates a minimum value of cohesion in the direction of easy fracture, that is, normal to the cleavage plane itself. The ideal *cleavage-exposed surface* is a homoatomic plane of the crystal network. The *morphologic cleavage plane* is parallel to this crystal face and its position is described by the indices of the exposed surface. Let us demonstrate this for talc, a mineral with a layer structure. Every layer of this mineral is composed of the following seven planes of O, Si, (O/OH), Mg, (O/OH), Si, and O atoms. The O/OH plane is that plane in which O atoms are most closely packed together. Since two-thirds of the oxygen atoms in this plane are strongly bonded at the same time to atoms of the Mg plane and to atoms of the Si plane, this plane cannot become a plane of cleavage. The same is true for the Mg plane. The packing of atoms in the oxygen plane is less dense, but these atoms form strong covalent bonds only with Si atoms of the neighboring plane. Bonding between these atoms and oxygen atoms of neighboring layer are of the van der Waals type and are therefore weak. This O plane can readily become a cleavage-exposed surface (see page 10).

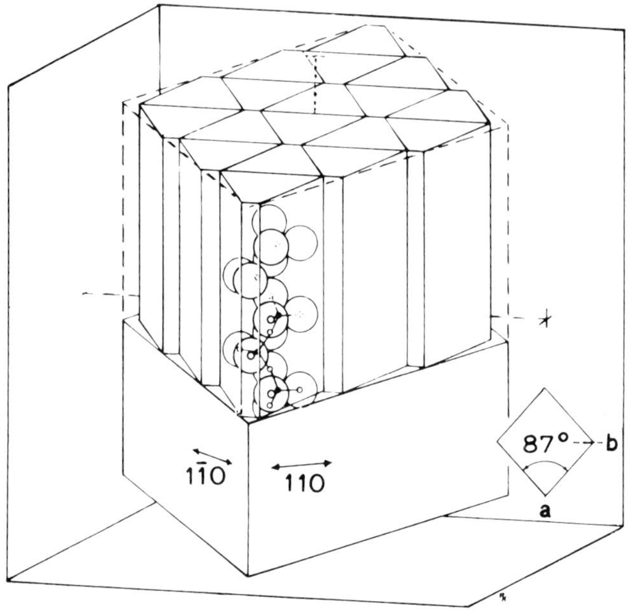

a

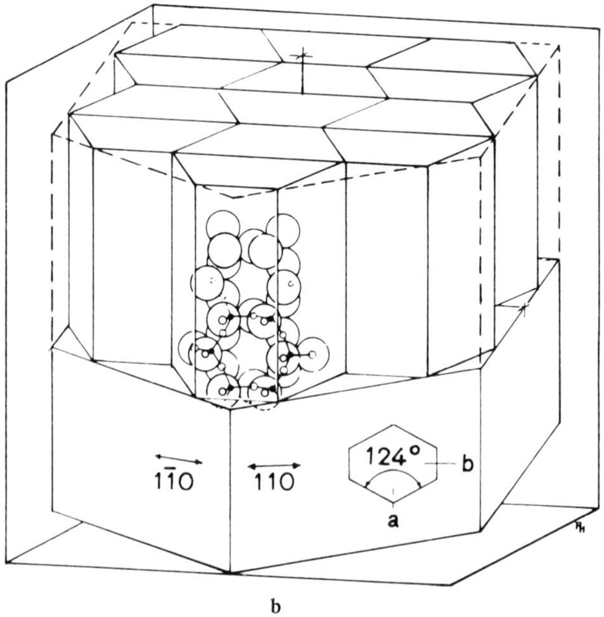

b

Fig. 2.2a and b. The relationship between atomic structure in **a** pyroxenes and **b** amphiboles. The linked Si–O chains are shown in cross section. The planes of weakness are shown by *heavy solid lines,* and resulting cleavage directions (110) by *broken lines.* (From Mason and Berry, W. H. Freeman and Company. Copyright 1968)

Talc has other network planes that can readily become cleavage-exposed surfaces but not parallel to the layers. The bonding between this plane and one of its parallel neighbors should be weak while that between this plane and the second neighbor should be strong. It is therefore to be expected that this plane will be composed of OH groups. This crystal plane has a simple relation to the crystallographic hexagonal axis and is the commonly occurring form of the crystal of talc. Similar cleavage planes are found for most clay minerals and micas.

In a tectosilicate crystal cleavage-exposed surfaces are comprised of O atoms. Since the oxidation number of this element is smaller than its coordination number, a crystallographic plane can be found in which the number of strong chemical bonds formed on one side of the plane is much greater than the number of bonds formed on the other side of the plane. From the many planes that can fulfil this requirement, the one with the highest density of oxygens will be the cleavage-exposed surface, since this plane has a high density of chemical bonds with only one of the neighboring crystallographic plane.

Cleavage does not split polyatomic ions and radicals in ionic crystals such as olivines, pyroxenes, and amphiboles (Fig. 2.2a, b). In crystals of monoatomic ions cleavage generally develops in such away as to expose planes of anions. An amorphous body obviously shows no cleavage. In short, the pattern of atomic or molecular packing determines the angles between faces that should be identical in all specimens of a particular mineral.

1.4 Capillary Pressure

A convex surface is associated with an increase of the vapor pressure of the liquid. Such convexity, with high curvature, appears in emulsions, aerosols, and foams. The associated special properties are important for the study of these colloid systems. An understanding of phenomena of curved liquid surfaces is also basic to the treatment of several processes of importance in systems of water in sediments and soils. The treatment of this problem is similar to that carried out for vapor pressure in a cylindrical capillary tube, where the liquid–gas interface is curved.

The pressure on one side of a curved surface differs from that on the other side. It is higher on the concave side than on the convex side. Suppose that a curved interface between two phases whose volumes are V_1 and V_2 is displaced into a new position. If it is required that the displaced surface be parallel to its original position, the surface area must be changed so that the curvature will be the same in both positions. Any change in surface area is associated with work done against the surface tension. Bubbles do not in fact contract because the surface tension forces are balanced by the force exerted by the internal excess pressure Δp.

Suppose that ΔO is the area of a sufficiently small section of a curved interface, which can be described by two radii of curvature r_1 and r_2, and that both radii are the same for all points of this section. The temperature T and the volume $V = V_1 + V_2$ are constant. When the section is displaced by dx the volume of one of the phases will increase by $dV_1 = \Delta O dx$ while the other will decrease by dV_2. Since the total volume is constant, $dV_1 = -dV_2$ (Fig. 2.3). The displacement of the section results in a change of its area by $d\Delta O$.

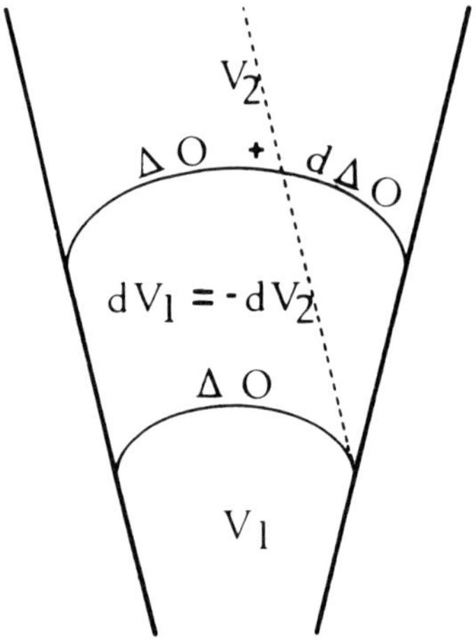

Fig. 2.3. Change of surface area of a bubble from a change of the volume of the bubble

Geometric considerations yield the following relation[3]

$$d\Delta O = \Delta O(r_1^{-1} + r_2^{-1})dx.$$

In the equilibrium condition $dA = 0$ and according to Eq. (2.17) for a one-component system at constant temperature we have

$$dA = -P_1 dV_1 - P_2 dV_2 + \sigma d\Delta O = 0,$$

where P_1 and P_2 are pressures in phases 1 and 2, respectively, i.e., on both sides of the interface. Substituting $-dV_1$ for dV_2 and $\Delta O(r_1^{-1} + r_2^{-1})\,dx$ for $d\Delta O$ into the expression we obtain

$$dA = -P_1 dV_1 + P_2 dV_1 + \sigma\Delta O\,(r_1^{-1} + r_2^{-1})\,dx = 0$$

or

$$dV_1(P_1 - P_2) = \Delta O\sigma\,(r_1^{-1} + r_2^{-1})\,dx.$$

Since $\Delta O dx = dV_1$ we obtain

$$\Delta P = P_1 - P_2 = \sigma\,(r_1^{-1} + r_2^{-1}). \qquad (2.30)$$

which is known as the *Laplace equation.* ΔP is the pressure drop associated with the

---

3 This is illustrated for a sphere, where $r_1 = r_2$ as follows. Since $O = 4\,\pi r^2$, $O + dO = 4\,\pi(r + dx)^2 = 4\,\pi r^2 + 8\,\pi r dx + 4\,\pi\,(dx)^2$. Since dx is very small, $4\,\pi\,(dx)^2 \to 0$ and $dO = 8\,\pi r dx = 4\,\pi r^2(r^{-1} + r^{-1})\,dx = O(r^{-1} + r^{-1})\,dx$.

curvature of the surface and is called *capillary pressure.* In the case of a spherical element Eq. (2.30) becomes

$$\Delta P = 2\,\sigma/r. \tag{2.31}$$

In the case of a plane surface $r_1 = r_2 = \infty$ and the capillary pressure is equal to zero.

Let us examine the change in vapor pressure accompanying the pressure change in a liquid that occurs when its surface is curved. Consider a one-component system consisting of two phases, a liquid (phase 1) and its vapor (phase 2). Let us compare two states of equilibrium: In the first, the surface is planar ($r = \infty$) and in the second, the surface is curved. When dn_1 moles of substance are transferred from phase 1 to phase 2 at constant temperature and pressure we have, according to Eq. (2.22), $dG = \mu_1 dn_1 + \mu_2 dn_2$. Since the system is closed, $dn = dn_1 + dn_2 = 0$ and $dn_1 = -dn_2$. Also, at equilibrium $dG = 0$. Hence, $dG = (\mu_1 - \mu_2)dn_1 = 0$. However, $dn_1 \neq 0$. Hence $\mu_1 - \mu_2 = 0$ and $\mu_1 = \mu_2$, which means that in a state of equilibrium the chemical potentials of the liquid and its vapor are identical.

On changing the curvature of the liquid surface, there will be a change in the chemical potential of any of the phases. However, as the system reaches a state of equilibrium, again the chemical potential will be the same for both phases. If $\Delta\mu_1$ and $\Delta\mu_2$ represent the changes in μ_1 and μ_2, respectively, due to the variation in curvature, at the new equilibrium state $\mu_1 + \Delta\mu_1 = \mu_2 + \Delta\mu_2$ and $\Delta\mu_1 = \Delta\mu_2$. The relation between the change of the chemical potential of the liquid phase and the change of its pressure is given by Eq. (2.26) $\Delta\mu_1 = \bar{v}\Delta p$ and this relation for the vapor phase is given by Eq. (2.27) $\Delta\mu_2 = RT \ln p_r/p_\infty$ where p_r is the vapor pressure over a curved surface of radius r, and p_∞ is the vapor pressure over a plane surface. Since $\Delta\mu_1$ is equal to $\Delta\mu_2$ we obtain $\bar{v}\Delta p = RT \ln p_r/p_\infty$, and with Eq. (2.31) for spherical surfaces we obtain

$$\Delta p = 2\,\sigma/r = (RT/\bar{v}) \ln (p_r/p_\infty). \tag{2.32}$$

This is the *Thomson-Gibbs* or *Kelvin equation.* This equation enables the calculation of the vapor pressure over a small spherical surface of a liquid

$$p_r = p_\infty \exp (2\sigma\bar{v}/RTr). \tag{2.33}$$

From Eq. (2.33) it is obvious that the vapor pressure over a drop will be greater than the vapor pressure over a plane surface of a large mass of the same liquid. Smaller drops should possess a higher vapor pressure than larger ones. Since the transformation of any substance from the state in which its vapor pressure is high into that state in which its vapor pressure is low is spontaneous, in emulsions or aerosols substance is transferred from finer droplets to the coarser ones and the coarse ones grow at the expense of the fine droplets.

The same considerations are applicable to solid crystals. In this case $\ln (p_r/p_\infty)$ attains measurable values only in the case of very small particles. These conclusions are applicable also to solubility. Solubility of a body with curved surfaces is greater than that with flat surfaces. Small crystals are more soluble than large ones. This effect can be observed only with very small particles.

Capillary pressure is one of the parameters considered in many equations that have been derived to describe migration of water in soils and sediments (e.g., Childs, 1969; Morel-Seytoux and Khanji, 1974).

1.5 Contact Angle of a Liquid at a Boundary of Three Phases and the Wetting Process

The following discussion of solid–liquid interfaces is concerned with the wetting of rocks and soils by water, oil, or molten magma and the spreading of the liquid. Effects due to gravity and curvature of solid rock have been excluded.

If a drop of liquid is placed on a plane surface of a dry solid, it can partly spread, displacing the gas that covers the solid surface. Along its periphery three phases are simultaneously in contact–liquid drop, vapor, and solid base–forming three interfaces. As can be seen from Figure 2.4a, b the three phases meet on a bidimensional projection in point A, and the projections of three interfaces form three angles, the sum of which is 360°. The angle formed between the projections of the liquid–solid and gas–solid interfaces is 180°. The angle formed between the projection of the liquid–solid and liquid–gas is denoted by α and is defined as the *contact angle*. The angle formed between the projection of the liquid–gas and gas–solid interface is $180 - \alpha$. The angles formed are obviously not arbitrary.

The liquid molecules may either interact with other molecules of the same liquid (cohesion) or they may interact with the solid surface and gas molecules, forming interfaces (adhesion). The work of gas-liquid or gas-solid adhesion is very low in comparison with solid–liquid adhesion or with liquid–liquid cohesion. If solid–liquid adhesion exceeds liquid–liquid cohesion, the contact angle α is small and the liquid is said to be wetting the solid. A contact angle of zero means a collapse of the drop and a complete wetting of the solid. If the contact angle approaches 180°, the result is a very strong liquid–liquid cohesion and a low liquid–solid adhesion, the solid surface remains absolutely unwetted, and the drop maintains its spherical shape.

When a drop on a plane solid surface is in a state of equilibrium the vectorial sum of the three forces acting to increase the three interfaces will be zero. According to Eq. (2.2) these forces can be expressed by the respective interface tensions. From Figure 2.4a, b it is obvious that

$$\sigma_{GS} = \sigma_{LS} + \sigma_{LG} \cos \alpha,$$

where σ_{LS}, σ_{GS}, and σ_{LG} are the tensions of liquid–solid, liquid–gas, and gas–solid interfaces, respectively. Hence

$$\cos \alpha = (\sigma_{GS} - \sigma_{LS})/\sigma_{LG}. \tag{2.34}$$

Each of these interface tensions tends to decrease the appropriate surface. The contact angle plays an important role in the migration of liquids through solids, some examples of which will be given in later chapters. The presence of certain organic compounds (such as detergents) in natural waters results in a decrease of their surface tensions σ_{LG} and σ_{LS}. According to Eq. (2.34) the contact angle α decreases, the result of which is a better spreading of the water drop and a better wetting of the rock or soil.

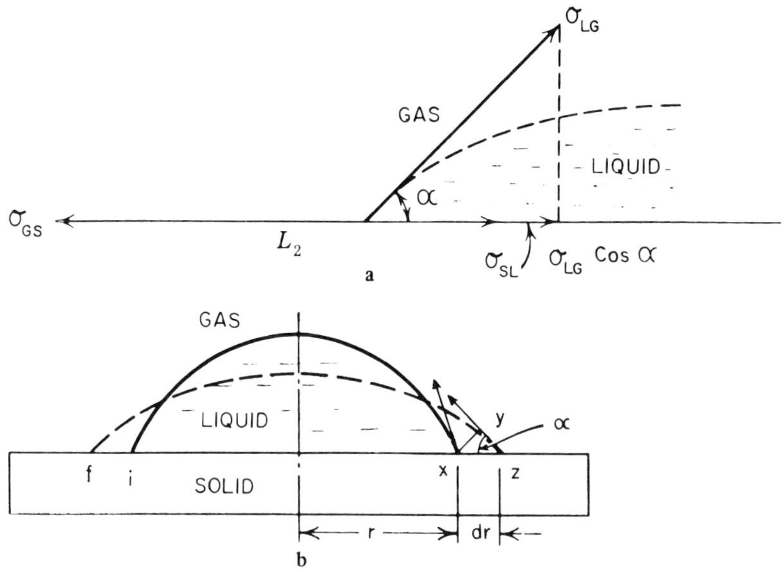

Fig. 2.4a and b. Contact angle at the air–liquid–solid phases contact: **a** the effect of surface tensions σ_{LG}, σ_{SG}, and σ_{LS} on the contact angle; **b** the effect of contact angle on the spreading of a drop on a solid

The type of mineral and its solvation energy have great influence on the contact angle and spreading of the drops. A small angle is obtained for water on clays but a larger one is obtained for water on quartz sediments.

When a liquid flows on a plane solid two characteristic contact angles are obtained, one at the side where the liquid advances and one at the side where the liquid retreats. Under specific physical conditions these angles are constant for a specific liquid–solid system. These angles are determined by the ratios of the surface tensions described in Eq. (2.34). These angles are also influenced by the roughness of the surface.

One of the results of the capillary pressure is the capillary rise of liquids in porous systems, an important phenomenon in the behavior of water in soils and sediments. When a cylindrical capillary tube is partly immersed in liquid, the liquid rises in the tube and a meniscus is formed at the upper edge of the liquid, as a result of the contact angle which the liquid makes with the wall of the tube. If the angle of contact is less than 90°, the meniscus is concave and the liquid pressure, p_1, just below the meniscus, will be less than the pressure p_O applied on the bulk liquid body, which may be the atmospheric pressure, or the hydrostatic pressure of the surrounding rock, if the liquid is exposed or covered, respectively. The bulk liquid, being under a higher pressure, will force the liquid in the capillary tube to rise, until an equilibrium is obtained when the hydrostatic pressure applied by the liquid column in the tube is equal to the capillary pressure. If the capillary is of a sufficiently small bore the meniscus will be spherical in profile. The projection of the capillary is shown in Figure 2.5. When the contact angle is zero, the meniscus is hemispheric, projected as half a circle, and the radius of curvature is equal to the radius of the capillary tube. On the

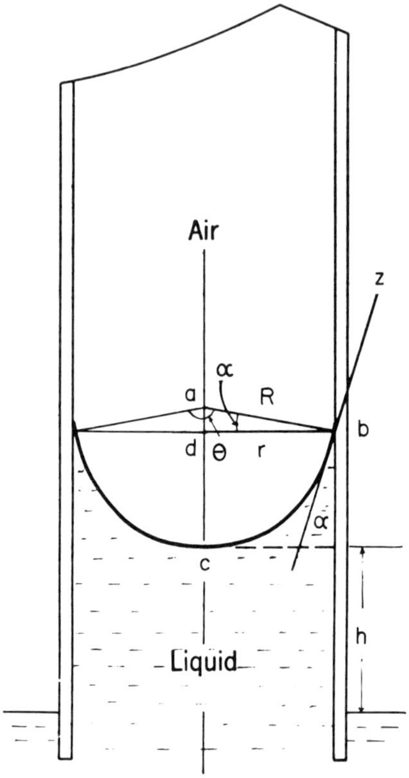

Fig. 2.5. A schematic projection of a capillary. For explanation see text

other hand, if the liquid forms a contact angle with the walls of the tube which differs from zero, the diameter of the tube becomes a chord that cuts an arc Θ of the circle, the projected meniscus. If R and r are the radii of the curvature of the meniscus and the capillary tube, respectively, and α is the contact angle, geometric considerations yield the following relations: $\Theta = \pi - 2\alpha$ and $R = r/\cos\alpha$.

According to Eq. (2.31), the pressure just below the meniscus is less than atmospheric by an amount $2\,\sigma_{LG}\cos\alpha/r$. The capillary pressure determines the height h of the capillary rise. If ρ and ρ_O are the densities of the liquid and the surrounding fluid, respectively, the weight of the liquid column in the capillary tube is $\pi\,r^2gh(\rho - \rho_O)$ and the force per one unit area is $gh\,(\rho - \rho_O)$, where g is the acceleration coefficient. Under equilibrium $2\,\sigma_{LG}\cos\alpha/r = gh(\rho - \rho_O)$. Thus, the capillary rise is given by

$$h = \frac{2\,\sigma_{LG}\cos\alpha}{rg\,(\rho - \rho_O)} = \frac{\sigma_{GS} - \sigma_{LS}}{rg\,(\rho - \rho_O)}. \tag{2.35}$$

From Eq. (2.35) it is obvious that the capillary rise increases with the increase of solid–liquid adhesion. The amount of capillary water in soils and sediments will depend on the hydrophilic properties of the minerals and the organic materials present. The drying of soils and sediments is associated to some extent with capillary rise. Conversion of hydrophilic surfaces of soil minerals into hydrophobic surfaces by treating them with suitable organic compounds results in a decrease in the degree of drying of the soils (Hergenhan, 1972). According to Hollis et al. (1977) in some soils (from the west midlands of England) organic matter may become the most important factor controlling retained water capacity and available water.

One of the processes in which wetting plays an important role is *infiltration.* This process is defined as "entry of water into soil" (Richards, 1952). The rate at which water entry proceeds into a soil depends upon factors such as wetting and transmission characteristics of the soil, soil layering, soil structure, the initial soil water content, and the kind and quantity of material in solution (Nielsen and Vachaud, 1965). Soils that exhibit resistance to wetting with water are commonly termed "water-repellent".

Aerophilic sandy soils exhibit water-repellency (Jamison, 1946). Infiltration of water vertically downward into water-repellent soil is markedly different from that occurring in wettable soils (De Bano, 1969). Infiltration velocity for wettable soil normally decreases with time, whereas infiltration velocity for repellent soil increases with time. High initial infiltration velocities are common in wettable soils, but very low initial velocities are typical in repellent soils. De Bano suggests that the presence of a water-repellent layer in an otherwise wettable soil profile decreases infiltration when the advancing wetting front reaches the water-repellent layer. A theoretical analysis by Mansell (1969) indicates that a water-repellent layer should be more effective in decreasing surface evaporation by reducing upward capillary flow than in retarding lateral or downward infiltration.

2. Liquid Surface

A liquid surface is usually well defined and therefore, in principle, easier to treat than a system having the less well-defined solid surface. It may be regarded as a region whose thickness, although indeterminate, is of the order of a few molecular diameters unless the temperature is close to the critical value.

2.1 Liquid Surface Tension

The present section is based on the treatment of Jasper (1967) supplemented by data more specific for geochemical systems. Within both gaseous and liquid phases the molecules are subjected to a spherically symmetric force field with an average zero resultant interaction. The molecular constituents of the surface region, however, are subjected to a tensile stress as the consequence of an asymmetrically distributed intermolecular attraction that has a resultant normal to the surface and is directed toward the center of mass of the liquid phase. This produces an effect, manifests at all points and in all directions in the liquid surface, which is equivalent to that of a horizontal tangential force along the liquid–vapor interface. This effect is known as the *surface tension.*

The increase of surface area takes place as a result of the appearance of new molecules coming up from the bulk of the liquid, and not because of an increase of distance between molecules in the surface layer. Therefore, during the formation of a liquid surface, work must be done against attraction forces within the interior of the liquid. This means that surface tension of pure liquids does not depend on the magnitude of the area. If n is the number of molecules that form one cm^2 of surface (molecular density), and $(w_g - w)$ is a measure of the work performed when one molecule is transported from the bulk to the surface (molecular attraction), then Eq. (2.2) becomes

$$\sigma = \delta W/\delta O = n(w_g - w). \tag{2.36}$$

In the presence of their vapors and at room temperature pure liquids usually have surface tensions in the range 10–80 mNm^{-1} (dyne/cm), water (72.75 mNm^{-1} at 20 °C) being in the upper and organic liquids in the lower part of this range. If the system consists of two different liquid phases that meet to form an interfacial region, the interfacial tension is intermediate between the surface tensions of the components.

The surface molecular density appearing in Eq. (2.36) is temperature dependent. Under orthobaric conditions the effect of increasing the temperature of a liquid would be to decrease the surface density and consequently the surface tension.

Assuming that the surface tension is a linear function of the temperature over a wide range, the following equation was suggested

$$\sigma = K(T_c - T). \tag{2.37}$$

The constant K is specific for the liquid involved and is independent of the temperature, and T_c is the critical temperature. This equation is valid for narrow temperature ranges. It is not valid in the vicinity of the critical temperature nor is it valid for a wide temperature range.

The surface tensions of a few molten metals and of some liquid crystals as well as of some basalts, andesites, and rhyolite obsidians above their liquidus temperatures in air, appear to have positive temperature coefficients.

The positive temperature dependence of molten rocks is the result of thermal dissociation of silicate polymers. The probability of obtaining dense surface packing of branched huge polymers is small due to steric hindrance, unless the polymers are completely linear. With decreasing size and branching of polymers the density of the surface layer increases. Molecular attraction of the polymeric species decreases only slightly with increasing temperature. This attraction depends on two factors: it decreases with decreasing size of species and increases with decreasing branching of the polymer. Since the number of species that form one cm^2 of surface increases with temperature, it follows from Eq. (2.36) that the surface tension also increases with the temperature. The effect of increasing the silica content of the melt would be to increase the polymer size and consequently to reduce the surface tension.

For liquid–gas systems it is possible to vary isothermally either the component of attraction or that of density of Eq. (2.36) or both. The former might be accomplished by selecting a series of gases whose molecular species differ in varying degrees from those of the liquid. The work performed when one molecule is transported from the bulk of the liquid to the surface decreases with increasing strength of bonds between liquid and gas molecules. Increasing the pressure of the gas leads to a higher adsorption of gas molecules onto the interface (see Sect. 3). Under these conditions the density of liquid molecules in the interface decreases and consequently the surface tension also decreases.

The surface tension of solutions will depend on the nature of the component present in excess in the surface phase. Since the free surface energy is determined by the nature and arrangement of the molecules in the surface phase, the greatest reduction in this property will be attained if the molecules of the component that has the least intrinsic free surface energy accumulate in the surface phase in excess of the molecules of the other component. Water, for example, is a polar compound with a relatively large free surface energy and therefore, at any constant temperature, the surface tension of most aqueous solutions of nonelectrolytes decreases with increasing concentration. This is obvious from Eq. (2.36) in which $(w_g - w)$ is much greater for water than for nonelectrolytes. The surface tension of a solution decreases as a result of a decrease of n_{H_2O}.

If the surface tension of the solution is less than that of the pure solvent at the same temperature and continues to decrease with increasing concentration of the solute, the solute is said to be *surface active* (or a *surfactant*). Most organic compounds that are soluble in water are surface active and will lower the surface tension. If the compound is miscible with water in all proportions, the surface tension in most cases first decreases rapidly at lower concentrations and then, very slightly as the mole fraction of the organic compound approaches unity. This sequence appears to be general for completely miscible systems, including magmas and volatiles, especially when the surface tensions of the pure components are considerably different in their magnitudes.

The surface tension of an aqueous solution of an electrolyte may be greater than that of the pure solvent at the same temperature and increases with increasing concentration of solute. In this case a *negative sorption* occurs, the concentration of the solute being less in the surface than in the bulk phase. As a result of the electrostatic field induced by the ions, stable solvation atmospheres around the ions are formed with polar solvents, and $(w_g - w)$ increases with increasing electrolyte concentration and consequently the surface tension increases.

The surface tension of dilute irreversible or unstable aqueous sols is usually similar to that of the solvent. In the majority of stable colloidal sols the surface tension is lower than that of pure water. Humic substances, proteins, soaps, and most detergents belong to the latter group. However, the surface tension of sols of silicic acid, metal hydroxides, and starch is higher than that of water. The fraction of hydrophobic moieties in compounds belonging to this group is low. Some silicates slightly raise and others slightly lower the surface tension of water (McBain, 1950).

2.1.1 Surface Tension of Natural Waters

Many geologic phenomena are associated with the migration of aqueous solutions through rocks, sediments, and soils. From the basic thermodynamic equations that deal with the propagation of liquids in solids, (2.34) and (2.35), it is obvious that such a migration depends on the surface tension of the solution.

Natural waters contain organic surface-active substances in various concentrations. These compounds are primarily decomposition products of plants, animals, and microorganisms. They are obtained by biologic activity mainly in the upper horizons of soils, by thermal decomposition in the lower layers of the sediments and by chemical reactions catalyzed by inorganic colloids in all horizons. The presence of fulvic and humic acids as well as of pyrolyzed fulvic acid, leaf extracts and saturated soil extracts may lower the surface tension of the aqueous solution from 72.75 mNm^{-1} up to values less than 45 mNm^{-1} at 20 °C, depending on their concentration and the pH of the solution (Chen and Schnitzer, 1978). An additional source of organic surfactants in natural water is the high industrial activity, which increases from year to year, giving rise to ever-increasing amounts of sewage and agricultural pesticides. Solutions rich in organic surface-active material will penetrate more easily into regions of low porosity, and chemical weathering may increase.

Organic surfactants, e.g., amphipathic monomers such as fatty or amino acids and their salts, contain hydrophilic groups, such as COOH or NH_2, as well as nonpolar hydrophobic moieties, such as alkyl chains. The fact that the hydrophilic group is accessible to hydration contributes to the aqueous solubility of the compound. The hydrophobic moiety, which causes the water to build an unstable "hydrophobic hydration" structure, is easily repelled from the bulk solution. The work performed to transport such a molecule from the bulk to the surface is much lower than that required to transport a water molecule. Such an organic molecule or ion has, therefore, a greater concentration in the solution surface than within the bulk phase. This phenomenon is termed *positive sorption* and will be further discussed in the next section (see also Sect. 4 in the Introduction). Saline ground waters have surface tensions higher than that of pure water. As a result the penetration of such solutions into regions of low porosity will occur to only a small extent. The surface tension of seawater is slightly higher than that of pure water at the same temperature, but is less than that of a salt solution having a concentration similar to that of seawater. This is due to the fact that seawater contains organic surfactants in addition to the inorganic salts.

2.1.2 Stability of Magmatic Emulsions (Table 2.2)

If two mutually immiscible liquids are brought into contact, and through some mechanism one is dispersed throughout the other in the form of very small droplets, the total interfacial area is increased as the volume of the droplets is decreased; the interfacial free surface energy is increased and the system becomes unstable. This is evidenced by the tendency of the droplets to coalesce spontaneously into an indepen-

Table 2.2. Some systems of immiscible magma liquids. (After Carmichael et al., 1974)

1. Mixtures of SiO_2 with divalent cations at high silica concentration	(Kracek, 1933)
2. The system $KAlSiO_4$–Fe_2SiO_4–SiO_2	(Roedder, 1951)
3. Mixtures of magnetite, apatite, and liquids of dioritic composition	(Philpotts, 1967)
4. The system $NaAlSi_3O_8$–$NaCl$–H_2O	(Koster van Groos and Wyllie, 1969)
5. The system Na_2CO_3–$NaAlSi_3O_8$	(Koster van Groos and Wyllie, 1969)

dent homogeneous liquid phase of minimum free surface energy. The disperse system may reduce the total interfacial free surface energy as a result of the presence of a third substance in the interfacial region, which has the effect of reducing the degree of dissimilarity between the contiguous phases. The transition between the two liquids of unlike molecular species accordingly becomes less abrupt, and the total interfacial free surface energy is thereby reduced to a minimum value. The third substance is known as an *emulsifier* and in this capacity it exerts its effect as a *stabilizer* by preventing coalescence of the dispersed liquid phase.

According to Anderson (1974) the behavior of sulfur in the magmas appears to reflect preeruption saturation with respect to sulfide melt: The concentration of sulfur in the mantle source exceeds that of the basaltic and picritic magmas derived from it and only part of the sulfur is transferred into the surface reservoir. A magma rising from the mantle or near the base of the crust, will undergo a decrease in temperature and commence precipitating an immiscible sulfidic liquid. The sulfur content of basaltic liquids before effervescence rarely exceeds 0.16% by weight, probably as a result of saturation with respect to a sulfide melt under fairly constant conditions of temperature and oxygen fugacity. The sulfide–silicate interfacial tension is very high if the silicate melt contains high concentrations of SiO_2 but decreases with increasing concentration of FeO. Ferrous iron is an efficient emulsifier giving the sulfide-in-silicate emulsion. The stability of this emulsion decreases with decreasing concentration of ferrous iron, accomplished with the coalescence of the droplets. Sulfur is removed from the silicate melt by separation of a sulfide phase as differentiation proceeds. This is consistent with the low concentration of sulfur in plutonic igneous rocks and selective accumulation of sulfides in the lower, more mafic, crust.

Clear examples of liquid immiscibility are occasionally found among igneous rocks. According to Carmichael et al. (1974) liquid immiscibility on a small scale may be a much more widespread phenomenon than is generally acknowledged, particularly in acid rocks. In many of the systems, the liquid immiscibility depends on the temperature. At very high temperatures the system may become a single liquid phase. The temperature at which immiscibility ceases is defined as the *critical solution temperature.*

As the cooling magma reaches the critical solution temperature globules in a state of a high dispersion are obtained. As long as the temperature of the magma remains near this critical solution temperature, this emulsion is very stable. This is due to the fact that at this temperature the various liquid phases closely resemble each other chemically, since they are nearly mutually miscible in all proportions. Under these conditions, the densities of the liquids are nearly alike and the interfacial free surface energy is very low. Any decrease in the magnitude of the interfacial free surface energy as the result of coalescence of the dispersed globules would be negligible. Therefore the tendency for such a coalescence likewise would be small. On further cooling the chemical composition of the liquid phases will differ, and the interfacial free surface energy becomes high enough to make the emulsion unstable and to effect the coalescence of the dispersed globules.

2.1.3 Magma Droplets and Frothy Liquids

Volcanic eruptions are associated with the formation of magma droplets (liquid in gas system) and vesicular liquid (gas in liquid system). All these systems have a discontinuous magma surface. The disruption of magma depends not only on the relative volumes of liquids and gases but also on the relation of gas pressure to the liquid surface tension (McBirney and Murase, 1971). Exsolution of the magmatic volatiles results in nucleation and growth of gas bubbles. When the total forces exerted by entrapped gases exceeds the surface tension of the liquid fraction over the same cross-sectional area, the liquid will be disrupted.

The cooling and differentiation of a magma and the exsolution of the volatiles have two opposing effects on its surface tension. The surface tension of molten rocks decreases with increasing silica content. Usually molten rocks have a positive temperature dependence; with falling temperature surface tension decreases. Water and dissolved gases are surface active and have the effect of lowering surface tension. It appears likely that exsolution of volatiles would increase surface tension to varying degrees in different magmas and different volatiles, so that bubbles would tend to expand to differing extents in basalt, andesite, or rhyolite without disrupting the liquid.

According to McBirney and Murase the effect of this relation on frothy liquids can serve to explain why certain liquids froth and others do not, or why others even form ash flows. If surface tension is constant or decreases during liquid frothing, bubbles that are formed by exsolution will be of a definite size and at a definite internal pressure at which the surface tension is exactly balanced. The relation between the surface tension of the liquid, the size of the bubble, and the internal pressure is given in Eq. (2.31). Further growth of the bubble will cause adjacent bubbles to coalesce and disrupt the continuity of the liquid while small bubbles will go back into solution. If, however, the surface tension varies in such a fashion that it increases as the radius of the bubble increases, individual bubbles will retain their walls and their separate identities. The large adjacent bubbles will not coalesce, but the liquid will continue to expand as a coherent froth.

2.2 Evaporation

Evaporation is of importance in those systems having water as one of the components. We will first deal with a one-component, two-phase fluid system, which is the simplest of all. In a liquid-vapor system in orthobaric equilibrium at any temperature between the triple point and critical value, the phases are physically distinct; the gas phase is characterized by complete internal disorder and absence of structure and the liquid is characterized by a structure of short-range order and transience. A liquid is a condensed state of matter and the molecules must collide very frequently as a consequence of their velocity and proximity. A temperature-dependent fraction of the molecules will possess kinetic energies considerably greater than the mean. There is a definite minimum value of the kinetic energy that will enable molecules to just reach the surface region. If a molecule acquires sufficient additional energy through collision, it may enter this region or even pass through it into the vapor phase. When molecules enter the interfacial region, their kinetic energies are partially transformed into potential energy concomitant with the position which they occupy. Thus, the energy required for this interphase transfer is not lost but transformed into a different form.

Molecules in the surface layer of pure liquids are involved in thermal motion like that of molecules in the bulk of the liquid or in the gaseous phase. This can be demonstrated as follows. The number of collisions of gas phase molecules per second, on a surface area of one cm^2 is, according to the kinetic theory, $p/\sqrt{2\pi mkT}$, where p is the vapor pressure of the liquid at temperature T, and m is the mass of a molecule. Not every colliding molecule passes from the gaseous phase into the surface layer. At equilibrium the number of condensing molecules is equal to the number of molecules that evaporate. If α is the condensation coefficient, then n_m, the number of molecules that evaporate during one second from one cm^2 of the liquid surface is given by

$$n_m = \alpha p/\sqrt{2\pi mkT} \tag{2.38}$$

For water at 20 °C, $\alpha = 0.03$ and $p = 17.5$ mmHg. This yields $n_m = 0.3 \times 10^{21}$ molecules/s from 1 cm^2 of surface. Taking into account the size of water molecules, one obtains that 1 cm^2 can accommodate simultaneously not more than 10^{15} molecules. It follows that a molecule can remain in the surface layer, on the average, for only 3×10^{-6} s. Equation (2.38) can be experimentally evaluated if evaporation takes place in a vacuum and the quantity of vapor evolved under these conditions is measured.

Evaporation from natural water surfaces involves several processes occurring simultaneously. Some of these will be described here. If a partly saturated air mass moves from the land to a water surface, all the lower boundary conditions change abruptly. The surface humidity greatly increases and water vapor flux is suddenly considerably larger than it is over the land. Air masses saturated by water move in directions of winds. Water vapor diffuses vertically and horizontally due to molecular and turbulent diffusion. Under conditions of fully turbulent flow, molecular diffusion is negligibly small (Weisman and Brutsaert, 1973).

It was suggested that an increase of gross evaporation from sea surfaces as well as intense evaporation from spray droplets in the atmospheric surface layer occur at high wind velocities (Wu, 1974). Such an increase, if it is appreciable, changes the heat balance near the sea surface and influences, consequently, the interaction between the sea and the atmosphere. According to Wu the contribution of water vapor in the atmosphere surface layer from evaporation increases with increasing wind velocity. Evaporation would be mainly from droplets when the wind velocity is greater than 15 m/s.

The formation of a monolayer (a film that is one molecule thick) by surface-active compounds, at the air–liquid interface of lakes and seas, suppresses the evaporation of their water. The mechanism of evaporation in this case will involve penetration by water vapor molecules from the water surface through the monolayer film and then diffusion of these molecules across the air diffusion sublayer just above the monolayer film. The movement of water vapor molecules through the monolayer can be regarded as a flow of molecules through flexible channels or pores, the walls of which are the hydrophobic hydrocarbon chains. The movement of water molecules through these channels or pores involves collisions between water molecules and monolayer molecules (Lou and Rasmussen, 1973).

Langmuir and Schaefer (1943) proposed an equation, analogous to Ohm's law of electricity, in which they related the evaporation suppression of a monolayer to a *resistance r,* where r is the total resistance of the combined monolayer and air sublayer. The rate of evaporation of mass N from one unit of surface area is given by the equation

$$N/Ot = (c_L - c_V)/r, \tag{2.39}$$

where O is the surface area of the water body, N is the evaporated mass, t is the time of evaporation, necessary for the system to reach equilibrium, c_L and c_V are concentrations of water vapor at the water surface (below the monolayer film) and in the gaseous phase, respectively.

Total resistance can be thought of as two resistances in series, the resistance of the monolayer r_f and that of the air sublayer just above the monolayer r_b. By consideration of the kinetics involved in the evaporation–condensation process, at equilibrium the following equation is obtained for the monolayer film resistance

$$r_f = (1/\alpha)\,(2\,\pi M/RT)^{1/2} \tag{2.40}$$

where α is the condensation coefficient, M is the molecular weight of water, R is the specific gas constant, and T is the absolute temperature. To relate the dependence of film resistance r_f to the structure of the hydrocarbon chain, Langmuir and Schaefer derived an empirical correlation

$$\log_{10} r_f = -3.08 + 0.0425\,F + 0.122\,n \tag{2.41}$$

where F is the surface tension in the boundary monolayer and n is the number of carbon atoms in the chain. From Eq. (2.39, 2.40, 2.41) it follows that the rate of evaporation through a monolayer decreases with increasing length of hydrocarbon chain of the surface-active compound. Evaporation suppression is an important method for conserving water.

3. Sorption

Sorption is a process in which species such as molecules, ions, or radicals are partitioned between a bulk phase and an interface. The sorbed species move from liquid or gaseous to solid–liquid, solid–gas, liquid–liquid, and liquid–gas interfaces. Sorption involves a number of phenomena, such as the accumulation of one or more substances in the interface, orientation of the molecules in the interface, ion exchange, capillary condensation[4], migration of fluids in porous media, catalysis, hydrolysis, and many others (McBain, 1950)–the most important reactions in geologic–colloid systems.

In systems comprising solids, sorption can be the result of high activity of the solid surface, involving either short-range chemical valence bonds, termed *chemisorption,* or long-range physical interactions through electrostatic or van der Waals forces, termed *physical-type sorption* or *adsorption.* The former is confined to surface atoms, but the latter may involve many successive layers of molecules or ions.

The terms absorption and adsorption are used to differentiate between the sites of sorption. If the substance penetrates into the interior, as in the picking up of organic compounds by coals and by smectites or the dissolution of a gas in a liquid, the process is known as absorption. If the substance is concentrated at the surface of a solid or liquid the process is known as adsorption (Remy, 1956).

In systems comprising liquids sorption can be the result of repulsive forces acting to remove the soluble species from the bulk solution due to lyophobic solvation. Solutes may be "structure breakers" of solvent associations in the liquid phase, giving rise to new solvation associations. If the newly formed solvent–solute assemblage results in a fine-structure more stable than that of the pure solvent, the solution is stable and the solvent is said to have a "lyophilic solvation" structure. If it is less stable than the original fine structure of the liquid, the solvent is said to have a "lyophobic solvation" structure and the solution is unstable. The low solubility of lyophobic solutes is linked with a large entropy decrease that accompanies dissolution. A major part of the entropy decrease is attributed to enhancement of solvent–solvent interactions in the liquid around solute particles. This enhancement results in a lyophobic solvation structure and reduced degree of randomness of solvent molecules: These solutes are regarded as "structure makers". The lyophobic solvation structure does not necessarily resemble the structure of the bulk liquid.

If the solute is an *amphipathic* molecule, comprised of lyophilic and lyophobic moieties, the solute solvation atmosphere may comprise some regions where a high stability of solvation is obtained due to solvent–solute interaction and some regions of a low stability, where there is no solvent–solute interaction and a lyophobic solvation atmosphere is obtained. The fraction of solvent forming the lyophobic solvation structure can be reduced either through the formation of micelles of solute molecules or by withdrawing solute molecules to the surface of the solvent. In both cases the lyophilic moiety of the solute remains in contact with the solvent.

The body on which sorption takes place is called the *sorbent,* and the substance which is sorbed is the *sorbate.* The process of removal of sorbed substances from or out of the sorbent is called *desorption.*

Sorption is accompanied by the release of heat, called the *heat of sorption,* since the surface energy of the sorbent is lowered. It follows from Le Chatelier's Principle that sorption diminishes with rise of temperature, unless the sorption is associated with a very high activation energy, which is sometimes the case with chemical sorption.

---

4 It is possible for gases or vapors to be condensed by the action of capillary forces in the void spaces of porous substances. This is kown as capillary condensation.

At constant temperature sorption is proportional to the surface area of the sorbent. It also depends on the concentration of the sorbate in the gas phase or in the solution from which the process occurs. Physical sorption generally rises very rapidly at low concentrations of the sorbate and increases relatively little at higher concentrations.

3.1 Sorption of Gases and Vapors by Solids

Gases are adsorbed by all sorts of solids. The amount of the physical sorption increases with the condensability of the vapor. Heat is evolved during sorption just as during the liquefaction of vapor. The physical type of sorption, having a low activation energy, is pronounced at low temperatures and disappears at very high temperatures, but chemisorption can sometimes occur at the lowest temperatures as well as at a white heat. Chemisorption is of essential importance in the catalysis of gas reactions and is much more specific than physical types of sorption.

Adsorption of molecules on solids can result either in a monomolecular or a multimolecular layer, named *monolayer* or *multilayer* respectively. A graph or a mathematical expression that shows the quantity of material sorbed as a function of the pressure or concentration of the sorbate in equilibrium with the sorbent at constant temperature is called a *sorption isotherm.* The adsorption is usually described by the *Langmuir isotherm*

$$X/M = abp/(1 + ap), \qquad (2.42)$$

where X is the weight of the substance sorbed by M grams of solid, p is the partial pressure of the gas or vapor employed, a and b are constants. The Langmuir isotherm (Fig. 2.6) is based on the postulation that adsorption in most cases results in a mono-

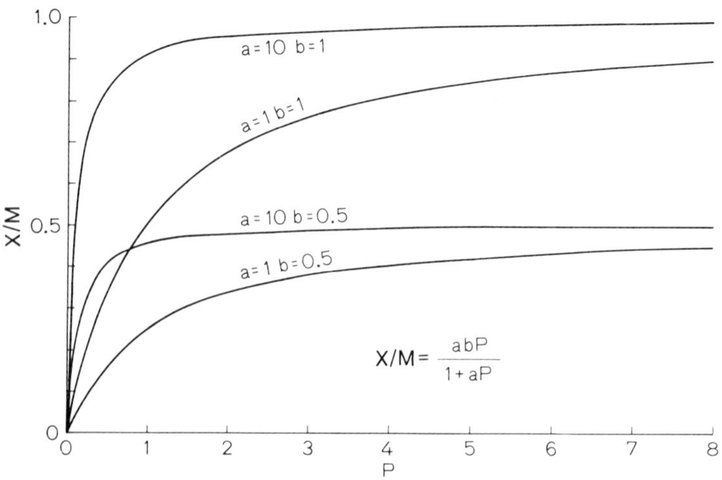

Fig. 2.6. Representative Langmuir isotherms. (From Dean, 1948)

layer and that there is no cohesion between the sorbed molecules either in the gaseous phase or after they have been adsorbed. The isotherm can be developed on the basis of molecular kinetic theory. Let us assume that a certain number N_s of adsorbed molecules is available in every square centimeter of a solid adsorbent, when the entire surface is occupied by the adsorbed molecules. In other words, N_s identical adsorption sites are available in every square centimeter of the adsorbent. At equilibrium, a certain fraction Θ of this monolayer is occupied by N_a adsorbed molecules, so that $\Theta = N_a/N_s$. According to kinetic studies, the number of molecules desorbed each second from each cm^2 will be $\Theta N_s \nu \exp(-\varphi/kT)$, where ν is a constant of proportionality determining the probability of desorption, φ is the activation energy of desorption, k is Boltzmann's constant, and $\nu \exp(-\varphi/kT)$ is the desorption probability. The number of molecules that fall from the gaseous phase onto one cm^2 is $p/\sqrt{2\pi mkT}$ where m is the mass of an adsorbed molecule. Only molecules that fall on the unoccupied active surface $(1 - \Theta)$ have a probability α of being adsorbed. The number of adsorbed molecules will then be $\alpha(1 - \Theta)p/\sqrt{2\pi mkT}$. At equilibrium

$$(1 - \Theta)(\alpha p)/\sqrt{2\pi mkT} = N_s \Theta \nu \exp(-\varphi/kT). \tag{2.43}$$

This equation can be written as follows

$$(1 - \Theta)p = \Theta N_s/ab,$$

where ab is constant and $\sqrt{2\pi mkT}(\nu/\alpha)\exp(-\varphi/kt) = 1/ab$. It follows that $p = \Theta p + \Theta N_s/ab = \Theta(p + N_s/ab)$ and

$$\Theta = p/(p + N_s/ab). \tag{2.44}$$

Since $\Theta = N_a/N_s$, we obtain

$$N_a = N_s p/(p + N_s/ab).$$

N_s is constant and may be designated by b. We obtain

$$N_a = bp/(p + 1/a) = abp/(1 + ap). \tag{2.45}$$

The Langmuir isotherm is obtained if N_a is expressed by X/M.

From the above it follows that the constant b of the Langmuir isotherm describes the *adsorption capacity* of the surface. It can be shown that the constant a describes the reciprocal of the gas pressure at which half of the adsorption capacity of the solid surface is saturated by the adsorbed gas. If $\Theta = 1/2$, Eq. (2.44) becomes $1/2 = p/(p + 1/a)$ or $p + 1/a = 2p$. Hence $a = 1/p_{1/2}$. The Langmuir isotherm is applied in studies of sorption of gases and vapors by most solids, including highly porous bodies. Corrections must be introduced if the adsorption results in a multimolecular layer or if the adsorbed molecules are elongated, and up to a certain extent of adsorption are parallel to the solid surface and to a certain extent are perpendicular. Corrections must also be applied when the adsorbing surface is heterogeneous and if there is more than one type of adsorbate.

Brunauer, Emmett, and Teller extended Langmuir's theory to multilayer adsorption. Their equation (known as the B. E. T. equation) is applicable only to the adsorption of vapors and can be written as

$$v = \frac{v_m\, cp}{(p_0 - p)\,[1 + (c - 1)\,(p/p_0)]}, \tag{2.46}$$

where v is the volume the adsorbed vapor would have if it were compressed to a liquid and v_m is the volume of the adsorbed molecules that would cover the surface with a monolayer; p_0 and p are the saturation pressure of the vapor and the equilibrium pressure at which the adsorption is measured, c is a measure of the relative magnitude of the forces between the adsorbed molecules and the forces between the adsorbent and the adsorbed molecules (Brunauer, 1944). From v_m the surface area of the solid phase can be calculated. The B. E. T. sorption isotherms of inert gases (such as N_2 and the noble elements) are used to determine the surface areas of mineral species in the laboratory.

3.2 Sorption by Solids from Solutions

Sorption of solutes is much more common in geologic processes than sorption of gases and is perhaps the most important phenomenon in hydrosphere and lithosphere colloid systems. It governs the mutual interaction between organic compounds and inorganic minerals, especially clays. It determines many of the properties of soils.

Sorption from solution, in contradistinction to sorption of gases and vapors from the gaseous phase, is not greatly affected by temperature or even by extreme pressure. It involves mainly chemical interactions of high sorption energies, and solvent and solute compete for the available surface. It enables the interaction of substances having a very low vapor pressure, like salts, with the surface of the solid.

The so-called *Freundlich isotherm* is a purely empirical representation of sorption at intermediate concentrations. It takes the form

$$X/M = kC^{1/n}, \; (n \geqslant 1) \tag{2.47}$$

where C represents the concentration of the solution in equilibrium with the sorbent and does not refer to the original concentration of the solution used and k and n are constants. A large value of k means a very high sorption (Fig. 2.7). Neither of the constants in the isotherm has any theoretical significance.[5]

---

5 The Freundlich isotherm is sometimes used to describe sorption of gases onto solids or liquids. It then takes the form $X/M = kp^{1/n}$, where p is the partial pressure of the sorbed gas.

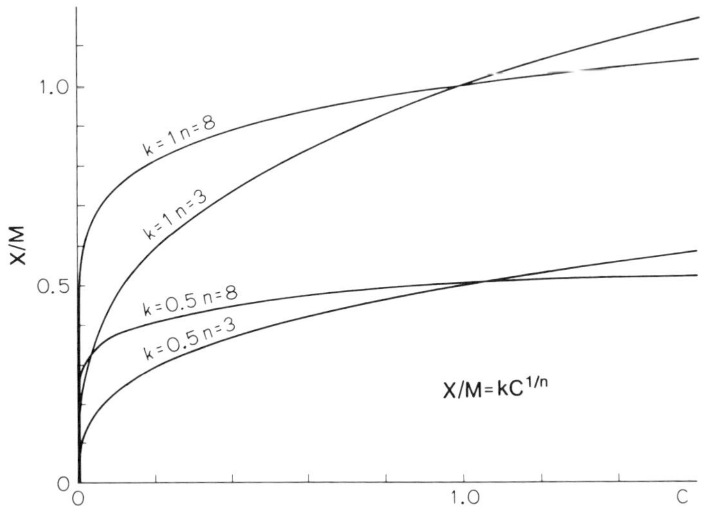

Fig. 2.7. Representative Freundlich isotherms. (From Dean, 1948)

In many cases, where the sorption onto the solid–liquid interface results in a monomolecular or monoionic layer, investigators make use of the Langmuir isotherm, Eq. (2.42), substituting concentration for pressure.

3.3 Sorption onto a Liquid–Gas Interface

In the following section the adsorption of surface-active agents from a solution onto the liquid–gas interface and the formation of a monomolecular layer will be discussed. The Langmuir adsorption isotherm obtains the form

$$N_a = abC/(1 + aC), \tag{2.48}$$

where N_a is the number of molecules adsorbed on one cm^2, b the maximum possible value of N_a, and $1/a$ is equal to the concentration $C_{1/2}$ at which $N_a = (1/2)\, b$. It is obvious that on a liquid surface all sites are equally susceptible to adsorption, and therefore b corresponds to the state when the entire surface is occupied by the adsorbed molecules, and is determined by the surface area occupied by the sorbed molecule. If this molecule is nonspherical and has a highly elongated shape, the area it occupies on the surface depends on its position, whether it is horizontal or vertical, the latter orientation giving a greater density of adsorbed molecules. That means that b is not necessarily constant and can increase with concentration (Fig. 2.8a, b).

As adsorption on a liquid surface increases, the mean separation of sorbed molecules can decrease to some extent, until dense packing is attained, resulting in interaction between neighboring sorbed molecules through their lyophobic moieties. This interaction may result in a greater tendency for soluble molecules to be sorbed onto the

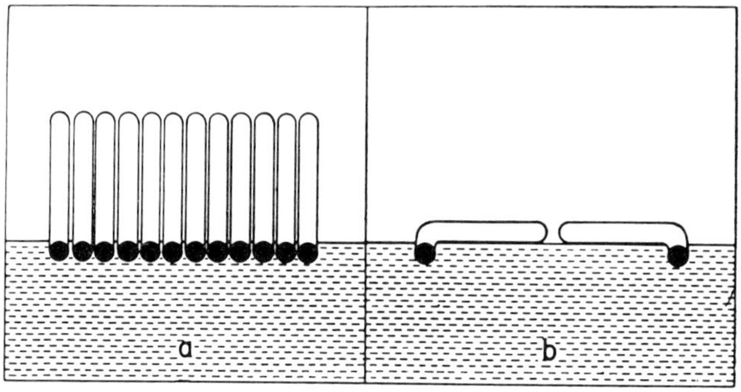

Fig. 2.8a and b. Monolayers of chain-structured molecules on liquid: **a** condensed, **b** expanded. (From Dean, 1948)

surface, which means a smaller probability of desorption (ν in Eq. 2.43). From the definition of a (Eq. 2.44), it is obvious that a increases with decreasing ν. It follows that a can increase with concentration.

3.4 Gas Exchange Across a Gas–Liquid Interface

This reaction is of fundamental importance in processes taking place in natural waters, oceans, lakes, rivers, etc. and in liquid–volatile interactions of the magma. The pathways of many environmental pollutants involve transfer across air–water interfaces.

The solubility of a gas in a liquid is proportional to its partial pressure. This may be formulated by the equation

$$C_L = (1/h)p = (1/H)C_G, \tag{2.49}$$

where h and H are constants, C_L and C_G are the gas concentrations in the liquid and gaseous phases respectively and p is the partial pressure of the gas in the gaseous phase. This relation is known as Henry's law and is valid when the dissolved substance neither combines with the solute nor undergoes dissociation in the process of dissolution. Henry's law is obtained from the Freundlich isotherm (2.47), for the special case when the power $1/n$ is equal to one.

The theory of gas exchange has recently been treated by Liss (1973). Let us assume a system comprising a liquid and a lyophobic gas. At a gas–liquid interface the flux F of gas through each boundary layer is given by the equation

$$F = \overline{k}\Delta C, \tag{2.50}$$

where F has dimension of mass $\times$ length^{-2} $\times$ time^{-1}. ΔC is the concentration difference across the particular layer and $\overline{k}$ is the corresponding *exchange constant*, also known as the *permeability coefficient* or *mass transfer coefficient*. Its value depends on many factors, of which the degree of mixing of the liquid and gas and the chemical reactivity of the gas concerned are some of the most important. The reciprocal of the exchange constant is a measure of the *resistance* to gas transfer and is designated by $\bar{r}$.

The total resistance to the exchange of gas will be due to a combination of the resistance due to the gas and liquid phases, $\overline{r_G}$ and $\overline{r_L}$, respectively. This resistance arises since the gas molecule has to diffuse through the boundary layers of a highly structured nature, one layer composed mainly from the gas molecules and the other composed mainly from the liquid molecules. The interface gaseous boundary layer has a density higher than the bulk gaseous phase, and the mobility of the molecules through this layer is slower than that in the bulk gas. The interface liquid boundary layer has a lyophobic solvation structure in which the molecular mobility is less than that in the bulk liquid. The exchange coefficient is related to the diffusion coefficient D by the equation $\bar{k} = D/J$, where J is the width of the layer. The diffusion of a species through a mobile phase depends on the mobility of the molecules that comprise the bulk phase. Since molecular mobilities of both boundary layers are less than those of the bulk phases, the exchange coefficients in each of these boundary layers are also smaller than in both the gaseous and the liquid phases.

Let us consider the changes in gas concentration near the interface between gas and liquid where gas exchange is taking place. In the bulk of the gas, which is assumed to be well mixed, the gas concentration is C_G. At the surface the gas concentration is C_{SG}, where the surface excess $\Gamma_{SG} = (\int_0^J dC_{SG}/dJ - C_G)J$, and J is the width of the gas boundary layer. Similarly, the concentration of dissolved gas in the liquid is C_{SL} at the liquid boundary layer and C_L in the bulk of the well-mixed liquid. On the assumption that the transport of gas across the interface is a steady-state process, and that net amount of gas entering the interface from the gaseous phase per one unit of time is equal to the net amount that passes from the interface into the bulk liquid phase, the flux of the gas can be described according to Eq. (2.50) by

$$F = \overline{k_G}(C_G - C_{SG}) = \overline{k_L}(C_{SL} - C_L), \qquad (2.51)$$

and $C_{SL} = (k_G/k_L)(C_G - C_{SG}) + C_L$. If the exchanging gas obeys Henry's law then

$$C_{SG} = HC_{SL}, \qquad (2.52)$$

where H is the Henry's law constant for molecules migrating between the gaseous boundary layer and the liquid boundary layer. Eliminating C_{SG} and C_{SL} between Eqs. (2.51) and (2.52) gives $C_{SL} = (k_G C_G + k_L C_L)/(k_L + Hk_G)$ and

$$F = (C_G - HC_L)/(1/k_G + H/k_L) = (C_G/H - C_L)/(1/k_L + 1/Hk_G). \qquad (2.53)$$

Since $\overline{k_G}$, $\overline{k_L}$, and H are constants, the equation can be written as

$$F = \overline{K_G}(C_G - HC_L) = \overline{K_L}(C_G/H - C_L) = \overline{K_L}(C_G - HC_L)/H, \qquad (2.54)$$

where $\overline{K_G} = \overline{k_G}\,\overline{k_L}/(\overline{k_L} + H\overline{k_G})$ and $\overline{K_L} = H\overline{k_G}\,\overline{k_L}/(H\overline{k_G} + \overline{k_L})$. Equation (2.54) shows the relation between the gas flux and the concentrations in the bulks of the phases without regard to the gas concentration in the interface. It follows that $\overline{K_G} = \overline{K_L}/H$.

Total resistance, $\overline{R_t}$, which can be expressed either on a gas phase $1/\overline{K_G}$, or a liquid phase $1/\overline{K_L}$ basis, depends on the exchange constants of the individual phases and the values of the Henry's law constants for the gas concerned:

$$\overline{R}_{t(G)} = 1/\overline{K_G} = 1/\overline{k_G} + H/\overline{k_L} \tag{2.55}$$

$$\overline{R}_{t(L)} = 1/\overline{K_L} = 1/\overline{k_L} + 1/H\overline{k_G}. \tag{2.56}$$

In using Eq. (2.56), $1/\overline{k_L}$ may be written as $\overline{r_L}$ and $1/H\overline{k_G}$ as $\overline{r_G}$, which gives

$$\overline{R}_{t(L)} = \overline{r_L} + \overline{r_G}. \tag{2.57}$$

Relationships such as Eq. (2.57) can be used to ascertain the relative importance of gas and liquid phase resistances for the exchange of any particular gas. However, except in special circumstances it is difficult to measure individual resistances, and most field and laboratory studies have measured only the total resistance.

3.4.1 Air–Sea Gas Exchange

During the passage of a typical storm the atmospheric pressure may fall and rise about 1–2%. During the pressure change, gas will transfer from one phase to another at a rate dependent on the partial pressure difference and wind velocity. The normal seasonal heating and cooling of the water column causes gas to move across the interface. The flux is dependent on the amount of heating and cooling.

The main body of each fluid is assumed to be well mixed. The main resistance to gas transport comes from the gas and liquid interfacial layers, across which the exchanging gases transfer by molecular diffusion. Under more turbulent conditions the thickness of the liquid film decreases and its resistance to gas exchange is reduced.

The Exchange of Oxygen: The oxygen is assumed to be only physically dissolved in the water and chemically unreactive. The diffusion coefficient for oxygen is about 10^4 times greater in air than for the gas dissolved in water. Hence diffusion processes in the liquid rather than in the gaseous phase are likely to control its transfer across an air–water interface. This, coupled with the low solubility of oxygen in water ($H \sim 30$), implies that $\overline{r_L} \gg \overline{r_G}$.

The Exchange of Carbon Dioxide: Although carbon dioxide is considerably more soluble in water than oxygen ($H \sim 1$), diffusion processes in the liquid dominate transport of this compound across the interface. At certain pH values the gas exists not only as dissolved CO_2 but also as ionic species, according to

$$CO_2 + H_2O \rightleftharpoons H_2CO_3 \rightleftharpoons H^+ + HCO_3^- \rightleftharpoons 2\,H^+ + CO_3^{2-}$$

and

$$CO_2 + OH^- \rightleftharpoons HCO_3^- \rightleftharpoons H^+ + CO_3^{2-}. \tag{2.I}$$

Below pH 5 all the carbon dioxide in the water will be present as dissolved CO_2 and H_2CO_3 molecules. At higher pH values, where some of the dissolved carbon dioxide forms ionic species, the exchange is more complex than for oxygen. At the water surface there is not only the usual concentration gradient of CO_2 molecules in molecular solution but also a similar gradient for bicarbonate and carbonate ions. In addition to the normal diffusion transport of gas molecules near the water surface, reequilibration of the dissolved molecular and ionic species of carbonic acid occurs, thus facilitating the transfer of this gas.

In a system containing oxygen and carbon dioxide at a pH below 5, one would expect that $\gamma = \overline{K_{L(CO_2)}}/\overline{K_{L(O_2)}} = 1$, but becomes greater than 1 at higher pH values. This has been proved experimentally. The exchange constants for both oxygen and carbon dioxide increase approximately as the square of the wind velocity.

The Exchange of Sulfur Dioxide: According to Liss (1971) the passage of sulfur dioxide across natural air–water interfaces, unlike O_2 and CO_2, is likely to be controlled by resistance in the air. This is in part due to the greater solubility of sulfur dioxide ($H \sim 3.8 \times 10^{-2}$), which increases the value of $\overline{r_G}$. Another factor is the rapid rate of hydration of SO_2 relative to CO_2. This leads to a low value of $\overline{r_L}$, the value of which varies with pH. At pH below 2.8 $\overline{r_L} > \overline{r_G}$; at greater pH values $\overline{r_G}$ becomes increasingly dominant.

The Effect of a Monomolecular Surface Layer of Organic Soap Molecules: There is some evidence that such a layer can decrease the rate of exchange of dissolved gases, such as oxygen and carbon dioxide. To obtain this effect, the film has to be in the closepacked condition. In large bodies of water depletion of oxygen beneath an impermeable surface film is unlikely because dissolved oxygen levels are near or above saturation due to wind-induced mixing. Surface expansion caused by waves reduces the closepacked character of the film molecules. Oxygen produced by marine organisms during photosynthetic periods would also compensate for oxygen losses (Ch. 1, Sect. 6).

The Effect of Oil Film on Oxygen Exchange: Such a film, which comprises nonpolar molecules, has a low structured nature, the molecules being mobile. The oil film has little effect on oxygen exchange unless its thickness is greater than 1 μm. Films of greater thickness tend to reduce the exchange constant. It is assumed that except in a region of accidental oil spillage, the amount of oil likely to be found at the sea surface would be insufficient to have any effect on oxygen exchange rates (Ch. 1, Sect. 6).

The Effect of Air-Bubble Solution on Air–Sea Gas Exchange: Bubbles contribute significantly to gas exchange (Atkinson, 1973). Air bubbles are carried down the water column by vertical turbulence. Under normal circumstances the bubbles will go into solution because of hydrostatic pressure and surface free energy. They represent a source of dissolved gas distributed throughout the water column. Craig and Weiss (1971) found values of up to 10% for the amount of air injected into the water by bubble solution. Bubbles and their dissolution cause a transfer of gas into the water column. The bubbles, once having entered the main part of the water column, will never reach the surface. Only bubbles that penetrate just the upper water or so have any chance of reaching the surface before disappearing by dissolution. Thus the effect of bubble solution on gas transfer from air to water is more pronounced than that from water to air. Aqueous bubble dispersion retards degassing of the water column. There is a time lag of about 5 h in the response of the water column to an atmospheric pressure change.

3.4.2 Gas Exchange in Pyroclastic Flows

The amount of gas that can be dissolved in the silicate melt depends on the temperature and pressure, and on the composition of the remainder of the liquid fraction. Following Henry's law (2.49) and Le Chatelier's Principle, the solubility decreases with decreasing pressure and increases with decreasing temperature. It increases also with increasing silica content (Burnham and Jahns, 1962).

The escape of a gas from a magma involves penetration of volatile molecules from the bulk liquid through both boundary layers of the liquid–gas interface to the gaseous phase. Because of the low molecular mobility of melted silicate species, diffusion processes in the liquid rather than the gas phase are likely to control the transfer of volatiles across a magma–gas interface. At a low concentration of volatiles the liquid boundary layer is composed primarily of the silicate species, forming a net structure that can be regarded as having channels or pores in which the silicate species form the walls. The diffusion of volatiles through these channels or pores involves collisions between these molecules and silicate polymers. The density of the liquid silicate boundary layer increases with temperature, up to a certain extent. The denser this structure, the smaller are the diameters of the pores and channels, and the greater is the number of collisions between volatile molecules and silicate polymers; consequently $\overline{r_L}$, the resistance to diffusion, increases.

The original volatiles of a magma exsolve from the melt during magma ascent as a result of reduction of confining pressure. At this stage the temperature is high enough to cause a far too great resistance $\overline{r_L}$ for the diffusion through the interface to occur, equilibrium in gas exchange is not attained, and the droplets of molten silica become highly supersaturated with volatiles. On cooling, the resistance to gas transfer decreases. Although the solubility should be the lowest at the high temperature of the explosion, the release of gas from the melt is not restricted to the initial explosion and the droplets continue to give off gas for some time as they are carried along in the avalanche. Each fragment is cushioned from the adjacent fragments by an envelope of expanding gas that is continuously being added to by escape of more gas from the supersaturated fragments themselves. Macdonald (1972) gives evidence indicating that gas must still have been coming out of solution and expanding in the interiors of the blocks during the very last stages of movement of the avalanche and after it comes to rest. A dramatic demonstration of the continued evolution of gas is furnished by the explosion of blocks after the avalanche has stopped.

There are three sources of volatiles in a pyroclastic flow: (1) those engulfed by the moving ash flow in the base of the eruption column; (2) those released from supersaturated flowing droplets of silicate melt; and (3) those released during the solidification of the liquid droplets. Solidifying droplets are not necessarily supersaturated with the volatiles because the temperature of solidification may be sufficiently low that the resistance of the interface to the gas transfer is low and the droplet may be in a state of equilibrium with the surrounding gases. Only a small fraction of the volatiles is released from the melt during the solidification since the solubility of the volatiles in the melt under the conditions of the avalanche is very low.

3.4.3 Gas Exchange During the Migration of Magma to the Earth Surface

Carbon dioxide and water are the most abundant volatiles in the lower crust and the upper mantle, where many magmas are generated. They can exist as dissolved species in magma and can form a coexisting free CO_2–H_2O fluid phase. Magmas migrate from regions of high temperatures and high hydrostatic pressures to regions of lower temperatures and hydrostatic pressures. The degree of polymerization of silicate species in the magma varies with pressure and temperature. It rises with pressure but decreases when the temperature is raised (Kushiro, 1975), giving rise to changes in the dissolved volatile content of the magma and in the composition of the CO_2–H_2O fluid.

A volatile of low solubility, which has the properties of a polymerizing agent, such as CO_2, causes the magma to begin to boil at depths much greater than those at which it would do so without CO_2 if it contained the equivalent amount of water. Even if there is insufficient H_2O in the system to saturate the magma, part of the H_2O enters the coexisting fluid phase. Furthermore, the partitioning of H_2O and CO_2 between magma and fluid is pressure dependent and consequently will change as a magma rises (Holloway, 1976).

Dissolution of volatiles in magma is associated with physical and chemical processes. In the physical process the dissolved molecule enters vacancies in the silicate melt. The physical dissolution increases with the degree of polymerization of the silicates, namely, with the decrease in temperature and increase in pressure. In the chemical process the dissolved species interacts with the silicate melt (Ch. 1). Water affects the silicate structure by breaking Si–O–Si bonds through (1) the substitution of bridging oxygens with nonbridging hydroxyls and (2) the exchange of metallic cations with protons (Burnham, 1975).

The mechanism of dissolution of carbon dioxide in magma differs from that of water. Its solubility in silicate melts is about one-half to one order of magnitude less than that of H_2O. Total carbon dioxide contents of magmas increase with temperature and hydrostatic pressure. It dissolves in silicate melts as both discrete CO_2 molecules and CO_3^{-2} anions. The equilibrium between the two carbon oxygen species in the melt can be epxressed by the following equation

$$CO_2\ (\text{melt}) + O^{2-}\ (\text{melt}) \rightleftharpoons CO_3^{2-}\ (\text{melt}). \qquad (2.\text{II})$$

The activity of O^{2-} is directly related to melt structure and in effect is an expression of the number of nonbridging oxygens in the melt. It expresses the degree of polymerization in the magma. A melt with no polymers (and no bridging oxygens) has by definition O^{2-} activity of 1. The more polymerized the melt, the more bridging oxygens, the lower the O^{2-} activity, and the reaction described in (2.II) shifts to the left side of the equation. Consequently, the positive correlation between temperature and CO_2 dissolution stems from the chemical process, whereas the physical process is responsible for the positive correlation between pressure and CO_2 dissolution (Mysen, 1976).

The presence of H_2O enhances the solubility of CO_2 to a maximum at a molar $CO_2/(CO_2 + H_2O)$ ratio of the volatile fraction of the system less than 1. The role of H_2O in silicate melts in forming cages for excess molecular CO_2 and the role of CO_2 as a polymerizing agent for silicates were discussed in Chapter 1. The difference between CO_2 solubilities in hydrous and H_2O-free melts decreases with increasing temperature and with decreasing pressure. Under these conditions the dissolution of CO_2 through the chemical process of (2.II) becomes more important than through adsorption in water cages. According to Mysen for every melt there is a temperature above which H_2O does not increase the amount of carbon dioxide that can be dissolved.

The molar ratio $CO_2/(CO_2 + H_2O)$ in the melt decreases with increasing pressure. This is because under the enhanced pressure the chemical dissolution of water increases whereas that of carbon dioxide decreases. This means that the separation of the fluid from the magma moving toward the surface should be characterized by reduction in the CO_2 content of the fluid as the total pressure is reduced. If the system is closed and the separated vapor remains in equilibrium with the melt, the composition of the fluid will change as the total pressure is reduced in such a way that the ratio of water to CO_2 will approximate the initial value for the system. If the rising magma is an open system and if there is a volatile exchange between the magma and the surrounding medium, the boiling depth and vapor compositon will be determined by many other factors. One possibility is that the H_2O and CO_2 released by boiling are lost completely from the system. In such a case the composition of the vapor phase will vary over a wide range up to the compositon of pure water (Kadik and Lukanin, 1973).

3.5 Ion Exchange

The present section is based on the treatments of Bolt (1967), Thomas (1967), and Laudelout (1972). Many solid minerals, when placed in contact with an aqueous solution, show a marked tendency to replace certain ions with ions of the same sign from solution. When the species lost and gained are positively charged, the phenomenon is called *cation exchange.* With negatively charged ions, *anion exchange* is obtained. In geologic studies cation exchange has recieved far more attention than anion exchange and in the present section will be considered almost exclusively. The processes involved are essentially the same, so that the detailed treatment of cation exchange will provide the principles required for analogous treatment of anion exchange. Exchangeable cations need not necessarily have the same oxidation numbers. The solid is termed *ion exchanger* and the ions, besides being called *exchangeable ions*, are sometimes termed *counterions* (or the German term *Gegenionen*).

Clay minerals and zeolites are the most effective ion exchangers among the minerals, and together with amorphous phases and organic substances, such as humics and kerogens, they are responsible for most ion exchange reactions in geologic systems.

In the presence of two ionic species, an equilibrium distribution is established that may be expressed by the relation between the ratio of the concentration of the two ionic species in the bulk solution and their ratio in the exchanger.

In several instances the exchange reactions between ions and minerals may be partly irreversible and often the exchange capacity of the solid substance depends, among other factors, on the pH value of the system. Exchange irreversibi with the ionic strength of the equilibrium solution and vanishes when ments are extrapolated to zero electrolyte concentration. A thermodyn... tion may thus only be applied to values obtained from extrapolating the experim... results to vanishing ionic strength, unless the reversibility of the exchange reaction has been demonstrated.

Our discussion here will be limited to reversible reactions and to cation pairs. Let us assume a system containing an amount of exchanger with a constant capacity to sorb reversibly certain cations and unable to sorb anions (which may be repelled due to the presence of an electric field). The exchange reaction can be expressed by

$$j\text{Mineral}(i+) + i\text{Salt}(j+) \rightleftharpoons i\text{Mineral}(j+) + j\text{Salt}(i+), \tag{2.III}$$

where i and j are the charges of the exchangeable cations, respectively. The system is suspended in an aqueous solution. The following *experimentally accessible* quantities are recognized:

1. Liquid content of the system, V, in ml per gram exchanger
2. Equilibrium concentrations, $c_0^{(i+)}$, $c_0^{(j+)}$, $c_0^{(-)}$ where $c_0^{(-)} = c_0^{(i+)} + c_0^{(j+)}$
3. Total amounts present in the system of the different ions, $C^{(i+)}$, $C^{(j+)}$, $C^{(-)}$ in milliequivalents (meq) per gram exchanger, where $C^{(i+)} + C^{(j+)} > C^{(-)}$.

Since some of the water molecules are positively sorbed by the solid to form the liquid–solid interface and anions are negatively sorbed, the electrolyte concentration in the equilibrium solution may be higher than in the original solution.

Using these experimental quantities one may define the surface excess of the different ions with respect to the solvent as:

$$\begin{aligned} \Gamma^{(i+)} &= C^{(i+)} - Vc_0^{(i+)} \\ \Gamma^{(j+)} &= C^{(j+)} - Vc_0^{(j+)} \\ -\Gamma^{(-)} &= Vc_0^{(-)} - C^{(-)} \end{aligned} \tag{2.58}$$

in which $\Gamma^{()}$ indicates the excess of cation i, j, or of the anion in meq/g exchanger. For anions the surface excess tends to be a negative number due to the anion repulsion. Summation of the preceding quantities gives the definition *of the cation exchange capacity* in meq/g of exchanger, Γ, as

$$\Gamma = \Gamma^{(i+)} + \Gamma^{(j+)} + (-\Gamma^{(-)}) = C^{(i+)} + C^{(j+)} - C^{(-)}. \tag{2.59}$$

This means that the cation exchange capacity is the excess of cations over anions present on the surface of a solid.

The anion present in the bulk and in the interface is partly balanced by the ith cation and partly by the jth cation. It is convenient to aportion the total amount of anions present in excess of the cations, according to the equilibrium concentration of the latter. The total amounts (in meq/g exchanger) of the i-valent and j-valent salts are $[c_0^{(i+)}/(c_0^{(i+)} + c_0^{(j+)})]\,C^{(-)}$ and $[c_0^{(j+)}/(c_0^{(i+)} + c_0^{(j+)})]\,C^{(-)}$ respectively. The *amount sorbed,*

n, constitutes the excess of the i- or j-valent cation with respect to the amount of its salt present in the system in meq/g exchanger and is equal to

$$n^{(i+)} = C^{(i+)} - \frac{c_0^{(i+)}}{c_0^{(i+)} + c_0^{(j+)}} C^{(-)} = \Gamma^{(i+)} - \frac{c_0^{(i+)}}{c_0^{(i+)} + c_0^{(j+)}} \Gamma^{(-)}; \tag{2.60}$$

$$n^{(j+)} = C^{(j+)} - \frac{c_0^{(j+)}}{c_0^{(i+)} + c_0^{(j+)}} C^{(-)} = \Gamma^{(j+)} - \frac{c_0^{(j+)}}{c_0^{(i+)} + c_0^{(j+)}} \Gamma^{(-)}.$$

Taking into consideration N_z, the fractional amount of the i- or j-valent component sorbed, which is defined:

$$N_i = n^{(i+)}/\Gamma \text{ and } N_j = n^{(j+)}/\Gamma, \tag{2.61}$$

for the present system we obtain

$$1 = N_i + N_j = [\Gamma^{(i+)} + \Gamma^{(j+)} - \Gamma^{(-)}]/\Gamma = [C^{(i+)} + C^{(j+)} - C^{(-)}]/\Gamma. \tag{2.62}$$

The exchange equation is then a functional relationship between N_i, N_j, $c_0^{(i+)}$ and $c_0^{(j+)}$ at equilibrium.

At equilibrium, $\Delta G_{P,T} = 0 = j\mu(i+)_S + i\mu(j+)_L - i\mu(j+)_S - j\mu(i+)_L$ in which ΔG is the Gibbs free energy change, μ is the appropriate chemical potential, and the subscripts S and L of the chemical potentials indicate the exchanger and the solution respectively. By choosing suitable standard states for the different components, the standard Gibbs free energy of exchange can be defined in terms of the equilibrium values of the *activities* of reactants and products.

$$\Delta G^0_{P,T} = j\mu^0(i+)_S - i\mu^0(j+)_S - j\mu^0(i+)_L + i\mu^0(j+)_L$$

$$= -RT \ln \frac{a(i+)^j_S}{a(j+)^i_S} \cdot \frac{a(j+)^i_L}{a(i+)^j_L} = -RT \ln K, \tag{2.63}$$

where a is the appropriate activity and $\mu = \mu^0 + RT \ln a$.

Many investigators have raised arguments against the application of activities to colloid systems. The activity concept of solutes was introduced as a concentration correction for interaction with reference to homogeneous systems. In solutions the mean activity of a salt has a specific meaning, both ions being present in roughly equal numbers, the activity thus representing a salt concentration corrected for mutual interaction between the ions of the salt. In colloid systems the effect of the giant colloid particle is quite the opposite. It accumulates large numbers of the much smaller ions in its surroundings, thus creating inhomogeneity. To assign one value to the activity of the ions in such a system does seem inconsistent with the principles underlying the activity concept in solution chemistry.

To render Eq. (2.63) applicable Gaines and Thomas (1953) selected the following standard states. For the exchanger they chose the homoionic state, in equilibrium with the solvent at unit activity, whereas for the salts the conventional standard state in solution was accepted (an ideal solution of the salt at unit molarity). It follows that for mono–divalent exchange, the standard Gibbs free energy of exchange is defined for 2 moles of monovalent ion in the exchange position of the solid in homoionic form in equilibrium with an infinitely dilute salt solution of the monovalent ion, reacting reversibly with one mole of salt solution of the divalent ion at unit activity to give one mole of divalent ion in the exchange position of the solid and 2 moles of the dissolved monovalent cation, both in their respective standard states. By this arbitrary choice of the standard state of the exchanger, ΔG^o becomes a measure of the relative affinity between exchanger and the two cations concerned, and thus lends itself to a comparison between different cation pairs and the same exchanger or between different exchangers and the same cation pair.

Applying the foregoing choice of standard states, Gaines and Thomas introduced activity coefficients applying to the exchanger, which are defined as $a(i+)_S = f_i N_i$ and $a(j+)_S = f_j N_j$, thus coupling the activity of the exchanger with a particular cation to the equivalent fraction adsorbed of the latter. A *selective coefficient* K_N of the exchange reaction can be defined as follows:

$$K_N = \frac{(N_i)^j}{(N_j)^i} \cdot \frac{[a(j+)_L]^i}{[a(i+)_L]^j} \tag{2.64}$$

This coefficient is obtainable experimentally. From Eqs. (2.63) and (2.64) we obtain $K = K_N(f_i^j/f_j^i)$, or

$$\ln K = \ln K_N + j \ln f_i - i \ln f_j \tag{2.65}$$

Gaines and Thomas showed that if K_N values obtained at vanishing ionic strengths are used, for the thermodynamic equilibrium constant K

$$\ln K = \int_0^1 \ln K_N dN_i$$

where N_i is the equivalent ionic fraction of a monovalent cation i exchanged with another monovalent cation j.

If the selective coefficients at a finite electrolyte concentration are known, then the preceding formula is no longer rigorous and should be replaced by the following:

$$\ln K = \int_D \ln K_N dN_i + \int n_w d \ln a_w,$$

where n_w is the number of moles of water found in the surface phase and a_w is the water activity. The right-hand integral indicates the change in Gibbs free energy of the solvent sorbed per unit exchanger. D is an integration path going from a standard initial state of a pure homoionic exchangeable i to a standard final state of homoionic exchangeable j. It involves concentrating the infinitely dilute solution of the i salt to a given finite concentration, carrying the exchange at that concentration by increasing the proportion of j in the solution and finally diluting the j solution to zero j salt concentration.

For the mono–divalent exchange ln K becomes

$$\ln K = 1 + \int_D \ln K_N dN_i + \int n_w d \ln a_w .$$

The right-hand integral is rather difficult to determine, but according to Gaines and Thomas (1955), this term is usually negligible in comparison to the other terms.

The *thermodynamic excess functions* express the deviation from ideality of the heteroionic exchanger with respect to the pure homoionic forms. The excess thermodynamic functions for the formation of a heteroionic exchanger are estimated from the activity coefficients of the adsorbed ions. The change in free energy of mixing, ΔG_m, is the difference between the observed value of the thermodynamic functions and the sum of the free energies of the pure constituents. For a system containing two monovalent ions, i and j, where N_i and N_j ($N_j = 1 - N_i$) are the equivalent fractions of i and j ions in the exchange position and f_i and f_j are the corresponding activity coefficients of the ions at the surface, ΔG_m can be written as follows:

$$\Delta G_m = RT\,[N_i \ln N_i f_i + (1 - N_i) \ln (1 - N_i) f_j].$$

ΔG_m^{ID} is the change in free energy of mixing in an ideal surface phase that obeys Raoult's law. For an equivalent fraction it is expressed as follows:

$$\Delta G_m^{ID} = RT\,[N_i \ln N_i + (1 - N_i) \ln (1 - N_i)].$$

The change in excess free energy of mixing, ΔG_m^E, is defined as:

$$\Delta G_m^E = \Delta G_m - \Delta G_m^{ID}.$$

Vansant and Uytterhoeven (1972) showed that in some systems the change in excess free energy can be positive, as in the exchange of Na montmorillonite with ethyl-, propyl-, and butyl-ammonium, whereas in others it can be negative, as in the exchange of Na-montmorillonite with inorganic ammonium ion. When the change in the excess function is negative the heterogeneous surface phase (which in that case contains a mixture of ammonium and sodium ions) is more stable than it would be if the mixing was ideal or if the system contained a mixture of homoionic Na- and NH_4-montmorillonites. A positive change in the excess function implies a deviation from ideality and a less stable mixture of the exchangeable ions compared to an ideal mixture.

3.6 Sorption and Surface Tension

Surface tension at the interface may be greatly altered by sorption, and the external properties of a liquid or solid are correspondingly modified, sometimes to a surprising extent. The Gibbs adsorption equation is a thermodynamic expression that relates the surface excess Γ of a species to both surface tension σ and the bulk activity of the adsorbate. In systems where σ is directly and simply measurable (i.e., liquid–liquid and liquid–gas systems) the Gibbs adsorption equation may be used to determine the surface concentration. In other systems, such as those including a solid phase, where the surface concentration can be measured directly but σ cannot, the Gibbs equation can be used to calculate the lowering of the interface tension.

Let us now determine this relationship for the relatively simple case of a two-component system, a solvent and a solute at a bulk concentration of c. When a small quantity dn_g of the solute is adsorbed by the solvent surface at constant temperature ($dT = 0$), the increase of the Helmholtz free surface energy is, according to Eq. (2.17), $dA_g = \sigma dO + \mu_g dn_g$. At the same time, since free energy is an extensive property, the surface free energy is composed of two components, the product of the surface area and the surface tension and the product of the number of adsorbed molecules and their chemical potential, and it is described by the sum $A_g = \sigma O + n_g \mu_g$ and $dA_g = \sigma dO + \mu_g dn_g + O d\sigma + n_g d\mu_g$. By equating both expressions for dA_g, we obtain $O d\sigma + n_g d\mu_g = 0$. Since at equilibrium $d\mu_g = d\mu$, the above equation can be written as $n_g/O = -d\sigma/d\mu$. Hence, for the surface excess we obtain

$$\Gamma = n_g/O = -(\delta\sigma/\delta\mu)_T. \tag{2.66}$$

From this equation it follows that positive adsorption lowers the surface tension. In sufficiently dilute solutions $d\mu = (RTdc)/c$ and Eq. (2.66) becomes

$$\Gamma = -\frac{c}{RT}\left(\frac{\delta\sigma}{\delta c}\right)_T. \tag{2.67}$$

These equations are called *Gibbs adsorption isotherms.* Since $(\delta\sigma/\delta c)$ in Eq. (2.67) is negative, it follows that adsorption increases with the increase in the bulk concentration of the solute.

In many geologic colloid systems, adsorption occurs on the surfaces of insoluble solids. It can be demonstrated that thermodynamic equations similar to (2.66) and (2.67) relate the surface concentration of a species to both surface tension and the bulk activity of the adsorbate. If the adsorption occurs from a solution, the surface excess increases with increasing concentration of the adsorbate in solution. If the adsorption occurs from the gaseous phase, the surface excess increases with increasing partial pressure of the adsorbate. This positive adsorption lowers the surface tension of the solid. Sorption therefore plays an important role in the stabilization of geologic colloid systems (see Ch. 4).

4. Electrochemistry of Heterogeneous Systems

At almost all interfaces there is a segregation of positive and negative charges in a direction normal to the phase boundary. The charges may be in the forms of ions and electrons or dipolar molecules and polarized atoms attached to the particles. This charge originates by the preferential adsorption of either positive or negative ions at the interface, by the adsorption and orientation of dipolar molecules, or by a transfer of electrons from one phase to another. In addition to specific adsorption, the net particle charge of some solids may arise either from the preferential dissolution of some ions of the crystal or from imperfections within the interior of the crystal. Electric potential differences are set up across the interface and an *electric double layer* of positive and negative charges is formed. For charged particles the chemical potentials are actually *electrochemical potentials*. Furthermore, the system as a whole must remain electrically neutral. Thermodynamic relationships do not help in the understanding of the fine structure of interfaces, and to this end an examination of the nature of electrochemical potentials is required (Sonntag and Strenge, 1972).

4.1 Electric Double Layer at a Solid–Liquid Interface

In this section the distribution of ions in an aqueous electrolyte solution in the vicinity of a solid surface will be described. Ions having a charge of the same sign as that of the solid surface are known as *co-ions* while those having a charge of an opposite sign are known as *counterions*. The aqueous phase is divided into four regions of distinct dielectric behavior (Bell and Levine, 1972). Region A (dielectric constant ϵ_1) consists of a layer of preferentially oriented water molecules in contact with the boundary. This is the *inner Helmholtz layer* where the specific adsorbed "bare" ions are without their hydration shells. Region B (dielectric constant ϵ_2) is a region of both free water molecules and molecules attached to hydrated ions. This is the *outer Helmholtz layer,* defined by the closest approach of a fully hydrated charge ion to the boundary (Fig. 2.9). The counterions are arranged in the outer Helmholtz layer at a distance δ from the interface. Region C (dielectric constant ϵ_3) is the *Gouy-Chapman diffuse layer* where the concentration of the counterions decreases with increasing distance

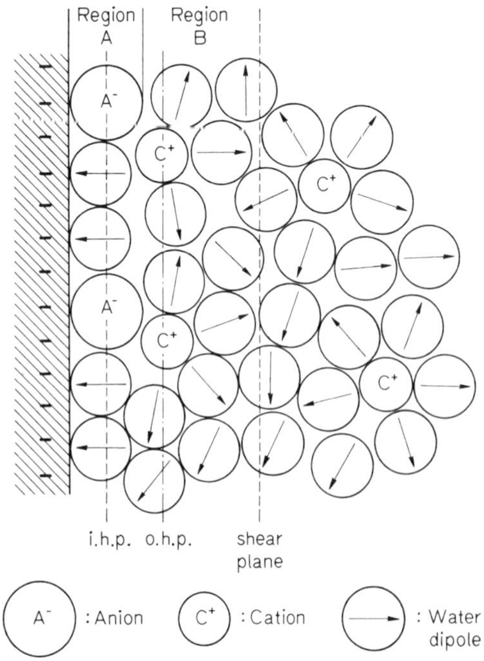

Fig. 2.9. A schematic representation of the electric double layer. (From Eagland, 1975. Used with permission of Plenum Press)

from the plane interface. ϵ_4 is the dielectric constant of the bulk solution (region D). The dielectric constant of the water increases from region A to region D.

4.1.1 Diffuse Double Layer

Counterions are electrically attracted by the oppositely charged solids. At the same time, however, these ions have a tendency to diffuse away from the surface toward the bulk of the solution, where their concentration is lower. The net result of the two competitive tendencies is an equilibrium distribution of ions in which their concentration gradually decreases with increasing distance from the solid surface. Simultaneously, there is a deficiency of co-ions in the vicinity of the surface, since these ions, which have an electric charge with the same sign as the particle, are repelled. There is a gradual increase of co-ion concentration with the distance from the solid surface. This diffuse character of the ions was recognized by Gouy in 1910 and by Chapman in 1913, and is often referred to as *Gouy-Chapman diffuse layer.*

To evaluate the distribution of ions in the diffuse double layer, the *space charge density* must be taken into account, which is defined as:

$$\rho = \Sigma F z_i c_i, \tag{2.68}$$

where z_i is the valency and c_i is the concentration of the ith ion respectively (at any given point within the diffuse double layer, c_i is variable and changes according to the distance from the solid surface) and F is the Faraday constant.[6]

In the bulk of the solution, where the concentration of cations $+z_i$ is equal to that of the anions $-z_i$, this charge density is zero. Furthermore, the condition of electroneutrality of the entire system makes it necessary for the surface charge of the solid phase, e_s, to have the same absolute value as $\int_{x=0}^{x=\infty} \rho dx$, namely,

$$|e_s| = \left| \int_{x=0}^{x=\infty} \rho \, dx \right|. \tag{2.69}$$

The diffuse distribution of ions is associated with a smooth variation of potential from its value φ_0 at the solid surface to zero in the bulk solution (zone D). If the surface charge results from the adsorption of *preferred ions,* the electric potential at the particle surface is determined solely by the concentration of these ions in solution. The ions are called *potential-determining ions* and the potential is given by Nernst equation

$$\varphi_0 = (kT/ze_s) \ln c_\infty / c_{\infty 0}, \tag{2.70}$$

in which k is the Boltzmann constant, c_∞ is the actual concentration of the potential-determining ions in solution, and $c_{\infty 0}$ is their bulk concentration in a solution whereby

---

6 The Faraday constant is the electrostatic charge of one gram-equivalent of an ion according to $F = N_a \bar{e}$, where N_a is the Avogadro number and $\bar{e}$ is the charge of one electron.

the net charge of the surface becomes zero, i.e., $\varphi_0 = 0$. If the double layer is the result of interior crystal imperfections or diadochic replacement of ions, the charge per unit area is determined by these imperfections and not by the properties and concentration of the solute and is therefore a fixed quantity. The electric potential is constant and does not depend on salt concentration.

The relation between the potential φ at any given point and the space charge density ρ is given by the Poisson equation

$$\nabla^2 \varphi = (4\,\pi/\epsilon)\,\rho, \qquad (2.71)$$

where ∇^2 is the Laplace operator,

$$\nabla^2 = \frac{\delta^2}{\delta x^2} + \frac{\delta^2}{\delta y^2} + \frac{\delta^2}{\delta z^2}$$

and ϵ is the dielectric constant. If $c_{i\infty}$ is the bulk concentration and $Fz_i\varphi$ is the energy required to transfer one gram-equivalent of the positive ions from the bulk of solution with $\varphi = 0$ to a point with potential φ, the concentration of the ith ion is given by the Boltzmann equation:

$$c_i = c_{i\infty} \exp\left(-Fz_i\varphi/RT\right), \qquad (2.72)$$

which can also be written,

$$\ln c_i/c_{i\infty} = -\,Fz_i\varphi/RT.$$

A solution of these equations becomes simpler in the case of a plane surface, where the Laplace operator ∇^2 is reduced to d^2/dx^2. Equation (2.71) becomes

$$(d^2\varphi/dx^2) = -(4\,\pi/\epsilon)\,\rho. \qquad (2.73)$$

The boundary conditions for the solution of Eq. (2.73) are

$$x = 0,\ \varphi = \varphi_0,\ (d\varphi/dx) = (d\varphi/dx)_0 \neq 0;$$

$$x = \infty,\ \varphi = 0,\ (d\varphi/dx) = 0.$$

Let us analyze the simple case of a binary electrolyte for which $|z_1| = |z_2| = z$. Substituting the equivalent of ρ from Eq. (2.68) in Eq. (2.73) we obtain

$$\frac{d^2\varphi}{dx^2} = \frac{4\,\pi}{\epsilon}\,\rho = -\,\frac{4\,\pi}{\epsilon}\,\Sigma\,Fz_ic_i = -\,\frac{4\,\pi}{\epsilon}\,[Fzc_{(+)} - Fzc_{(-)}].$$

Substituting the value of c_i from (2.72) we obtain

$$\frac{d^2\varphi}{dx^2} = -\,\frac{4\,\pi}{\epsilon}\,Fzc_\infty\left[\exp\left(-\frac{Fz\varphi}{RT}\right) - \exp\left(\frac{Fz\varphi}{RT}\right)\right] = \frac{8\,\pi\,Fz}{\epsilon}\,c_\infty\,\sinh\left(\frac{Fz\varphi}{RT}\right). \qquad (2.74)$$

Since $(dA/dx) = (dA/d\varphi)\,(d\varphi/dx)$, and if we assume that $d\varphi/dx = A$ then the following substitution can take place

$$\frac{d^2\varphi}{dx^2} = \underbrace{\frac{d(d\varphi/dx)}{dx}}_{(dA/dx)} = (d\varphi/dx) \cdot \underbrace{(d/d\varphi)\,(d\varphi/dx)}_{(dA/d\varphi)}. \tag{2.75}$$

Let us multiply both sides of Eq. (2.74) by $d\varphi$. Let us also multiply and divide the right side of the equation by RT/RT. Substituting (2.75) in (2.74) we obtain

$$\frac{d^2\varphi}{dx^2}\,d\varphi = \frac{d\varphi}{dx}\,d\left(\frac{d\varphi}{dx}\right) = \frac{8\,\pi\,RT}{\epsilon}\,c_\infty \sinh \frac{Fz\varphi}{RT}\,d\left(\frac{Fz\varphi}{RT}\right). \tag{2.76}$$

Let us integrate Eq. (2.76). The integration of the central part of the equation is as follows. Assuming that $A = d\varphi/dx$, then, the central part can be written as AdA and its integral is equal to $(1/2)\,A^2$ + Const. = $1/2\,(d\varphi/dx)^2$ + Const. To integrate the right-hand side, we assume that ϵ is independent of x, and since at $x = \infty$, $\varphi = 0$, $d\varphi/dx = 0$ and $\cos h\, 0 = 1$,[7] the first integration of Eq. (2.76) between $x = 0$ and $x = \infty$ gives

$$(d\varphi/dx)^2 = (16\,\pi\,RT/\epsilon)\,c_\infty\,[\cosh(Fz\varphi/RT) - 1].$$

Since

$$\cosh x - 1 = 1/2\,[\exp(x) + \exp(-x)] - 1 = 1/2\,[\exp(x) + \exp(-x) - 2] =$$

$$1/2\,[\exp(x/2) - \exp(x/2)]^2$$

we obtain

$$\frac{d\varphi}{dx} = -\sqrt{\frac{8\,\pi RT}{\epsilon}}\,c_\infty\,[\exp(Fz\varphi/2\,RT - \exp(-Fz\varphi/2\,RT)]. \tag{2.77}$$

Some important conclusions may be drawn from this equation:

1. The minus sign before the square root indicates that φ decreases exponentially with the distance from the solid surface. The maximum value of the potential is at the solid surface.

2. The slope $d\varphi/dx$ increases in proportion to $\sqrt{c_\infty}$. This applies to ions that do not affect the potential φ_0 and to potential-determining ions. In the first case, when the concentration of such ions is increased, the curve $\varphi(x)$ starts again from φ_0(at $x = 0$); for large c_∞ the curve descends more steeply and approaches $\varphi = 0$ at shorter distances from the surface.

3. Experimentally it is impossible to determine φ_0. An electric potential ζ of a plate lying in the diffuse double layer, at a certain distance Δ from and parallel to the particle surface can be determined by electrokinetic means. This electrokinetic

---

7 $\cosh x = 1/2\,[\exp(x) + \exp(-x)]$ $\cosh(0) = 1/2\,[\exp(0) + \exp(-0)] = 1/2\,(1 + 1) = 1$.

potential is named *zeta potential.* In accordance with the principles of hydrodynamics the velocity of the liquid in an electric or a mechanical field is zero at the solid surface and increases toward the bulk liquid. For the electrokinetic phenomena to take place it is necessary to have a certain drift of the charged liquid with respect to the solid surface. The seat of the zeta potential is a *shearing plane* (or a *slippling plane*) between the bulk liquid and an envelope of water that moves with the particle at a distance Δ from the solid surface. The zeta potential will then be $\zeta = \varphi_\Delta$. Equation (2.77) establishes a relationship between ζ and c_∞. At maximum dilution, ζ approaches φ_0 whereas at higher c_∞ it drops down and approaches zero. With the present state of hydrodynamic techniques Δ cannot be precisely determined.

From Eqs. (2.72) and (2.77) the concentration of the counter- and co-ions at any point in the diffuse double layer can be computed. Their distributions as a function of the distance from the surface are shown in Figure 2.10. The concentrations of both cations and anions at the bulk solution are indicated by the line BD. The concentration distribution of the ions of opposite sign is given by curve DA, that of the ions of the same sign as the surface charge by curve DC. The total excess of counterions is represented by surface area BAD, the total deficiency of co-ions, known as "Donnan exclusion", is represented by surface area BCD. The total net diffuse layer charge, which is equivalent to the surface charge, is represented by the surface area CAD.

When the bulk electrolyte concentration is increased (line B′D′ in the Figure), the distribution of the ions in the diffuse layer changes, as can be seen from the curves A′D′

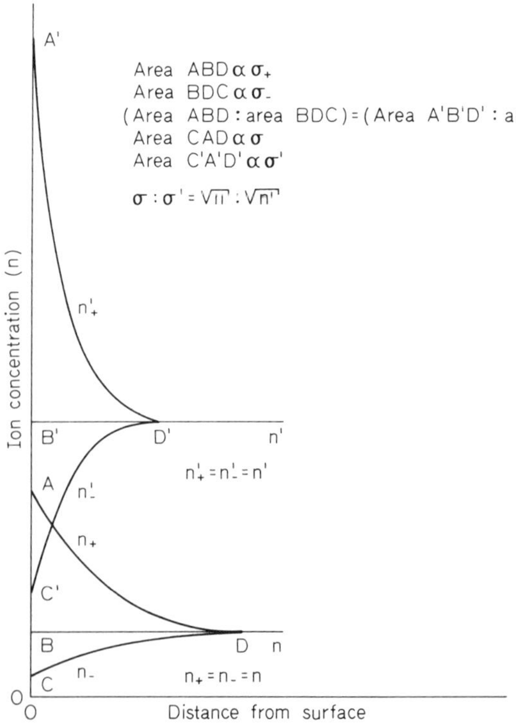

Fig. 2.10. A schematic representation of charge distribution in the diffuse double layer of a negative particle at two electrolyte concentrations. (From van Olphen, 1963. Used with permission of John Wiley and Sons)

and C′D′ for counter and co-ions, respectively. All the concentrations of the cations and of the anions everywhere in the diffuse layer are now higher. The diffuse counterion atmosphere is compressed toward the surface, and the diffuse charge is concentrated in regions closer to the surface. Some quantitative information about the computed thickness of the diffuse layer in solutions of various electrolyte concentrations and about the effect of the oxidation number of the counterion is given in Table 2.3.

Table 2.3. Approximate "thickness" of the electric double layer as a function of electrolyte concentration at constant surface potential. (After van Olphen, 1963)

Concentration of electrolyte	Thickness of the double layer in μm	
mol/l	Monovalent ion	Divalent ion
0.01	1000	500
1.0	100	50
100.0	10	5

4.1.2 Calculation of the Surface Charge

From Eqs. (2.69) and (2.73) we obtain

$$e_s = \int_{x=0}^{x=\infty} \rho\, dx = \int_{x=0}^{x=\infty} \frac{\epsilon}{4\,\pi} \cdot \frac{d^2\varphi}{dx^2}\, dx = \int_{x=0}^{x=\infty} \frac{\epsilon}{4\,\pi} \cdot \frac{d(d\varphi/dx)}{dx}\, dx$$

$$= \int_{x=0}^{x=\infty} \frac{\epsilon}{4\,\pi}\, d\left(\frac{d\varphi}{dx}\right) = \frac{\epsilon}{4\,\pi}\left[\left(\frac{d\varphi}{dx}\right)_\infty - \left(\frac{d\varphi}{dx}\right)_0\right].$$

Since $(d\varphi/dx)_{x=\infty} = 0$ we obtain $e_s = -(\epsilon/4\,\pi)/(d\varphi/dx)_{x=0}$. It should be remembered that e_s and $\int\rho\, dx$ are of opposite signs. By substituting the value of $d\varphi/dx$ found from Eq. (2.77) for $x = 0$ and $\varphi = \varphi_0$, we obtain

$$e_s = -\sqrt{\frac{\epsilon^2}{16\,\pi^2} \cdot \frac{8\,\pi\, RT}{\epsilon} c_\infty}\, \left[\exp\left(\frac{Fz\varphi_0}{2RT}\right) - \exp\left(\frac{-Fz\varphi_0}{2\,RT}\right)\right]$$

$$= \sqrt{\frac{\epsilon RT}{2\,\pi} c_\infty}\, \left[\exp\left(\frac{Fz\varphi_0}{2RT}\right) - \exp\left(-\frac{Fz\varphi_0}{2RT}\right)\right]. \tag{2.78}$$

For sufficiently small φ_0, where $Fz\varphi_0/2\,RT$ tends to zero, the exponentials may be expanded into a series which, as a first approximation, yields $Fz\varphi_0/2\,RT$ and $-Fz\varphi_0/2\,RT$. Equation (2.78) becomes

$$e_s = \sqrt{\frac{\epsilon RT}{2\,\pi} c_\infty}\, \frac{Fz\varphi_0}{RT} = \frac{\epsilon}{4\,\pi} \sqrt{\frac{8\,\pi\, F^2 z^2 c_\infty}{\epsilon RT}}\, \varphi_0. \tag{2.79}$$

A value γ can be defined as:

$$\gamma = \sqrt{\frac{8\,\pi\, z^2 F^2 c_\infty}{\epsilon\, RT}}. \tag{2.80}$$

The expression $1/\gamma$ is parallel to the concept of the "radius of ionic atmosphere" in the theory of strong electrolytes. In the case of colloids it represents the "thickness of the diffuse double layer", which decreases with increasing concentration and charge of the counterion. Substituting (2.80) in (2.79) gives

$$e_s = \frac{\epsilon}{4\pi} \gamma \varphi_0 \tag{2.81}$$

The following conclusions are derived from Eq. (2.81):
(1) The surface charge is proportional to the surface potential.
(2) The double layer capacity is equivalent to the capacity of a plane condenser with plate distance $1/\gamma$.[8]

4.1.3 Stern's Model of the Double Layer

Stern (1924) considered the ions forming the double layer to be distributed not only because of their thermal kinetic motion and the surface electric field, but also because of specific chemical interactions between the ions and the outer phase. The nonelectrostatic forces of attraction between ions in solution and the outer phase can be considerable at short distances when ions penetrate into region A. These *short-range* interactions are characterized by a strong dependency upon distance and comprise hydrogen and covalent bonds and π or hydrophobic interactions, which do not occur when the ions are in the diffuse layer (Region C).

If a layer is formed by the specific sorption of ions of a certain type, the distance of closest approach of the ions to the charged surface is limited by the size of these ions, and a *molecular condenser* is formed by the solid surface charge and the charge in the plane of the centers of the closest ions. The formation of such a layer is equivalent to the accumulation of a certain quantity of electric charge at the interface, at a distance β from the solid's surface, where β is the radius of the ion, so that the electric potential changes linearly with the distance, from value φ_0 at the solid surface to a value φ_β in the immediate vicinity of the interface. These ions in the inner Helmholtz layer are known as *potential-determining ions.*

As was explained in the Introduction, the penetration of hydrated ions into the water boundary layer is not favored. The crowding of the hydrated counterions at the solid surface (where $x = 0$) is prevented. In this way, the outer Helmholtz layer is obtained, in which the closest approach of the centers of the fully hydrated ions to the solid boundary is δ. The *inner Helmholtz plane,* where the centers of all specific adsorbed ions lie, is situated in region A, while the *outer Helmholtz plane* is in region B. By considering regions A and B as a parallel plate condenser, the potential is shown to drop linearly from value φ_β to a value φ_δ, which is called the *Stern potential.* Beyond the outer Helmholtz plane, the electric potential decreases roughly exponentially with increasing distance. The inner and outer Helmholtz layers are termed the *Stern double layer* (Fig. 2.11).

---

8 A condenser with charge e_s and plate distance δ has a potential $\varphi = (4\pi\delta e_s)/\epsilon$ where ϵ is the dielectric constant of the material in the condenser.

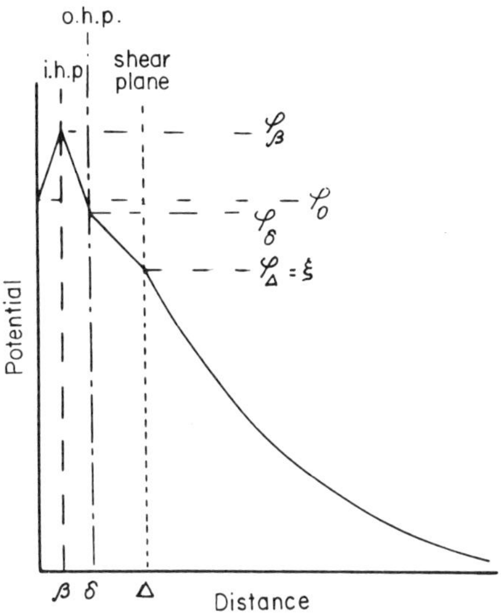

Fig. 2.11. Variation of potential corresponding to Figure 2.9. (From Eagland, 1975. Used with permission of Plenum Press)

If e_1 is the charge of the Stern layer, e_2 the charge of the diffuse layer, and e_s the charge of the solid surface, the condition for electroneutrality of the system requires that $e_s = -(e_1 + e_2)$. The *interface charge* is defined as the solid-surface charge plus the Stern layer charge and is equal to $-e_2$, $|e_s + e_1| = |e_2|$. It is calculated as in Eq. (2.78), with the integration carried out from $x = \delta$ rather than from $x = 0$, and correspondingly from φ_δ rather than φ_0. Equation (2.78) becomes

$$e_2 = \sqrt{\frac{\epsilon RT}{2\pi} c_\infty} \, [\exp (Fz\varphi_\delta/2\,RT) - \exp (-Fz\varphi_\delta/2\,RT)]. \tag{2.82}$$

Assuming that the Stern layer is monoionic, and making use of Langmuir's equation, by an approximate statistical treatment one obtains

$$e_1 = NzF/\left(1 + \frac{1}{c_\infty} \exp \frac{zF\varphi_\delta + \theta_+}{RT}\right) - NzF/\left(1 + \frac{1}{c_\infty} \exp - \frac{zF\varphi_\delta - \theta_-}{RT}\right), \tag{2.83}$$

in which N is the number of available sorption sites per 1 cm^2 of the surface and θ_+ and θ_- are the sorption potentials of cation and anion, equivalent to the energy of sorption without the electric energy $zF\varphi_\delta$. The value of e_s can now be computed.

Stern's correction amends Gouy and Chapman's theory, which led to high values for the charge of the double layer, and consequently for its capacitance. According to Gouy and Chapman the integration of Eq. (2.69) is carried out from $x = 0$ and not from $x = \delta$.

4.2 Electrokinetic Phenomena

There are four main phenomena in which the relationship between motion and electric charge or potential is demonstrable, two of which depend upon the existence of an external electromotive force, namely, *electroosmosis* and *electrophoresis*, and two upon the application of mechanical force, namely, *streaming* and *sedimentation potentials*. The first two are only minor in geologic systems but are of general importance in the laboratory study of colloid properties of minerals and in the geophysical research of surface properties of rocks and of conductivities in fluid-bearing rocks.

4.2.1 External Electromotive Forces

1. Electroosmosis occurs when the solid is not free to move, while the liquid can move. It is observed with immersed porous materials, compacted powders, and with gels and jellies. It consists in the movement of a liquid through a capillary or a porous diaphragm under the influence of an electric field. The two chief rules of electroosmosis are: (1) for any voltage applied a proportional hydrostatic pressure is produced and (2) for any amount of current passed, a proportional volume of liquid is transported. For a constant current intensity it is independent of the area and thickness of the porous diaphragm. The contribution of electric potentials to the migration of water in the lithosphere has not yet been studied thoroughly.

2. Electrophoresis or cataphoresis is the movement of particles, including colloid particles, drops, globules, and bubbles in an electric field. Electrophoresis of colloid particles is similar to the phenomenon called ionic mobility, which occurs with small ions consisting of one or a few atoms. It can be demonstrated with a U-tube fitted with two platinum electrodes located near the top of the two limbs, with a colloid brown ferric hydroxide solution at the bottom of the U-tube, covered by a colorless solution of very dilute electrolyte. The ferric hydroxide is positively charged, the corresponding negative charges in the diffuse layer being sorbed chloride ions. During electrophoresis the band of brown color is seen to move uniformly away toward the cathode. The chloride ion moves toward the anode. The migration may be measured quantitatively by analysis of the substance in the anode half or the cathode half, or in both, before and after a certain amount of current has passed. Sometimes the movement can be observed under the microscope or ultramicroscope.

Hydrogen and hydroxyl are the fastest moving ions known. They move at a rate of 35 and 18×10^{-4} cm/s, respectively, in a field of 1V/cm. Ordinary ions are those of potassium and chloride, which move at a rate of 6.8×10^{-4} cm/s. The heaviest and slowest ions move at a rate of at least 2×10^{-4} cm/s. Bubbles, drops, suspensions, and colloid particles move at the same rate as ordinary ions, that is, between 2×10^{-4} and 20×10^{-4} cm/s/V/cm. Thus a negative charge riding on a drop of paraffin oil may move much faster than if it were being carried by an atom of chlorine and is therefore a better conductor than a chloride ion. However, the total conductivity of the colloid solution is usually low because the total number of particles and charges is small compared with the total number of ions in an ordinary solution of an electrolyte. The charge on a ferric hydroxide sol has been determined as one positive charge to 206 chemical equivalents of ferric iron, the rate of movement and conductivity being 57% that of the same charge on potassium ions. The corresponding free chloride ion therefore carries 1.00/1.57 or 64% of the current, leaving 36% of the current to be carried by the colloid particles (McBain, 1950).

Clay minerals have two electric polarizabilities. One is a permanent dipole moment originating from structure; it depends on the atomic masses and is oriented parallel to the long axis of the clay particle. The second, an induced dipole moment, is perpendicular to the first and is caused by the external electric field. The induced dipole depends on the polarizability of the electric double layer of the particle. The orientation of clay particles in an electric field depends on the interaction between the electric moments of the particle and the strength of the applied field (Petkanchin and Brückner, 1976). Because of this preferred orientation, electrophoretic mobilities of clay particles are usually small, in the range of 1 to 3×10^{-4} cm/s/V/cm (Marshall, 1949).

The zeta potential, ζ, of a suspension is calculated from the electrophoretic velocity, ν_{el}, using the Helmholtz-Smoluchowski equation

$$\nu_{el} = \epsilon \zeta E / 4 \pi \eta, \tag{2.84}$$

in which E is the applied field strength, and ϵ and η are the dielectric constant and viscosity of the medium, respectively.

3. Electrodialysis is specific to the electrophoresis system. It is carried out by applying an electric field to a colloid solution placed in a siphon, the ends of which are closed by a membrane. The small ordinary ions migrate through the membrane toward the electrodes while the particles that cannot move through the membrane tend to accumulate on the side toward which they are moving. This is a widely used method of purifying colloid systems from electrolytic impurities.

Much research has been devoted to the study of purification of clay minerals by this method. It would be expected that during electrodialysis exchangeable cations would migrate to the cathode and the negative clay particle would react with water molecules, which are in great excess, and become protonated, resulting in the migration of hydroxyls to the anode. The interaction of the negative clay particle with water is described by the following chemical equation:

$$\text{Clay}(-) + H_2O \rightarrow \text{Clay.H} + OH^-. \tag{2.IV}$$

Electrodialysis has frequently been used to prepare H clays for cation exchange study. This procedure must be used with caution since electrodialysis of some clay minerals may cause their decomposition because of their instability in acid solutions. The process of electrodialysis does not involve only the removal of mobile ions (counterions and excess electrolyte) to the cathode and anode chambers, although this predominates at the beginning of the process. Basic constituents, such as iron and magnesium, are frequently found at the anode, while silicic acid moves in considerable quantities with the bulk of the bases to the cathode (Marshall, 1949). Kelley (1935) demonstrated that the replacement of exchangeable ions by H^+ results in the movement of Al from octahedral to exchange positions. Unexchangeable cations, other than Al, are also removed from the interior of the clay crystal. The amount of loss varies with the clay mineral. Magnesium-rich smectite minerals seem to be most susceptible to the removal of unexchangeable cations, a process that is accompanied by a breakdown of structure. Other smectites are quite susceptible to such alteration, but less so than the magnesium-rich smectites. Those rich in aluminum are the most stable smectites while the iron-rich smectites are more susceptible to alteration than the aluminum-rich smectites but still more stable than the magnesium-rich smectites (Caldwell and Marshall, 1942).

A similar sequence of stability was found during the electrodialysis of various micas. Those rich in aluminum (muscovite) are the most stable while those rich in magnesium are the most unstable. Roy (1949) studied the effects of electrodialysis on biotite and found that within a few hours the biotite lost an appreciable number of cations and that 80–90% were lost at the end of 28 days. Despite this great loss the structure of the biotite was not completely destroyed. After a similar length of time, muscovite and phlogopite micas lost only a small amount of their iron and alkalies. There was almost no loss of aluminum.

Attapulgite-sepiolite, vermiculite, chlorite, and some of the biotite micas are quite susceptible to disintegration by electrodialysis while kaolinite, an aluminum-rich clay mineral, is only minimally affected. The stability of the various clay minerals in acid solutions will be further discussed in Chapter 7.

4.2.2 Mechanically Caused Motion

1. Streaming Potential. As is the case with electroosmosis, the solid is not free to move but the liquid can move. The potential arises from forcing a liquid through any capillary system. It occurs in the seepage of water in soil (Ravina and Zaslavsky, 1968) or in streaming of water through porous rocks. The potential difference is directly proportional to the pressure difference that causes the flow, and is independent of the area and thickness of the porous, permeable rock. Swartzendruber and Gairon (1975) observed electric potentials during the upward entry of water into columns of air-dry mixtures of sand and clays.

2. Sedimentation Potential and the Dorn Effect. As with electrophoresis, the solid charged particles are free to move in the dispersing medium. The electric potential arises along the direction of sedimentation during the sedimentation of particles under the force of gravity. A similar effect is obtained from falling drops or from rising bubbles and whenever charged particles are moving under the influence of mechanical forces. This effect was first described by Dorn in 1880 and in its wider application is called the *Dorn effect.* Atmospheric electricity is largely a result of this effect.

4.3 Theory of Electrokinetic Phenomena

4.3.1 Liquid Flow Through a Capillary or a Porous Diaphragm

The liquid adjacent to the walls of a capillary has an excess diffuse electric charge compensated by a corresponding excess of the opposite charge on the capillary wall. If an electric field is applied along the capillary, the charges will tend to migrate within the capillary with respect to its walls, forcing the polarized liquid molecules to flow, thus bringing about electroosmosis. If, however, the liquid flows through the capillary as a result of a pressure difference between the two ends, the counterions are forced to flow with the flowing liquid along the capillary, giving rise to the streaming potential.

For the present treatment let us assume that the liquid charge does not comprise a diffuse layer, but a thin layer with a charge density of $+e_s/\mathrm{cm}^2$ at a very short distance Δ from the solid surface of the capillary, which has the charge density of $-e_s/\mathrm{cm}^2$.

This is electrically equivalent to a condenser with charge e_s, plate distance Δ, and a potential

$$\zeta = 4\,\pi\Delta\, e_s/\epsilon, \tag{2.85}$$

where ϵ is the dielectric constant of the material in the condenser. In an electric field of strength E directed parallel to the interface, each cm^2 of the surface of the liquid will be acted upon by the force tending to move the liquid with respect to the wall

$$f_1 = E e_s. \tag{2.86}$$

The frictional force f_2/cm^2 between the liquid and the wall, is equal to (see Ch. 9)

$$f_2 = \eta\, \overline{U}/\Delta, \tag{2.87}$$

where η is the viscosity of the liquid in the double layer and $\overline{U}/\Delta$ is the velocity gradient perpendicular to the direction of the migration. The frictional force increases with the velocity gradient until force f_1, which causes this velocity gradient, becomes equal to the frictional force. At this stage a constant velocity of the liquid in the capillary will be established. For this *steady-state flow* we obtain

$$E e_s = \eta\, \overline{U}/\Delta. \tag{2.88}$$

According to Eqs. (2.85) and (2.88) the rate of flow of the bulk liquid is given by

$$\overline{U} = (\epsilon\zeta/4\,\pi\eta)E. \tag{2.89}$$

Since the amount ν of the liquid transferred through a capillary of radius r during one second is

$$\nu = \overline{U}\,\pi\, r^2 \tag{2.90}$$

where $\pi\, r^2$ is the cross section of the capillary, then, according to Eq. (2.88)

$$\nu = (\pi\, r^2\;\Delta\, e_s/\eta)\, E \tag{2.91}$$

or, making use of (2.89)

$$\nu = \pi\, r^2\;(\epsilon\zeta/4\,\pi\,\eta)\, E = (\epsilon\zeta r^2/4\,\eta)\, E. \tag{2.91*}$$

Electroosmosis is usually carried out through a porous diaphragm or plug of thickness L with a large number of capillaries, and it occurs simultaneously through this large number of capillaries whose total cross section is $\Sigma\pi r^2$. For a porous diaphragm made of nonconducting material the total cross section can easily be determined from the electric resistance. If the capillaries are filled with a solution of specific conductivity $K = 1/\rho$, the system can be regarded as a conductor, the resistance of which, R, increases with the thickness of the diaphragm and decreases with its cross section, according to the equation

$$R = \frac{\rho L}{\Sigma \pi r^2} = \frac{1}{K} \cdot \frac{L}{\Sigma \pi r^2}. \tag{2.92}$$

According to Ohm's law $R = EL/i$ where EL is the potential difference between both sides of the diaphragm and i is the current intensity. After substituting R for EL/i in Eq. (2.92) we obtain

$$\Sigma \pi r^2 = i/KE. \tag{2.93}$$

Equations (2.91) and (2.91*) become

$$v = \Delta e_s (\Sigma \pi r^2/\eta) E = (\Delta e_s/\eta K)\, i \tag{2.94}$$

and

$$v = (i/KE) \cdot (\epsilon \zeta/4\, \pi \eta) E = (\epsilon \zeta/4\, \pi \eta K)\, i. \tag{2.94*}$$

Near the walls of the capillary the values of ϵ, η, and K are not necessarily the same as in the bulk liquid, because of the excess of surface charge-determining ions. For capillaries that are not extremely narrow, the following relation holds: $K = K_\infty + (S/O)K_S$, where S is the perimeter of the cross section O of the capillary, K_S is the surface conductance, and K_∞ is the bulk conductance. For a cylindrical capillary $S/O = 2\,\pi r/\pi r^2 = 2/r$, so that the correction for the surface conductance $(K - K_\infty)$ decreases with increasing r. The electric current and the electroosmotic stream flow parallel to the capillary walls. Equations (2.94) and (2.94*) may therefore be applied to any arrangement of capillaries in a porous diaphragm and in rocks.

Let us now analyze the streaming potential. If there is a pressure gradient dP/dh along a capillary filled with a liquid, where h is the distance from the end of the capillary, this will result in a flow. The maximum flow velocity is along the capillary axis. The velocity $\overline{U}_x$, decreases with a parabolic profile, reaching the value of zero at the walls of the capillary. At a distance $x < r$ from the capillary axis, where r is the capillary radius, we have

$$\overline{U}_x = (1/4\,\eta) \cdot (dP/dh)\,(r^2 - x^2). \tag{2.95}$$

For a cylindrical capillary at constant section $(dP/dh = P/h)$ the velocity $\overline{U}_\Delta$ at a small distance $\Delta \ll r$ from the wall is given by

$$U_\Delta = \frac{1}{4\,\eta}\,\frac{P}{h}\,[r^2 - (r - \Delta)^2] \approx \frac{1}{2\,\eta}\,\frac{P}{h}\, r\Delta. \tag{2.96}$$

The excess charge of the liquid comprising a thin layer at a distance Δ from the capillary wall will be transported at the same speed, resulting in a convection current. If $e_{s\Delta}$ is the excess charge per one cm^2 of the thin liquid layer at a distance Δ the convection current i_c is given by

$$i_c = 2\,\pi r \overline{U}_\Delta e_{s\Delta}. \tag{2.97}$$

The charges are transferred from one end of the capillary to the other, generating a potential difference Eh, which increases with increasing accumulation of charges, causing an increasing conduction current i through the solution in the capillary of resistance R. When this conduction current becomes equal to the convection current, a constant steady-state value of $Eh = V_{st}$ will be established, where V_{st} is the streaming potential. According to Ohm's law

$$i = V_{st}/R = V_{st}(K\pi r^2/h) \tag{2.98}$$

where $R = (1/K) \cdot (h/\pi r^2)$. When i_c is equal to i we obtain from Eqs. (2.97) and (2.98)

$$V_{st}K\,(\pi r^2/h) = 2\,\pi r \overline{U}_\Delta\, e_{s\Delta}. \tag{2.99}$$

From Eqs. (2.85), (2.96), and (2.99) we obtain for the streaming potential

$$V_{st} = (\Delta e_{s\Delta}/\eta K)\,P = (\epsilon\zeta/4\,\pi\eta K)\,P. \tag{2.100}$$

The dimensions of the capillary do not enter into Eq. (2.100). This equation is valid also for porous diaphragms independently of the shape of their capillaries, as long as the same conditions are fulfilled for which Eq. (2.100) has been derived, i.e., small thickness of the double layer with respect to the diameter of the capillaries, and the absence of surface conductance of any significance with respect to the bulk liquid conductance in the capillaries. The reasons for these requirements are the same as those for Eqs. (2.94) and (2.94*). The dependence of K on pore or capillary radius also remains valid.

In many rocks and soils these conditions are often not fulfilled. In the region of the double layer both viscosity and dielectric constants may change owing to the strong electric field in the vicinity of the surface. A decrease in the dielectric constant and an increase in the viscosity occur in the double layer. Such changes lead to a reduction of the streaming potential. In compacted sediments the capillaries are so narrow that the double layer is distorted by the curvature of the wall and the streaming potential becomes smaller than expected. If surface conductance is high the streaming potential further decreases. Apparently, according to van Olphen (1963), taking into account all refinements leads to smaller streaming potentials than those predicted by Eq. (2.100). It would seem that streaming potentials in porous formations are low.

4.3.2 Electrokinetic Phenomena of Dispersed Systems

The mechanism of electrophoresis and sedimentation potential is analogous to that of electroosmosis and streaming potential, with the difference that in the former case the liquid phase remains stationary and the particles are displaced owing to their small size and consequent high mobility. It would be useful to define a magnitude that may characterize the mobility of a colloid particle during electrophoresis, the *electrophoretic mobility*, $u_e = u/E$, where E is the applied electric field and u is the actual migration velocity of the particles. An equation analogous to (2.88) is used to describe the relationship between the applied electric field and the migration velocity of the particles. According to Eqs. (2.85) and (2.88) the electrophoretic mobility is given by

$$u_e = e_s \Delta / \eta = \epsilon \zeta / 4 \pi \eta. \tag{2.101}$$

This magnitude is analogous to the ionic mobility in an electric field. Experimentally it was found that the electrophoretic velocity depends on the thickness of the diffuse double layer $1/\gamma$ and on the radius of the particle as well as on the geometric shape of the particle.

Sedimentation potential E_{sed} can be calculated as follows. If the charge of a colloid particle is q and the number of such particles per cm^3 is c, then the convection current per cm^2 of a cross section normal to the direction of sedimentation will be

$$I_c = quc. \tag{2.102}$$

As a result of the transfer of the charges, a conduction current of a density I is generated. When this conduction current becomes equal to the convection current a steady state is obtained, and

$$I = E_{sed} d/R, \tag{2.103}$$

where $E_{sed}d$ is the steady-state potential difference between electrodes separated vertically by a distance d, and $R = d/K$ is the resistance.

Under steady-state conditions the electric force qE acting on the particle should be equal to the frictional force $u_e E 6\pi\eta r$, which arises when the particle moves at a speed $u = u_e E$, where r is the particle radius. Since $qE = u_e E 6\pi\eta r$ and after substituting the value of u_e from Eq. (2.101) there will be obtained

$$q = (3/2)\, \epsilon\, \zeta\, r. \tag{2.104}$$

As will be shown in Chapter 5, the sedimentation rate of a spherical particle is given by $u = (2/9) \cdot (r^2/\eta)\,(\rho - \rho_0)\, g$, where ρ and ρ_0 are the densities of the disperse phase and dispersing medium, respectively, and g is the acceleration due to gravity. Equating I with I_c (Eqs. 2.102 and 2.103) gives

$$E_{sed} d/R = quc \tag{2.105}$$

and substituting the values of u, R, and q (Eq. 2.104) gives

$$E_{sed} = (\epsilon\zeta/3\, \eta K)\, r^3\, (\rho - \rho_0)\, c\, g. \tag{2.106}$$

Refinements to this equation have been made by several investigators (see Sheludko, 1966) to deal with solid particles moving in an electrolyte solution and with the upward movement of gaseous bubbles. In these cases the effect of ionic diffusion is associated with the electrokinetic phenomena and must be taken into account.

4.4 Electric Properties of Dust Storms

Dust storms are electrically active and produce remarkable electric perturbations in the electric state of the atmosphere. Negative values of atmospheric electric potential gradients of -10 kV m^{-1} were found by Rudge (1914) in his measurements in South Africa. Positive potential gradients of up to 15 kV m^{-1} were found in the Northern Sahara in Algeria by Demon *et al.* (1953). Negative potential gradients of up to -600 V m^{-1} have been observed by Uchikawa (1951) in dust storms in Japan. Harris (1967, 1969) observed negative potential gradients of up to -5 kV m^{-1} and negative air–earth currents associated with the Harmattan dust haze, which extends over most of West Africa each year between November and March.

The mechanism of charge generation in dust storms seems to be closely associated with the dispersion of dust into the atmosphere. The net electrification may be further modified by the subsequent impact of dust particles as they are blown along the ground. Most of the dust storms whose major constituents are clay minerals produce only negative space charge and negative potential gradient at the earth's surface. Sand storms or dust storms whose major dust constituent is silica produce particles having both types of space charge and potential gradient at ground level.

Charging of particles by collisions results from proton or electron transfer (see Ch. 4). Ionization due to cosmic rays and radioactivity in the atmosphere will generate bipolar ions. An equilibrium charge distribution will develop in the aerosol as a result of the random thermal motion of the ions and the frequent collisions between ions and particles. The size distribution of the dust particles and the charge distribution on them are important factors in the mechanism of charge generation and its transportation in the dust storms. Transportation of charge in dust storms can occur by both convection and conduction processes, that is, migration of charged particles and migration of ions and electrons. The relative efficiency of the two processes is determined by the nature of the charge carriers and the charge distribution over dust particles of different sizes.

Intense electrification has been observed in dust devils.[9] According to Freier (1960), who studied the potential gradient due to a dust devil in the Sahara, there should be negative charge up and positive charge down in dust devils. According to Crozier (1970) the negative space charge densities in the dust devils range from 6×10^5 to 9×10^6 $\bar{e}/cm^3$, where $\bar{e}$ is the electron charge.

According to Kamra (1972) there are two factors that affect the atmospheric electric parameters at ground levels during a dust storm: (1) the turbulent component and (2) the background component. The turbulent component is generally associated with wind gusts that raise excessive amounts of local dust into the lower atmosphere. Such puffs and whirls of dust caused by wind gusts often produce strong electric perturbations at ground levels. These electric perturbations last for only a few minutes, and their frequency depends on the wind velocity. Kamra (1969a, b) observed that

9 As a consequence of strong heating of the near-surface air, destabilization can result in the formation of ***dust devils,*** which are swirling plumes of rising dusty air, and ***wild dusty winds,*** which are larger local dust movements. These phenomena are small in scale and short-lived. The dust tends to settle out in the surrounding environment.

with increasing wind velocity, there was a systematic deviation of the potential gradient and its electric agitation from their fair weather values. Whenever the wind begins to blow and raises appreciable amounts of dust into the atmosphere, the potential gradient begins to decrease and within a few minutes becomes negative. If the strong winds continue, the potential gradient remains negative. It attains its fair-weather value only when the winds subside. Negative potential gradients of many thousands of volts per meter are common whenever the winds are sufficiently strong.

The background component results from the widely scattered dust carried to high altitudes and is a function of the average wind velocity and the convection in the atmosphere. The electric charge needed to create steady deviations of electric parameters need not be generated locally but can be brought from great distances, provided the rate of charge production and its transportation are strong enough to prevent the electric charge in the dust cloud from being neutralized because of the limited relaxation time of the atmosphere. This effect is apparently what happens during the Harmattan season in West Africa, whereby large negative potential gradients occur even in the absence of strong winds, during the Harmattan dust haze (Ette, 1971; Harris, 1969).

In a widespread dust storm, the finer particles are carried to high altitudes, whereas a wide spectrum of particle sizes can be observed close to the ground. The electric charge being transported to high altitudes by these fine particles or by air will affect the electric field on the ground but may or may not significantly affect the space charge at altitudes close to the ground. Although the clouds of dust raised by wind gusts produce strong turbulent perturbations in potential gradient and space charge, there seems to be a steady background effect of the general dustiness of the atmosphere that tends to keep the potential gradient of the atmosphere negative.

Measurements by Kamra (1972) over sand dunes indicate a positive dipole type of charge distribution in sand storms. This positive dipole, with positive charge up and negative charge down, fluctuates vertically with varying wind velocity and has its center at a height of about one meter. At low wind velocities space charge at the one-meter level is overwhelmingly positive, and the corresponding negative charge of the dipole remains below the one meter level. Hence fluctuations are produced in the potential gradient only. At higher wind velocities both positive and negative space charge regions of the dipole are raised such that the mixed region of the two space charges fluctuates up and down at about the one-meter level. Under such conditions the space charge of the one-meter level fluctuates around zero, but the potential gradient remains predominantly positive.

References

Anderson, A. T.: Chlorine, sulfur and water in magmas and oceans. Geol. Soc. Am. Bull. *85*, 1485–1492 (1974)

Atkinson, L. P.: Effect of air bubble solution on air-sea gas exchange. J. Geophys. Res. *78*, 962–968 (1973)

Babcock, K. L.: Theory of the chemical properties of soil colloidal systems at equilibrium. Hilgardia *34*, 417–542 (1963)

Bell, G. M., Levine, P. L.: Diffuse layer effects due to adsorbed ions and inner layer polarizability. J. Colloid. Interface Sci. *41*, 275–286 (1972)

Bolt, G. H.: Cation exchange equations used in soil science–a review. Neth. J. Agric. Sci. *15*, 81–103 (1967)
Brunauer, S.: Physical adsorption of gases and vapors. Oxford: University Press 1944
Burnham, C. W.: Water and magmas, a mixing model. Geochim. Cosmochim. Acta *39*, 1077–1084 (1975)
Burnham, C. W., Jahns, R. H.: A method for determining the solubility of water in silica melts. Am. J. Sci. *260*, 721–745 (1962)
Caldwell, O. G., Marshall, C. E.: A study of some chemical and physical properties of the clay minerals nontronite, attapulgite and saponite. Missouri Univ., Coll. Agr. Res. Bull., No. 354 (1942)
Carmichael, I. S. E., Turner, F. J., Verhoogen, J.: Igneous petrology. New York: McGraw-Hill 1974
Chapman, D. L.: A contribution to the theory of electrocapillarity. Philos. Mag. *25*, 475–481 (1913)
Chen, Y., Schnitzer, M.: The surface tension of aqueous solutions in soil humic substances. Soil Sci. *125*, 7–15 (1978)
Childs, E. C.: An introduction to the physical basis of soil water phenomena. New York: John Wiley 1969
Clark, W. E.: Prediction of ultrafiltration membrane performance. Science *138*, 148–149 (1962)
Craig, H., Weiss, R. F.: Dissolved gas saturation anomalies and excess helium in the ocean. Earth Planet. Sci. Lett. *10*, 289–296 (1971)
Crozier, W. D.: Dust devil properties. J. Geophys. Res. *75*, 4583–4585 (1970)
Dean, R. B.: Modern colloids. New York: D. van Nostrand Co. 1948
De Bano, L. F.: Water movement in water repellent soils. In: Proc. Symp. Water-repellent soils, held at the Univ. Calif., Riverside, on May 6-10, 1968, and quoted by Mansel, 1969 (1969)
Demon, L., DeFelici, P., Gondet, H., Kast, Y., Pontier, L.: Premiers résultates obtenus au cours du printemps, J. Rech. Cent. Nat. Rech. Sci., *24*, 126–137 (1953)
Eagland, D.: The influence of hydration on the stability of hydrophobic colloid systems. In: Water, a comprehensive treatise. Franks, F. (ed.). New York: Plenum Press, 1975, Vol. 5, pp. 1–74
Ette, A. I. I.: The effect of the Harmattan dust on atmospheric electric parameters. J. Atmos. Terr. Phys. *33*, 295–300 (1971)
Freier, G. D.: The electric field of a large dust devil. J. Geophys. Res. *65*, 3504–3508 (1960)
Gaines, G. L., Thomas, H. C.: Adsorption studies on clay material. II. A formulation of the thermodynamics of exchange adsorption. J. Chem. Phys. *21*, 714–718 (1953)
Gaines, G. L., Thomas, H. C.: Adsorption studies on clay minerals. V. Montmorillonite-cesium-strontium at several temperatures. J. Chem. Phys. *23*, 2322–2326 (1955)
Gibbs, W.: The collected works. New York: Longmans, Green and Co., 1928, Vol. 1
Gouy, G.: Sur la constitution de la charge électrique à la surface d'un électrolyte. Ann. Phys. (Paris), Série 4 *9*, 457–468 (1910)
Harris, D. J.: Electrical effects on the Harmattan dust storms. Nature *214*, 583 (1967)
Harris, D. J.: Atmospheric electric field measurements during the Harmattan dust haze in Northern Nigeria. In: Planetary electrodynamics. Coronity, S. C., Hughes, J. (eds.). London: Gordon and Beach, 1969, Vol. 1, pp. 39–46
Hergenhan, H.: Hydrophobization of a soil layer to interrupt capillary water movement. Arch. Acker. Pflanzenbau Bodenkd. *16*, 399–409 (1972)
Hollis, J. M., Jones, R. J. A., Palmer, R. C.: The effect of organic matter and particle size on the water-retention properties of some soils in the west midlands of England. Geoderma *17*, 225–238 (1977)
Holloway, J. R.: Fluids in the evolution of granitic magmas. Consequences of finite CO_2 solubility. Geol. Soc. Am. Bull. *87*, 1513–1518 (1976)
Jamison, V. C.: The penetration of irrigation and rain water into sandy soils in Central Florida. Soil Sci. Soc. Am. Proc. *10*, 25–29 (1946)
Jasper, J. J.: Measurement of surface and interfacial tension. In: Treatise on analytical chemistry. Part I. Theory and practice. Kolthoff, I. M., Elving, P. J. (eds.), New York: Interscience Publishers, 1967, Vol. 7, pp. 4611–4760

Kadik, A. A., Lukanin, O. A.: The solubility dependent behavior of water and carbon dioxide in magmatic processes. Geochem. Intern. 115–129, (Trans. from Geokhimiya, No. *2,* 163–179), (1973)
Kamra, A. K.: Short term variations in atmospheric electric potential gradient. J. Atmos. Terr. Phys. *31,* 1273–1279 (1969a)
Kamra, A. K.: Effect of wind on diurnal and seasonal variations of atmospheric electric field. J. Atmos. Terr. Phys. *31,* 1281–1286 (1969b)
Kamra, A. K.: Measurements of electrical properties of dust storms. J. Geophys. Res. *77,* 5856–5869 (1972)
Kelley, W. P.: The agronomic importance of calcium. Soil Sci. *40,* 103–109 (1935)
Koster van Groos, A. F., Wyllie, P. J.: Melting relationship in the system $NaAlSi_3O_8$–NaCl–HOH at one kilobar pressure with petrological applications. J. Geol. *77,* 581–605 (1969)
Kracek, F. C.: Ann. Rept. Geophys. Lab. (Carnegie Inst. Yrbk. 32), pp. 61–63, (Quoted by Carmichael et al. 1974 (1933)
Kushiro, I.: On the nature of silicate melt and its significance in magma genesis: regularities in the shift of liquidus boundaries involving olivine, pyroxene, and silica minerals. Am. J. Sci. *275,* 411–431 (1975)
Langmuir, I., Schaefer, V. J.: Rate of evaporation of water through compressed monolayers on water. J. Franklin Inst. *235,* 119–162 (1943)
Laudelout, H.: Cation exchange in soils. S. C. I. Monograph No. 37, Sorption and transport process in soils, pp. 33–39 (1972)
Liss, P. S.: Exchange of SO_2 between the atmosphere and natural waters. Nature (London) *233,* 327–329 (1971)
Liss, P. S.: Processes of gas exchange across an air-water interface. Deep-Sea res. *20,* 221–238 (1973)
Lou, Y. S., Rasmussen, G. P.: Evaporation retardation by monomolecular layers. Water Resour. Res. *9,* 1258–1263 (1973)
Macdonald, G. A.: Volcanoes, Englewood Cliffs, New Jersey: Prentice-Hall 1972
Mansell, R. S.: Infiltration of water into soil columns which have a water-repellent layer. Soil Crop Sci. Florida Proc. *29,* 92–102 (1969)
Marshall, C. E.: The colloid chemistry of the silicate minerals. New York: Academic Press 1949
Mason, B., Berry, L. G.: Elements of minerology. San Francisco: W. H. Freeman and Co. 1968
McBain, J. W.: Colloid science. Boston: Heath 1950
McBirney, A. R., Murase, T.: Factors governing the formation of pyroclastic rocks. Bull. Volcanol. *34,* 372–384 (1971)
Morel-Seytoux, H. J., Khanji, J.: Derivation of an equation of infiltration. Water Resour. Res. *10,* 795–800 (1974)
Mysen, B. O.: The role of volatiles in silicate melts: solubility of carbon dioxide and water in feldspar, pyroxene and feldspatoid melts to 30 KB and 1625 °C. Am. J. Sci. *276,* 969–996 (1976)
Nielsen, D. R., Vachaud, G.: Infiltration of water into vertical and horizontal soil columns. J. Indian Soc. Soil Sci. *13,* 15–23 (1965)
Petkanchin, I., Brückner, R.: Orientation mechanism of palygorskite and kaolinite derived from the electric reversing pulse technique. Colloid Polymer Sci. *254,* 433–435 (1976)
Philpotts, A. R.: Origin of certain iron-titanium oxide and apatite rocks. Econ. Geol. *62,* 303–315 (1967)
Ravina, I., Zaslavsky, D.: Nonlinear electrokinetic phenomena. I. Review of literature. Soil Sci. *106,* 60–66 (1968)
Remy, H.: Treatise on inorganic chemistry (translated by Anderson, J. S., Kleinberg, J.). Amsterdam: Elsevier 1956
Richards, L. A.: Report on the subcommitte on permeability and infiltration of the committee on terminology of the soil science. Soil Sci. Soc. Am. Proc. *16,* 85–88 (1952)
Roedder, E.: Low temperature immiscibility in the system K_2O–FeO–Al_2O_3–SiO_2. Am. Mineral. *36,* 282–286 (1951)
Roy, R.: Decomposition and resynthesis of micas. J. Am. Ceram. Soc. *32,* 202–210 (1949)

Rudge, W. A. D.: On some sources of disturbance of the normal atmospheric potential gradient. Proc. R. Soc. London, Ser. A *90,* 571–582 (1914)

Sheludko, A.: Colloid chemistry. Amsterdam: Elsevier 1966

Sonntag, H., Strenge, K.: Coagulation and stability of disperse systems. Konder, R. (translator). Jerusalem: Israel Program for Scientific Translations 1972

Stern, O.: Zur Theorie der elektrolytischen Doppelschicht. Z. Elektrochem. *30,* 508–516 (1924)

Swartzendruber, D., Gairon, S.: Electrical potentials during water entry into an air-dried mixture of sand and kaolinite. Soil Sci. *120,* 407–411 (1975)

Thomas, H. C.: The thermodynamics of ion exchange on colloidal materials with applications to silicate minerals. Techn. Rept. U. S. Atomic Energy Commission Div. Reactor Dev. Technol. Univ. N. Carolina, Chapel Hill (1967)

Uchikawa, K.: The disturbance of the electric field in the atmosphere at dust storm, Tateno. J. Aer. Observ. *5,* 10–15 (1951)

van Olphen, H.: An introduction to clay colloid chemistry. New York: Interscience Pub. 1963

Vansant, E. P.: Uytterhoeven, J. B.: Thermodynamics of the exchange of n-alkylammonium ions of Na-montmorillonite. Clays Clay Miner. *20,* 47–54 (1972)

Weisman, R. N., Brutsaert, W.: Evaporation and cooling of a lake under unstable atmospheric conditions. Water resour. Res. *9,* 1242–1257 (1973)

Wu, J.: Evaporation due to spray. J. Geophys. Res. *79,* 4107–4109 (1974)

Chapter 3
Formation of Aqueous Solutions and Suspensions of Hydrophobic Colloids

The two general ways in which suspensions are formed are by building up of particles from molecular or atomic units, or by breaking down particles of macroscopic dimensions, termed respectively condensation and dispersion. Both phenomena will be discussed in this chapter.

1. Condensation Process (Formation of New Phases)

In molecular or ionic solutions, conditions can be created in which individual molecules or ions combine to form particles of colloid dimensions. This requires as a first step the formation of a supersaturated solution. The supersaturation is then relieved by the formation of *nuclei,* or by condensation upon nuclei already present, followed by growth to larger particles. The colloid solution is stable as long as the particles are not too large.

A solution is defined as supersaturated when it contains a higher concentration of solute than that which can remain at equilibrium with the precipitate. The degree of supersaturation is expressed by the *supersaturation ratio, S. R.,* which is defined as the square root of the ratio of the ion activity products in the supersaturated state, K^0_{ss}, to those in the saturated state, K^0_{sp}, and is given by

$$S.\,R. = (K^0_{ss}/K^0_{sp})^{1/2} \tag{3.1}$$

The precipitation of ionic crystals occurs by formation of ion clusters that grow into nuclei, and finally into macroscopic crystals. First, ions come together to give clusters of gradually increasing size. In the vapor state such clusters are more stable than the separate ions. This is not true for aqueous solutions, in which ions cause the formation of stable water structures and the dielectric constant becomes high. Clusters are therefore more prone to break apart than they are to remain whole. Only a few exist long enough to combine into crystal nuclei.

Nuclei that are larger than clusters may contain hundreds or thousands of ions. Though still unstable, they have a good chance of growing into macroscopic crystals. Nuclei are formed at supersaturation ratio of a value higher than a certain critical value, which may vary for different substances, and to effect nucleation, it is necessary that the solution be supersaturated. However, it is known that solutions are not absolutely homogeneous but have certain concentration fluctuations of the solute with local supersaturations, enabling the formation of ionic or molecular crystal embryos. When the environmental conditions are such that the equilibrium is shifted in the direction of the formation of these embryos of a certain critical size, crystal nuclei are formed and crystal growth begins. The crystal size may vary somewhat for different solutes.

The degree of nucleation increases very rapidly with supersaturation. When the supersaturation ratio is very small, only a few nuclei are formed, and hence the solutions are rather stable, precipitation often requiring several days or months. As the ratio increases, the stability of the solution decreases. Finally, at the critical supersaturation ratio the spontaneous formation of crystalline particles results.

In the presence of impurities that act as seeds, particularly if the impurity is isomorphous with the compound to be precipitated, and that may provide a surface on which the precipitated ions can deposit, nucleation may occur at supersaturation ratios lower than the critical ratio.

Temperature will also affect the value of the critical supersaturation ratio. This ratio decreases with rising temperature probably because of an increase in the number of collisions of atoms or molecules per unit concentration.

The critical supersaturation ratio decreases when a system is subjected to turbulent effects, because of the probability of increasing concentration fluctuations in the form of some minute volumes of the solution with very high concentrations. Nucleation can be induced by seeding with crystallites of the substance.

The rate of formation of the nuclei, V_1, may be expressed by

$$V_1 = k_1' \exp\left[-\frac{k_1''}{(\ln C/C_0)^2}\right] \tag{3.2}$$

where k_1' and k_1'' are constants, C is the total concentration of the precipitating substance, and C_0 is its equilibrium solubility, hence C/C_0 denotes the supersaturation ratio.

Nucleation is ultimately superseded by growth of existing nuclei to macroscopic size. The rate of particle growth is determined by diffusion of substance from supersaturated volumes within the bulk toward those particles that are surrounded by very thin envelopes of solution in equilibrium with the solid. Hence, it may be expressed in analogy to a diffusion equation by

$$V_2 = k_2(C - C_0)/C_0 \tag{3.3}$$

where V_2 is the rate, k_2 is the proportionality constant, and $(C - C_0)$ denotes the initial supersaturation.

The nature of the initial disperse particles depends upon the *aggregation velocity* and the *orientation velocity*. The former is a function of the degree of supersaturation, whereas the latter is dependent on the characteristics of the substance being precipitated. Strongly polar compounds, such as calcite or gypsum, have a high orientation velocity. When the aggregation velocity greatly exceeds the orientation velocity, amorphous precipitates such as hydrous oxides and sulfides are formed, and conversion to a crystalline modification is extremely slow. When the orientation velocity is high compared to the aggregation velocity, crystalline particles are easily formed.

Growth by orientation is a slow complex process occurring when supersaturation is low. Ion pairs do not deposit at random on the surface of the crystal. In slow growth each face of the crystal grows a layer at a time. The greatest energy is released when an ion-pair deposits on the edge of a growing layer. It may come directly from the solution, by diffusion along the surface of the crystal, or from an adsorbed layer of hydrated ions. Initiation of a new layer is more difficult than completion of an old one and begins only with the creation of a nucleus on top of the old layer. By building layer on layer the crystalline particle gradually grows to macroscopic size.

When supersaturation is very high, growth occurs in a haphazard, less systematic way. The particles are made up of less regular crystals and may often be amorphous. In rapid growth foreign ions momentarily adsorbed on the surface may be covered over by the next crystal layer. Droplets of the solution may even be trapped in the crystal. Some of these faults are eliminated as the particle ages. Amorphous particles become more crystalline on standing. This is often associated with a decrease in solubility and stability of the suspended system.

The formation of highly dispersed particles is assisted by increase in the concentrations of the reactants, in other words, by supersaturation of the solution. In such a case the rate of formation of the nuclei increases faster than the growth rate of the particles formed. This can be seen from Figure 3.1, which graphically represents Eqs. (3.2) and (3.3). The nucleation rate may be so fast that most of the excess dissolved mass is precipitated as critical nuclei, leaving little material for further growth. If, by chance, a minute volume of very high supersaturation persists after the spontaneous nucleation, nuclei in the vicinity of this quantity will grow very fast. To sum up, some of the particles condensed from a highly supersaturated solution will be of a huge size and some of extremely small size. A colloid system having particles of various sizes is called a *polydisperse system.*

Crystallization nuclei and small crystallites (primary particles) combine together to form an aggregate (a secondary particle). The secondary particle may also contain liquid droplets. The mechanism of aggregation will be discussed in more detail in Chapter 8. An increase in the number of crystallization nuclei and microcrystals brings about an increase in the rate of particle aggregation and growth of aggregates. Large aggregates do not form stable disperse systems. Therefore, for each reaction associated

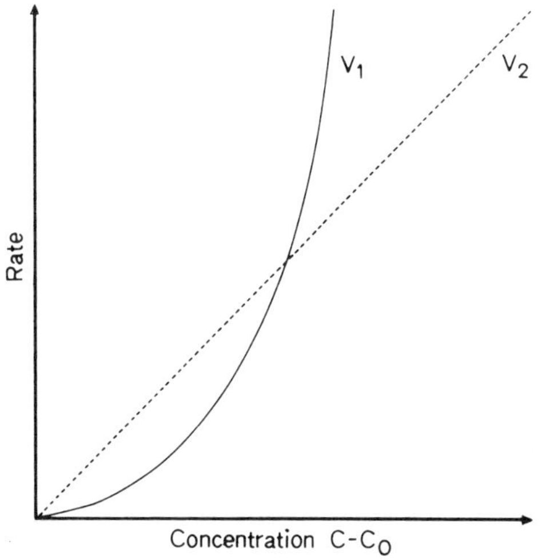

Fig. 3.1. The rate of nuclei formation, V_1, and the rate of particle growth, V_2, versus supersaturation, $C - C_0$.

with the formation of a new disperse phase, there exist certain optimum conditions for obtaining the colloid solution.

Condensation is a common phenomenon in geologic systems. Dissolved metallic cations occur in the hydrospheric constituents either in the form of hydrated charged ions or complexed with organic or inorganic ligands. An increase in the concentration of anions such as silicate, phosphate, or carbonate or a change in the pH of the solution may result in the formation of an insoluble species. A change in the Eh of the system (oxidation–reduction potential) may lead to the decomposition of the organic ligand and the oxidation of the metallic cation with the formation of an insoluble species. If the formation of the new solid phase occurs in a dynamic system, the growth of crystalline particles may be prevented, and a colloid dispersion is formed. The particles will settle only when the colloid dispersion reaches a basin suitable for aggregation and sedimentation of the particles.

The best known condensation process occurring in the hydrosphere is the hydrolysis of polyvalent cations and the formation of sparingly soluble hydroxides and hydrous oxides. The solubility of the metallic cation at any pH value can be calculated from the appropriate solubility product. In Figure 3.2 the solubilities of the more important components of geologic systems, in the absence of precipitating or complex forming anions other than OH^-, are plotted against pH.

In nature, observed pH values lie mostly between 4 and 9 (Krauskopf, 1967). Streams in humid regions generally show values between 5 and 6.5, in arid regions between 7 and 8. Soil water, especially if decaying vegetation is abundant, may have pHs down to 4 or a little lower. Ocean water normally shows a pH range of 8.1 to 8.3. Soil water and playa-lake water in desert regions may have pHs of 9 or even higher. The lowest recorded pHs in nature are found in solutions in contact with oxidizing

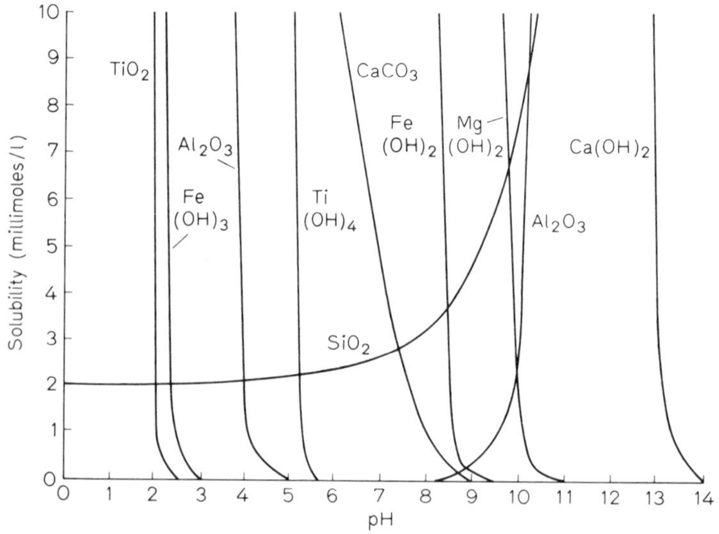

Fig. 3.2. Solubility in relation to pH for some components released by chemical weathering. (From Loughnan, 1969. Used with permission of American Elsevier Publishing Company, Inc.)

pyrite; values even less than zero have been recorded in such environments (Mason, 1952).

Clay minerals may be precipitated directly from solutions. This process is defined as *neoformation* (Keller, 1952; Tarr and Keller, 1937). According to Millot (1970) neoformation takes place to a considerable extent in basin areas where the neighboring land area has undergone intense weathering. The material supplied to the basin is largely in solution with little or no detrital material. Smectite, palygorskite, and sepiolite are believed to be precipitated under such conditions, with the relative Mg, Si, and Al content of the waters determining which of these clay minerals is formed.

The geochemistry of neoformation of clay minerals may throw light on the mechanism of clay mineral formation during weathering of rock-forming silicate, which to this time has not been clarified. Unfortunately, the theory of this process and its experimental investigations are far from complete.

1.1 Chemistry of Aluminum in Natural Waters

Aluminum is an abundant constituent of rock and soil minerals. However, it is a very minor constituent of natural waters, occurring in concentrations of less than 0.1 mg/l because of its low solubility. High concentrations occur rarely and usually are associated with waters having a low pH.

Al ion is amphoteric, so that when the pH of an acidic solution of this ion is raised, a precipitate of the composition $Al(OH)_3.xH_2O$ is formed, and if the pH is raised further, this precipitate is redissolved, forming a complexed Al anion, in which the ratio between OH^- and Al^{+3} is greater than three. These amphoteric properties manifest themselves in a wide variety of hydroxy complexes that may be present in varying proportions in an aqueous solution depending upon the pH. These include Al^{+3}, $AlOH^{+2}$, $Al_2(OH)_2^{+4}$, $Al(OH)_2^+$, and $Al(OH)_4^-$, all hydrated. The proportion of each complexed ion was calculated from thermodynamic data of chemical equilibria by Huang and Keller (1972) and is shown in Figure 3.3a–c.

The hydrogeochemistry of Al is determined by the high *polarizing power* of this ion, which is related to the ratio *e*/*r* between *e*, the charge, and *r*, the radius of the "bare" Al ion. This ion is a water structure maker. When dissolved in water it is surrounded by a tightly bound shell of oriented water molecules. The aluminum lies at a center of an octahedron, at each vertex of which is a water molecule. The coordination number is six, at least for solutions of which the pH is near or below neutrality. There is a tendency towards a four coordination arrangement in alkaline solutions.

The coordinated water molecules are polarized and consequently a hydrolysis of the Al occurs resulting in a decrease in the pH of the solution according to the equations

$$Al^{3+} + 2\,H_2O \rightleftharpoons Al(OH)^{2+} + H_3O^+ \qquad pK_{11} = 4.89$$
$$Al^{3+} + 4\,H_2O \rightleftharpoons Al(OH)_2^+ + 2\,H_3O^+ \qquad pK_{12} = 10.32$$
$$Al^{3+} + 6\,H_2O \rightleftharpoons Al(OH)_3 + 3\,H_3O^+. \qquad pK_{13} = 16.28.$$

This process leads to the hydrolytic precipitation of Al and the formation of a colloid solution, the mechanism of which will be discussed later.

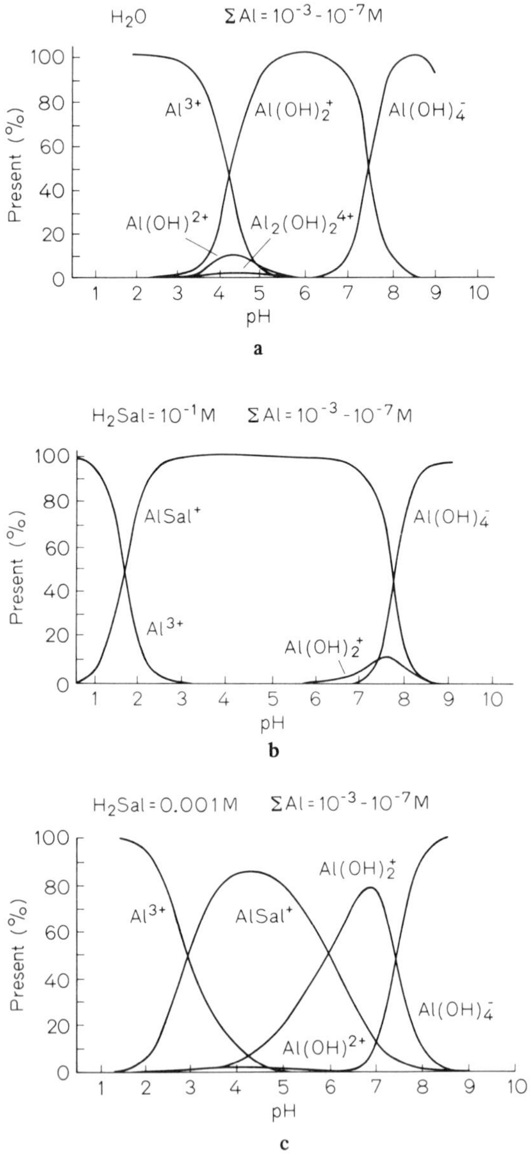

Fig. 3.3a–c. Distribution of Al ionic species (all hydrated) from an Al-hydrate mineral in **a** distilled water, **b** 0.1 M, and **c** 0.001 M solutions of salicyclic acid. Total Al ranges between 10^{-3} and 10^{-7} M. (From Huang and Keller, 1972)

Water in well-drained soil horizons containing abundant organic matter, such as the podsols, may attain pH values below 4. Under these conditions, Al is soluble and may migrate down the soil profile to less acid regions where precipitation and enrichment take place. Lateritic soil, despite the fact that it is well leached and contains clay minerals similar to those of the podsols, never reaches such a low pH because of the low organic content. The low acidity does not allow the Al ion to migrate (Loughnan, 1969). This is well illustrated in Figure 3.4, in which the silica:alumina ratios for the clay colloids from a podsol and a lateritic soil are plotted against depth.

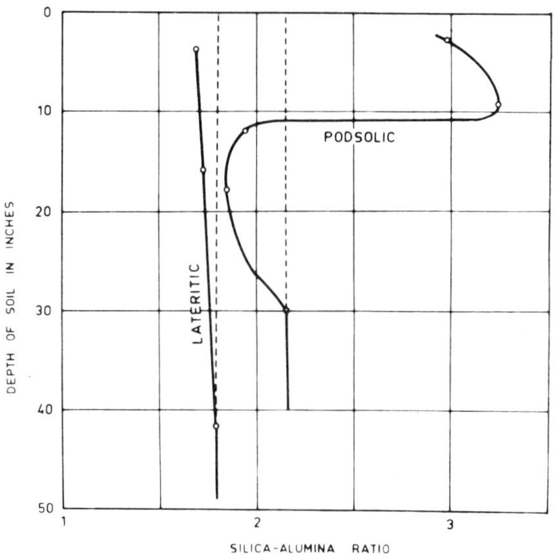

Fig. 3.4. SiO_2 to Al_2O_3 ratios for a lateritic and a podsolic soil. (From Loughnan, 1969: Used with permission of American Elsevier Publishing Company, Inc.)

The geochemical behavior of Al in aqueous systems is strongly influenced by the tendency of this ion to form soluble complex ionic species with various organic ligands. Huang and Keller (1972) showed by laboratory experiments that significantly large amounts of Al become soluble in the presence of dilute organic acids at room temperature. Salicyclic acid (3.I) is one of the acids they studied. This acid and many of its derivatives are present in the volatile oil of the blossoms and leaves of various plants. When this acid interacts with Al, a soluble cationic complex (3.II) is obtained that is stable in the pH range 3 to 7.5. The high stability of the complex results from the

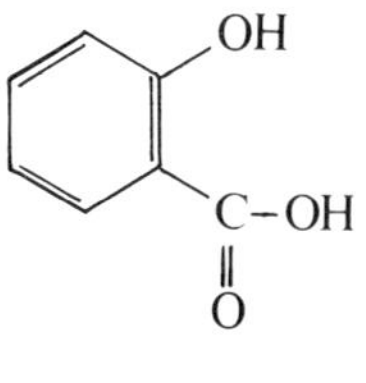

(3.I) H_2Sal

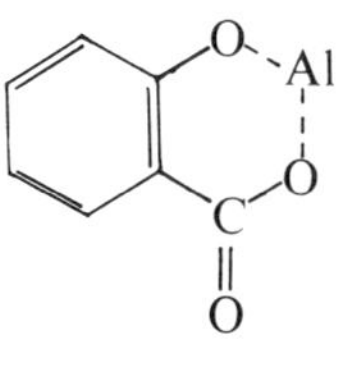

(3.II) $[Al\ Sal]^+$

six-member ring formed between Al and the two functional groups of the salicylic acid. Such a structure is defined as *chelated* and a five or six-member ring is defined as a *chelate.* The distribution of Al ionic species in the absence and presence of salicylic acid is given in Figure 3.3. Huang and Keller also examined Al solubility in the presence of other organic compounds that represent amino- aliphatic and aromatic acids present in humic acids and found that the presence of these acids can greatly increase the solubility of Al to 70–85 mg/l.

Al forms a six-coordinated series of soluble complexes with F ions. Although fluoride is a minor constituent of most natural waters, the complexing action is strong enough to have considerable influence on the form of dissolved Al even when very little F is present (Hem, 1968a). In natural water containing 2 mg/l of F, Al would probably be present mostly as AlF_2^+ and AlF_3. Nearly all the Al would be complexed even when only a few tenths of a milligram per liter of F are present, provided that the total Al is always considerably less than the total F. In many natural waters, however, Al content is relatively large compared to the F content. Hem examined the effect of F ions on the concentration of Al in two samples of natural waters collected from Mammoth Spring and Big Horn Spring in Wyoming, with pH values of 6.6 and 6.2, and found Al concentrations of 0.2 and 0.4 and F concentrations of 2.4 and 3.5 mg/l respectively. Both of these waters contain moderate amounts of silica, resulting in a low activity of free Al. He concluded that Al present in these waters is in the form of the undissociated complex AlF_3, and is therefore in concentrations higher than those expected for free Al in the presence of silica, and increases with increasing concentrations of F.

Two sulfate complexes of aluminum, $AlSO_4^+$ and $Al(SO_4)_2^-$, may occur. The complexing effect of sulfate is much weaker than that of fluoride. Aluminum sulfate complex species will not predominate except in low pH waters with high sulfate content. This may occur in regions where sulfide minerals are oxidized. In systems where sulfate complexing is potentially important, there rarely will be enough dissolved aluminum to use up more than a small fraction of the available sulfate (Hem, 1968a).

According to Hem (1968b), in addition to the monomeric ions of aluminum, natural waters contain polymerized hydroxide aggregates of colloid or subcolloid size. The occurrence of polymerized hydroxides will depend on the past history of the water and on kinetics rather than on equilibrium conditions. Some standard procedures for determination of aluminum are insensitive to the hydroxide polymer, and many published analyses probably reflect this fact.

Soil solutions contain soluble polymeric hydroxy Al ions, which behave like normal exchangeable cations and are adsorbed by soil particles. Their concentration in the soil solution increases with the electrolyte concentration and with the valency of the desorbing cation (Bache and Sharp, 1976).

The transport of aluminum in common natural waters (pH = 4–9) occurs either as dissolved complexed ions or as colloid solutions and suspensions. Transport continues until either a strongly precipitating anion is encountered, or, until the complex has been destroyed. The aluminum complex is destroyed when the solution is enriched with another metallic cation that forms more stable complexes with the ligand.Organo-Al complexes may also be destroyed as a result of the oxidation or reduction of the organic ligand. Significant residual deposits of aluminum-rich minerals, such as hydroxides, phosphates, and silicates, indicate that these anions are strong precipitating agents for aluminum under field conditions (Huang and Keller, 1972).

1.1.1 Mechanism of Polymerization of Hydrated Aluminum

According to Mattock (1954) the tendency for metal cations to form polymeric hydroxides in solution is most pronounced when the metal-oxygen bond is partly covalent. This tendency decreases both when the bonding is strongly electrostatic, as in NaOH, and when it is strongly covalent, as in $HClO_4$ or H_2SO_4. The aluminum–oxygen bond is partly covalent and hence aluminum tends to form polymeric hydroxy complexes.

The polymerization process of hydrated aluminum ions involves the following steps: (1) hydrolysis of the monomer and dimerization; (2) further hydrolysis, condensation by the orientation of aluminum species along *a* and *b* crystallographic axes and the formation of a bidimensional polymeric ionic species of a single gibbsite-type layer (sheet structure); (3) condensation by the parallel orientation of gibbsite-type sheets perpendicular to the *c* crystallographic axis and the formation of a tridimensional *tactoid*; and (4) condensation by the aggregation of the various Al species and the formation of amorphous particles (Hem, 1968b). Monomeric ions of Al are designated by Al^a. Polymeric ions having a sheet structure are designated by Al^b whereas the tridimensional tactoids are designated by Al^c.

Both polymerization and depolymerization are controlled by two factors, the activity of ions in solution and the pH of the environment. The polymerization reaction rates are extremely slow in dilute solutions but become faster in more concentrated solutions. The reactions are catalyzed by various soluble salts and solid surfaces and are therefore fast in many geologic environments. Changes in the acidity of the medium result in transfer of protons and not in the exchange of OH^- ions by H_2O molecules in the hydration atmosphere of the aluminum species.

The ion $Al(H_2O)_6^{3+}$ is the predominant form in acidic solutions. When the pH is raised the hydroxy complex ion $AlOH(H_2O)_5^{2+}$ is obtained. Two monomeric ions may condense to form the dimer $Al_2(OH)_2(H_2O)_8^{4+}$ (Fig. 3.5) with a four-membered planar ring comprising a double OH bridge between adjacent aluminum ions (Johansson, 1963a, b). In a solution containing only the monomeric and dimeric ions the gram-formula ratio of OH to aluminum ($g = [OH]/[Al^{3+}]$) cannot exceed one. At concentra-

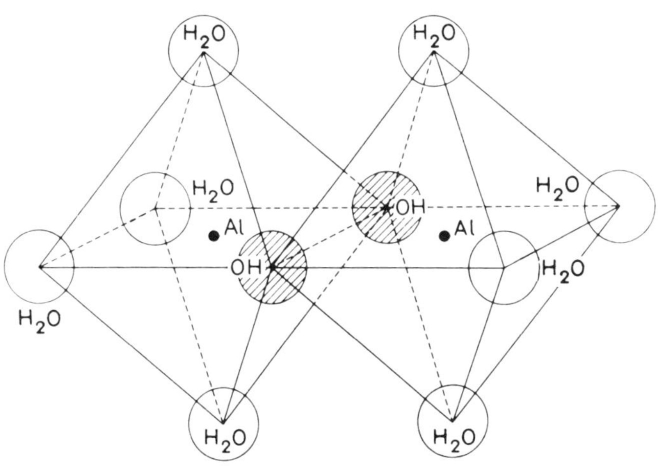

Fig. 3.5. Schematic representation of dimeric cation $Al_2(OH)_2(OH_2)_8^{4+}$. [Al^{3+} in the centers of the condensed octahedra. Oxygen (OH^- and H_2O) in the corners] (From Hem and Roberson, 1967; source: U. S. Geol. Surv.)

tions of aluminum greater than 10^{-2} M, where g = 1, the dimer predominates whereas the monomer predominates at concentrations below $5 \times 10^{3-}$ M.

The equilibrium constant for the dimerization in distilled water is given by

$$2\,Al^{3+} + 4\,H_2O \rightleftharpoons Al_2(OH)_2^{4+} + 2\,H_3O^+ \qquad pK_{22} = 6.2.$$

The dimer concentration increases with the ionic strength (Stol et al., 1976).

As the acidity of the solution decreases, further deprotonation of the aluminum-hydrated species occurs and condensation continues. For this type of condensation to occur an aluminum ion must participate in only one double OH bridge in one plane. If this requirement is fulfilled, then the largest polymeric species obtained is the gibbsite sheet. In the gibbsite sheet each of the aluminum ions participates in three double OH bridges. In the gibbsite the gram formula ratio g of OH to Al is 3.

When 3 > g > 1, one may anticipate several combinations that depend on the ratio g and may represent various steps in the formation of the gibbsite sheet. For instance, when g = 5/3, the ion $[Al_6(OH)_{10}(HOH)_{16}]^{8+}$ is obtained, which is thought to have a branched chain structure. When g = 2, the ion $[Al_6(OH)_{12}(HOH)_{12}]^{6+}$ is obtained, which has a ring structure formed by six aluminum hydroxy octahedra. This unit is apparently the most stable configuration that can be built up from six AlOH octahedra. For g = 2.52, an ion consisting of three rings and two additional tails is obtained, namely, $[Al_{16}(OH)_{36}(HOH)_{24}]^{12+}$. All these structures are based on the assumption that the OH groups are in the form of double OH bridges (Fig. 3.6).

According to Hsu and Bates (1964) polymers consist of multiples of six octahedra rings. When g = 2.3, three condensed rings give $[Al_{13}(OH)_{30}(HOH)_{18}]^{9+}$. When g = 2.5, seven condensed rings give $[Al_{24}(OH)_{60}(HOH)_{24}]^{12+}$. When g = 2.7, nineteen condensed rings give $[Al_{54}(OH)_{144}(HOH)_{36}]^{18+}$. That is, when the value of g increases the size of the polymeric cation also increases. The Al ion at the edge of such a polymeric cation is not fully coordinated by OH groups and is more labile than aluminum inside the polymeric ion. The stability of these polymers depends on the thickness of

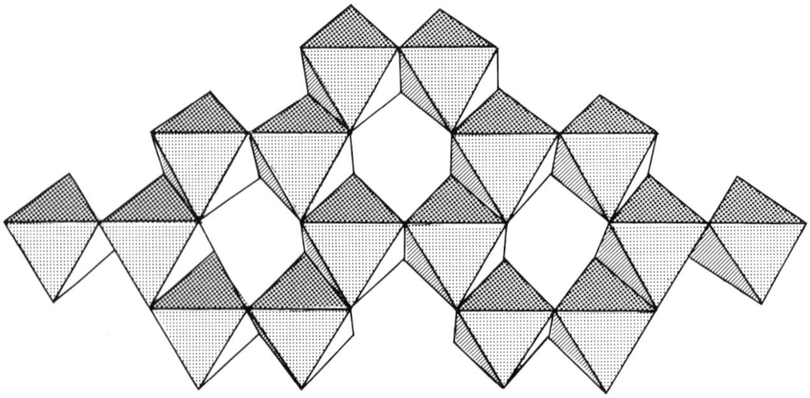

Fig. 3.6. Schematic representation of network of aluminum hydroxide octahedra (gibbsite-type sheet). (From Hem and Roberson, 1967; source: U. S. Geol. Surv.)

the diffuse double layer, which in turn is dependent on the net positive charge per aluminum atom and also on the nature of the counteranion. According to Stol et al. (1976) relatively small polynuclear complexes are formed when g does not exceed 2.2. Above this ratio there is a rapid increase in the size of the polymer and at g = 2.5 most of the building blocks form cyclic structures.

Miceli and Stuer (1968) point out that the polymers are usually present in high ionic strength solutions containing a high concentration of aluminum. Hem and Roberson (1967) and Smith (1969) found evidence of polymeric species in low ionic strength solutions containing not more than 13 mg Al per liter. When g does not exceed 3, it may require ten days or more for the ionic species to reach a diameter of 100 nm. The polymers are important intermediates occurring during the precipitation of aluminum hydroxide. Their concentration depends upon the conditions of mixing and rate of hydrolysis. When once formed, they remain even if the solution is highly diluted because there is no rapid equilibrium between $Al(HOH)_6^{3+}$ and polymers. The hydration energies of species of Al^b are very high and these behave as hydrophilic colloids.

Several gibbsite sheets can combine to give a tactoid in which the sheets lie parallel to one another perpendicular in the *c* axis. The forces between the sheets are mainly due to hydrogen bondings and van der Waals interactions. The total surface area of such a particle becomes very small in comparison to the sum of the areas of the surfaces of the single gibbsite sheets that comprise the tactoid, hence the hydration energy per unit mass of particle of Al^c is much lower than that per unit mass of particles of Al^b. Particles of Al^c are hydrophobic colloids.

1.1.2 Aging of Aqueous Solutions of Aluminum and Precipitation of Hydroxides

From the preceding discussion it follows that the geologic aqueous systems may contain three different types of aluminum species: Al^a (monomers), Al^b (hydrophilic colloids), and Al^c (hydrophobic colloids). For a particular g the amount of Al^a decreases from the moment that aluminum has dissolved, until it reaches a constant value after an incubation period, the time of which increases with decreasing initial pH of the system. The constant value of Al^a at equilibrium is independent of how long the solution has aged. At the same time the total amount of Al^b plus Al^c is also constant. This indicates that the minus Gibbs free energies of formation of polymeric hydroxy complexes of type Al^b are not much smaller than that of solid Al^c (Smith, 1971).

The growth of polymeric hydroxy complexes of type Al^b during the aging of aluminum salt solutions results in the release of protons from hydrated aluminum species and a downward drift in pH. Most changes of pH occur during the first 10–14 days of contact between the ion and water, but this decrease in pH continues even for several months. If $g \leq 1$, equilibrium will not be achieved even after several years of aging.

Initially, the solid material may be partly or totally amorphous, but rapidly becomes crystalline and takes on the structure of gibbsite. A precipitate of aluminum hydroxide is obtained immediately, as g becomes higher than 3. Under these conditions supersaturation of the solution is very high and the aggregation velocity becomes very high compared to the orientation velocity of gibbsite sheets perpendicular to the *c* axis.

Such a precipitate is composed of large amorphous aggregates of $Al(OH)_3$. After one day of aging, the precipitate is still amorphous to X-rays. After seven more days of aging, bayerite and small amounts of gibbsite are found in the precipitate. With further aging transformation of bayerite to gibbsite takes place by the mechanism of reorientation. Gibbsite is the more stable form of $Al(OH)_3$ and predominates in nature. As a result of the transformation of the hydroxide from amorphous into a crystalline modification the solubility of aluminum decreases with increasing time of aging.

In an acidic environment, when the supersaturation of the solution is very low, the aggregation velocity becomes very small and may be slower than the orientation velocity of gibbsite sheets perpendicular to the c axis. Under these conditions the formation of the solid phase is very slow and may take several days or weeks. Very fine hexagonal hydrophobic colloid gibbsite particles with a diameter smaller than 100 nm can be separated from such a solution. The minus standard free energy of formation of these small gibbsite particles at 25 °C is 272.3 kcal per gram-formula and that of the amorphous $Al(OH)_3$ obtained from alkaline solutions is even less. The minus standard free energies of formation of crystalline gibbsite and bayerite are 275.3 and 274.6 kcal per gram-formula, respectively (Parks, 1972). These free energy data indicate that the amorphous $Al(OH)_3$ tends to recrystallize and that the small gibbsite crystallites tend to grow when they are subjected to aging.

The precipitation of aluminum with hydroxyls is exemplified geologically by the ores of bauxite. Despite the seemingly residual nature of deposits of bauxite, Huang and Keller (1972) state that contradictory field, petrographic, and mineralogic evidence, e.g., crystals of gibbsite and boehmite in oolitic and pisolitic structures, and trails of those minerals in joints and channels through which solutions have moved, indicates that precipitation of aluminum from solutions had occurred.

1.1.3 Precipitation of Aluminum Phosphate

Precipitation of aluminum by phosphate is exemplified by large-scale deposits in the "leached aluminum phosphate zone" occurring in the Pliocene Bone Valley Formation of the phosphate deposits of Florida (Altschuler, et al., 1956).

Hsu (1968) studied the chemical mechanism of this reaction. He showed that different products are formed, depending upon whether hydroxy polymeric complexes of type Al^b or the monomer Al^a react with the phosphate anion. When small amounts of phosphate are added to a solution of the hydrophilic colloid of the type Al^b, the anion is sorbed by the polymer, replacing HOH molecules outside the core, whereas structural OH^- groups in the polymer are not replaced. The following two assumptions should be considered: (1) The relative attraction force of polymers for phosphate is proportional to their charge density, or net positive charge per aluminum atom, and (2) the phosphate ion tetrahedron tends to link two aluminum polymers together rather than to complex with only one polymer. In an ideal system all polymers will have the same attraction force for phosphate. The first phosphate ion added to the system may be adsorbed by any two polymers it first encounters. The polymers that interact with this phosphate ion become less positively charged. Consequently, the second phosphate ion added will not bond to the first pair of polymers but rather to its neighbors. As long as the amount of phosphate added is not sufficient to neutralize all the positive charges of the polymers present, the colloid suspension is stable. The system appears turbid because the phosphated polymers are sufficiently large to cause light scattering.

At the *isoelectric point,* that is, when all Al^b species are completely neutralized as a result of the phosphate adsorption, there is no electric repulsion between the particles and the colloid suspension is not stable. Particles cluster together and are precipitated from the solution. This flocculation occurs at a P/Al atomic ratio of 0.3.

With phosphate concentrations higher than that occurring at the isoelectric point, the excess phosphate in the solution competes with the structural hydroxyl for aluminum and may break the polymer into smaller units. The hydroxy-aluminum-phosphates obtained are negatively charged, and thus are capable of adsorbing cations. There is again an electric repulsion between the charged small particles and again a stable colloid suspension is obtained.

Hsu also studied the reaction of the monomer Al^a with phosphate. When fresh solutions of $AlCl_3$ were mixed with NaH_2PO_4, clear solutions were obtained up to a P/Al atomic ratio of 8. Precipitation occurred when this ratio in solution was higher than 8. However, all precipitates had a nearly constant P/Al atomic ratio of about 1.2 and also contained small amounts of sodium. Although the solutions were not colloid the precipitates showed a low degree of crystallinity.

1.1.4 Precipitation of Aluminum Silicate

The presence of silica can greatly alter the nature of the hydrolytic reaction products of aluminum even at initial concentrations as low as those found in natural waters (Hem et al., 1973; Luciuk and Huang, 1974).

When small amounts of silicic acid are added to a solution of the hydrophilic colloid of the type Al^b, the acid is *specifically* sorbed by the polymer. During this *specific sorption* one OH bridge is exchanged by a $SiO_4H_3^-$ group, and one of the four oxygen atoms then serves as a bridge between two aluminum and one silicon atom. A second molecule of silicic acid may react in a similar way with an OH group of another bridge having one aluminum atom in common with the first bridge. This exchange is followed by a condensation that may occur between the two silicate groups resulting in a six-membered ring, comprised of one aluminum, two silicon, and three oxygen atoms. The structure obtained by this type of sorption is shown in scheme (3.III).

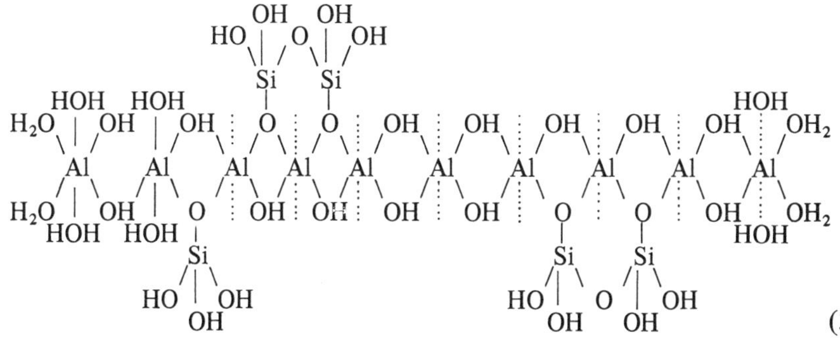

(3.III)

In connection with polymeric species, oxy-metal rings in general are more stable than their hydrate analogs, if there are no steric hindrances. Octahedral centered aluminum, with an O–Al–O angle of 90°, is able to participate in a planar four-membered ring, while a tetrahedral centered silicon, with an O–Si–O angle of 105°, is able to participate in a six-membered ring, which is not planar. As a result of ring formations, the angles are slightly changed, and the stability of the rings decreases with increase of the size of angles from the theoretical values mentioned above.

Stabilization due to ring formation is explainable in terms of favorable entropy change. Qualitatively, one can understand the favorable entropy as follows. The replacement of a coordinated water molecule by either an $[Al(OH)]^{2+}$ or $[Al(OH)_2]^{1+}$ should have about equal probability. The replacement of a second water molecule by the other OH group in the coordinated $[Al(OH)_2]^{+}$, however, is much more probable than its replacement by a free $[Al(OH)]^{2+}$ from solution, since the $[Al(OH)_2]^{+}$ is already tied to the metal ion and the OH group is in the immediate vicinity of the HOH it will replace. Thus the formation of $[(HOH)_4Al(OH)_2Al(HOH)_4]^{4+}$ is more probable than the formation of the less stable $[(HOH)_5Al(OH)Al(HOH)_5(OH)Al(HOH)_5]^{7+}$. The same applies for the formation of the six-membered oxy-aluminosilica ring.

Another way to visualize the more favorable entropy change is to consider that a process in which the number of independent particles increases proceeds with an increase in entropy (the larger the number of discrete molecules or ions, the greater is the possible disorder). The formation of the six-membered ring can be formulated as follows:

$$[Al(HOH)_6]^{3+} + 2\,H_4SiO_4 \rightarrow [(HOH)_4Al \cdot (OSi)_2O(OH)_4]^{+} + 3\,HOH + 2\,H^{+} \quad (3.IV)$$

This formula demonstrates that during the condensation of three species, six species are freed and therefore this process should proceed with a favourable entropy.

Two-, three-, and more polycondensed six-membered oxy-aluminosilica rings can be formed on both sides of the sheet polymer of type Al^b. It can be realized that if specific sorption of silicic acid occurred onto 2/3 of hydroxyl sites of a gibbsite type sheet, followed by intrasilica condensation, a clay mineral structure is obtained. The orientation velocity in this type of reaction is extremely slow. Examination of the hydrolytic reaction products of aluminum formed in the presence of silicic acid reveals that materials amorphous to X-ray and electron diffraction are the only products even at a very low initial aluminum concentration (e. g., 1.39×10^{-4} mol/l) and even in systems aged over four years.

At silica concentrations below about 10^{-4} mol/l, microcrystalline gibbsite is formed at pH below 6 and crystalline bayerite at pH above 7, but the solid phase appears only after a much longer aging time than that required for crystallization in silica-free solutions. The solid phase contains only a small fraction of the total aluminum, whereas the aqueous phase contains polymeric hydrophilic colloid particles of Al^b, which sorb silica. At OH/Al ratios below 3, the reaction products form a stable colloid solution not removable by ultrafiltration.

The stabilization of colloid solutions of aluminum hydroxide by very small amounts of silicic acid are explained in the following way. The six-membered rings may appear on both sides of the gibbsite-type sheet. Their presence results in steric hindrance, and consequently the orientation velocity of aluminum species along the *a* and *b* directions becomes very small. Furthermore, van der Waals interactions between parallel sheets are now very weak, there being only very few contact points between them, and hydrogen bondings are almost nonexistent. The transformation of the hydrophilic colloid into a hydrophobic one, which should result from the condensation of several microlayers into one cluster, is extremely slow. At a high saturation of gibbsite-type sheet with silicic acid, van der Waals interaction and aggregation velocity increase, resulting in an amorphous precipitate.

The transformation of this amorphous phase into a crystalline clay mineral phase can occur by either of the following two mechanisms: (1) continuous silicification of the gibbsite-type sheet up to saturation, by specific sorption of silicic acid after it has been desorbed from the amorphous phase, and (2) by the condensation through orientation in the *a* and *b* directions of the six-membered oxy-aluminosilica rings to form a phyllo-aluminosilicate. Six-membered rings are obtained from the monomers according to Eq. 3.VII after the amorphous aluminosilicate has been depolymerized. As a result of the steric hindrance existing during crystallization via mechanism (1), mechanism (2) requires a lower activation energy. It is therefore to be expected that the rate of growth of crystalline clay minerals depends on the rate of depolymerization of the amorphous material. Such transformations require very long aging periods.

Laboratory syntheses of clay minerals from amorphous gels are complicated and occur under very special conditions. Most of the syntheses described in the literature were performed under hydrothermal conditions. In this Section, the synthesis of kaolinite, the most thoroughly studied, will be described (see, e.g., DeKimpe, 1967, 1969; DeKimpe et al., 1961; Rodrique et al., 1972; La Iglesia Fernandez and Martin Vivaldi, 1972; Wey and Siffert, 1961). These workers combined solutions of organically complexed aluminum with an organically complexed form of orthosilicate or with very dilute solutions of silicic acid. This procedure provides gels free of cationic and anionic impurities. Organic acids are also important because they serve as a hexa-coordinating agent for aluminum. Since they form stable complexes with aluminum, activity of Al^a is greatly reduced. Consequently, the concentration of Al^b is reduced and also supersaturation of the solution becomes low with respect to the solubility product of clay minerals, resulting in an abrupt fall in the rate of aggregation. In nature, humic substances catalyze kaolinite formation (La Iglesia Fernandez and Martin Vivaldi, 1972).

The optimal pH suitable for crystallization of kaolinite is close to 4. At this pH, the binding of alumina to silica is favored. At lower pH values the dissociation of Al–O–Si group is favored with the formation of Al^a whereas at higher pH values the self-association of alumina is favored. In alkaline solution, in which aluminum has the tendency to be fourfold coordinated, it may substitute a silicon in silica polymeric species. After prolonged aging of aluminum hydroxide containing some Mg ions, in solutions containing very low concentrations of silica but very high concentrations of potassium, illite and a mixed-layer illite-smectite were obtained (Harder, 1975).

Under neutral conditions Mg may be substituted for Al in the gibbsite-type sheet, and it is possible to co-precipitate Mg with aluminum hydroxide. The Mg content of such a precipitate depends on the Mg concentration in the initial solution and the pH during the precipitation process. Smectites were synthesized by Harder (1972) at room temperature by the aging of Mg-containing aluminum hydroxides in solutions of very low silica concentrations (15–30 mg SiO_2/l). Polymerization of the silicic acid inhibits the formation of smectites, and the reaction is therefore favored at pH 10. At neutral pH values smectite is formed only if the concentration of Mg in the $Al(OH)_3$ is at least 6%. Lower Mg concentrations are necessary for smectite formation at higher temperatures. According to Harder the tetrahedral coordination of Al in alkaline solution inhibits the formation of the octahedral sheet in phyllosilicates at low temperatures. Some of the Al enters into an amorphous tectosilicate structure, and the remainder forms metastable bayerite. At a certain Mg content, the participation of Al in sixfold coordination, the formation of oxy-aluminosilica rings, and the condensation of these rings through orientation in the *a* and *b* directions seem to be facilitated.

1.2 Chemistry of Iron and Manganese in Natural Waters

Iron is an abundant and widespread constituent of rocks and soils. Although manganese is much less abundant than iron in the earth's crust, it is one of the most common elements and is widely distributed in rocks and soils. Both elements are transition metals and are capable of existing in more than one valence state. The hydrogeochemistry of these elements is determined by the polarizing power of the respective ions, which greatly increase with ionic charge. The ionic radius, which is important in crystal geochemistry and is also an important factor in determining the polarizing power of the ion, decreases with increasing oxidation number of the element. As a consequence, there is a difference in solubility of hydroxides between these valence states, the solubility decreasing with increasing oxidation number of the metal. Hydrolysis and precipitation of Fe(III) and Mn(IV) occur at very low pH values (< 3) and the precipitates tend to be amorphous and to form colloid solutions, while those of Fe(II) and Mn(II) are crystalline, and hydrolysis and precipitation occur at pH values above 8.5.

The most common form of dissolved iron in natural waters is the ferrous ion Fe^{2+}. The complex $FeOH^+$ may occur in solutions very low in concentrations of dissolved carbon dioxide species. Ferrous complexes are formed by many organic molecules especially by those called *chelating agents,* which have more than one electron donating group, being able to form ring structures incorporating the iron ion. Almost all iron occurring in soil solutions is complexed with organic matter (Olomu et al., 1973). Above pH 11 the anion $HFeO_2^-$ can exist in appreciable quantity, but this high pH is rarely attained in natural systems.

Ferric iron occurs in acid solutions as Fe^{3+}, $FeOH^{2+}$, $Fe(OH)_2^+$ and as some polymeric forms (hydrophilic colloids), the sizes and structures of which depend on the pH. Above pH 4.8, however, the solubility of ferric species is less than 0.01 mg Fe/l. Colloid ferric hydroxide is commonly present in surface waters, and small quantities may persist even in water that appears to be clear. Fe(III) forms stronger complexes than Fe(II) with various organic substances, and these complexes play an important role in the mobility of this element.

The most common form of dissolved manganese in natural waters is the manganous ion Mn^{2+}. The complex $MnOH^+$ may occur to some extent if the solution is not too acidic. Manganese also forms soluble complexes with many organic natural compounds, but these complexes are less stable than those of iron. Brockamp (1976) studied the leaching of Mn and Fe by aqueous solutions of various organic acids and showed that Mn is more concentrated than Fe in solutions from weathered rocks, where Fe is oxidized, but that more Fe than Mn is dissolved from fresh basalt, where the Fe is divalent.

The solubilities of iron and manganese in natural waters and soil solutions and the predominant dissolved species can be effectively summarized by means of Eh–pH diagrams (Figs. 3.7 and 3.8). The oxidation potentials for the transition of ferrous to ferric iron and of manganous to manganic manganese fall within the range found in natural waters, and consequently all these oxidation states are common in geologic systems. Both elements can be transported by flowing water along chemical potential gradients from acid and/or reducing environments poor in sulfide, and are precipitated as

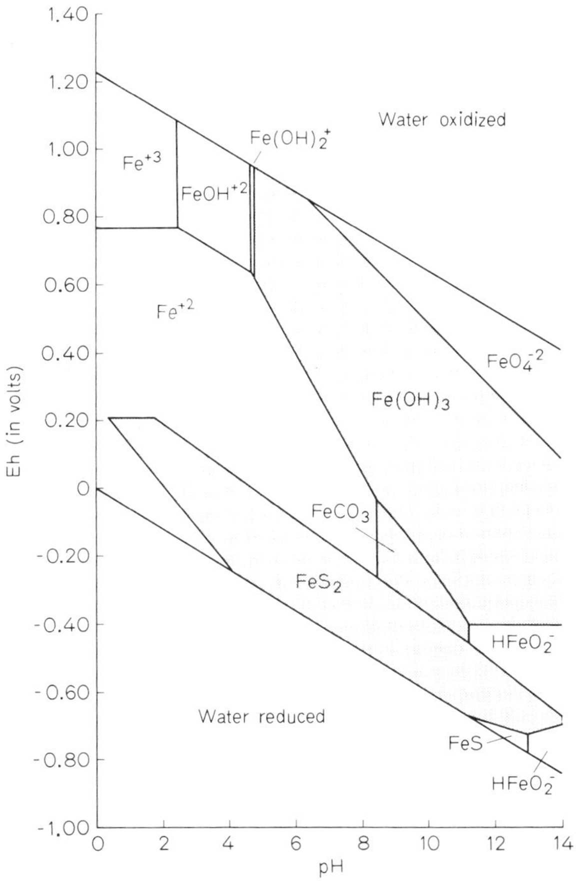

Fig. 3.7. Fields of stability for solid and dissolved forms of iron as a function of Eh and pH at 25 °C and 1 atmosphere pressure. Activity of sulfur species 96 mg/l as SO_4^{-2}, carbon dioxide species 100 mg/l as HCO_3^-, and dissolved iron 0.0056 mg/l (From Hem, 1970; source: U. S. Geol. Surv.)

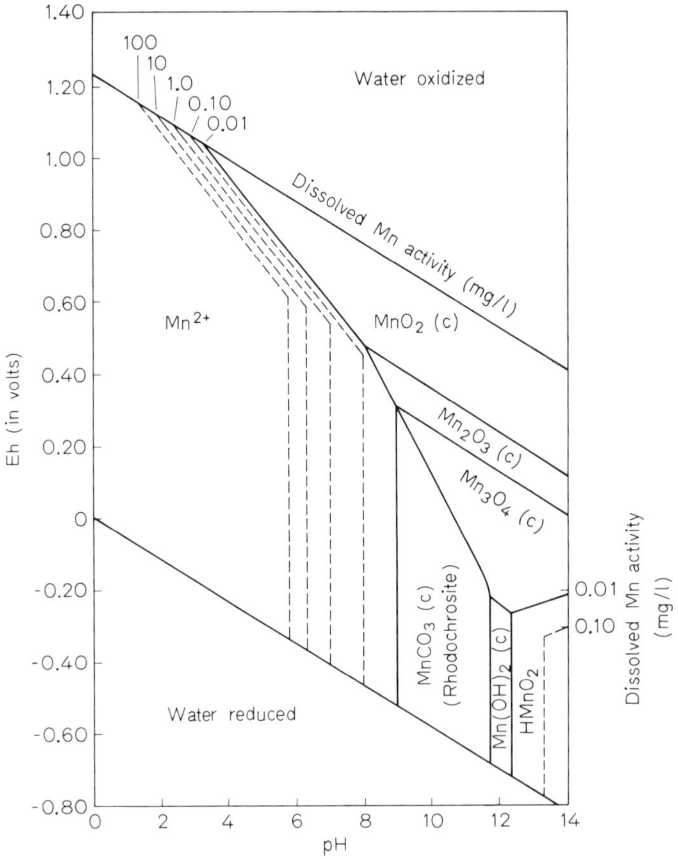

Fig. 3.8. Fields of stability of solids and solubility of manganese as a function of Eh and pH at 25 °C and 1 atmosphere pressure. Activity of dissolved carbon dioxide species 100 mg/l as HCO_3^- species. Sulfur species absent. Used with permission of the publisher. (From Hem, 1970; source: U. S. Geol. Surv.)

hydrous oxides on reaching oxidizing and/or basic environments. They are also precipitated in reducing environments rich in sulfide ions, as iron or manganese sulfide. For both oxidic and sulfidic prepcipitates, orientation velocities are extremely low, and during the first stages of the formation of the new solid phases the solid particles will remain in a colloid state until they reach a basin where conditions are suitable for coagulation and sedimentation.

Reduction of the Mn(IV) and Fe(III) oxides in geologic systems occurs under anaerobic conditions, whereby bacterial oxidation of organic matter proceeds at a rate such that the dissolved oxygen supply is depleted. The bacteria then use the higher oxides as a source of oxygen. Organic compounds, and especially soluble sugars, such as starch and glucose, are responsible for the drop of the Eh. Hydrogen that is formed by the decomposition of the soluble sugars, is the reducing agent for Fe and Mn. Nitrates lessen the drop of Eh whereas sulfates have only minimal effect on the Eh drop (Yamane and Sato, 1968). Meek et al. (1968) showed that flooding a specific soil in the presence of organic matter caused a reduction in Eh to –0.1 V, and if the same soil was flooded without organic material, the Eh was similar to that of the unflooded soil. Solubility of Mn was 46 mg/l in the former case, but only 3.7 mg/l in the latter case.

The rates of oxidation of Fe(II) and Mn(II) and or precipitation of the hydrous oxide of the oxidized metal are greatly increased with increase of the pH: Stumm and Lee (1961) found the rate of oxidation of ferrous iron in aerated water to increase 100-fold by raising the pH one unit. This oxidation is extremely slow in strongly acid solutions.

From a comparison of the corresponding diagrams for the iron–water and manganese–water systems, it is obvious that manganese is considerably more soluble than iron in the Eh–pH range of natural waters.

In general, the oxygen of the air is the agent for the oxidation of iron and manganese. As a result of the low solubility of hydrous oxides of the oxidized metals this process is spontaneous in alkaline solutions. Bacteria may also influence the rates of oxidation. It has been hypothesized that the largest manganese deposits in Russia were formed with the aid of manganese-oxidizing bacteria (Zajic, 1969). According to Ivarson and Heringa (1972) the bacteria are able to alter and control their environmental pH, which is a contributing factor in the oxidation of iron and manganese.

Oxidation is catalyzed by the presence of colloid materials in suspension. Manganic oxide, MnO_2, is a good catalyst for the oxidation of Mn(II). Clay minerals serve as good catalysts for the oxidation of iron and manganese.

Organic materials that form stable complexes with Fe(II) and Mn(II) may prevent the oxidation of these elements. If the complex obtained has a low dissociation constant, the activity of the divalent cation in solution is very low, and the oxidation potential is reduced to such an extent that no Fe(III) or Mn(IV) hydroxides are precipitated. Bodenheimer et al. (1963) examined the precipitation of manganic oxide from alkaline solutions of Mn(II) in the presence of montmorillonite and various amines, by air oxidation. The presence of tetramethylenepentaamine, which gives a very stable complex with Mn^{2+}, prevented its oxidation and precipitation. The stability of this complex results from the formation of four chelate rings in which every Mn atom participates (3.V). Other polyamines, giving rise to less stable chelates, did not prevent the oxidation of manganese.

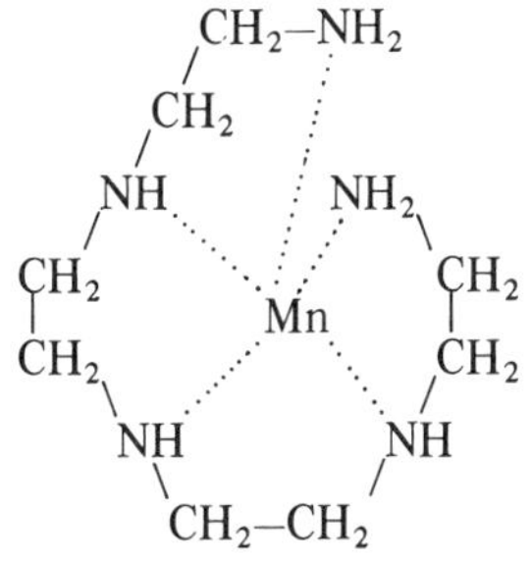

(3.V)

In addition to the inorganic controls on the precipitation of the transition metals, certain organic acids may form hydrophobic colloids of insoluble salts with these elements. For example, hydrophilic colloid humic acid may become hydrophobic due to the adsorption of Mn (Crerar et al., 1972).

Deep sediments are sufficiently reducing to effect diagenetic remobilization. Concentrations of dissolved Fe and Mn as high as 10 mg/l have been measured within interstitial waters of reducing Baltic Sea sediments (Debyser and Rouge, 1956). The pore water of the uppermost oxidizing layer of the marine sediments contains Fe and Mn in the same concentrations as the ocean bottom water (Strakhov, 1976). Total Mn is most concentrated in the uppermost oxidizing beds and decreases with depth (Li et al., 1969; Van der Weijden et al., 1970). In the uppermost oxidizing beds, Mn oxides reprecipitate, hence Mn(II) concentrations fall. A typical variation with depth of Eh, pH, total Mn, and dissolved Mn in marine sediments, as summarized by Crerar and Barnes (1974), is shown in Figure 3.9. Because concentrations of soluble Mn reach a maximum toward the top of the reducing zone, Crerar and Barnes concluded that the soluble Mn is mobilized upward. In nearshore and other environments where such diagenetic rates are high, Mn can be supplied to surface deposits by upward diffusion along chemical gradients. There is a strong organic control on the upward diffusion of manganese along sedimentary Eh–pH gradients (Shanks, 1971). The Eh does not play an important role in the deposition of Fe from seawaters since they are supersaturated with respect to $Fe(OH)_3$. It is therefore expected that the upward diagenetic mobility of Fe along sedimentary Eh gradients of marine sediments should not be as pronounced as that of Mn. This has been shown by Van der Weijden et al. (1970).

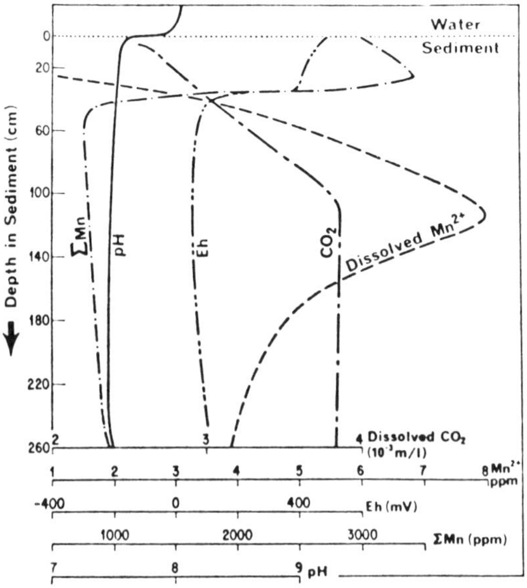

Fig. 3.9. Geochemical profile of a marine sediment, showing vertical distribution of total Mn, dissolved Mn, Eh, pH, and dissolved CO_2. (From Crerar and Barnes, 1974. Used with permission of the publisher)

Changes of Eh and pH at various zones of a sedimentation basin are responsible for geochemical separations of Fe from Mn (Krauskopf, 1957). Deposits from shallow water and nearshore deposits usually have low Mn/Fe ratios whereas deep-water deposits have higher Mn/Fe ratios. Mart and Sass (1972) showed from geologic evidence that the manganese ore of Um Bogma, Sinai, was formed by the settling out of the colloid ore material from an *accumulation basin,* in which finely particulate material, such as oxides of Fe and Mn, either sorbed on clay minerals or complexed with organic substances, was trapped in tidal pools of low mechanical energy. A differentiation process took place when the particulate material was suspended in the basin, during which the cores of the lenses were enriched in Mn, whereas the margins were enriched in Fe. This differentiation was explained by Eh–pH differences between the margins and the core. Although the Eh of the margins was high, the pH was low and not sufficient for the precipitation of manganese oxides. It was sufficiently high to precipitate ferric hydroxide. The Eh of the core is low, but the pH is sufficiently high for Mn(II) to precipitate. Under these conditions the organic complexes of Fe(II) are sufficiently stable and this element does not settle out. As a consequence, manganese migrates away from the rim and precipitates in the core whereas iron migrates in the opposite direction.

1.2.1 Condensation of Hydrated Ferric Iron

When ferric iron ion is dissolved in water the initial reaction product is hydrated monomeric species. The first stage, hydrolysis, which follows the hydration process of the ion, is very fast and results in the acidic pH of the solution. The initial pHs of 0.1, 0.01, and 0.001 M solutions of the salt $Fe(ClO_4)_3$ immediately after their preparation are 1.42, 2.32, and 3.21, respectively (Hsu and Ragone, 1972). The first-stage hydrolysis is followed by dimerization and formation of

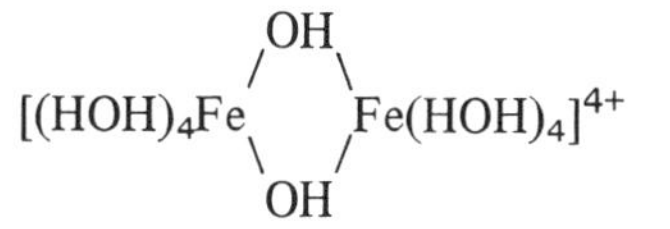

(3.VI)

According to Mulay and Selwood (1954) the monomeric species is paramagnetic whereas the dimer is diamagnetic. This indicates that there is a Fe–Fe interaction within the dimer. An aqueous solution of the dimer absorbs light at the wavelength 335 nm while the monomers Fe^{3+} or $Fe(OH)^{2+}$ absorb at the wavelength 240 nm. The dimerization is an endothermic reaction, the dimer being more stable at high temperatures. Equilibrium constants for the reactions occurring during the initial stage of hydrolysis have been estimated by Hedström (1953) as follows:

$$Fe^{3+} + HOH \xrightleftharpoons{k_{11}} FeOH^{2+} + H^+ \qquad k_{11} = (9.0 \pm 1.0) \times 10^{-4}$$

$$Fe^{3+} + 2\,HOH \xrightleftharpoons{k_{12}} Fe\,(OH)_2^+ + 2\,H^+ \qquad k_{12} = (4.9 \pm 1.0) \times 10^{-7}$$

$$2\,Fe^{3+} + 2\,HOH \xrightleftharpoons{k_{22}} Fe_2(OH)_2^{4+} + 2\,H^+ \qquad k_{13} = (1.22 \pm 0.1) \times 10^{-3}$$

In the second stage, condensation of monomeric hydrolyzed iron species to hydrophilic polymeric species occurs (Lamb and Jacques, 1938). The continuing condensation of hydrolyzed species induces further hydrolysis of hydrated species, known as a *secondary hydrolysis,* resulting in a further decrease in solution pH. The time necessary for the secondary pH drop is called the "induction period". The higher the iron concentration the longer will be the induction period. Induction periods for 0.0002, 0.0005, and 0.02 M solutions of $Fe(ClO_4)_3$ are 2 and 10 min and 4 months, respectively. Solutions with concentrations higher than 0.04 M did not show a decrease in pH even after 10 months of aging, and solutions appeared a clear yellow (Hsu, 1973).

The polymeric species obtained during the second stage differ from those obtained during the parallel stage in the aging of aluminum salt solutions. A great variety of structures is obtained at the end of the induction period, some of which may grow further and precipitate and some of which achieve a metastable structure and do not grow further, thus forming stable colloid solutions. The structures of the hydrophilic colloids depend on the anions present, the initial ratio between Fe and OH in the solution, the initial hydroxyl concentration, and the time of aging of the solution. When the gram-formula weight ratio $g = [OH^-]/[Fe^{+3}]$ exceeds 1, a metastable yellow solution of hydrolyzed and of dimeric iron is obtained, which after an induction time of several weeks will precipitate α FeOOH. This precipitate contains only a small portion of the total iron. When g exceeds 2 the solution appears dark brown and after an induction time of several days α FeOOH and γ FeOOH will precipitate from the colloid solution. Here again only a small portion of the total iron appears in the solid phase. When g exceeds 3, amorphous $Fe(OH)_3$ will precipitate immediately. On aging, this precipitate is partly converted into α Fe_2O_3 and partly into α FeOOH, but the greatest fraction remains in an amorphous state (Feitknecht and Michaelis, 1962).

The hydrolysis of $Fe(NO_3)_3$ in the presence of bicarbonate was investigated by Spiro et al. (1966). The hydrolysis results in precipitation of amorphous ferric hydroxide and formation of hydrophilic polymers. The polymers were separated by gel-filtration and dialysis. The average molecular weight of these polymers is 1.4×10^5. The empirical composition is $[Fe(OH)_x(NO_3)_{3-x}]_n$, where x is 2.3–2.5 and n is approx. 900. Unlike polymeric Al^b, this iron polymer is not a microcrystal of ferric hydroxide, and the aging of the polymer will not result in the precipitation of ferric hydroxide or oxyhydroxide. It is a positively charged cation composed of hydrated ferric doubly bridged by hydroxyls. Electron-microscopic study and viscosity measurements of colloid solutions indicate that the polymer has a spherical structure. The rate of formation of this polymer is very fast while the rate of its dissociation is very slow. Aqueous solutions of this polymer are stable and may persist for many years, unless the polymers are decomposed by acids.

Biedermann and Chow (1966) studied the hydrolysis of ferric iron (> 0.001 M) in NaCl solution (0.5 M). The hydrolysis was slow and led to the precipitation of an insoluble compound having the formula $Fe(OH)_{2.7}Cl_{0.3}$. X-ray study of this compound showed a similar picture to that of α FeOOH (goethite), which is common in sea sediments. It seems that a similar mechanism of precipitation occurs in seawater. Polymeric hydroxy-ferric cations, which may act as nuclei for continuing condensation of monomeric species, are probably similar to those described previously for aluminum. When they become large they may settle under gravity, forming amorphous hydroxide.

During prolonged aging, the fresh $[Fe_n(OH)_{3(n-x)}]^{3x+}$ gradually dehydrates to the less soluble species $[Fe_nO_{n-x}(OH)_{n-x}]^{3x+}$ involving the conversion of *ol-* into *oxo*-linkages (Schindler et al., 1963). Settling of these polymers results in the formation of α FeOOH.

According to Hsu and Ragone (1972), in a solution supersaturated with respect to iron hydroxide, the initial number of nuclei relative to the concentration of monomeric iron species is the key factor governing particle size distribution, which in turn governs the color, turbidity, and stability during aging. A solution of 0.001 M $Fe(ClO_4)_3$, with an initial pH of 3.21, is highly supersaturated with respect to ferric hydroxide, and a large number of nuclei spontaneously form in situ shortly after the preparation of the solution. The pH begins to drop after an induction period of 40 min. There are not sufficient ferric ions left in the solution to enable the further growth of the nuclei. A clear, stable, brown solution is developed within 6 h, which does not show any precipitate even after 10 months. The pH of this sol is 2.64. A solution of the same salt at a higher concentration (0.01 M), gives an initial pH of 2.32. Supersaturation with respect to ferric hydroxide is not longer so high and only a small number of nuclei are formed in situ. The pH begins to drop after an induction period of 5 days and the solution becomes slightly turbid within 9 days as a result of the nuclei growth. After 10 months the solution is very turbid, becomes dark yellow and impervious to light. The pH drops to 1.75. The precipitate consists mainly of α FeOOH, indicating crystal growth by the mechanism of orientation. When the same freshly prepared solution is acidified, the number of nuclei formed in situ is even smaller and a much higher induction period is necessary before any turbidity is observed. No sol is obtained at this high acidity.

By increasing the number of nuclei one can prevent their growth to a macroscopic settleable form even in acid solutions. Hsu (1972) added clear brown sols of iron hydroxide as nuclei to freshly prepared acidified ferric iron solution and studied their acceleration effects on the hydrolysis and sedimentation of iron. Depending upon the quantity of seeding solution added, samples differed greatly in appearance and in stability during aging. With a small amount of seeding solution added and following a relatively long induction period, the sample rapidly developed to a dense, cloudy yellow suspension, from which FeOOH precipitate settled shortly afterward. With an increased amount of seeding solution added, the sample gradually became more brownish and less turbid, and eventually a clear brown sol was obtained. Although α FeOOH was the major hydrolyzed species in all samples, a yellow precipitate that settled under gravity occurred only in those solutions that contained small quantity of seeding solution.

Particle size of the polymeric iron hydroxide increases with increasing initial iron concentration in solutions (Hsu, 1973). The solubility of iron hydroxide increases with decreasing particle size. The solubility product of freshly precipitated $Fe(OH)_3$ is 2.5×10^{-39} (Biederman and Schindler, 1957). The solubility products of $Fe(OH)_3$ obtained by hydrolysis and polymerization of iron in 0.01 and 0.001 M solutions of $Fe(ClO_4)_3$ are 6.6×10^{-38} and 3.9×10^{-37} respectively (Hsu and Ragone, 1972).

According to Dousma and de Bruyn (1976) the polymeric iron hydroxides can be classified by their size and structure into two groups and are referred to as small and large polymers. The large polymers are characterized by the presence of oxo-, Fe–O–Fe linkages, which are responsible for the greater stability of these polymers. The large polymers are obtained when the pH of the solution exceeds a certain value, which depends on the total concentration of the ferric iron. For example, at pH 2.0, 2.5, and 3.2, the concentrations are 6.25×10^{-2}, 6.25×10^{-3}, and 6.25×10^{-4} gram-atom Fe per liter, respectively. For the same Fe concentrations the small polymers are obtained at pH values of 1.5, 2.2, and 2.9 respectively. The small polymers react rapidly with acids to depolymerize whereas the large polymers undergo a very slow depolymerization reaction with acids.

1.2.2 Iron and Manganese in Soil Solution

In soil materials, organic redox systems produced by microbial fermentation processes have great effects on the measured oxidation potentials and on the reduction of manganese and iron oxides. Tri- and tetravalent manganese oxides are reduced first, giving rise to soluble ions of Mn(II). Next, ferric oxide is reduced giving rise to soluble ferrous ions, and finally sulfates are reduced to sulfides. A reduction of the insoluble oxides by microbially formed sulfides, which themselves are oxidized by this process, also seems to occur. Therefore, sulfides do not occur as a stable sulfur phase in considerable amounts before all available iron oxides are reduced to Fe(II) (Brümmer, (1974). Then, formation of the iron sulfide phase takes place by the condensation of Fe^{2+} with S^{2-}. The solubility product of FeS is 5×10^{-18}. Manganese forms a sulfide phase to a much smaller extent, and the concentrations of both ions should be much higher before condensation of Mn^{2+} with S^{2-} occurs. The solubility product of MnS is 1×10^{-11} and this phase has only little significance in soil mineralogy.

Many transition metals form stable soluble complexes with organic soil material. Aqueous soluble ferrous iron is found in many soils in oxidizing environments. The oxidation of this iron is prevented by the presence of organic compounds coordinated to this metal, forming stable chelates (Olomu et al., 1973). Most soluble manganese in soil solution is divalent and is complexed with organic material. Soil solutions from various localities in the USA contained 13 mg Mn/l at pH 7, of which 84–99% were complexed. The presence of such large concentrations of Mn(II) in a soil solution of a neutral soil was still less than that predicted from measured values of the Eh of the soil solution and the Mn(II)/MnO_2 half-cell potential. This may be the result of the oxidation of Mn(II) and the precipitation of MnO_2 by soil microorganisms (Geering et al., 1960). Soluble manganese comprises lateral transportation downslope and reorganization in the lower slope profiles mainly as concretions. Yaalon et al. (1972) found that in Israeli soils subjected to the moderate weathering environment of the Mediterranean climate, the total Mn increased downslope by 50–80% and became immobilized in the lower slope soils in the form of Mn-rich nodules. The process of nodule formation is strongest in the B horizon of poorly drained pedons. Miehlich and Zöttl (1973) obtained maximum concentration of Mn and formation of Mn concretions in the B2g horizon of pseudogleys from loess in Bavaria under spruce and beech forests.

The reduction of ferric iron to the ferrous state under anaerobic conditions is induced by water-logging. This soil process is known as the gleying process. The ferrous ion is leached from the profile by lateral or ground water. Readily soluble "free" and surface coating iron hydroxides are the predominant phases that contribute iron to the soil solutions, and they are removed from the system by the gleying process. The dissolution of iron from crystalline clay minerals seems to be of minor significance during this process (Mitchell et al., 1968).

Lepidocrocite, the orange-colored γ form of FeOOH, is normally formed from the oxidation of Fe(II) compounds produced by microbiologic reduction. From a consideration of the energy relationship, lepidocrocite should not persist but should transform to goethite, the more stable polymorph, especially in the presence of ferrous iron. Nevertheless, it occurs as a constituent of hydromorphic soils of many regions, particularly in clayey or loamy noncalcareous gleys and pseudogleys of various countries (Schwertmann and Thalmann, 1976). According to Oosterhout (1967) the transformation from lepidocrocite to goethite proceeds through the process of dissolution and reprecipitation. The rate of transformation is controlled by three steps: (1) the dissolution of lepidocrocite; (2) the formation of goethite nuclei, and (3) crystal growth of goethite. Schwertmann and Taylor (1972a) showed that any of these processes can be rate-determining under appropriate conditions. Soil constituents may be capable of retarding or even inhibiting the transformation. Hiller (1966) showed that the transformation of lepidocrocite to goethite can be completely inhibited in the presence of traces of silicate, aluminate, and stannate. According to Schwertmann and Taylor (1972b) the presence of Si merely retards the nucleation stage of the transformation. They did not find any decrease in the dissolution rate of the lepidocrocite due to surface adsorption of silica. They also did not find any retardation of the transformation if the silica was introduced after the nucleation stage. They showed that Si is adsorbed and incorporated into the goethite structure. Due to its retarding effect on the nucleation, larger crystals of goethite are formed in the presence of Si, many of which are twinned.

Concentrations and nodules in which Fe and Mn are concentrated relative to the soil matrix are common in some soils, and they result from dissolution–precipitation processes of these elements in the soil solutions. They are typical of hydromorphic soils of humid temperate climates, particularly in soils with impeded internal drainage. Their highest concentrations and largest sizes appear in horizons that are subjected to seasonal water logging, under conditions in which there is only a slow removal of the end products of weathering. In the first stage Mn becomes mobile by reduction. It is then concentrated in various forms and during drying out of the soil under aerobic conditions, it is reoxidized and precipitates around localized nuclei. Fe undergoes a similar process but Mn appears to be concentrated first, associated with and followed by Fe at a stage where Mn is nearly exhausted (Schwertmann and Fanning, 1976). The following four properties are discernible: (1) Concretions and nodules have higher Fe and lower Mn concentrations in the outer layers than in the core (Blume and Schwertmann, 1969). (2) The ratio of concentration in the nodules to concentration in the whole soil is generally much greater for Mn than for Fe (Taylor et al., 1964). (3) The larger concretions contain higher Mn/Fe ratios than the smaller ones (Phillipe et al., 1972). (4) With increasing depth there is a general trend for the Mn/Fe ratio in the concretions to increase.

1.2.3 Fate of Iron in a Limestone Aquifer

Edmund (1973) studied the relation between concentrations of various trace-metal ions in a limestone aquifer and the oxidation-reduction potential, pH of the solution, and concentrations of dissolved oxygen and dissolved sulfide (which is present mainly as HS^-). He conducted his research on ground waters travelling down-dip in the Lincolnshire Limestone (Middle Jurrasic) of Eastern England. The composition of the recharge water is modified by cation exchange, oxidation-reduction reactions, and mixing with connate water. There is a sharp oxidation-reduction barrier, some 10–12 km from the onset of confined conditions along the line of traverse. The Eh on the oxidizing side of the barrier is around +400 mV, apparently buffered by the presence of dissolved oxygen, and it then falls sharply to around +150 mV, corresponding to the complete reaction of the oxygen. The dissolved oxygen falls to zero. The water becomes steadily more reducing down-dip, coincident with the presence of sulfide species. In the deepest part of the aquifer there is an indication that in waters of connate origin, the Eh is relatively constant, around −100 mV.

Sulfate concentration is relatively constant (around 120 mg SO_4/l) in the oxidizing ground water. There is a sudden decrease in the amount down-dip some 3 km beyond the oxidation-reduction barrier, as a result of sulfate reduction. Even at depth, however, sulfate is still present.

Sulfide species are measurable (0.001 mg S/l) when the Eh is around +125 mV, although there is a section of the aquifer where neither sulfide nor oxygen was detected.

Two buffer sections can be recognized in the aquifer, which are separated by a third section in which a steady change in pH occurs, producing an alkaline barrier. For some 20 km from the onset of confined conditions, the pH is maintained at 7.2 but then rises to a level of about 8.3. This change is an expression of concomitant dissolution of carbonate minerals and ion exchange reactions.

The behavior of iron in the aquifer depends on the oxidation-reduction potential. Light brown residues, indicating ferric species, were found in water with Eh greater than +200 mV. The highest total dissolved iron values occur at low pH when the Eh is also low. Such conditions obtained in the aquifer between the Eh barrier and the alkaline barrier. The remaining water produces fine-grained black residues of FeS. The system shows very low values for soluble sulfide species, indicating that the condensation of Fe^{2+} with S^{2-} is rapid. Equilibrium between ferrous minerals other than sulfides and soluble Fe^{2+} is never reached, since FeS will be precipitated. The reaction will then proceed in dependence on the rate of production of HS^- and on the availability of ferrous materials. Precipitation of other trace metals is also governed by the precipitation of their sulfides.

1.2.4 Iron and Manganese in Seawater

The Fe concentration of ocean waters ranges up to 7.3 μg/l and the average is 3.0 μg/l (Crerar and Barnes, 1974). According to Byrne and Kester (1976) the complexes $Fe(OH)_2^+$ and $Fe(OH)_3$ are the most common soluble iron species in the normal pH range of seawater, whereas $Fe(OH)_4^-$ and $Fe(OH)^{2+}$ are present in negligible concentrations. There is at least one other significant ferric complex, possibly with bicarbonate,

carbonate, or borate ions in seawater above pH 8. From data on solubility constants Byrne and Kester concluded that iron in the ocean can exist in soluble form in concentrations not in excess of 2×10^{-8} gram-atom/l. Measurements of natural concentrations of iron in ocean water generally show dissolved iron concentrations not less than 2×10^{-8} gram-atom/l (Spencer et al., 1970; Brewer et al., 1972). The oceans are weakly supersaturated with respect to $Fe(OH)_3$. Byrne and Kester suggest that the observed iron concentration in excess of the solubility value is due to differences in the particle size distribution or chemical composition of naturally occurring colloid iron. They showed that 450-nm filters do not remove the smallest colloid particles present under natural conditions. It was observed in a simple filtration experiment at pH 7.87 that the concentration of Fe(III) passing a 50-nm filter was 5.9×10^{-10} gram-atom/l and that of Fe(III) passing a 450-nm filter was 3.9×10^{-9} gram-atom/l. Organic complexes of Fe(III) are probably insignificant equilibrium species of iron in seawater (Kester et al., 1975).

The Mn content of ocean waters ranges between 0.5 and 8.0 μg/l (Rona et al., 1962). Approximately 10% of the total Mn in seawater is colloidal or particulate. An additional 10% is organically complexed and the remainder consists of simple soluble inorganic ions and complexes. Mn^{2+} is the most common soluble species in seawater. $MnCl^+$ and $MnSO_4^0$ are present in considerable concentrations at the normal pH of seawater, whereas $MnHCO_3^+$, $MnCl_2^0$, $MnCl_3^-$, $MnOH^+$, MnF^+, and $Mn(OH)_3^-$ are present in negligible concentrations (Crerar and Barnes, 1974). Ehs of marine bottom waters generally lie within the limits 0.20–0.45 V. Based on these data, Crerar and Barnes concluded that the oceans are undersaturated with respect to manganic oxides in contrast to the supersaturation state of ferric oxides. Spontaneous precipitation of Mn from seawater is impossible.

2. Dispersion Process

Dispersion involves two stages: (1) weathering of the parent rock material, which can be either physical disintegration or chemical decomposition followed by the formation of new minerals, or both, and (2) suspension of the small particles in the liquid phase, which involves detachment of the particles from the upper layer of the sediment and peptization. In many cases the two stages occur simultaneously. In some rare cases the two stages are separated by long geologic periods.

2.1 Physical Weathering

If a stress is applied to a consolidated material it leads to strains and eventually to rupture. *Disintegration* is a breakdown whereby the rock loses its coherence, and aggregates are broken down to smaller units. The chemical composition of the materials in the rock changes only minimally, but the mineralogic constitution may be changed. Crystal defects and amorphous surface layers are formed, and if the stress is applied over long periods, crystalline minerals may become amorphous.

In principle it should be possible to break down any macroscopic solid into colloid particles by sufficient mechanical grinding. In practice a definite limit is reached, depending on the rock and the weathering processes, owing to the tendency of the small particles to aggregate again by virtue of their high surface energy.

There are many agents of physical weathering that lead to disintegration of rocks and crushing of fragments. The principles of these processes are similar and include (1) explosion of solid material as a result of a bursting pressure exerted by the expansion of fluids and (2) mechanical disintegration caused either by geologic activity or by influences of living organisms. The following may be regarded as representative agents of physical weathering.

1. Condensation of Solids in Cracks: When liquid water freezes it expands 9%, thereby exerting a bursting pressure on its surroundings. The many repeated alternations of freezing and thawing in cold climates is much more effective in breaking rocks than the fewer, but deeper, freezes of the extreme Arctic.

Fine dirt particles enter into small crevices in a boulder and put the rock under stress. Expansion of the particles widens the cracks and during subsequent cycles of contraction and expansion the boulder splits. (Ollier, 1965). According to Yaalon (1970) cracking of boulders on desert surfaces is the result of gypsum and salt crystal growth in the fine crevices. Dew, together with airborne aerosols and the infrequent rains, supply the required moisture and salts for the process to proceed.

2. Wind: About three-quarters of the material moved by the wind does not rise as a dust cloud but comprises grains of silt and sand and even bigger rock fragments that roll or skip along the ground. Collisions between these grains and the ground may cause its pulverization.

3. Running Water: Streams, and such slope processes as rill wash and downslope movements carry with them heavy boulders and cobbles. In the process of sliding along the bed of the stream these fragments may grind minerals thereon.

4. Sea Waves: Waves move sand and gravel even if they are moderate. Besides the fact that the wave itself is a powerful weathering agent, storm-driven pebbles and sand act as a gigantic horizontal saw biting ceaselessly into the land.

5. Glaciers: These are slow-moving, thick masses of ice. As a glacier slowly creeps and slides downhill impelled by its own weight, it drags with it the material underneath. Thus glaciers transport immense quantities of coarse rock waste. The moving glacier, with its tremendous weight, is one of the most powerful rock-grinding agents known. Hummocky ridges of boulders, sand, and silt, distributed without appreciable sorting or stratification, are piled along the ice front. Most pebbles are sharply angular. Some fragments are faceted, with nearly flat, grooved, and polished surfaces formed as the boulder scrapes along the bed rock. Such unsorted debris, deposited directly by the ice, is called "tills".

The melt water is dirty milky white. The suspended particles are not necessarily clay minerals, but may be finely ground rock-forming minerals. Rubble and coarse sand are deposited in front of the glacier. The finer debris moves on into rivers to build flood plains and into lakes and seas to form bottom deposits. The colloid suspensions obtained with this fine debris are stable for long periods. The famous "green" coloration of several lakes in regions of glaciers is due to this colloid solution of the pulverized rock. This color is observed during summer and autumn. In winter the upper layer of the lake freezes. The layer of ice prevents the mixing of the lake water by the wind. Slowly, the small particles settle out. When the ice layer melts, the lake is observed to have a blue color. This phenomenon is cyclic.

Rock flour is also widely blown about in late summer when the glacial streams shrink and expose their channels to the wind. Unstratified deposits, constisting chiefly of silt particles carried by the wind, appear near glaciated areas. These are known as "loess".

6. Rolling Rocks: The impact of a rock falling or rolling from above can lead to the disintegration of the ground.

7. Plant, Animal and Human Activities: Organisms help to break rock down into soil, both mechanically and chemically. Growing plant roots powerfully wedge soil and rock. They open cracks in rocks and loosen the bondings between mineral grains. Burrowing animals move and mix the soil effectively.

8. Explosive Activity of Volcanoes: This has been discussed in Chapter 1.

During the geologic history of the earth more than 70% of weathered magmatic or metamorphic rocks have been reduced in grain size from diameters between 0.1 to several millimeters, to diameters of less than 20 μm. The physical weathering of rock debris is a process in which the rock material undergoes transformation in several cycles. That is to say that with the passage of time the same rock material is being ground over and over again (Wedepohl, 1971).

Quartz has a very high resistance to crushing processes. It has a high surface tension and no cleavage. According to Kuenen (1960) the transport distance necessary to bring about a perfect rounding of a quartz grain is of the order of several earth circumferences. Average mineral composition of igneous rocks shows 18% quartz. It rises to 30–44% in graywackes in which physical weathering has reached a significant level. On further weathering the sorting according to grain size also becomes very important. The average quartz content of sandstones is 65–67%, while that found in shales and clays is only 20%. Even deep-sea sediments frequently contain 10–15% quartz, part of which has probably been transported as a suspension that was stabilized due to the many crystalline defects and the amorphous coatings of the crushed quartz crystals. Some of the quartz has probably been transported by wind, in the form of aerosols (see, e.g., Jackson et al., 1971; Rex and Goldberg, 1958).

Feldspars are almost as hard as quartz, but cleavage is well marked, especially in the plagioclase feldspars, and this allows rapid attrition. They occur in most igneous rocks (64%) and metamorphic rocks. They are abundant in arkoses and graywackes (35%) but are uncommon in clays and shales, because of their chemical weatherability. Special conditions are required to form feldspar-rich suspensions of the sort that might settle and lithify into arkoses and graywackes. Weathering in arid regions, particularly by granular disintegration, may produce transportable debris with little chemical alteration. Nile muds contain fresh microcline and orthoclase, indicating that the suspension is derived from an arid region. Glacial regions also show much physical and little chemical weathering and so can also provide feldspar-rich suspensions (Ollier, 1969).

The layered structure of mica-type minerals results in easy disintegration of mica grains into flakes. Micas occur abundantly in igneous and metamorphic rocks and are also found in smaller amounts in numerous sedimentary rocks. They are usually absent in wind-blown sediments or rocks derived from them, as micas are very prone to attrition during wind transport (Ollier, 1969).

2.2 Chemical Weathering

Most minerals of igneous and metamorphic rocks, which are formed under high temperatures and high pressures and in chemical environments differing from those of the surface, are thermodynamically unstable when exposed at the earth's surface. They react slowly with oxygen, carbon dioxide, and moisture from the atmosphere and at a faster rate with liquid water containing some organic and inorganic dissolved compounds, forming new minerals. Most of the newly formed minerals are clay minerals, which tend to form small crystals and are capable of forming colloid systems. This type of weathering, whereby the chemical composition of the newly formed solid phase differs from that of the parent phase (before the process had started) is known as *chemical weathering.* Clay minerals are also subjected to chemical weathering if they reach an environment that has a chemical constitution different from that of the environment of their formation. This type of chemical reaction will be discussed further in Chapter 7.

From a study of the weathering of granite gneiss, diabase, and amphibolite Goldich (1938) concluded that the weathering sequence for the common rock-forming minerals is:

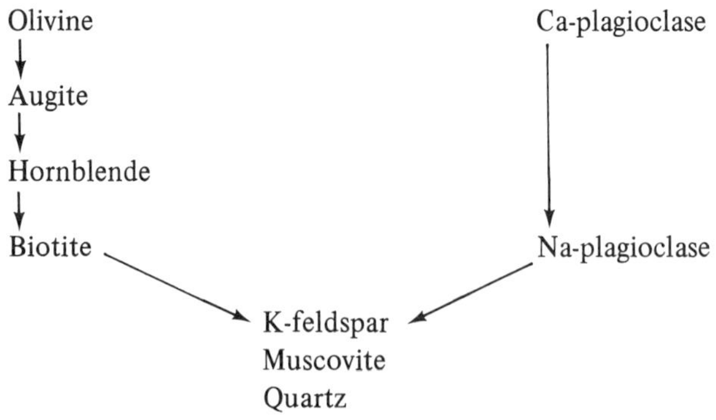

The weathering of silicates involves at least three classes of chemical reactions: (1) reaction of the mineral with an aqueous solution with which it is not in equilibrium; (2) reactions among soluble species in the aqueous phase; and (3) reactions between the aqueous solution and stable and/or metastable species sorbed on the residual primary particles or forming an independent solid phase. In general, reactions among species in an aqueous solution are rapid compared to those among condensed phases and are reversible through the entire process. The first and third classes of these reactions are the rate-limiting steps in the weathering process, whereas the remaining class is a step that can be represented by reversible reactions and evaluated with the aid of equilibrium thermodynamics. Following this model, Helgeson (1971, 1972) carried out a kinetic study of mass transfer between K-feldspar and aqueous solutions. In the initial stages of reaction of K-feldspar with an aqueous phase, H^+ is exchanged for K^+ on the surface of the reactant feldspar. A congruent dissolution of the mineral causes the concentration of aluminum in solution to increase until the bulk solutions reaches saturation with respect to gibbsite. If no supersaturation occurs, a surface layer of gibbsite then begins to precipitate from the solution at the expense of K-feldspar while the concentrations of SiO_2 and K^+ continue to increase in the bulk solution, and dissolution becomes incongruent. At a certain stage, the interstitital solution in the surface layer of gibbsite reaches saturation with respect to kaolinite at the interface of the gibbsite zone and the reacting feldspar. A zone of kaolinite begins to precipitate from the solution between the gibbsite zone and the reacting feldspar. The rate-limiting step of the kaolinite formation is the slow diffusion of material from the reacting mineral through surface layers of intermediate reaction products into the bulk solution, but the mineralogy of the reaction product is determined solely by the reversible chemical reaction in the aqueous phase and their equilibrium thermodynamics.

Taking this into consideration, the activities of ions in the solid phase are defined as being equal to 1. Phase diagrams describing the *fields of stability* of minerals at various ratios between the activities of ions dissolved in the weathering solutions are useful in describing the minerals formed in various weathering zones. These diagrams are a basis for quantitative conclusions regarding silicate mineral alterations and modifications (Kramer, 1968).

Although thermodynamic calculations indicate that silicate minerals are sparingly soluble, the sequence of aqueous dissolution of the common rock-forming silicate minerals obtained by computation of free energy of aqueous dissolution is in good agreement with Goldich's sequence of weathering (Yariv, 1971). This may indicate that chemical weathering is governed mainly by aqueous dissolution. Water usually percolates in the weathering system, and its contact with the surface of a definite crystal is very brief. It may have time to react only with the surface of the mineral. In closed systems, after the surface of the crystal has reacted with water, further reaction becomes more difficult and weathering may eventually stop. In other words, these heterogeneous reactions have very high activation energies and are therefore very slow. On the other hand, in the geologic systems, the weathering process requires a large amount of percolating water and the soluble weathered products are continuously removed. The chemical equilibrium is, therefore, displaced in the direction of dissolution. After the ions have passed into aqueous solution, either by ion exchange or by heterogeneous hydrolysis (weathering reactions of class 1), or by the breakdown of the silicate surface layer (reaction of class 3) they may form a new mineral phase (reaction of class 2) and precipitate. The whole process may occur in situ, or after the ions have been removed elsewhere.

The kinetics of reactions in which silicates are involved are usually slow because of the complexity of the crystal structure of the minerals. The major components to be considered in silicate weathering are Na_2O, K_2O, MgO, CaO, FeO, Fe_2O_3, Al_2O_3, SiO_2, and CO_2. These normally exist in the aqueous phase as Na^+, K^+, Mg^{2+}, Ca^{2+}, Fe^{2+}, H_4SiO_4, HCO_3^-, H_2CO_3 and monomeric, hydrophilic polymers and organic complexes of Al^{3+} and Fe^{3+}.

It is debatable whether chemical weathering should be treated in a section that deals with condensation rather than with dispersion. However, there is a very strong connection between parent rock, percolating water (which is the dispersing agent), and the weathering product, which goes beyond chemical composition (see, e.g., Ch. 7, Sect. 4). It is therefore more convenient to deal with chemical weathering together with dispersion processes. In addition to precipitation reactions from solutions and aging of precipitates, which were discussed in Section 1, the following are the most common chemical reactions occurring during the weathering of common rock-forming minerals.

2.2.1 Ion Exchange

This reaction is best demonstrated from the behavior of micas that have a crystal structure similar to that of clay minerals. Studies of the artificial weathering of micas show that their interlayer potassium is replaced by a hydrated metallic cation and that the micas are eventually converted into a swelling mineral resembling vermiculite (Barshad, 1948). Of the inorganic cations, Ba^{2+} is the most effective exchanging cation (Rausell-Colom et al., 1965). The *c*-spacing of the ultimate product is 1.22 nm for the Ba form, 1.48 for the Ca form, and 1.43 for the Mg form, indicating the adsorption of one and two water layers into the interlayer space of Ba-type and of Ca- or Mg-type weathered mica, respectively (Kodama and Ross, 1972; Kodama et al., 1974).

Hydrobiotites, in which potassium is removed only from alternative layers, and which are characterized by a regular alternation of 1.0 nm biotite and 1.4 nm vermiculite type unit cells, are common in nature. They have been synthesized by using very dilute solutions to exchange the potassium and by oxidizing the iron of biotite during exchange (Farmer and Wilson, 1970).

Ion exchange of K by H is obtained by treating mica with organic acids. A vermiculite is thereby obtained with exchangeable Al and H in the interlayer space of the mineral. The Al is obtained from the partial dissolution of the mineral. If the organic acid forms complexes with Al and the Al concentration in the solution thereby increases, a vermiculite is obtained with complexed hydroxy-aluminum in the interlayer space (Vicente Hernandez and Robert, 1975).

As interlayer potassium is removed from a mica, various other irreversible changes occur in the unit layer, such as loss of charge, loss of octahedral cations, and structural distortions, so that if potassium is reintroduced into the weathered mica the original structure is not always obtained. Reentry of K in the K-depleted muscovite results in a *c*-spacing of 1.04 nm and not of 1.0 nm given by the original muscovite (Kodama and Ross, 1972).

As the particle size of the mica is reduced, exchange is more rapid, but when the particle size is very small, and is of the order of one μm, the exchange becomes very small and decreases further as the particle size further decreases (Scott, 1968; Reichenbach and Rich, 1968). The exchange is controlled by a diffusion process and normally the boundary between the altered and unaltered mica is well defined and proceeds inwardly from the edge of the particle (Mortland, 1958; Quirk and Chute, 1968).

The rate of K release is much slower in muscovite than in biotite or phlogopite (Rausell-Colom et al., 1965; Reichenbach and Rich, 1968). In phlogopite and biotite, where the OH group is normal to the silicate sheet the proton is near the interlayer potassium and exchange is rapid. In muscovite the OH group is at an acute angle to the silicate sheet. The increased distance between the proton and the potassium causes the cation to be held more firmly and consequently the exchange reaction is slower. While an ion exchange reaction occurring in expanding clay minerals is very rapid and can be identified within a few seconds, several weeks are necessary before ion exchange in micas can be detected.

2.2.2 Heterogeneous Hydrolysis

The reaction between silicate minerals particles and water acidified by CO_2 or oganic acids involves an exchange of metal ions for H^+ on the surface of the solid, leading to an increase in the pH of the aqueous phase. The protonated residual anionic silicate phase forms a weak acid. Physical sorption of organic acids appears to be of only minor importance in weathering reactions. The degree of the exchange reaction depends directly on the oxygen:silicon ratio in the silicate structure, being greatest for the olivines and least for the quartz. As the oxygen:silicon ratio increases from quartz to olivines, a greater percentage of oxygen bonding power is available to form bonds with cations other than silicon, including protons. The reaction seems to be dependent on the total surface area of the solid, that is, on the number of exchange sites available (Deju and Bhappu, 1966; see pages 213–216).

Deju (1971) constructed a mathematical model for the chemical weathering of rocks by percolating water in which the mathematical description of water flow includes terms for the ion exchange reactions. The application of the model to field data shows that protonation of silicate surfaces through the ion exchange mechanism is the chemical reaction that determines the rate of chemical weathering and that the rate of metallic ion desorption from the bed subjected to weathering increases with the oxygen:silicon ratio of the minerals and with the solubility of the corresponding metal hydroxide but decreases with the rise pf pH of the percolating solutions. The rate of desorption decreases in the order from monovalent to trivalent cations and from smaller to larger cations. High oxygen:silicon ratio silicates weather more extensively than feldspars and quartz sands. Desorbed species are transported in solution, being deposited at a point where the porosity of the bed is reduced and where ultrafiltration governs the geologic process (see Ch. 10). Certain organic compounds that may form stable soluble complexes with di- and polyvalent cations increase the desorption of these metals from the rock surface.

2.2.3 Structural Collapse of the Residual Anionic Lattice of the Surface Layer

As a result of proton sorption a high acidity is obtained at the solid-liquid interface, and Si–O–Si and Si–O–M groups (where M is a metal atom such as Mg, Al, or Fe) dissociate, forming small polymers and monomers. The reaction is catalyzed by the adsorbed proton, and it can be regarded as the reverse reaction of the condensation and polymerization described in the previous section. If the dissociation products form small species with a high hydration energy, they may go into solution whereas the larger species form the surface residual layer, the structure of which differs from that of the bulk crystal. The surface residual layer reaches a steady-state thickness that depends on the temperature and acidity of the aqueous solution (Paces, 1973).

The breakdown of the silicate framework can also be regarded as a hydrolysis reaction, in which water molecules dissociate into hydrogen and hydroxyl ions, permitting the hydroxyl ions to bind to exposed Si or M and the hydrogen ions to oxygens. This type of hydrolysis has almost no effect on the pH of the solution whereas the hydrolysis discussed in Section 2.2.2 causes the pH of the liquid phase to rise considerably.

2.2.4 Oxidation–Reduction in Minerals

Oxidation–reduction reactions either in the solid phase or in solution of leached species, may be important in the weathering of minerals containing considerable concentrations of iron. Ferromagnesium minerals are known to be unstable and usually are the first to undergo chemical weathering. Ferrous iron is easily leached by percolating water. Thermodynamic data indicate that magnesium should be more easily leached. If, however, iron in the solution is oxidized and the oxidation is followed by precipitation of $Fe(OH)_3$, ferrous iron-bearing silicates are then less resistant to leaching than magnesium-bearing silicates (Yariv, 1971).

If iron comprises a major element of a mineral, oxidation–reduction reactions may result in a complete destruction of the original mineral structure, and the formation of a new phase, for example, oxidation of pyrite, results in the formation of iron oxides. Ferric oxides are more soluble than pyrite and consequently are more subjected to further weathering.

Oxidation–reduction of minerals, of which iron comprises a minor element, e.g. amphiboles, micas, and clay minerals, may result in no change or in only slight changes in the crystal structure. The oxidation of structural ferrous iron in these minerals is associated with (1) a decrease in the net negative charge of the silicate anion; (2) a reversible deprotonation of structural OH; and (3) an irreversible ejection of octahedral metallic cations (Veith and Jackson, 1973). Most vermiculites, if they are formed from micas, have a layer charge less than the parent mica. The simplest explanation for the charge reduction on weathering to vermiculite would be ferrous iron oxidation to ferric iron (Walker, 1949). Barshad and Kishk (1970) showed a reversible increase in cation exchange capacity of vermiculites after reduction of structural ferric iron. Ismail (1970) was able to reverse the process of oxidation and restore a weathered biotite to nearly its original charge by reducing its ferric iron.

Air oxidation is accompanied by the loss of hydrogen from the silicate hydroxyl plane (Farmer et al., 1971) and thereby the oxidation of Fe is not accompanied by any loss of layer charge. The reaction proceeds as follows:

$$2\,Fe^{2+} = 2\,Fe^{3+} + 2\,e^-$$

$$2\,[OH]^-_{lattice} + 2\,e^- = 2\,[O]^{2-}_{lattice} + H_2. \qquad (3.VII)$$

A deficiency in sturctural OH and an excess of O^{2-} was found in many trioctahedral micas from various localities and was attributed to deprotonation of hydroxyls during Fe oxidation (Rimsaite, 1970).

Gilkes et al. (1973) showed that oxidation results in stabilization of biotite against weathering. They porposed that during the oxidation of lattice ferrous iron, ejection of some octahedrally coordinated cation occurs so that the charge balance can be retained and thus adjacent hydroxyls may develop an inclined orientation characteristic of dioctahedral micas. This in turn results in a more stable environment for interlayer K, so that losses to weathering agents such as salt solutions become very small. The rate of dissolution of biotite in dilute acid solutions also decreases as the proportion of ferric iron in octahedrally coordinated sites increases.

Rosenson and Heller-Kallai (private communication) showed that air oxidation of lattice ferrous iron in montmorillonite is more rapid in aqueous suspensions than in the dry state. The rate of oxidation increases with increasing surface acidity and surface area of the clay mineral.

Iron-bearing olivines become less stable after oxidation. Oxidation of iron in olivine crystal results in increasing concentration of lattice defects (Nitsan, 1974). Dissolution, and consequently weathering, is a defect-sensitive property of olivines and is therefore likely to show an increase after oxidation of the solid mineral.

2.2.5 Role of Biologic Activity in Chemical Weathering

Plant roots and associated microorganisms possess potent mineral-weathering capabilities. The variety of organic weathering agents produced by decomposing plant and animal tissue, as excretions from living root cells, or as metabolites from microbes, creates an environment in which primary minerals can be attacked and degraded to their ionic components and residual compounds. Many potential weathering agents have been identified in plant root exudates and many more are produced in organic matter decomposition. Ions released from mineral surfaces by one of the mechanisms previously described are incorporated almost immediately by ion substitution and chelate formation into microbial or higher plant systems. In biotite, mineral alteration has been related by Mortland et al. (1956) to oxidation of octahedral ferrous iron by root material with subsequent removal of potassium. In acid soil environments where many organic chelating agents are present, biotite weathering may result in an amorphous metal and alkali-depleted product rather than in crystalline clays containing significant amounts of exchangeable cations. Higher plant and microbial utilization of nutrient ions or leaching could rapidly remove released ions from weathering sites, allowing the weathering processes to continue unhindered by end-product accumulation (Boyle and Voigt, 1973). In environments rich in biogenous products the Goldich weathering sequence does not necessarily hold due to the interaction of the minerals with organic substances.

2.2.6 Role of Percolating Water in Chemical Weathering

Leaching is the selective removal of chemical constituents which are soluble or which form stable hydrophilic colloid solutions from rocks, rock debris, or soils, by the action of percolating water. Species (mainly ionic) of low formula weights, are transferred from the rock into the percolating water by one of the following chemical processes, characteristic of chemical weathering: ion exchange, heterogeneous hydrolysis, depolymerization of aluminosilicate networks, formation of soluble organometal complexes, and oxidation–reduction reactions. Leached species can leave the aqueous percolating phase by the following reactions: condensation and formation of sparingly soluble species, by being adsorbed (or ion-exchanged) on surfaces of the solid noneroded phase or by the flocculation and settling out of the colloid species. The chemical and mineralogic composition of the residual geologic system depends on the degree of leaching.

From field observations, Mohr et al. (1972) and Tedrow and Wilkerson (1953) concluded that alkali and alkaline earth metals are the first to be leached from igneous rocks during the early stages of weathering in tropical and temperate regions. These metals reside mainly in plagioclase and in ferromagnesium minerals of relatively low stability toward weathering. These minerals disintegrate in the early stages of weathering.

Because of their high solubility these metals are removed from the weathering region. Pyroxene phenocrysts are relatively more resistant to weathering than are matrix minerals found in andesite, and Hendricks and Whittig (1968) noted a tendency of Mg to resist leaching during the earliest weathering stages of this rock. Navrot and Singer (1976) found a nearly complete leaching of Na, Mg, and Ca during the transformation of a basic igneous rocks into kaolinite soil clays in humid Mediterranean conditions. The leaching of K was incomplete and the residual K was associated with the formation of some illite. Butler (1954) and Johnson and Likens (1968) found that under temperate humid weathering conditions, Mg released by the breakdown of primary minerals is also retained in pedogenic clays. This element is capable of entering the octahedral sheet of the clay minerals resulting in the formation of smectites.

The leaching of Fe from parent rocks is limited, both by the low solubility of Fe(III) and by the high stability of its host minerals. In basic igneous rocks a fraction of Fe is contained in the divalent form in relatively unstable minerals. Navrot and Singer (1976) found that only a moderate part of Fe was leached from basic igneous rocks during clay formation and that a large fraction of this metal, after it had been released from olivines and pyroxenes, was then incorporated into the clay fraction.

Certain organic compounds, which may form stable soluble complexes with di- and polyvalent cations, bring about a reduction in the activity of the polyvalent cations in solution and increase the leaching of these metals. The bulk of dissolved organic matter in natural waters consists of highly oxidized and chemically and biologically stable polymeric compounds closely resembling soil humic substances. Average concentrations of these aquatic humics were estimated to be 3–45 mg/l in major rivers of the United States. Fractional elution of soil organic matter by meteoritic waters is the main process contributing to the presence of humic matter in rivers. The aquatic humic polymers participate in complex formation through ionizable functional groups, such as carboxylic and phenolic, having a wide range of acidity (Reuter and Perdue, 1977).

Low-molecular-weight organic acids are effective in mineral degration and metal leaching through complex formation with metallic cations (Huang and Keller, 1970).

These acids constitute only a minor fraction of the organic fraction of soils and sediments. Humic substances, which are of high molecular weight, also play a significant role in the leaching processes of rock weathering, because of their capacity to extract considerable amounts of major and transition metals. Singer and Navrot (1976) showed that after a rapid initial rise, the leaching rates of metallic cations from basalts decrease rapidly. Fe is the most easily leached metal, followed by Al, Ca, and Mg. The transition metals are leached in trace amounts, relative to their contents in basalts. Many metal humates, being hydrophobic colloids, display low solubility in water, but they are readily converted into hydrophilic colloids and become mobile in the presence of excess humic acids (Baker, 1973). The mechanism of the conversion of these colloid species from hydrophobic to hydrophilic is discussed in Chapter 8. According to Bloomfield et al. (1976) transition metals are mobile in natural waters partly in association with colloid humified organic matter and partly in true solution in the form of anionic complexes.

The limiting factor for the leaching of silica during weathering of aluminosilicate rocks is the low solubility of this element. For the losses in Si to become marked, intensive leaching must have taken place, and this is common only under tropical conditions. Protons may serve as catalysts for the depolymerization of silica species and consequently, the removal rate of silica may be accelerated by an acid weathering environment (Hendricks and Whittig, 1968). This will be further discussed in Chapter 6.

2.2.7 Acidity of Groundwater in Aluminosilicate Rocks

Shvartsev (1975) measured pH values of ground waters collected from various acid igneous rocks and compared them to theoretically calculated pH values. For the calculations he assumed that during the weathering of the aluminosilicate, clay minerals are produced and cations pass into solution. The system is balanced by the adsorption of protons onto the solid clay phase and the presence of hydroxyl ions in the ground waters. An example is the weathering of anorthite to give kaolinite according to the chemical equation:

$$2\,Ca\,[Al_2Si_2O_8] + 6\,H_2O \rightarrow Al_4Si_4O_{10}(OH)_8 + 2\,Ca^{2+} + 4\,OH^- \qquad (3.VIII)$$

The cation content of the ground water should allow the calculation of a theoretical pH for water interacting with the aluminosilicates. The actual pH values for leaching ground waters from nepheline syenites, granite gneisses and schists, and granodiorites were found to be between 5.3 and 7.4 pH units. They are 4 or 5 units lower than the calculated values, which ranged between 9.9 and 11.5. This occurs because the excess alkalinity is neutralized by carbon dioxide according to the chemical equation

$$OH^- + CO_2 \rightarrow HCO_3^-. \qquad (3.IX)$$

Water passing through the soil in the upper horizons in the earth's crust becomes enriched in carbon dioxide, probably of biologic origin.

2.3 Detachment of Particles Caused by Rainfall

The term *detachment* is used to describe the removal of transportable fragments of rock and soil materials from their resting place by physical eroding agents, usually falling raindrops, running water, or wind. As a result of detachment, particles or aggregates are made available for dispersion in colloid suspensions.

Ellison (1952) and Young and Wiersma (1973) determined the relative importance of raindrop impact and flowing water as contributors to the erosion process and concluded that the major force initiating soil detachment is derived from the impact of falling raindrops. Not all particle sizes are equally susceptible to detachment (Fig. 3.10). Because of their physical mass, gravel and very coarse sand may be very resistant to detachment. Coarse and medium-size particles detach quite readily under raindrop impact. From the coarse sand sizes through the silt sizes, resistance to detachment increases (Farmer, 1973).

In the present section the mechanism of detachment of particles caused by rainfall is described. The detachment of particles by flowing fluids will be treated in Chapter 5. The model used in the present section deals with three extreme stages: (1) dry soil; (2) soil–water mixture (fluidized soil); and (3) soil and overland flow. In all three stages the decrease in relative detachability is determined in part by the two factors, particle size and hydration ability (Yariv, 1976). Most minerals that are included in the category of small-size particles have a high energy of hydration. Sorption of water from falling rain is rapid due to the high kinetic energy of the reacting molecules (water molecules in the rain drops).

2.3.1 First Stage–Dry Soil

When rain begins to fall, the soil is dry, and collisions between raindrops and soil particles may result in detachment. If the soil particle does not sorb water, the collision may be regarded as a collision of two elastic bodies. If this collision leads to

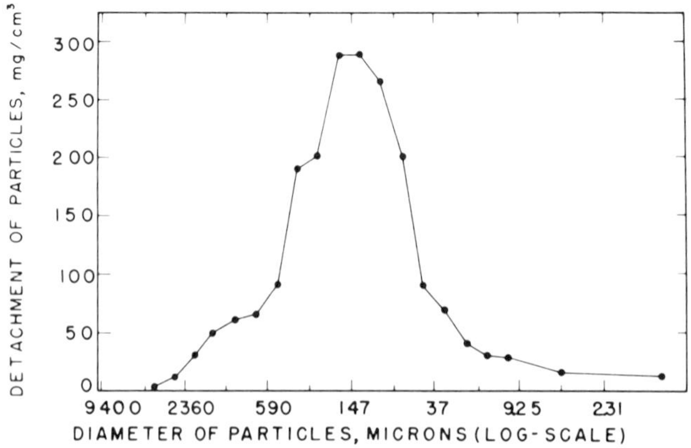

Fig. 3.10. Detachment of soil particles (mg/cm³ of water) in simulated rainfall. (From Mazurak and and Mosher, 1968. Reproduced by permission of Soil Sci. Soc. Am.)

the sorption of water molecules by the particle, the mechanical force by which the drop is acting to detach the particle is thus effectively smaller. Such a collision can also lead to a breakdown of the soil aggregates into smaller units. This causes a decrease in the efficiency of raindrops in detaching soil particles (Rose, 1961).

The detachment of a nonhydrated individual grain of weight G due to force F acting on the grain has been described by Yalin (1972). The detachment of a grain occurs when the value of the ratio F/G exceeds a certain limit determined by the force direction, the friction coefficient, and the disposition of the contacts between neighboring grains. There is a distribution in sizes and impact velocities of individual raindrops that depends on the range of drop sizes and on P, the intensity of rainfall. This distribution can be expressed as D, and is a function of P and of K_m, the kinetic energy of a drop having that size, s_{50}, at which 50% of the water falls in drops whose diameters are larger than s_{50} and 50% in drops of diameters smaller than s_{50}. Only those drops with a kinetic energy high enough to fulfil Yalin's stipulation are effective in detaching particles. Hence, the probability, A, of detachment of particles with a diameter $2r$, density d, and a weight of $G = 4/3\, d\pi r^3$ can be expressed by the equation

$$A = xyz'z''D^{-r^3} \tag{3.4}$$

where x and y are, respectively, probability constants ($x \leqslant 1$) that depend on the concentration and density of particles of radius r. z' and z'' are, respectively, probability constants ($z' \geqslant 1 \geqslant z''$) that depend on the concentration of aggregates which tend to disintegrate to give particles of radius r, and aggregates of radius r which tend to disintegrate.

If the particle tends to hydrate as the direct result of the impact of a raindrop, then a correction of Yalin's stipulation can be made as follows: $F/(G + H) \geqslant$ constant, where H is defined as the "hydration force" and is equal to the amount of the mechanical force that was lost by the raindrop due to hydration. The hydration force is proportional to the surface area of the particle (πr^2) and the hydration energy of the mineral. Because $H = h\pi r^2$, where h is a coefficient depending on the mineral type (i.e., its hydration energy), the probability, A, of particles with a diameter of $2r$ being detached is given by the equation

$$A = xyz'z''D^{-(r^3 + hr^2)}. \tag{3.5}$$

Since H decreases with increasing particle size, a possible solution to Eq. (3.5) results in a detachment behavior similar to that described in Fig. 3.10.

2.3.2 Second Stage–Soil–Water Mixture

As the rain continues the upper layer of the soil is fluidized. Impacting raindrops may cause the splash of drops of this fluidized soil. The size of the splashed drop depends on the kinetic energy of the impacting raindrop. Large particles can be detached only as large drops of fluidized soil. It is therefore to be expected that the probability of detachment will decrease with increasing particle size.

For a given amount of energy supplied by an impacting raindrop the size of the splashed drop will decrease with increasing surface tension of the fluidized soil. Although the surface tension of a liquid, in general, is unaffected by the presence of relatively large quantities of inert or inactive substances, it is markedly influenced by active impurities, even in trace quantities. Grains of soil minerals that form irreversible or unstable sols (e.g., quartz) have almost no effect on the surface tension of water. However, amorphous oxides and clay minerals increase the surface tension. Surface tension can be related to the ratio between immobile and mobile water molecules in the fluidized bed. For a given concentration of solid material the surface tension will increase with increasing hydration energy of minerals.

Not all drops that may be formed in the fluidized soil will splash. Splashing occurs only after the drop has moved through a viscous fluid, the viscosity of which depends also on the kinds of minerals, and increases greatly for those minerals that have high surface charge densities and high hydration energies (Ch. 9). Small drops, having high vapor pressure, are not stable and may disintegrate rather than splash. It may be concluded from the preceding discussion that for fluidized soil the relative detachability decreases with increasing particle size or with increasing hydration energy of the detaching minerals. By the adoption of suitable coefficients, Eq. (3.5) and a figure similar to Fig. 3.10 describe the probability of detachment of particles during the second stage.

When water is able to infiltrate to greater depths, the concentration of mobile water molecules in the fluidized upper layer decreases. Surface tension and viscosity of this layer increase and the probability of detachment of particles decreases.

2.3.3 Third Stage–Soil and Overland Water

The third stage occurs when the fluidized soil is covered by a layer of free water or a dilute soil suspension. Drag forces acting on the fluidized soil are due to the turbulence of the layer of free water and may be the result of either raindrops impacting this layer or of overland flow. The model previously described for the formation of drops of fluidized soil followed by their migration through this fluid is also useful in the present case. By the adoption of suitable coefficients, Eq. (3.5) may serve to describe the probability of detachment of particles during this stage. In addition, the drop of fluidized soil must migrate through the overland water layer before it splashes. Depending on the thickness of the overland water layer, its flow rate and its viscosity, the fluidized soil drops may disintegrate so that soil particles are transported by flowing water. In this case, although detachment of soil particles occurs, it does not result in splashing, but in the formation of a colloid solution or a suspension. According to Young and Wiersma (1973), the amount of soil transported in surface flow is higher than that transported in splash.

In practice, there is no clear distinction between the three stages. A part of the soil layer may be fluidized and another part may still be dry, while with yet another part overland water may exist. The contribution of each stage depends on the duration and intensity of the rain and the infiltratability of the soil.

2.4 Peptization

Peptization is the term used by Thomas Graham (mid 19th century) to describe the process of separating from each other and suspending in a liquid particles of a colloid that have aggregated to form a gelatinous or finely granular mass. This definition clearly does not apply to the physical or chemical weathering of the parent rock. Analytical chemists use this term to describe the process of washing out precipitates until they pass through the filter paper after the removal of coagulating salt, followed by suspension of the colloid size particles by the flowing water.

In the present book this term is used to describe the process of the conversion of unconsolidated particulate rock material into colloid solutions and suspensions, due to the interaction with flowing water. This may include the removal of soluble coagulating salt and the addition of a soluble foreign material named *peptizer*, which may increase the ease of dispersal and the stability of the suspension. Peptization differs from detachment in that the former results in the formation of an aqueous suspension whereas the latter results in particles lying on a basement. However, the detached particles are readily peptizable.

Peptization progresses as follows. In the first stage water molecules penetrate into the aggregate, forming monolayers on the surface of the primary particles and water bridges between adjacent particles. In the second stage sufficient water penetrates into the aggregate for the surface of the solid particles to become electrically charged and for the diffuse electric double layers surrounding primary particles to be fully developed (Ch. 4). Two particles interact only when their diffuse double layers interpenetrate. Close approach of charged surfaces together with their associated double layers will generate repulsive forces between them (Ch. 8) that may result in swelling and further, in disintegration of the aggregate into smaller aggregates and primary independent particles. Repulsion and consequently the degree of disintegration of the aggregate, increase with increasing thickness of the double layer and depend on the type and concentration of the electrolyte. They increase with increasing electrolyte concentration up to a few micromoles per liter, but decrease with still further increase in concentration. A very small amount of electrolyte is necessary so that co-ions and counterions may form a diffuse double layer of a considerable thickness. As more electrolyte is added the thickness of the double layer is greatly reduced and repulsion between the primary particles decreases. Swelling and the degree of disintegration of the aggregate are also reduced. The degree of disintegration will increase after the excess of the electrolyte has been washed out by the flowing water.

In the third stage the particles are suspended by drag forces acting on them due to the turbulence of the flowing water (Ch. 5). Further disintegration of aggregates occurs as a result of shearing applied by the flowing water (Ch. 9). The thermal movement of water molecules may be sufficient to suspend very small particles in the range of colloid size.

Rapid peptizations occur and stable suspensions are formed in water if there are sufficiently great repulsive forces to overcome the interparticle attractions caused by van der Waals and electric forces. Small particles form more stable suspensions than big particles and peptization of small particles is much more rapid than that of big particles.

2.4.1 Peptizers

A colloid solution of independent primary AB particles can become more stable by the addition of a compound C which is easily sorbed on the surface of the particle, leading to weakening of interparticle links inside an aggregate. The rate of peptization of AB becomes higher in the presence of C, and C is therefore kown as a peptizer. The effect of the peptizer is demonstrated as follows:

1. There are strong interparticle attractions between AB particles. Let us assume that A and B are surfaces that have charges of opposite signs. A stable aggregate is then formed due to the electrostatic attractions between oppositely charged surfaces according to the equation

$$AB + AB \rightarrow ABAB \ldots \qquad (3.X)$$

2. When C is added to the system it is sorbed onto one of the surfaces according to the equation

$$AB + C \rightarrow ABC \qquad (3.XI)$$

3. C can either block the charge of B or even convert the sign of the charge of this surface to that of A.

Only weak interparticle attractions exist between ABC articles. An unstable aggregate is formed that is easily disintegrated during the peptization process according to the equation

$$ABC - ABC - \ldots \rightarrow ABC + ABC \qquad (3.XII)$$

The concentration of C should be small to avoid thinning of the electric diffuse double layer, which in turn results in aggregation.

Peptization is an important process occurring during the erosion of rock material, and natural peptizers are important chemical agents for erosion. The interaction of some natural peptizers with some clay minerals will now be described. It should be noted that most of the information obtained in the laboratory applies to minerals of the smectite group.

Oxygen planes comprising faces of layers of clay minerals are negatively charged and the edges of the layers are positively charged. In the absence of a peptizer an aggregate can be formed by face-to-edge interaction. Carbonate or hydroxyl ions are specifically sorbed onto the edges of the layers, changing their charges from positive to negative. Their presence in small amounts prevents aggregation (Szántó, 1962). The presence of small amounts of phosphates in soils has the same effect, and the same argument is true for those phenolates (which are decomposition products of humic substances) that contain at least three phenol groups attached to one benzene ring, two of which should be adjacent, e.g., (3.XIII) (van Olphen, 1963). Certain aromatic cations,

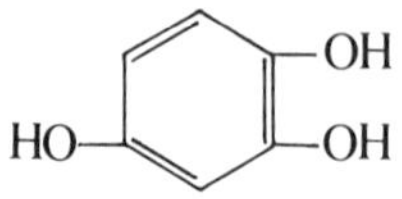

(3.XIII)

which are specifically sorbed by the negative oxygen planes, due to π interactions of the aromatic ring with this plane, may change the net charge of the face from negative to positive. At this stage, and as long as the thickness of the diffuse double layer is great, there will be an electrostatic repulsion between faces and edges as well as between faces and faces (Yariv and Lurie, 1971).

Small amounts of amphiphathic molecules may increase the rate of peptization. The polar molecules are adsorbed on the clay surfaces and prevent close approach of the clay particles (Neumann and Sansom, 1970). Further adsorption of these molecules decrease the thickness of the double layer, until van der Waals bonding takes place and the rate of peptization decreases.

The adsorption of uncharged nonionic polymers and surfactants is also known to peptize and stabilize lyophobic colloid particles. Depending on the surface concentration of the adsorbed molecules, there are two different approaches to the explanation of the stability of such dispersions (Bagchi, 1972, 1974).

1. When two particles with low density clouds of adsorbed polymer layers approach each other at distances of separation of their surfaces that are less than twice the thickness of the adsorption layer, mixing of the two adsorption layers takes place. If this process occurs with a net increase in the free energy, repulsion occurs between the particles, and if the repulsion energy is larger than the van der Waals attraction, a stable dispersion will result. Such a mechanism is probable when the concentration of the polymer in the adsoprtion layer is low. This is known as the "mixing" mechanism or "osmotic effect". The total repulsive interaction between two particles is the sum of the *entropic* and the *enthalpic repulsions.* The entropic repulsion is the increase in the free energy of the polymer chains due to a restriction of the possible number of configurations of the adsorbed polymer on close approach of the two particles and mixing of the two layers. When two surfaces with adsorbed polymer molecules approach each other at close distances, some solvent has to be squeezed out from the overlap volume of the solid-liquid interface and the polymer concentration in the adsorbed layer increases. The enthalpic repulsion is the increase in the free energy of the adsorbed polymer chains due to an increase in enthalpy arising from an increase in the polymer concentration in the overlap volume.

2. The polymer-adsorbed layers around the particles are quite dense because the surface coverage is very high. When the particles approach each other to a distance less than twice the thickness of the adsorbed layer, a denting of the two adsorption layers is produced. This process resembles the squeezing of two pieces of wet sponge together, resulting in the squeezing out of some solvent from the adsorbed layer. The squeezing out of solvent will also cause the concentration of the polymer in the interaction region to increase. In this type of interaction the particles may be regarded as if they undergo pseudoelastic collisions on approach rather than mixing of their adsorption layers. Such elasticity associated with the adsorption layers may be considered to be similar to the rubber elasticity of polymers. This mechanism is called the "denting mechanism" or the "volume restriction effect".

2.4.2 Effect of Ion-exchange Reactions on Peptization

The particle size distribution of clay minerals is assumed to depend on several factors, among which is the type of the exchangeable cation. This has been thoroughly investigated for smectites and will be discussed here. An aqueous suspension of any smectite mineral contains single unit layers dispersed in the system together with *tactoids* in which face-to-face interaction between unit layers results in the formation of a cluster of parallel layers separated by water layers of a small thickness (less than 1.0 nm). This water layer comprises the *interlayer space* of the tactoid. For any exchangeable ion an equilibrium exists between smaller and bigger tactoids in the suspension, and any discussion must relate to an average size of the tactoid.

A smectite tactoid can be regarded as an association colloid and the average number of unit layers that comprise a tactoid depends on the exchangeable cation present (Table 3.1). The most reliable methods for measuring tactoid size are those based on light transmittance (Banin and Lahav, 1968) and surface area measurements (Edwards, et al., 1965). Li-montmorillonite was chosen as the reference form since it shows the lowest light scattering, and thus it presumably has the smallest average tactoid size, assumed to be made up of one unit layer. In a water solution of low electrolyte concentrations, most particles of Na-montmorillonite comprise single unit layers. The number of tactoids having more than a single unit layer is greater in Na-montmorillonite than in Li-montmorillonite suspension. The average size of the tactoids increases with increasing charge of the exchangeable ion. Ca-montmorillonite suspensions contain tactoids averaging several (3–9) clay unit layers.

The history of the system greatly affects the average size of the tactoid, which depends not only on the cation present at the exchange site, but also on the cation that was previously sorbed on the clay. Drying of the clay results in larger tactoids (Lahav and Banin, 1968). During the exchange of Ca by Na in montmorillonite suspensions, the breakdown of the tactoids occurs when the equivalent fraction of Na in the montmorillonite increases from 0.2 to 0.5 and does not go further with further exchange of Ca by Na (Shainberg and Otoh, 1968). On the other hand, in the reverse exchange, during a titration of Na-montmorillonite by $CaCl_2$, formation of the tactoids occurs mainly when the equivalent fraction of Na is much higher than 0.5 (Lurie and Yariv, 1968).

Ion-exchange reactions occur widely in geologic systems but they become significant peptization and erosion agents when sodium exchanges di- and polyvalent cations and

Table 3.1. Tactoid size of montmorillonites containing various echangeable cations

Exchangeable cation	Average number of unit layers comprising a tactoid	
	Calculated from optical measurements (Banin and Lahav, 1968)	Calculated from surface area measurements (Edwards et al., 1965)
Li	1.00	1.00
Na	1.47	1.15
K	2.00	1.43
Rb	2.15	–
NH_4	2.60	2.34
Cs	4.60	4.00
Mg	9.60	8.45
Ca	10.88	5.68
Ba	11.15	9.76
Al	11.40	–

the excess electrolyte is washed out. The appearance of cracks filled with gypsum in clay sediments results from the percolation of sodium sulfate solutions through calcium-saturated clay sediments. Calcium is exchanged by sodium and gypsum precipitates according to the equation

$$Ca - clay + Na_2SO_4 \rightarrow CaSO_4 + Na_2 - clay \qquad (3.XIV)$$

The sodium clay consists of very small particles and is readily peptized. It is eroded from the system by percolating water, leaving space for further precipitation of gypsum.

References

Altschuler, Z. S., Jaffe, E. B., Cuttitta, F.: The aluminum phosphate zone of the Bone Valley formation and its uranium deposits, U. S. Geol. Survey Profess. Paper *300*, 495–504 (1956)

Bache, B. W., Sharp, G. S.: Soluble polymeric hydroxy-aluminum ions in acid soils. J. Soil Sci. *27*, 167–174 (1976)

Bagchi, P.: Enthalpic repulsion between two identical sperhical particles coated with a polymeric adsorption layer. J. Colloid Interface Sci. *41*, 380–382 (1972)

Bagchi, P.: Theory of stabilization of spherical particles by nonionic polymers. J. Colloid Interface Sci. *47*, 86–99 (1974)

Baker, W. E.: The role of humic acids from Tasmanian podzolic soils in mineral degradation and metal mobilization. Geochim. Cosmochim. Acta *37*, 269–281 (1973)

Banin, A., Lahav, N.: Particle size and optimal properties of montmorillonite in suspensions. Isr. J. Chem. *6*, 235–250 (1968)

Barshad, I.: Vermiculite and its relation to biotite as revealed by base exchange reactions, X-ray analyses, DTA and water content. Am. Mineral. *33*, 655–678 (1948)

Barshad, I., Kishk, F. M.: Factors affecting potassium fixation and cation exchange capacities of soil vermiculite clays. Clays Clay Miner. *18*, 127–137 (1970)

Biederman, G., Chow, T. J.: Studies on the hydrolysis of metal ions. Part 57. The hydrolysis of the iron and the solubility product of $Fe(OH)_{2.70}Cl_{0.30}$ in 0.5 NaCl medium. Acta Chem. Scand. *20*, 1376–1388 (1966)

Biederman, G., Schindler, P.: On the solubility product of precipitated iron hydroxide. Acta Chem. Scand. *11*, 731–740 (1957)

Bloomfield, C., Kelso, W. I., Pruden, G.: Reactions between metals and humified organic matter. J. Soil Sci. *27*, 16–31 (1976)

Blume, H. P., Schwertmann, U.: Genetic evolution of profile distribution of aluminum, iron, and manganese oxides. Soil Sci. Soc. Am. Proc. *33*, 438–444 (1969)

Bodenheimer, W., Kirson, B., Yariv, S.: Intensification of colour reactions between copper ions and polyamines by montmorillonite. Anal. Chim. Acta *29*, 582–583 (1963)

Boyle, J. R., Voigt, G. K.: Biological weathering of silicate minerals–implications for tree nutrition and soil genesis. Plant Soil *38*, 191–201 (1973)

Brewer, P. G., Spencer, D. W., Robertson, D. W.: Trace elements profiles from the GEOSECS–II test station in the Sargasso Sea. Earth Planet. Sci. Lett. *16*, 111–116 (1972)

Brockamp, O.: Dissolution and transport of manganese by organic acids and their role in sedimentary Mn ore formation. Sedimentology *23*, 579–586 (1976)

Brümmer, G.: Redoxpotentiale und Redoxprozesse von Mangan, Eisen und Schwefelverbindungen in hydromorphen Böden und Sedimenten. Geoderma *12*, 207–222 (1974)

Butler, J. R.: The geochemistry and mineralogy of rock weathering. 2. The Nordmarka area, Oslo. Geochim. Cosmochim. Acta *6*, 268–281 (1954)

Byrne, R. H., Kester, D. R.: Solubility of hydrous ferric oxide and iron speciation in seawater. Mar. Chem. *4*, 255–274 (1976)

Crerar, D. A., Barnes, H. L.: Deposition of deep-sea manganese nodules. Geochim. Cosmochim. Acta *38*, 279–300 (1974)

Crerar, D. A., Cormick, R. K., Barnes, H. L.: Organic controls on the sedimentary geochemistry of manganese. Acta Univ. Szegediensis *20*, 217–226 (1972)

Debyser, J., Rouge, P. E.: Sur l'origine du fers des eaux interstitielles. Compt. Rend. Acad. Sci. (Paris) *243*, 2111–2113 (1956)

Deju, R. A.: A model of chemical weathering of silicate minerals. Geol. Soc. Am. Bull. *82*, 1055–1062 (1971)

Deju, R. A., Bhappu, R. B.: Surface properties of silicate minerals. Soc. Mining Eng. Trans., March 1966, 67–70 (1966)

De Kimpe, C. R.: Hydrothermal aging of synthetic aluminosilicate gels. Clay Miner. *7*, 203–214 (1967)

De Kimpe, C. R.: Crystallization of kaolinite at low temperature from aluminosilicic gels. Clays Clay Miner. *17*, 37–38 (1969)

De Kimpe, C. R., Gastuche, M. C., Brindley, G. W.: Ionic coordination in aluminosilicic gels in relation to clay mineral formation. Am. Mineral. *46*, 1370–1381 (1961)

Dousma, J., de Bruyn, P. L.: Hydrolysis-precipitation studies of iron solutions. I. Model for hydrolysis and precipitation from Fe(III) nitrate solutions. J. Colloid Interface Sci. *56*, 527–539 (1976)

Edmund, W. M.: Trace element variations across an oxidation-reduction barrier in a limestone aquifer. Proc. Symp. Hydrogeochem. Biogeochem., Tokyo, 1970, Washington D. C.: The Clarke Co., *1*, 500–526 (1973)

Edwards, D. G., Posner, A. M., Quirk, J. P.: Repulsion of chloride ions by negatively charged clay surfaces. Parts 1–3. Trans. Faraday Soc. *61*, 2808–2823 (1965)

Ellison, W. D.: Raindrop energy and soil erosion. Emp. J. Exp. Agric. *20*, 81–97 (1952)

Farmer, E. E.: Relative detachability of soil particles by simulated rainfall. Soil Sci. Soc. Am. Proc., *37*, 629–633 (1973)

Farmer, V. C., Russell, J. D., McHardy, W. J., Newman, A. C. D., Ahlrichs, J. L., Rimsaite, J.: Evidence for loss of protons and octahedral iron from oxidized biotites and vermiculites. Mineral. Mag. *38*, 121–137 (1971)

Farmer, V. C., Wilson, M. J.: Experimental conversion of biotite to hydrobiotite. Nature (London) *226*, 841–842 (1970)

Feitknecht, W., Michaelis, W.: Über die Hydrolyse von Eisen III Perchloratelösungen. Helv. Chim. Acta *45*, 212–224 (1962)

Geering, H. R., Hodgson, J. F., Sdano, C.: Micronutrient cation complexes in soil solution: IV. The chemical state of manganese in soil solution. Soil Sci. Soc. Am. Proc. *33*, 81–85 (1969)

Gilkes, R. J., Young, R. C., Quirk, J. P.: Artificial weathering of oxidized biotite. Potassium removal by sodium chloride and sodium tetraphenylboron solutions. Soil Sci. Soc. Am. Proc. *37*, 25–28 (1973)

Goldich, S. S.: A study of rock weathering. J. Geol. *46*, 17–58 (1938)

Harder, H.: The role of Mg in the formation of smectite minerals. Chem. Geol. *19*, 31–39 (1972)

Harder, H.: Illite mineral synthesis at surface temperatures. Proc. Int. Clay conf., Mexico, Bailey, S. W. (ed.). Welnette, IL.: Applied Publishing Ltd. 1975, pp. 305–306

Hedström, B. O. A.: Studies on the hydrolysis of metal ions. VII. The hydrolysis of the iron III ion. Ark. Kemi *6*, 1–16 (1953)

Helgeson, H. G.: Kinetics of mass transfer among silicates and aqueous solutions. Geochim. Cosmochim. Acta *35*, 421–469 (1971)

Helgeson, H. G.: Kinetics of mass transfer among silicates and aqueous solutions: correction and clarification. Geochim. Cosmochim. Acta *36*, 1067–1070 (1972)

Hem, J. D.: Graphical methods for studies of aqueous aluminum hydroxides, fluoride, and sulfate complexes. U. S. Geol. Survey Water-Supp. Paper 1827-B (1968a)

Hem, J. D.: Aluminum species in water. In: Trace inorganics in water. Adv. Chem. Ser., *73*, 1968b, pp. 98–114

Hem, J. D.: Study and interpretation of the chemical characteristics of natural water 2nd ed. U. S. Geol. Survey Water-Supp. Paper 1473 (1970)

Hem, J. D., Roberson, C. E.: Form and stability of aluminum hydroxide complexes in dilute solution. U. S. Geol. Survey Water-Supp. Paper 1827-A (1967)

Hem, J. D., Roberson, C. E., Lind, C. J. Polzer, W. L.: Chemical interactions of aluminum with silica at 25 °C. U. S. Geol. Survey Water-Supp. Paper 1827-E (1973)

Hendricks, D. M., Whittig, L. D.: Andesite weathering. II. Geochemical changes from andesite to saprolite. J. Soil Sci. *19,* 147–153 (1968)

Hiller, J. E.: Phasenumwandlungen im Rost. Werkst. Korros. *17,* 943–951 (1966)

Hsu, Pa Ho: Interaction between aluminum and phosphate in aqueous solution. In: Trace inorganics in water. Adv. Chem. Ser., *73,* 1968, pp. 115–127

Hsu, Pa Ho: Nucleation, polymerization and precipitation of FeOOH. J. Soil. Sci. *23,* 409–419 (1972)

Hsu, Pa Ho: Appearance and stability of hydrolyzed $Fe(ClO_4)_3$ solutions. Clays Clay Miner. *21,* 267–277 (1973)

Hsu, Pa Ho, Bates, T. F.: Formation of X-ray amorphous anc crystalline aluminum hydroxides. Mineral. Mag. *33,* 749–768 (1964)

Hsu, Pa Ho, Ragone, S. E.: Aging of hydrolyzed iron solutions. J. Soil Sci. *23,* 17–31 (1972)

Huang, W. H., Keller, W. D.: Dissolution of rock-forming silicate minerals in organic acids: simulated first-stage weathering of fresh minerals surfaces. Am. Mineral. *55,* 2076–2094 (1970)

Huang, W. H., Keller, W. D.: Geochemical mechanism for the dissolution, transport and deposition of aluminum in the zone of weathering. Clays Clay Miner. *20,* 69–74 (1972)

Ismail, F. T.: Oxidation-reduction mechanism of octahedral iron in mica type structures. Soil Sci. *110,* 167–171 (1970)

Ivarson, K. C., Heringa, P. K.: Oxidation of manganese by microorganisms in manganese deposits of Newfoundland soil. Can. J. Soil Sci. *52,* 401–416 (1972)

Jackson, M. L., Levelt, T. W. M., Syers, J. K., Rex, R. W., Clayton, R. N., Sherman, G. D., Uehara, G.: Geomorphological relationships of tropospherically derived quartz in the soils of the Hawaiian Islands. Soil Sci. Soc. Am. Proc. *35,* 515–525 (1971)

Johansson, G.: On the structure of some hydroxo salts of Al, Ga, In and Tl. Sven. Kem. Tidskr. *2,* 6–7 (1963a)

Johansson, G.: On the crystal structure of the basic aluminum sulfate. Ark. Kemi *20,* 321 (1963b)

Johnson, N. M., Likens, G. E.: Rate of chemical weathering of silicate minerals in New Hampshire. Geochim. Cosmochim. Acta *32,* 531–545 (1968)

Keller, W. D.: Observations on the origin of Missouri High-Alumina clays. In: Problems of clay and laterite genesis. New York: Am. Inst. Mining Engineer. 1952, pp. 115–134

Kester, D. R., Byrne, R. H., Liang, Y. J.: Redox reactions and solution complexes of iron in marine systems. In: Marine chemistry in the coastal environment. Symp. Ser., No. 18. Church, T. M. (ed.). Washington, D. C.: Am. Chem. Soc., 1975, pp. 56–79

Kodama, H., Ross, G. J.: Structural changes accompanying potassium exchange in a clay-size muscovite. Proc. Int. Clay Conf., Madrid. Serratosa, J. M. (ed.) Madrid: Div. Ciencias C. S. I. C. 1973, pp. 481–492

Kodama, H., Ross, G. J., Liyama, J. T., Robert, J. L.: Effect of layer charge location on potassium exchange and hydration of micas. Am. Mineral. *59,* 491–495 (1974)

Kramer, J. R.: Mineral-water equilibria in silicate weathering. XXIII Int. Geol. Cong. Proc. *6,* 149–160 (1968)

Krauskopf, K. B.: Separation of manganese from iron in sedimentary processes. Geochim. Cosmochim. Acta *12,* 61–84 (1957)

Krauskopf, K. B.: Introduction to geochemistry. New York: McGraw-Hill 1967

Kuenen, Ph. H.: Experimental abrasion. 4. Aeolian action. J. Geol. *68,* 427–449 (1960)

Lahav, N., Banin, A.: Effect of various treatments on the optical properties of montmorillonite suspensions. Isr. J. Chem. *6,* 285–294 (1968)

La Iglesia Fernadez, A., Martin Vivaldi, J. L.: A contribution to the synthesis of kaolinite. Proc. Int. Clay Conf., Madrid. Serratosa, J. M. (ed.) Madrid: Div. Ciencias C. S. I. C., 1973, pp. 173–185

Lamb, A. B., Jacques, A. G.: The slow hydrolysis of ferric chloride in dilute solutions. II. The change in hydrogen ion concentration. J. Am. Chem. Soc. *60,* 1215–1225 (1938)

Li, W. H., Bischoff, J., Mathieu, G.: The migration of manganese in the Arctic Basin sediments. Earth Planet. Sci. Lett. *7,* 265–270 (1969)

Loughnan, F. C.: Chemical Weathering of the silicate Minerals. New York: American Elsevier 1969

Luciuk, G. M., Huang, P. M.: Effect of monosilicic acid on hydrolytic reactions of aluminum. Soil Sci. Soc. Am. Proc. *38,* 235–244 (1974)

Lurie, D., Yariv, S.: Heterometric titration of sodium montmorillonite with calcium nitrate. Isr. J. Chem. *6,* 203–211 (1968)

Mart, J., Sass, E.: Geology and origin of the manganese ore of Um Bogma, Sinai. Econ. Geol. *67,* 145–155 (1972)

Mason, B.: Principles of geochemistry. New York: John Wiley & Sons, 1952

Mattock, G.: Factors affecting the hydrolysis and dimerization of metal ions. Acta Chem. Scand. *8,* 777–787 (1954)

Mazurak, A. P., Mosher, N.: Detachment of particles in simulated rainfall. Soil Sci. Soc. Am. Proc. *32,* 716–719 (1968)

Meek, B. D., MacKenzie, A. J., Grass, L. B.: Effects of organic matter, flooding time and temperature on the dissolution of iron and manganese from soils in situ. Soil Sci. Soc. Am. Proc. *32,* 634–638 (1968)

Miceli, J., Stuer, J.: Ligand penetration rates into metal ion coordination spheres. Aluminum, gallium and indium sulfate. J. Am. Chem. Soc. *90,* 6967–6972 (1968)

Miehlich, G., Zöttl, H. W.: Einfluß des Fichtenreinanbaus auf die Eisen- und Mangandynamik eines Lösslehm-Pseudogleys. In: Pseudogley and gley. Soil Sci. Soc. Trans. Comm. V and VI, Stuttgart-Hohenheim, 1972. 1973, pp. 51–55

Millot, G.: Geology of clays. Farrand, W. R., Paquet, H. K. (translators), New York-Berlin-Heidelberg: Springer-Verlag 1970

Mitchell, B. D., Bracewell, J. M., de Endredy, A. S., McHardy, W. J., Smith, B. F. L.: Mineralogical and chemical characteristics of a gley soil from North-East Scotland. 9th Int. Cong. Soil. Sci. Adelaide, Trans. Holries, J. W. (ed.) New York: Am. Elsevier, 1968, pp. 67–77

Mohr, E. C., van Baren, F. A., van Schuylenborgh, F. A.: Tropical soils. A comprehensive study. The Hague: Mouton 1972

Mortland, M. M.: Kinetics of potassium release from biotite. Soil Sci. Soc. Am. Proc. *22,* 503–508 (1958)

Mortland, M. M., Lawton, K., Uehara, G.: Alteration of biotite to vermiculite by plant growth. Soil Sci. *82,* 477–481 (1956)

Mulay, L. N., Selwood, P. W.: The hydrolysis of Fe(III) ion, magnetic and spectrophotometric studies on ferric perchlorate solutions. J. Am. Chem. Soc. *76,* 6207–6208 (1954)

Navrot, J., Singer, A.: Geochemical changes accompanying basic igneous rocks-clay transitions in humid Mediterranean climate. Soil Sci. *121,* 337–345 (1976)

Neuman, B. S., Sansom, K. G.: The study of gel formation and flocculation in aqueous clay dispersions by optical and rheological methods. Isr. J. Chem. *8,* 315–322 (1970)

Nitsan, U.: Stability field of olivine with respect to oxidation and reduction. J. Geophys. Res. *79,* 706–711 (1974)

Ollier, C. D.: Dirt-cracking–a type of insulation weathering. Aust. J. Sci. *27,* 236–237 (1965)

Ollier, C. D.: Weathering. Edinburgh. Oliver and Boyd. 1969

Olomu, M. O., Racz, G. J., Cho, C. M.: Effect of flooding on the Eh, pH and concentration of Fe and Mn in several Manitoba soils. Soil Sci. Soc. Am. Proc. *37,* 220–224 (1973)

Oosterhout, V. G. W.: The transformation of γ FeOOH to α FeOOH. J. Inorg. Nucl. Chem. *29,* 1235–1238 (1967)

Paces, T.: Steady state kinetics and equilibrium between ground water and granitic rock. Geochim. Cosmochim. Acta *37,* 2641–2663 (1973)

Parks, G. A.: Free energies of formation and aqueous solubilities of aluminum hydroxides and oxide hydroxides at 25 °C. Am. Mineral. *57,* 1163–1189 (1972)

Phillippe, W. R., Blevins, R. L., Barnhisel, R. I., Bailey, H. H.: Distribution of concretions from selected soils of the inner blue grass region of Kentucky. Soil Sci. Soc. Am. Proc. *36,* 171–173 (1972)

Quirk, J. P., Chute, J. H.: Potassium release from mica-like clay minerals. 9th Int. Cong. Soil Sci., Adelaide, Trans. Holmes, J. W. (ed.) New York: Am. Elsevier, 1968, pp. 671–687

Rausell-Colom, J., Sweatman, T. R., Wells, C. B., Norrish, K.: Studies in the artifical weathering of mica. In: Experimental pedology. Hallsworth, E. G., Crawford, D. V. (eds.) London: Butterworths, 1965, pp. 40–72

Reichenbach, H., Graf von, Rich, C. I.: Preparation of dioctahedral vermiculites from muscovite and subsequent exchange properties. 9th Int. Cong. Soil Sci., Adelaide, Trans. Holmes, I. W. (ed.) New York: Am. Elsevier, 1968, pp. 709–719

Reuter, J. H., Perdue, E. M.: Importance of heavy-metal organic matter interactions in natural waters. Geochim. Cosmochim. Acta *41,* 325–334 (1977)

Rex, R. W., Goldberg, E. D.: Quartz contents of pelagic sediments of the Pacific Ocean. Tellus *10,* 153–159 (1958)

Rimsaite, J.: Structural formulas of oxidized and hydroxyl deficient micas and decomposition of the hydroxyl group. Contrib. Mineral. Petrol., *25,* 225–240 (1970)

Rodrique, L., Poncelet, G., Herbillon, A.: Importance of the silica subtraction process during the hydrothermal kaolinitization of amorphous silicoaluminas. Proc. Int. Clay Conf., Madrid. Serratosa, J. M. (ed.). Madrid: Div. Ciencias C. S. I. C., 1973, pp. 187–198

Rona, E., Hood, D. W., Muse, L., Buglio, B.: Activation analysis of Mn and Zn in sea water. Limnol. Oceanogr. *7,* 201–206 (1962)

Rose, C. W.: Rainfall and soil structure. Soil Sci. *91,* 49–54 (1961)

Schindler, P. Michaelis, W. Feitknecht, W.: Solubility products of metal oxides and hydroxides. Solubility of aged iron (III) hydroxide precipitate. Helv. Chim. Acta *46,* 444–449 (1963)

Schwertmann, U., Fanning, D. S.: Iron-manganese concretions in hydrosequence of soils in Loess in Bavaria. Soil Sci. Soc. Am. Proc., *40,* 731–738 (1976)

Schwertmann, U., Taylor, R. M.: The transformation of lepidocrocite to goethite. Clays Clay Miner. *20,* 151–158 (1972a)

Schwertmann, U., Taylor, R. M.: The influence of silicate on the transformation of lepidocrocite to goethite. Clays Clay Miner. *20,* 159–164 (1972b)

Schwertmann, U., Thalmann, H.: The influence of Fe(II), Si and pH on the formation of lepidocrocite and ferrihydrite during oxidation of aqueous $FeCl_2$ solutions. Clay Miner. *11,* 189–200 (1976b)

Scott, A. D.: Effect of particle size on interlayer potassium exchange in micas. 9th Int. Cong. Soil Sci., Adelaide, Trans. *11,* 649–660 (1968)

Shainberg, I., Otoh, H.: Size and shape of montmorillonite particles saturated with Na/Ca ions, inferred from viscosity and optical measurement. Isr. J. Chem. *6,* 251–259 (1968)

Shanks, W. C.: Experimental study of iron and manganese migration in marine sediments. Geol. Soc. Am., 68th Ann. Meeting, Boulder, CO: Geol. Soc. Am. *4,* 235 (1972)

Shvartsev, S. L.: Volume and composition evolution for infiltrating groundwater in aluminosilicate rocks. Geokhimya *6,* 905–916 (1975) (English translation, Geochem. Intern. *12,* issue No. 3, 184–194)

Singer, A., Navrot, J.: Extraction of metals from basalt by humic acids, Nature (London) *262,* 479–481 (1976)

Smith, R. W.: The state of Al in aqueous solution and adsorption of hydrolysis products on Al_2O_3. Ph. D. thesis, Depart. Miner. Engineer., Stanford Univ., 1969

Smith, R. W.: Relations among equilibrium and nonequilibrium aqueous species of aluminum hydroxy complexes. In: Nonequilibrium systems in natural water chemistry. Adv. Chem. Ser., *106,* 1971, pp. 250–279

Spencer, D. W., Robertson, D. E., Turekian, K. K., Folsom, T. R.: Trace elements calibrations and profiles at the GEOSECS test station in the Northeast Pacific Ocean. J. Geophys. Res. *75,* 7688–7696 (1970)

Spiro, T. G., Allerton. S. G., Renner, J., Terzis, A., Bils, R., Saltzman, P.: The hydrolytic polymerization of iron(III). J. Am. Chem. Soc. *88,* 2721–2726 (1966)

Stol, R. J., van Helden, A. K., de Bruyn, P. L.: Hydrolysis-precipitation studies of aluminum(III) solutions. 2. A kinetic study and model. J. Colloid Interface Sci. *57,* 115–131 (1976)

Strakhov, V. M.: Formation conditions of iron-manganese nodules ores in modern water bodies. Lithol. Miner. Resour. (USSR) (English trans.) *11*, 1–13 (1976)

Stumm, W., Lee, G. F.: Oxygenation of ferrous iron. Ind. Eng. Chem. *53*, 143–146 (1961)

Szántó, F.: On the electrochemical properties and disaggregation of bentonites. Acta geol. (Budapest) *7*, 305–314 (1962)

Tarr, W. A., Keller, W. D.: Occurrences of kaolinite deposited from solutions. Am. Mineral. *22*, 933–935 (1937)

Taylor, R. M., McKenzie, R. M., Norrish, K.: The mineralogy and chemistry of manganese in some Australian soils. Aust. J. Soil Res. *2*, 235–248 (1964)

Tedrow, J. C. F., Wilkerson, A. S.: Weathering of glacial soil material. Soil Sci. *75*, 345–353 (1953)

van der Weijden, C. H., Schuiling, R. D., Das, H. A.: Some geochemical characteristics of sediments from the North Atlantic Ocean. Mar. Geol. *9*, 81–99 (1970)

van Olphen, H.: An introduction to clay colloid chemistry. New York: Interscience Pub. New York 1963

Veith, J. A., Jackson, M. L.: Iron oxidation and reduction effects on structural hydroxyl and layer charge in aqueous suspensions of micaceous vermiculites. Clays Clay Miner. *22*, 345–353 (1973)

Vicente Hernandez, M. A., Robert, M. M.: Transformation profonde des micas sous l'action de l'acide galacturonique. Problème des smectites des podzols. C. R. Acad. Sci. Ser. D *281*, 523–526 (1975)

Walker, G. F.: The decomposition of biotite in soils. Mineral. Mag. *28*, 693 (1949)

Wedepohl, K. H.: Geochemistry. Althaus, E. (translator), New York: Holt, Rinehart and Winston, Inc. 1971

Wey, R., Siffert, B.: Reactions de la silice monomoleculaire en solution avec les ions Al et Mg. Genese et synthese des argiles. Colloq. Int. C. N. R. S. *105*, 11–24 (1961)

Yaalon, D. H.: Parallel stone cracking, a weathering process on desert surfaces. Geol. Inst. Techn. Econ. Bull. (Bucharest), Ser. C, Pedology, *18*, 107–111 (1970)

Yaalon, D. H., Jungreis, C., Koyumdjisky, H.: Distribution and reorganization of Mn in three catenas of Mediterranean soils. Geoderma *7*, 71–78 (1972)

Yalin, M. S.: Mechanism of sediment transport. Oxford: Pergamon Press 1972

Yamane, I., Sato, K.: Initial rapid drop of oxidation-reduction potential in submerged air-dried soils. Soil Sci. Plant. Nutr. *14*, 68–72 (1968)

Yariv, S.: Some calculated standard free energy changes during chemical weathering of rock forming silicates. Isr. J. Chem. *9*, 695–710 (1971)

Yariv, S.: Comments on the mechanism of soil detachment by rainfall. Geoderma *15*, 393–399 (1976)

Yariv, S., Lurie, D.: Metachromasy in clay minerals. I. Sorption of methylene blue by montmorillonite. Isr. J. Chem. *9*, 537–552 (1971)

Young, R. A., Wiersma, J. L.: The role of rainfall impact in soil detachment and transport. Water Resourc. Res. *9*, 1629–1636 (1973)

Zajic, J. E.: Microbial biogeochemistry. New York: Academic Press 1969

Chapter 4
Surface Coatings on Rocks and Grains of Minerals

Rocks, rock debris, and grains are usually coated with X-ray amorphous layers formed by the following mechanisms: (1) Incongruent dissolution of rock material; (2) sorption of inorganic and organic species mainly from solutions and suspensions, but also from the atmosphere; (3) thermal diffusion of atoms from the core of the crystal to the surface resulting from abrasion and collision between solids in dry systems; (4) biologic activity, which is, however, outside the scope of this review and will not be considered.

Geologic processes involving solid rock and aqueous solution phases are affected by surface coatings. Indeed, the hydrogeochemistry of many elements is controlled by chemical reactions occurring between surface coating and solution rather than between crystalline mineral and solution. Surface coatings also play an important role in determining the geochemical properties of soils (Bracewell et al., 1970; Townsend and Reed, 1971) and sands, which have large surface areas (Heald and Larese, 1974) and in the dissolution characteristics and diagenesis of carbonate materials in seawater and in sediments (Weyl, 1967; see also pages 72–73).

The presence of an amorphous surface microstructure of quartz grains has been revealed by transmission and scanning electron microscopy (e.g., Krinsley and Doornkamp, 1973). Lin et al. (1974) used the same techniques to observe the same phenomenon in heavy sand minerals, zircon, tourmaline, magnetite, hornblende, and calcite.

1. Incongruent Dissolution of Silicates

The dissolution of aluminosilicate minerals in weathering environments differs from that of carbonates, sulfates, or chlorides in being incongruent, that is, the molar ratios of the chemical elements in the solution differ from those in either the rocks or rock fragments. Furthermore, in most of the laboratory dissolution experiments of silicate minerals and rocks in water, incongruency was observed (Correns, 1963; Keller et al., 1963; Marshall and McDowell, 1965). The rate of dissolution of elements varies greatly and those which dissolve relatively slowly enrich the near-surface region of the solid. Consequently, the structure of the mineral surface differs from that of the core and may react with water, producing a hydrated and/or swollen layer at the surface.

Two explanations have been proposed for the formation of the near-surface region by a dissolution process:

1. The dissolution is congruent, occurring by a dissociation process, the reaction mechanism being the reverse of condensation. Dissolved monomeric silica polymerizes

and is readsorbed on the solid surface. Alumina is more readily adsorbed than silica by negatively charged particles of silicate minerals and adsorption of alumina may result in the mineral particle becoming positively charged, as for example, when silica contains aluminum impurities (van Olphen, 1963). For further dissolution to occur the monomers formed by dissociation and hydration on the crystalline surface of the mineral have to diffuse through the sorbed layer, and thus the rate of solution depends on the diffusion rate, which decreases greatly.

2. Three separete processes are involved in the dissolution of the silicate (Luce et al., 1972): (a) removal of surface metallic cations by ion exchange (heterogeneous hydrolysis), (b) solid-state diffusion of atoms from the core to the crystal surface, and (c) collapse of the framework of the residual anionic polymeric lattice of the leached surface layer into soluble monomeric silica. It was demonstrated using the three magnesium silicates, forsterite, lizardite, and enstatite that the first process is the most rapid, metallic cations being removed at a faster rate than silicon and that the near-surface, partially leached region of the mineral particle is depleted in magnesium relative to silicon. For these three minerals diffusion coefficients for magnesium are greater than for silicon, resulting in incongruent dissolution over moderate periods of time. The diffusion flux becomes negligible compared with the rate of the third process, which is a slow dissolution of the surface. Because the near-surface, partially leached region in the mineral is more depleted in magnesium relative to silicon, its dissolution will cause the ratio of silicon to magnesium in solution to reach the stoichiometric ratio if the system is closed. Eventually the exchange rates decrease to that of the rate of dissolution of the material at the solid aqueous interface. Hence, over long periods the amount of silicon and magnesium dissolved is proportional to the time of the process and dissolution is congruent.

Some minerals may have very stable lattice structures, and collapse of the silicate framework with complete dissolution of the surface occurs very slowly. It has been shown (Nathan, 1968; Mendelovici, 1973) that breakdown of the palygorskite lattice occurs only after 80% of the magnesium has been leached by acid. It also has been demonstrated (Choi and Smith, 1972) that magnesium cations may be continuously liberated from chrysotile fibers leaving a stable silica skeleton with the original structure. The rate of dissolution of chrysotile in water is a function of its specific surface area, chrysotile dissolving incongruently.

After a prolonged acid attack the octahedral sheet of the palygorskite is dissolved completely, before the tetrahedral sheet dissolves, and surface Si–OH groups are formed (Mendelovici, 1973). The appearance of Si–OH groups has also been reported by Gastuche (1963) in HCl-treated biotite and by Fripiat and Mendelovici (1968) in HCl-treated chrysotile.

In the incongruent stage of the dissolution, silicates with chain, layer, or three-dimensional network structures become depleted of metallic cations, and provided that there is no specific adsorption of cations, they acquire a negative surface charge. Island structure silicates, however, acquire a positive surface charge when in contact with water. During the dissolution of akermanite, $Ca_2MgSi_2O_7$, the ions $Si_2O_7^{-6}$ dissolve rapidly and may be replaced by OH^-, forming a surface coating of MgOH (Siskens et al., 1975a). The incongruent stage of the dissolution of silicates having an island structure results in the depletion of the small silicate anion.

The complete dissolution process of the near-surface region becomes more complex if aluminum is present because of the lower solubility of aluminum compared to silica species. Silicon is preferentially liberated from the silica–alumina skeleton at the near-surface region, leaving polymeric and monomeric aluminum species, which in time reorganize to form an amorphous aluminous near-surface region. Positive aluminum network fragments, Al^{3+} and AlO^+, also dissolve, and part of the negative charge on anorthite surfaces has been attributed by Siskens et al. (1975b) to this process.

The dissolution of alkali feldspars in aqueous solutions is incongruent because alkali ions are preferentially leached from the feldspar, or because an aluminum-containing phase is precipitated. Petrovic et al. (1976) found from X-ray photoelectron spectroscopy study that during the incongruent dissolution of alkali-feldspar at 82 °C, which corresponds to deep diagenesis, no continuous precipitate coating is formed and the thickness of the leached layer does not exceed one feldspar unit cell. They concluded that when aluminum hydroxite and kaolinite accumulate on the surfaces of dissolving feldspar grains during the early stage of weathering, they do not form tightly adhering layers.

In their investigation of incongruency of dissolution of several rock-forming silicate minerals Huang and Keller (1970) found the following:

Olivine (Nesosilicate): Silicon dissolves preferentially and more rapidly than the other major cations in olivines. Magnesium dissolves somewhat slower than silicon and the near-surface region becomes iron rich.

Augite (Inosilicate): The surface is slightly enriched with silicon and very markedly with iron and aluminum.

Muscovite (Phyllosilicate): The potassium, sodium, and magnesium cations in micas are readily soluble, especially potassium. The solubility of silicon is significantly less and that of aluminum, iron, and calcium even lower. The near-surface region is consequently enriched in these elements.

Labradorite and Microcline (Tectosilicates): Dissociation of feldspars results in a surface zone greatly enriched in aluminum and to some extent in silicon but depleted in potassium, sodium, and calcium. In the presence of organic compounds that form stable complexes with aluminum and iron, according to Huang and Keller, the near-surface zone becomes enriched with aluminum and to some extent with iron, and the thickness of the depleted layer may increase by a factor of 30.

The weathering of the near-surface region by means of incongruent dissolution, in the area of grain contacts, was studied by Meunier and Velde (1976) from outcrops of granite in the early stages of its weathering. The phase that is formed in the grain contact region has a composition distinct from that of the initial mineral grains in the rock. This suggests solution–solid reactions that involve incongruent dissolution of the silicate minerals and the interactions between the surface coatings and the aqueous system. Three types of contact zones were identified: (1) Quartz is inactive during the early stages of the weathering process in granites, and contacts between quartz grains and all other minerals produce no new phases. (2) When muscovite (or sericite) and orthoclase are in contact, the mica is unaffected whereas the orthoclase is invaded by an illite-like phase growing at right angles to the grain contact. (3) When muscovite and plagioclase are in contact, the surface coatings of both minerals are unstable, and kaolinite or an expanding clay mineral grows at the grain contacts.

2. Sorption from Aqueous Solutions onto Mineral Surfaces

Since the nature of the forces acting between the mineral surface and the sorbates largely determines the type of process involved, it is appropriate to classify the various interactions as *long-range* and *short-range.* Long-range attractive interactions result from van der Waals and electrostatic forces and act not merely at the immediate surface but some distance beyond. Van der Waals forces vary inversely (usually to the seventh power) with the distance, and unless the sorbed molecules are large, are infinitesimally small beyond a few molecular diameters. The electrostatic forces also vary inversely with the distance, but only to the second power, and are therefore effective at much greater distances than the van der Waals forces. Short-range attractive interactions arise from various types of chemical bonds and intermediate bonds such as hydrogen bonds and π interactions. These require functional groups on the surface of the solid and in general are either acidic (proton donors or electron pair acceptors) or basic (proton acceptors or electron pair donors), determining the *surface acidity* of the rock particle. Most sorption reactions involving short-range interaction mechanisms are of the "acid–base" type of reaction.

2.1 Sorption by Long-range Interactions

Most silicate minerals have negatively charged surfaces at the pH values of natural waters. The formation of the diffuse double layer in the solid–liquid interface and the ion-exchange reactions discussed in Chapter 2 are examples of long-range interactions.

Ion-exchange reactions occur with both clayey and nonclayey fractions of soil, sediments, and of particulate material suspended in the hydrosphere. The nonclayey fraction reacts only to a very small extent compared to the clayey fraction. The pH has only a minimal effect on cation sorption by clay minerals but is very influential on cation sorption of the nonclayey fraction. For example, Shuman (1975) showed that lowering the pH of a soil–water system reduces Zn adsorption by sandy soils but not by soils rich in colloid-size material.

The composition of natural waters is often modified by ion-exchange reactions. The most common exchange in geologic systems is that of Na–K (water) for Ca–Mg (minerals). This effect is given in ground waters by the ratio [Cl – (Na + K)]/Cl. In the case of inverse exchange this ratio becomes negative and the ratio [Cl – (Na + K)]/ [SO_4 + HCO_3 + NO_3] is used (Swaine and Schneider, 1971). The interaction of ions with soil colloids lessens the mobility of the ions in soil solutions. Soil may therefore turn into a system that accumulates potentially hazardous metallic pollutants or at least attenuates their mobility (see, e.g. Fuller et al., 1976). Sorption by river and stream-suspended particles is an important factor in the transport of major cations (Lorrain and Souchez, 1972) and trace cations (Förstener and Müller, 1976) in rivers and springs.

Polymeric hydroxy cations of aluminum and iron (Ch. 3) are adsorbed in preference to monomeric species because (1) they have a higher positive charge than monomeric ions, and the electrostatic forces between them and the mineral particle surfaces become stronger; (2) they have a lower hydration energy than monomeric species, and only a small amount of energy is required to transfer the polymer from the aqueous solution to the interface; (3) since the polymer is large, the contribution of van der Waals forces is significant, in addition to the electrostatic interactions; (4) the exchange of a number of monomeric species for one polymeric cation is accompanied by an increase in the number of ions in the solution, and several adsorbed water molecules are also transferred from the interface to the liquid phase, thus increasing the degree of disorder and entropy of the system.

A thin amorphous layer of hydrated metal hydroxide can form a coating on rock and soil grains, and since the charge of amorphous aluminum and iron hydroxides is positive, the surface charge of the coated grain can therefore be positive (Blackmore, 1973). The surface properties of the coated grain differ from those of the uncoated in that it may now sorb anions by electrostatic interactions. The adsorption of phosphate and sulfate on the surface of mica minerals is greatly increased when the mineral is coated with amorphous hydrous oxides of aluminum or iron (Perrott et al., 1974).

Metallic cations, with radii similar to that of iron, such as Mn, Co, Ni, Cu, and Zn, may substitute for iron in the hydrophilic polymeric colloid. Furthermore, the metals incorporated in the polymeric species may be adsorbed simultaneously. Indeed, Carpenter et al. (1975) demonstrated that when ground water containing soluble iron and manganese enters the oxygenated flowing water in streams, iron and manganese oxides form coatings on rock surfaces because of the higher pH and Eh values independently of microbial activity.

Sorption of negatively charged soluble polymers such as humic material, silicates, and phosphates on the surfaces of positive particles of various metal oxides and the formation of negative coatings also results from long-range electrostatic attraction (Moshi et al., 1974).

Particles of low surface charge density may interact with large soluble polymers of equal charge by long-range van der Waals interaction. Jones and Uehara (1973) using transmission electron microscopy studied the formation of silica gel on surfaces of finely ground quartz. Freshly ground quartz appeared relatively free of gel coating but after three months in water the quartz particles were very thinly covered with a gel and after standing for six months in water the gel-covered particles coalesced to give a "coat of paint" effect.

The nonclayey soil minerals obtained from the weathering of basalts have higher negative surface charge densities than those obtained from granites and form surface coating of Fe and Al hydroxides more readily. The geochemistry of Mn in soils and its partition between clayey and nonclayey fractions are to a great extent determined by the nature of the coating oxides. Manganese is present mainly in the nonclayey fraction of soils originating from basalts (Biswas and Gawande, 1964; Yaalon et al., 1972). The main mafic minerals of basalt rocks, augite and olivine, are host minerals for Mn (Mitchell, 1964), which is responsible for the high average content of Mn in basalt-derived soils. Soluble Mn is adsorbed together with Fe and Al hydroxide onto the surfaces of the sandy particles. On the other hand, in granitic rocks weathering will gradually produce a well-drained quartz-rich soil poor in manganese, with most of the Mn associated in the clay and organic-matter fraction (Harris and Adams, 1966).

The nature of the hydrogeochemistry of Mn and the formation of nodules in marine environments are largely determined by the formation of surface coatings of Fe oxides. Ferromanganese nodules are present at almost all depths and latitudes both in open oceans, interior seas, and in fresh-water lacustrines. They range in size from microscopic grains to boulders of several meters in length. As was stated in Chapter 3, marine waters are undersaturated with Mn and manganese nodules should not form by precipitation. The manganiferric phases of the nodules originate by adsorption of hydroxy polymeric cations of Fe and Mn onto suitable submarine surfaces such as those of sea-floor silicates, oxyhydroxides, carbonates, phosphates, and biogenic debris. Detrital particulate matter may contain oxygen trapped in the interparticle space, and oxidizing Eh develops at the solid–liquid interfaces of the particles, which may reach values that are 200–400 mV greater than that of seawater. At the same time the surface of the sorbent, which acts as a catalyst, lowers the activation energy of the oxidation reactions. Once initiated, Mn deposits are self-perpetuating because both Fe and Mn are autocatalytically precipitated upon nodule surfaces (Crerar and Barnes, 1974).

Nodules and other marine sediments contain relatively high concentrations of various trace elements that are adsorbed by cation exchange, surface hydrolysis, and coprecipitation by substitution for the major metallic cation in the interior of the crystal. As a result of this adsorption, the elements Zn, Cu, Pb, Bi, Cd, Ni, Co, Hg, Ag, Mo, W, and V, although greatly undersaturated in seawater, have relatively short mean-residence times (Krauskopf, 1956).

The growth rates of nodules vary widely depending upon their occurrence: pelagic, shallow marine, or lacustrine. The pelagic have the lowest rate and are enriched with Mn, Ni, and Cu. According to Heye and Marchig (1977) fast-growing pelagic nodules are enriched in Mn, Ni, and Cu whereas slowly growing layers are enriched in Fe.

Many manganese nodules from the Atlantic, Pacific, and Indian Oceans show that the ferromanganese oxide layer is supported by a frame of material derived from the core, which is amorphous silica when the core consists of silicates, and calcium phosphate when the core consists of calcium phosphate (Lalou and Brichet, 1976). The ferromanganese oxide layer has a surface suitable for the adsorption thereon of silica and phosphate from solutions undersaturated with these compounds.

The leaching of minor elements from rocks during weathering processes depends on their capability of being incorporated into iron hydroxide particles as well as on their chemical properties. For example, Mn, Co, Ni, Cu, and Zn are contained in relatively unstable ferromagnesium minerals and pass readily into ionic solution during the initial stages of weathering. Their leaching from weathered rocks is moderated by their adsorption and by incorporation in ferric oxide particles and surface coatings. Yaalon et al. (1974), in their analysis of basalt soils from the Galilee, Israel, found Co and Ni to be among the least leached elements in the subhumid Mediterranean weathering environment. Manganese is only slightly leached in this environment. A minor element that is not adsorbed by the hydroxide phase, such as Sr, shows a very high leaching rate, which is similar to that of Ca. The leaching of minor elements, assessed according to their depletion rate from weathered basic igneous rocks of the

Upper Golan Heights, Israel, according to Navrot and Singer (1976) is Sr > Mn = Cu > Co = Ni > Zn > Cr. The relative capacities of the minor elements to be retained in the clayey fraction is Zn = Mn > Ni = Co = Cu > Cr = Sr.

Cation exchange of clay minerals is reduced when they are coated by amorphous Fe and Al hydroxides. An increase with depth in the cation exchange capacity in the upper horizons of anaerobic marine sediments is caused by the slow removal of amorphous oxide coatings from clay minerals (Sholkovitz, 1973). The exposed clay minerals may now more readily adsorb Mg from pore water by cation exchange and thereby profiles of decreasing Mg concentrations in pore fluids are obtained. According to Bischoff et al. (1975) this process may account for 5–10% of the magnesium brought by rivers to the sea.

2.2 Sorption by Short-range Interactions

Particular aspects of the chemisorption of water molecules on freshly exposed surfaces are dealt with in several chapters. In this section consideration will be limited to surfaces consisting of "broken bonds" of a mineral oxide type M_mN_nO, where M and N are either metallic or nonmetallic elements. Four types of broken bond sites are exposed, two of which are electron pair acceptors (Lewis acids), $-O-M^+$ and $-O-N^+$, two are electron pair donors (bases), $-M-O^-$ and $-N-O^-$ and are termed "functional groups". Ionic minerals, carbonates, sulfates, phosphates, and neso- or sorosilicates, do not yield $-O-N^+$ groups, where N is C, Si, P, or S, because the bond is strong and does not dissociate by mechanical disintegration. On the other hand, in phyllo- and tectosilicates with aluminum for silicon substitution, the crystals are preferentially cleaved along planes of high aluminum content, so that exposed surfaces are closer in character to alumina than to silica.

Depending on the functional group charge, water molecules present either a positive or negative pole and because of this polarization sorbed water can dissociate, becoming either proton or hydroxyl donor according to Eqs. (4.I) and (4.II).

$$-O-M^+ + HOH \rightleftharpoons -O-M-OH + H^+, \qquad (4.I)$$

$$-M-O^- + HOH \rightleftharpoons -M-OH + OH^-. \qquad (4.II)$$

In the first case a Lewis acid site is converted by the sorbed water to a Brønsted acid site, a property defined as "surface acidity".

2.2.1 Short-range Interaction of Water with Exposed Metallic Cations

Dissociation of an $-O-M+$ functional group followed by hydration of M produces an increase in the surface negative charge. When the cation is hydrated it moves into the outer Helmholtz layer or further into the Gouy-Chapman diffuse layer, the degree of dissociation depending on the polarizing power of the cation.

Dissociation occurs mainly with monovalent cations, less with di-, and much less with trivalent cations. It is related to the ratio of the coordination valency of the ion M and to the actual number of oxygens to which the surface ion M is linked, and the higher the ratio, the easier is the dissociation. Cations of nondissociable $-O-M^+$ groups are regarded as "specific sorbed" cations (see page 142).

Sorbed hydrated cations act as proton donors, protonation strength increasing with the cation polarizing power and with decreasing number of coordinated water molecules. The degree of dissociation of coordinated water molecules sorbed on solid surfaces and consequently the acid strength of the sorbed hydrated cation is always greater than the acid strength of the same hydrated cation in aqueous solutions. Also, the acid strength decreases when the particles are suspended in water.

Dissociable cations of $-O-M^+$ groups can exchange with other cations, and since exchange sites of this type are susceptible to heterogeneous hydrolysis they are pH dependent. Thus, when the pH of the suspension is reduced the exchange capacity of the mineral is also reduced.

The concentrations of cations such as K, Na, Ca, Mg, and Al in soil solutions and the acidity of the solutions are highly dependent on the exchange reactions taking place between soil colloids and solutions.

Aluminum in the form of $-O-Al$ groups is tightly held by soil colloids, is non-exchangeable, and does not undergo heterogeneous hydrolysis. Unless exchangeable aluminum comprises 60% of the total exchangeable sites on clay surfaces, soil solutions contain less than 1mg Al/l. With larger amounts of aluminum in the system, this ion may be found at other easily exchangeable sites and its concentration in the soil solutions is increased.

Many soils of the humid tropics have a high percentage of exchangeable aluminum and their pH is less than 5. Decrease of exchangeable aluminum and increase of calcium and magnesium, which undergo hydrolysis, raises the pH above 5; it rises to even higher values when the exchange positions are saturated with the more dissociable sodium and potassium. Soils saturated with sodium or potassium have higher ion exchange capacities than those saturated with aluminum (Kamprath, 1972).

2.2.2 Short-range Interaction of Water with Exposed Oxygens

Study of the effects of the hydrated functional oxide groups, $-M-O^-$ and $-N-O^-$, on the surface acidity of particles has established that for elements of the third row of the periodic table the polarizing power of M (or N) increases from sodium to chlorine because the cation charge increases as the radius decreases. The oxygen is a basic site, having at least one unshared electron pair and is able to accept a proton donated, for example, by a water molecule. The extent to which the oxygen is able to donate an electron pair to link the proton determines the *basic strength* of the $-M-O^-$ group.

The Na–O bond is predominantly ionic in character and four pairs of electrons in the outer shell of the oxygen can be regarded as under the influence of the oxygen nucleus field, which liberates one pair of electrons for the formation of a covalent bond with a proton. The basic strength of this group is very high, whereas the acid strength of the conjugated $-Na-OH$ group is very low.

Cl–O forms a covalent bond and the shared pair of electrons is considered to be equally under the influence of the oxygen and chlorine nucleic fields. Each unshared pair of electrons is held very strongly by the field of the oxygen nucleus and the basic strength is very low. Likewise, the acid strength of the conjugated surface –Cl–OH group, obtained by the protonation of the oxygen, is very high.

The effect on the basic strength of the oxygen induced by neighboring sodium or chlorine atoms is defined as *inductive effect,* that of chlorine affecting the ease with which a proton may leave the oxygen of the –Cl–OH group. In the same way electron-attracting properties of other groups and atoms attached to the oxygens can affect the surface acidity.

The ability of a chemically bonded atom to attract electrons is termed *electronegativity.* For atoms in the ground state it should be proportional to the sum of the energy required to remove an electron (ionization potential) and the energy evolved if the atom acquires an electron (electron affinity). Electronegativity increases from left to right in the periodic table and decreases from top to bottom. Values of electronegativity are normally quoted for atoms in the ground state and are not the same for atoms in compounds, termed the *valence state.* The electronegativity of an element in the valence state increases with the positive charge. For a particular valency the valence-state electronegativity increases from left to right in the periodic table and decreases from top to bottom. It therefore follows that as electronegativity increases from sodium to chlorine, surface acidity also increases in the same order.

2.2.3 Electric Charge of Surface Functional Groups

The electric charge of the hydrated functional groups is determined by the basic strength of the $-M-O^-$ group. The characteristics of groups exhibiting low, medium, and high basic strengths are now considered.

1. Medium Strength $-M-O^-$ Group. Hydration of the $-M-O^-$ group consists first of the oxygen donating an electron pair to water hydrogen.

$$-M-O'^- + H'O''H'' \rightleftharpoons -M-O'^- \ldots H'O''H''. \tag{4.III}$$

The electron pair common to O′ and H′ is close to the H′ nucleus for the greater part of the time, and since the electrons of the covalent bond between H′ and O″ are repelled by H′ they are for most of the time close to the oxygen. The H′–O″ bond becomes more polar than normal and dissociable. Because the electrons of the covalent bond between O″ and H″ are repelled by O″ they will be for the most part close to the H″ nucleus with the result that the polarity of the O″–H″ bond is less than that in pure water and the ability to dissociate is reduced. There is repulsion between positively charged H′ and H″ and also between negatively charged O′ and O″, so that when the surface is in contact with liquid water, dissociation occurs, the solution is alkaline, and the surface function group has no charge:

$$-M-O'H'O''H''^- + HOH \rightleftharpoons -M-O'H' + H''O'' \cdot HOH^-. \tag{4.IV}$$

2. High Strength $-M-O^-$ Group. When the inductive effect of M on O′ is very low, the oxygen atom is a very strong electron donor and coordinates readily with two protons:

$$-M-O'^{-} + 2\,H'O''H'' \rightleftharpoons -M-O'^{-}\begin{array}{l} \diagup H'O''H'' \\ \diagdown H'O''H'' \end{array} . \qquad (4.V)$$

When the surface is in contact with liquid water, dissociation occurs and the solution is much more alkaline than that of (1)

$$-M-M-O'^{-}\begin{array}{l} \diagup H'O''H'' \\ \diagdown H'O''H'' \end{array} + 2\,HOH \rightleftharpoons -M-O'^{+}\begin{array}{l} \diagup H' \\ \diagdown H' \end{array} + 2\,H''O'' \cdot HOH^{-} . \qquad (4.VI)$$

The surface functional group becomes positively charged, is termed *surface acid site,* and plays an important role in catalyzing certain heterogeneous reactions.

3. Low Strength –M–O⁻ Group. When M induces a very high effect on O′ the electron pair forming the O′–H′ bond (4.III) is for the greater part of the time close to the O′ nucleus and the electrons forming the bonds H′–O″ and O″–H″ are in positions similar to those of normal water and there is only slight polarization of the O″–H″ bond. However, there is a weak repulsion between H′ and H″, which increases with the inductive effect of M. When the surface is in contact with liquid water dissociation occurs and the solution is acidic:

$$-M-O'H'O''H''^{-} + H_2O \rightleftharpoons -M-O'H'O''^{2-} + H'' \cdot OH_2^{+} . \qquad (4.VII)$$

The negative charge on the functional group is higher, constituting a *surface basic site,* which like the acid site plays an important role in the catalysis of certain heterogeneous reactions.

The protonation–hydroxylation reactions of the $-M-O^{-}$ groups are reversible. The number of the oxide groups that are protonated or hydroxylated relative to the total number of the oxide groups depends not only on the inductive effect of the surface oxygen but also on the pH of the solution. Groups with a very high basic strength become negatively charged in strongly alkaline solutions (4.VII), and groups with a very low basic strength acquire a positive charge in strongly acidic solutions (4.VI).

The pH values of aqueous solutions and suspensions containing an equal number of equivalents of the oxides of elements of the third row of the periodic table, decrease in the order, Na_2O, MgO, Al_2O_3, SiO_2, P_2O_5, SO_3, and Cl_2O_7. The first three give alkaline solutions or suspensions, of which $Al(OH)_3$ is the weakest base, and the others give acid solutions or suspensions, H_4SiO_4 being the weakest acid.

2.2.4 Electric Charge of Surfaces

Positively, neutrally, and negatively charged groups are present simultaneously on the solid surface, the net charge density of which is the summation of the individual contributions of the positive and negative groups and the specific sorbed potential-determining ions. The charge depends on the chemical composition and the crystallographic structure of the solid. It is also a function of the pH of the solution in contact with the solid. For high pH values the surface concentration of OH^- is high and the net surface charge density of the solid is negative, becoming more negative with rise in pH. However, lowering of the pH produces a decrease in the surface OH^- concentration and the negative surface charge to a point where the surface concentrations of H^+ and OH^- are equal, termed the *point of zero charge* (p. z. c.). In the absence of constant surface charge and of specific sorbed ions other than H^+ and OH^-, it is the pH at which the electrokinetic potential is zero, and it is termed the *isoelectric point* (i. e. p.). This point does not necessarily occur at neutral pH values. The net charge becomes more positive with further decrease of pH. The p. z. c. and the i. e. p. are different only if specific adsorption of cations and anions other than H^+ and OH^- occurs at the p. z. c. The i. e. p. is a measure of the double layer potential to a first approximation and hence reflects only indirectly the surface conditions, whereas the p. z. c. represents them directly. By specific sorption of polyvalent cations the p. z. c. drops to lower pH values whereas the i. e. p. increases. Specific sorption of anions increases the pH of the z. p. c. and decreases the pH of the i. e. p.

The i. e. p. of a number of common oxides and hydrous oxides, calcite, and apatite are given in Table 4.1. For each group there is an obvious broad range of isoelectric points, and there are several reasons for this: (1) the degree of crystallinity and the crystal species produced depend very much on the method of preparation (2) dif-

Table 4.1. Isoelectric points of oxides and hydrous oxides and of some ionic minerals common in soils and sediments. (After Marshall, 1964; Parks, 1965)

Precipitate or mineral		pH
$Al(OH)_3$	Amorphous	7.1–9.45
α-$Al(OH)_3$	Crys. gibbsite	5.0
γ-$Al(OH)_3$	Crys. bayrite	6.2–9.25
$AlO(OH)$	Boehmite	9.1–9.4
α-Al_2O_3	Corundum	2.2(?)–9.2
γ-Al_2O_3		8.0–8.5
$Fe(OH)_2$		12.0
Fe_3O_4	Natural and synthetic magnetite	6.5
$Fe(OH)_3$	Amorphous	7.1–8.5
α-$FeO(OH)$	Goethite	3.2–6.7
$FeO(OH)$	Limonite	3.6
γ-$FeO(OH)$	Lepidocrocite	5.4–7.4
α-Fe_2O_3	Hematite	2.1(?)–8.6
γ-Fe_2O_3	Magnemite	6.7
$Mg(OH)_2$		~12
MgO		12.4
$Mn(OH)_2$		7.0
MnO_2		4.0–4.5
SiO_2	Quartz	2.2
SiO_2	Amorphous, sols or gels	1.8
TiO_2	Natural rutile	4.7
$CaCO_3$	Calcite	9.5
$Ca_5(PO_4)_3(F, OH)$	Fluoroapatite	6.0
$Ca_5(PO_4)_3(OH)$	Hydroxyapatite	7.0

ferent species of the same ultimate composition have different surface properties, (3) microcrystalline minerals have much lower i. e. p. than amorphous hydroxides, and (4) surface dehydration gives a more acidic i. e. p. than that for the hydrated material (Table 4.2). Also, hydration of the outer layer of an anhydrous oxide to an oxy-hydroxide or hydroxide may occur and spurious i. e. p. values are observed.

Table 4.2. Variations of isoelectric points with hydration. (After Parks, 1965)

Oxide	Anhydrous compound pH	Hydrous compound pH
Fe_2O_3	6.7	8.6
Al_2O_3	6.7	9.2
TiO_2	4.7	6.2

The hydration rate depends on the type of mineral. The oxides and hydrous oxides of aluminum are hydrated within a few days if only to the extent that they develop a film of $Al(OH)_3$. The i. e. p. value of oxides and hydroxides of aluminum at equilibrium with water is probably about 9.2. Values currently quoted for these compounds probably represent metastable states (Parks, 1965). Their surfaces in natural waters are almost invariably positively charged. Ferric oxides hydrate rapidly again, at least, to the extent that their surfaces have the properties of α-FeO(OH) (Parks, 1965). Although the i. e. p. of ferric oxides occurs at a lower pH than that of aluminum oxides, grains of iron oxide minerals are coated with amorphous iron oxide, which is positively charged in most natural waters. Breeuwsma and Lyklema (1973) estimated that the point of zero charge of hematite occurs at pH 8.5, dropping with specific sorption of divalent cation: For example, it is pH 6.5 in 0.3*N* $Ca(NO_3)_2$. Sorption of polyvalent anions increases the pH of the p. z. c.: it occurs at pH 9.5 in 0.3*N* K_2SO_4. Specific-sorbed cations and anions on aluminum hydroxide cause the pH of the p. z. c., to shift to lower and higher values respectively (Alwitt, 1972; Huang and Stumm, 1973).

Jepson et al. (1976) showed that with increasing silica adsorption the i.e.p. of gibbsite is reduced from pH 9 to a limiting value at about pH 3. This means that a pH above 3 the sorption of silica may result in a reversal of the surface charge of gibbsite. For a given pH, adsorption of Na cations increases and that of chloride anions decreases with increasing silica adsorption. A similar change in the net surface charge is obtained when phosphate is adsorbed onto hydrated alumina (Rajan, 1976). Lyklema (personal communication, 1975) has suggested that low isoelectric points recorded for some ferric oxides are due to thin layers of silicates covering their surfaces.

Incongruent dissolution of minerals should result in changes in their p. z. c. and i. e. p. values. Under the pH conditions of normal weathering solutions, the large, lower charged cations are leached preferentially to the small, more highly charged silicon. A change to a more siliceous surface should be accompanied by a change in the surface charge. An originally positive surface charge becomes less positive and a negative surface charge more negative. The p. z. c. of forsterite dispersed in aqueous suspension is initially 8.9. Since Mg is leached more than Si, the p. z. c. drops to 8.4

and 8.0 within 1 and 4 h, respectively. Longer dissolution periods lower this value toward that of silica gel (Luce and Parks, 1973).

Soils in general are characterized by surfaces bearing a constant charge, whereas soils of the humid tropics are characterized by their surface charge varying with changes in electrolyte concentration and in pH (van Raij and Peech, 1972). Soil organic matter, iron and aluminum hydroxides, and edges of clay minerals contribute to the pH dependency of soil surfaces, whereas the negative charge at low pH values is the result of the permanent negative charge of clay minerals. The p. z. c. of soils results from the overall composition of the soil system, giving an average of the organic and inorganic constituents. The presence of Fe and Al hydroxides tends to shift the p. z. c. to higher pH values whereas the presence of silica, clay minerals, and organic matter tends to shift the p. z. c. to lower pH values. Morais et al. (1976) found that the pH values of p. z. c. for Brazilian tropical soils ranged from 1.2–3.4 in the A horizon to 4.0–6.1 in the B horizon. At the prevailing field pH values the surface horizons have in general a net negative charge, while the subsurface horizons are either non- or positively charged. They attributed the negative charge of horizon A mainly to the presence of large amounts of organic matter compared with horizon B.

The surface coating of the mineral particle may form a porous layer. Lyklema (1971) found that porous oxides bear high surface charges without giving rise to particularly high electrostatic potentials. The actual charge of porous particles exceeds the surface charge produced by complete dissociation of the hydroxyl groups on the surface. This charge tends to be the higher the more porous the surface layer. These properties of the porous layer are a result of (1) potential-determining ions penetrating into the porous layer and allowing more charge to be accommodated per unit area, according to the facility of ion penetration, and (2) counterions also penetrating the surface, and the charge and potential at the solution side of the boundary consequently remaining at low levels.

3. Alteration of Minerals by Abrasion

Clelland et al. (1952) reported from chemical evidence a "high-solubility layer" on the surfaces of finely particulate "rock crystal", silica sand, fused amorphous silica, olivine, and orthoclase feldspar. It is a vitreous layer produced by mechanical crushing and grinding, the mean thickness of which varies from 30 to 150 nm.

Gordon and Harris (1955) stated that the amorphous layer on quartz does not have a precise thickness and simply represents a gradual transition from a noncrystalline exterior to a crystalline core. The conclusion was based mainly on X-ray diffraction examination, which indicated that particles smaller than 500 nm were essentially X-ray amorphous with virtually no crystalline core.

3.1 Abrasion of Silica and Silicate Minerals

3.1.1 Abrasion of Quartz

The D.T.A. curve of quartz exhibits an endothermic peak at 575 °C indicating the α to β structure inversion. Keith and Tuttle (1952) noted that particle size within the range

0.5–0.02 nm had no effect on the size of the peak. However, in studies of grinding quartz (Dempster and Ritchie, 1953; Clelland and Ritchie, 1952), it was shown that after a period of 17 h the peak size had decreased to an area equivalent to a quartz content of 50%. The diameter of the particles decreased from 250.0×10^3 nm to 1.3×10^3 nm and its density from 2.664 g/ml to 2.606 g/ml. The decrease in peak size was attributed to the formation on the surface of the particles of a thermally inert "non-quartz" layer. Barta and Bruthans (1962) observed that the decrease in size of the α to β inversion peak with the grinding of quartz was greater than that anticipated from the increase in specific surface area, and they suggested that mechanical grinding produces defects in the quartz lattice with a concomitant increase in reactivity.

3.1.2 Abrasion of Mica Minerals

The cation exchange capacity of unground muscovite is low (approximately 6 meq/100 g). It increases greatly with grinding. That of a muscovite from Goshen, Massachusetts increased after grinding for 1 h to 30 meq/100 g and to 208 meq/100 g after 24 h. X-ray evidence indicated that almost complete breakdown of the muscovite structure takes place in 8–9 h of dry grinding and is followed by recrystallization. The primary rupture occurs along the sheet surface resulting in delamination and rendering the interlayer K-exchangeable (Mackenzie and Milne, 1953).

Even short periods of grinding markedly affect the D. T. A. curves run on this mineral (Mackenzie, 1970). Mica minerals are usually anhydrous: however, the amorphous surface layer produced by grinding sorbs atmospheric moisture.

Muscovite usually recrystallizes to leucite, $KAl(SiO_3)_2$, at 1100 °C, but after grinding for 24 h, recrystallization occurs below 1000 °C because the increase in crystal defects enhances solid-state diffusion. Grinding under water is less destructive than dry grinding. The disaggregation by filing is the least destructive method.

The effect of grinding on trioctahedral micas appears to be much less than on dioctahedral species, presumably because the lack of vacancies in the octahedral layers of the format causes the diffusion of atoms, which occurs during grinding, to require a higher activation energy than the thermal effect produced by grinding. Vermiculite is also more resistant to dry grinding than muscovite because of the lubrication action of the water layers between the silicate sheets.

3.1.3 Abrasion of Clay Minerals

Of the clay minerals, kaolinite has received most attention because of its widespread use in ceramics. The grinding of kaolinite gives particles of lower crystallinity, with a lower proportion of structural hydroxyls and a lower temperature of dehydroxylation than unground kaolinite. The hydration energy increases with time of grinding (West, 1972). Grinding improves the plastic and dispersion properties of kaolinite.

The reactions occurring during the grinding of kaolinite can be classified in four groups: thermal diffusion, delamination, layer breakdown, and sorption of water (Yariv, 1975).

1. Thermal Diffusion of Atoms from Their Original Sites in the Crystal and the Formation of Lattice Defects. Although thermal diffusion occurs with all types of atoms, the best example is the migration of the small protons of hydroxyl groups. The process of proton migration within the mineral crystal is termed *prototropy*. Some of the migrating protons interact with structure hydroxyls to give water molecules, and may be desorbed by the clay at relatively low temperatures (Miller and Oulton, 1970). The mechanism is similar to thermal dehydroxylation and is represented as follows:

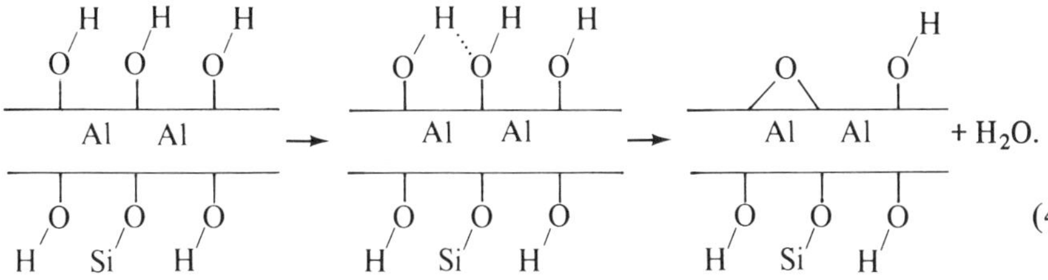

(4.VIII)

Other migrating protons may approach the fields of the *inner oxygens* common to the tetrahedral and octahedral sheets, as shown below (4.IX):

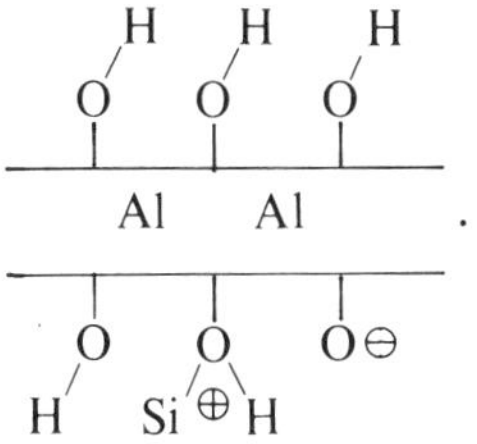

(4.IX)

Protons may migrate to the surface of the particle, the bulk particle becoming negatively charged and the surface gaining positive charge. The proton can be exchanged with other cations and so the cation exchange capacity increases by grinding. Sorption of anions, e.g., phosphate, is also increased by grinding (Murphy, 1939). The cation exchange reactions and the sorption of phosphate on the protonated kaolinite surface are shown below (4.X):

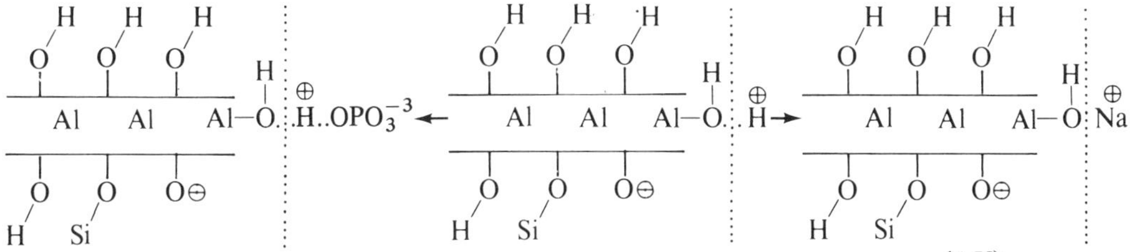

(4.X)

2. Separation of the Lamellae (Delamination). The primary rupture in kaolinite occurs along the sheet surface, and oxygen and hydroxyl planes become increasingly exposed with grinding. Small cations, such as Na, may penetrate the crystal through the hexagonal holes of the oxygen plane.

3. The Breakdown of the Kaolinite Layer. Fracture of the layer along other planes increases the number of exposed functional groups AlOH, SiOH, and Al–O–Si and thus increases the cation exchange capacity (Kelley and Jenney, 1936). There is a linear relationship between the cation exchange capacity and surface area of the mineral at the various stages of grinding. The surface exposed by this mechanism is termed the *broken bond surface.*

4. Sorption of Water. Freshly exposed kaolinite surfaces are very active and adsorb water from the atmosphere. Mechanical breakdown of the layer exposes more atoms, protons being attracted by exposed oxygens and hydroxyls by exposed aluminums and silicons. This is chemical sorption and in addition to this, physical sorption of water molecules occurs through a) the formation of hydrogen bonds between adsorbed molecules and hydroxyls of the broken bond surface; b) dipole–dipole interaction between water molecules and exposed hydroxyl planes; and c) hydration of cations exposed by thermal diffusion.

The four preceding groups of reactions result in the destruction of the kaolinite structure and the production of amorphous material. There is actually a gradual transition from a noncrystalline exterior to a crystalline core, the thickness of the former depending on the time and intensity of grinding. Takahashi (1959) ground dry kaolinite and halloysite for several hundred hours and noted in the initial stages a reduction in particle size, distortion, and ultimately destruction of the crystalline structure. However, prolonged grinding led to reaggregation of the amorphous material and the formation of spherical particles with a zeolite structure. The development of the zeolitic structure was accompanied by an increase in cation exchange capacity.

3.1.4 Interaction Between Surfaces of Silicate Minerals During Abrasion

Collisions between particles of different minerals may result in their surfaces interacting, and indeed repeated collision can have effects similar to grinding. Mishirky et al. (1974) investigated the reactions occurring during the grinding of kaolinite and calcined kaolinite (mainly mullite) mixtures. Kaolinite donated protons to calcined kaolinite, the kaolinite gaining a negative charge while the calcined kaolinite became positively charged. The electrostatic forces of attraction between the particles resulted in the adhesion of small particles of kaolinite to the large particles of the calcined kaolinite. This gives rise to a stable aggregate that inhibits the suspension of kaolinite in water. Erosion and sorting of kaolinite depends on the stability of this type of aggregate.

3.1.5 Interaction Between Organic Compounds and Surfaces of Minerals During Abrasion

The effects of grinding on the sorption of solid benzoic acid on surfaces of silica (calcined diatomaceous earth), alumina, and silica-alumina (calcined kaolinite) was studied by Yariv et al. (1967). Benzoic acid was chosen because it is a simple analogue of the more complex unsaturated acids that make up humic acids. Ground mixtures were examined by D. T. A. and infrared spectroscopy, and it was established that benzoic acid is chemically sorbed on the surfaces of alumina and calcined kaolinite as benzoate ion, whereas the functional surface groups Al–O–Al and Al–O–Si accept protons as follows:

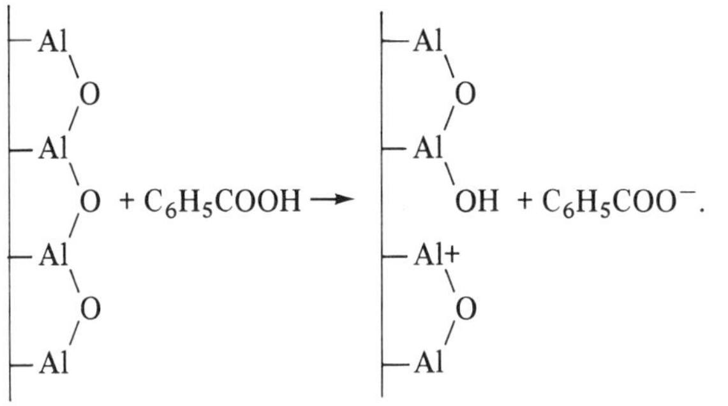

$$\left[\begin{array}{c}-\text{Al}\\\quad\diagdown\\\quad\quad\text{O}\\\quad\diagup\\-\text{Al}\\\quad\diagdown\\\quad\quad\text{O}\\\quad\diagup\\-\text{Al}\\\quad\diagdown\\\quad\quad\text{O}\\\quad\diagup\\-\text{Al}\end{array}\right] + C_6H_5COOH \longrightarrow \left[\begin{array}{c}-\text{Al}\\\quad\diagdown\\\quad\quad\text{O}\\\quad\diagup\\-\text{Al}\\\quad\diagdown\\\quad\quad\text{OH}\\\\-\text{Al}+\\\quad\diagdown\\\quad\quad\text{O}\\\quad\diagup\\-\text{Al}\end{array}\right] + C_6H_5COO^-. \qquad (4.\text{XI})$$

Neither benzoic acid nor benzoate ion was sorbed by the diatomaceous earth since the functional groups $Si-O^-$ and $Si-O-Si$ exposed upon breakdown of the diatomaceous earth particle are weaker bases than the benzoate ion and do not accept protons from benzoic acid.

3.2 Abrasion of Calcite

As with silicate minerals, the grinding of carbonate minerals involves partial crystal destruction and formation of crystal defects (Webb and Krüger, 1970). The effects of abrasion on the crystallinity and the stoichiometry of the calcite surfaces has been studied by Goujon and Mutaftschiev (1976) on the basis of the specific heat of immersion of calcite in water and the pH of saturated aqueous solutions. Four types of surface coatings on calcite particles were identified: 1) those composed of pure $CaCO_3$, 2) those composed of almost pure $CaCO_3$ and with a lower degree of crystallinity, 3) those containing up to 30% CaO groups, and 4) those containing $Ca(OH)_2$ groups. The third type has a high degree of crystallinity whereas the fourth type has the lowest degree.

Crystals of calcite particles form a nearly perfect rhombohedron. Crystals limited only by rhombohedral singular faces have a heat of immersion in water of about 480 ergs/cm^2. Their aqueous saturated solutions have a pH of 9.95, which is equal to the theoretical value for a pure saturated $CaCO_3$ solution. The surface of such crystals are of type (1), composed of pure calcium carbonate. Such surfaces do not adsorb CO_2 and do not dissociate below 420 °C, the temperature of the complete surface decomposition of calcium carbonate.

Ground powders are made up of sponge-like aggregates, and no evidence is found for a preferential cleavage along the rhombohedral faces. The grinding process destroys most of the surface crystallinity. Water is adsorbed on the amorphous surface during grinding, and surface groups of $CaCO_3$ hydrolyze, forming surface coatings with a high content of amorphous $Ca(OH)_2$ (type 4). The heat of immersion in water of the dry surface is as high as 960 ergs/cm^2.

Outgassing at temperatures below 150 °C eliminates the adsorbed water very slowly and incompletely. During dehydration, the surface slowly becomes reorganized. This dehydration and reorganization of the amorphous surface becomes very rapid and complete at temperatures above 150 °C. During the surface reorganization the surface dissociates to calcium oxide to an extent of about 30% according to

$$CaCO_3 \rightarrow CaO + CO_2 \tag{4.XII}$$

and

$$Ca(OH)_2 \rightarrow CaO + H_2O. \tag{4.XIII}$$

The surface obtained at this stage is well defined and reproducible. It corresponds to type (3) and is composed of randomly oriented surface $CaCO_3$ partially decomposed to CaO surface groups. The heat of immersion in water in this case is 760 ergs/cm^2. The aqueous saturated solutions have a pH of 10.06, which is equal to the theoretical value of a saturated solution of a mixture containing 70% $CaCO_3$ and 30% CaO. A thermal treatment of up to 420 °C of such powders does not lead to any further decomposition of the surface. At temperatures above 420 °C a bulk decomposition of the calcite leads to the formation of a CaO surface.

When ground powders are subjected to gaseous CO_2 treatment, a restoration of the surface calcium carbonate of type (2) occurs. The latter tends to occupy coherent positions with respect to the calcite substrate, thus increasing the degree of surface crystallinity. The heat of immersion in water in this case is 540 ergs/cm^2, which is somewhat higher than the heat of immersion of the type (1) surface. The saturated aqueous solution of this type of surface has a pH of 9.98. This type of $CaCO_3$ surface is very unstable, and the calcium oxide surface groups are restored by outgassing at a temperature as low as 50 °C.

3.3 Abrasion pH Values of Minerals

Abrasion pH is regarded by geochemists as the pH of suspension obtained by grinding particles of rock debris, soils, and mineral species under water (Stevens and Carron, 1948). The pH of the suspension may differ from the pH measured for the supernatant solution after the solid fraction has been separated. It may also differ from the pH of a system obtained by resuspending the previously separated particles in fresh water.

The pH of a suspension measured with the aid of a conventional glass electrode is determined primarily by the concentration of dissolved protons. Hydrogen and hydroxyl ions sorbed onto the solid–liquid interface and the surface acidity of the hydrous coating layer have a certain effect on the measured pH value of the system, as long as there are collisions between the electrode and suspended particles, and may lower or increase this value (see, e.g., Mattock, 1961; Lehmann and Lorenz, 1961).

The pH of the supernatant is determined by the hydrolysis of the dissolved species and the sorption of protons and hydroxyls on the freshly exposed solid surface. The rate of dissolution of the amorphous coating surface is high compared to that of the bulk crystalline core. The rate of dissolution greatly increases with grinding because both the specific surface area and crystalline defects (and internal solid-state diffusion of ions) greatly increase. Since silicic acid is a weak acid, minerals containing alkali and alkaline earth metals invariably yield abrasion pH values above 7. Minerals containing

Na and K have higher abrasion pH values than Mg and Ca minerals because the latter are partly hydrolyzed. Abrasion pH values of kaolinite, halloysite, and pyrophyllite are less than 7 because protons migrate from the mineral particle to the surface coatings during the abrasion process by the mechanism of prototropy. Prototropy also takes place during the grinding of other phyllosilicates and the abrasion pH values of these minerals are lower than those expected from silicate minerals containing alkali and alkaline earth metals.

When micas and alkali feldspars ground under dry conditions to pass a 200-mesh screen are placed in contact at room temperature with aqueous solutions low in alkali ions, alumina, and silica, a rapid ion exchange takes place during the first half-minute at the edges of the interlayers of the mica and on the surface of the feldspar with protons from the solution. The pH values of the resulting suspensions are greater than 7. In solutions in which the concentration ratios of $[K^+]/[H^+]$ is lower than 10^{9-10} or 10^{7-8}, an H-feldspar or H-mica structure is favored over K-feldspar or K-mica structure, respectively, at 25 °C (Garrels and Howard, 1959). This exchange is followed by the much slower dissolution of the mineral, whereby the elements of the crystal framework are also released into the solution and which results in a very slight decrease of the pH value. The latter process proceeds at room temperature for a period of several days.

Abrasion pH values of biotite decrease with increasing degree of weathering of the mineral, and pH values ranging between 9.5 and 6.5 were found for biotites from a gneiss in a warm humid climate, partly weathered under good drainage conditions (Grant, 1964). This is attributed to the leaching of interlayer K and its replacement by H ions. Abrasion pH values for some common minerals are given in Table 4.3.

Table 4.3. Abrasion pH values for some minerals. (After Stevens and Carron, 1948)

Mineral	Formula	Abrasion pH
Silicates		
Diopside	$CaMg(SiO_3)_2$	11
Olivine	$(Mg, Fe)_2SiO_4$	10, 11
Tremolite	$Ca_2Mg_5Si_8O_{22}(OH)_2$	10, 11
Augite	$Ca(Mg, Fe, Al)(Al, Si)_2O_6$	10
Hornblende	$(Ca, Na)_2(Mg, Fe, Al)_5(Al, Si)_8O_{22}(OH)_2$	10
Albite	$NaAlSi_3O_8$	9, 10
Talc	$Mg_3Si_4O_{10}(OH)_2$	9
Anthophyllite	$(Mg, Fe)_7Si_8O_{22}(OH)_2$	8, 9
Biotite	$K(Mg, Fe)_3AlSi_3O_{10}(OH)_2$	8, 9
Microline	$KAlSi_3O_8$	8, 9
Anorthite	$CaAl_2Si_2O_8$	8
Muscovite	$KAl_2AlSi_3O_{10}(OH)_2$	7, 8
Orthoclase	$KAlSi_3O_8$	8
Montmorillonite	$Al_2Si_4O_{10}(OH)_2.nH_2O$	6, 7
Halloysite	$Al_2Si_2O_5(OH)_4.nH_2O$	6
Pyrophyllite	$Al_2Si_4O_{10}(OH)_2$	6
Kaolinite	$Al_2Si_2O_5(OH)_4$	5, 7

Table 4.3 (continued)

Mineral	Formula	Abrasion pH
Oxides		
Boehmite	$AlO(OH)$	6, 7
Gibbsite	$Al(OH)_3$	6, 7
Hematite	Fe_2O_3	6
Carbonates		
Magnesite	$MgCO_3$	10, 11
Dolomite	$CaMg(CO_3)_2$	9, 10
Aragonite	$CaCO_3$	8
Calcite	$CaCO_3$	8

In certain playa lakes, where the rocks subjected to alteration are rich in minerals such as olivines, augite, and nepheline, with high abrasion pH values, the waters may become especially alkaline. Such an environment would be suitable for the formation of authigenic zeolites (Loughnan, 1966) and minerals of the palygorskite–sepiolite group (Loughnan, 1960). The highly alkaline environments obtained during the early weathering abrasion stages of basalts in Galilee, Israel, are attributable to the presence of olivines and pyroxenes (Singer, 1970).

4. Surface Structures of Gibbsite and Goethite

According to Russell et al. (1975) the surfaces of the crystalline hydroxides gibbsite and goethite are well defined and closely related to their bulk structures. A gibbsite tactoid consists of infinite planes of closely packed hydroxyl groups parallel to (001), the planes being bound together in pairs by Al ions to form a layer structure. Such a pair is defined as a gibbsite-type sheet. Gibbsite tactoids synthesized by Russell et al. were in the form of thin hexagonal plates having diameters of about 250 nm and thickness of about 9 nm. Their exposed surface consisted principally of (001) hydroxyl planes. The surface area of the particles was 104 m^2/g of which the edge surface comprised only 8 m^2/g and the (001) hydroxyl plane 96 m^2/g. Four types of inner OH groups are identifiable in gibbsite: 1) hydroxyls forming hydrogen bonds with hydroxyls of a neighboring gibbsite-type sheet; 2) and 3) two types of hydroxyls forming hydrogen bonds with hydroxyls within the same plane, and 4) hydroxyls that do not form hydrogen bonds. Types (2) , (3), and (4) are identifiable on the surface of gibbsite tactoid. They are the functional groups chiefly responsible for the surface reactions and sorption properties of this mineral (see e.g. page 273–276).

The goethite structure is built up from strips of condensed Fe(O, OH) octahedra, each strip being two octahedra in width. The strips share oxygens along their edges to give an open structure in which the gaps between the strips are bridged by hydrogen bonds involving OH groups shared between three Fe ions. The same hydroxyls are also exposed on the (100) surface together with two new types of hydroxyls, which arise from protonation of the oxide ions along the edges of the strips. Three types of surface hydroxyls can be identified: (1) hydroxyls, in which each OH is coordinated to only one Fe; (2) and (3) hydroxyls in which each OH group is coordinated to two and three Fe ions. Hydroxyls of type (1) are able to form hydrogen bonds with each other. They are also easily exchanged by anions such as phosphates and silicates during the specific sorption of these anions. Hydroxyls of type (2) act as proton donors, forming hydrogen bonds with adsorbed proton acceptors (pages 273–276).

References

Alwitt, R. S.: The point of zero charge of pseudoboehmite. J. Colloid Interface Sci. *40,* 195–198 (1972)

Barta, R., Bruthans, Z.: Contribution to the application of D. T. A. in the research of activity due to vibration grinding (English abstract). Silikaty *6,* 9–15 (1962)

Bischoff, J. L., Clancy, J. J., Booth, J. S.: Magnesium removal in reducing marine sediments by cation exchange. Geochim. Cosmochim. Acta *39,* 559–568 (1975)

Biswas, T. D., Gawande, S. P.: Relation of manganese in genesis of catenary soils. J. Indian Soc. Soil Sci. *12,* 261–267 (1964)

Blackmore, A. V.: Aggregation of clay by the product of iron (III) hydrolysis. Aust. J. Soil Res. *11,* 75–82 (1973)

Bracewell, J. M., Campbell, A. C., Mitchell, B. D.: An assessment of some thermal and chemical techniques used in the study of the poorly-ordered aluminosilicates in soil clays. Clay Miner. *8,* 325–335 (1970)

Breeuwsma, A., Lyklema, J.: Physical and chemical adsorption of ions in the electrical double layer on hematite. J. Colloid Interface Sci. *43,* 437–448 (1973)

Carpenter, R. H., Timothy, A. P., Smith, R. L.: Fe-Mn oxide coatings in stream sediments geochemical surveys. J. Geochem. Exploration *4,* 349–363 (1975)

Choi, I., Smith, R. W.: Kinetic study of dissolution of asbestos fibers in water. J. Colloid Interface Sci. *40,* 253–262 (1972)

Clelland, D. W., Cumming, W. M., Ritchie, P. D.: Physicochemical studies on dusts. I. A. high-solubility layer on silicious dust surfaces. J. Appl. Chem. *2,* 31–41 (1952)

Clelland, D. W., Ritchie, P. D.: Physicochemical studies on dusts. II. Nature and regeneration of the high-solubility layer on siliceous dusts. J. Appl. Chem. *2,* 42–48 (1952)

Correns, C. W.: Experiments on the decomposition of silicates and discussion of chemical weathering. Clays Clay Miner. *12,* 443–460 (1963)

Crerar, D. A., Barnes, H. L. Deposition of deep-sea manganese nodules. Geochim. Cosmochim. Acta *38,* 279–300 (1974)

Dempster, P. B., Ritchie, P. D.: Physicochemical studies on dusts. V. Examination of finely ground quartz by differential thermal analysis and other physical methods. J. Appl. Chem. *3,* 182–192 (1953)

Förstener, U., Müller, G.: Heavy metal pollution monitoring by river sediments. Fortsch. Miner. *53,* 271–288 (1976)

Fripiat, J. J., Mendelovici, E.: Derivatives organiques des silicate. I. Derivé methylé du chrysotile. Bull. Soc. Chim. *2,* 483–492 (1968)

Fuller, W. H., Korte, N. E., Niebla, E. E., Alesh, B. A.: Contribution of the soil to the migration of certain common and trace elements. Soil Sci. *122,* 223–235 (1976)

Garrels, R. M., Howard, P.: Reactions of feldspar and mica with water at low temperature and pressure. Clays Clay Miner. *6,* 68–88 (1959)

Gastuche, M. C.: Kinetics and dissolution of biotite. I. Interfacial rate process followed by optical measurements of the white silica rim. Proc. Int. Clay Conf., Stockholm. Rosenquist, I. Th. (ed.) Oxford: Pergamon Press, 1963, pp. 67–75

Gordon, R. Y., Harris, G. W.: Effect of particle size on the quantitative determination of quartz by X-ray diffraction. Nature (London), *175,* 1135 (1955)

Goujon, G., Mutaftschiev, B.: On the crystallinity and stoichiometry of the calcite surface. J. Colloid Interface Sci. *57,* 148–161 (1976)

Grant, W. H.: Chemical weathering of biotite-plagioclase gneiss. Clays Clay Miner. *12,* 455–463 (1964).
Harris, R. C., Adams, J. A. S.: Geochemical and mineralogical studies on the weathering of granitic rocks. Am. J. Sci. *264,* 146–173 (1966)
Heald, M. T., Larese, R. E.: Influence of coating on quartz cementation. J. Sediment. Petrol. *44,* 1269–1274 (1974)
Heye, D., Marchig, V.: Relationship between the growth rate of manganese nodules from the Central Pacific and their chemical constitution. Mar. Geol. *23,* M19–M25 (1977)
Huang, C. P., Stumm, W.: Specific adsoprtion of cations on hydrous $\gamma-Al_2O_3$. J. Colloid Interface Sci. *43,* 409–420 (1973)
Huang, W. H., Keller, W. D.: Dissolution of rock-forming silicate minerals in organic acids: simulated first-stage weathering of fresh mineral surfaces. Am. Mineral. *55,* 2076–2094 (1970)
Jepson, W. B., Jeffs, D. G., Ferris, A. P.: The adsorption of silica on gibbsite and its relevance to the kaolinite surface. J. Colloid Interface Sci. *55,* 454–461 (1976)
Jones, R. C., Uehara, G.: Amorphous coatings on mineral surfaces. Soil Sci. Soc. Am. Proc. *37,* 792–798 (1973)
Kamprath, E. J.: Soil acidity and liming. In: Soils of the humid tropics. Drosdoff, M. (ed.). Washington: Nat. Acad. Sci. 1972, pp. 136–149
Keith, M. L., Tuttle, O. F.: Significance of variation in the high-low inversion of quartz. Am. J. Sci., Bowen Volume, Pt. *1,* 203–252 (1952)
Keller, W. D., Balgord, W. D., Reesman, A. L.: Dissolved products of artificially pulverized silicate minerals and rocks: Part I. J. Sediment. Petrol. *33,* 191–204 (1963)
Kelley, W. P., Jenney, H.: Relation of crystal structure to base exchange and its bearing on base exchange in soils. Soil Sci. *42,* 376–382 (1936)
Krauskopf, K. B.: Factors controlling the concentrations of thirteen rare metals in seawater. Geochim. Cosmochim. Acta *9,* 1–32B (1956)
Krinsley, D., Doornkamp, J. C.: Atlas of quartz sand surface textures. Cambridge: Cambridge Univ. Press 1973
Lalou, C., Brichet, E.: On some relationships between the oxide layers and the cores of deep sea manganese nodules. Miner. Deposita *11,* 267–277 (1976)
Lehman, H., Lorenz, W.: pH-Messungen in Suspensionen tonmineralhaltiger Rohstoffe. Tonind. ztg. Keram. Rundsch. *85,* 325–333; 349–358 (1961)
Lin, I. J., Rohrlich, V., Slatkine, A.: Surface microstructures of heavy minerals from the Mediterranean coast of Israel. J. Sediment. Petrol. *44,* 1281–1295 (1974)
Lorrain, R. D., Souchez, R. A.: Sorption as a factor in the transport of major cations by meltwater from an Alpine glacier. Quat. Res. (N. Y.) *2,* 253–256 (1972)
Loughnan, F. C.: Further remarks on the occurrence of palygorskite at Redbank Plains, Queensland. J. Roy. Soc. Qld. *71,* 43–50 (1960)
Loughnan, F. C.: A comparative study of the Newcastle and Illawara Coal Measure sediments of the Sydney Basin. J. Sediment. Petrol. *36,* 1016–1025 (1966)
Luce, R. W., Bartlett, R. W., Parks, G. A.: Dissolution kinetics of magnesium silicates. Geochim. Cosmochim. Acta, *36,* 35–50 (1972)
Luce, R. W., Parks, G. A.: Point of zero charge of weathering forsterite. Chem. Geol. *12,* 147–153 (1973)
Lyklema, J.: The electric double layer on oxides. Croat. Chem. Acta *43,* 249–260 (1971)
Mackenzie, R. C.: Simple phyllosilicates based on gibbsite and brucite-like sheets. In: Differential thermal analysis. Mackenzie, R. C. (ed.). London: Academic Press, 1970, Vol. *1,* 497–537
Mackenzie, R. C., Milne, A. A.: The effect of grinding on micas: I. Muscovite. Mineral. Mag. *30,* 178–185 (1953)
Marshall, C. E.: The physical chemistry and mineralogy of soils, Vol. 1: Soil materials. New York: John Wiley & Sons 1964
Marshall, C. E., McDowell, L. L.: The surface reactivity of micas. Soil Sci. *99,* 115–131 (1965)
Mattock, G.: pH measurements and titrations. New York: The Macmillan Co. 1961
Mendelovici, E.: Infrared study of attapulgite and HCl treated attapulgite. Clays Clay Miner. *21,* 115–119 (1973)

Meunier, A., Velde, B.: Mineral reactions at grain contacts in early stages of granite weathering. Clay Miner. *11*, 235–240 (1976)
Miller, J. G., Oulton, J. D.: Protropy in kaolinite during percussive grinding. Clays Clay Miner. *18*, 313–323 (1970)
Mishirky, S. A., Yariv, S., Siniansky, W. I.: Some effect of grinding kaolinite with calcined kaolinite. Clay Sci. *4*, 213–224 (1974)
Mitchell, R. L.: Trace elements in soils. In: Chemistry of the soil. Bear, F. E. (ed.). New York: Reinhold 1964, pp. 320–368
Morais, F. I., Page, A. L., Lund, L. J.: The effect of pH, salt concentration and nature of electrolytes on the charge characteristics of Brazilian tropical soils. Soil Sci. Soc. Am. Proc. *40*, 521–527 (1976)
Moshi, A. O., Wild, L., Greenland, D. J.: Effect of organic matter on the charge of phosphate adsorption characteristics of Kikuyu red clay from Kenya. Geoderma *11*, 275–285 (1974)
Murphy, H. F.: The role of kaolinite in phosphate fixation. Hilgardia *12*, 341–382 (1939)
Nathan, J.: Dissolution of palygorskite by hydrochloric acid. Isr. J. Chem. *6*, 275–283 (1968)
Navrot, J., Singer, A.: Geochemical changes accompanying basic igneous rocks-clay transition in a humid Mediterranean climate. Soil Sci. *121*, 337–345 (1976)
Parks, G. A.: The isoelectric points of solid oxides, solid hydroxides and aqueous hydroxy complex system. Chem. Rev. *65*, 177–198 (1965)
Perrott, K. W., Langdon, A. G., Wilson, A. T.: Sorption of anions by the cation exchange surface of muscovite. J. Colloid Interface Sci. *48*, 10–19 (1974)
Petrović, R., Berner, R. A., Goldhaber, M. B.: Rate control in dissolution of alkali feldspars. I. Study of residual feldspar grains by X-ray photoelectron spectroscopy. Geochim. Cosmochim. Acta *40*, 537–548 (1976)
Rajan, S. S. S.: Changes in net surface charge of hydrous alumina with phosphate adsorption. Nature (London) *262*, 45–46 (1976)
Russell, J. D., Parfitt, R. L., Fraser, A. R., Farmer, V. C.: Surface structures of gibbsite, goethite and phosphated goethite. Nature (London) *248*, 220–221 (1975)
Sholkovitz, E.: Interstitial water chemistry of Santa Barbara basin sediments. Geochim. Cosmochim. Acta *37*, 2043–2073 (1973)
Shuman, L. M.: The effect of soil properties on zinc adsorption by soils. Soil Sci. Soc. Am. Proc. *39*, 454–458 (1975)
Singer, A.: Weathering products of basalts in the Galilee. I. Rock-soil interface weathering. Isr. J. Chem. *8*, 459–468 (1970)
Siskens, C. A. M., Stein, H. N., Stevels, J. M.: Surfaces of silicates in aqueous saline solutions. I. J. Colloid Interface Sci. *52*, 244–250 (1975a)
Siskens, C. A. M., Stein, H. N., Stevels, J. M.: Surfaces of silicates in aqueous saline solutions. II. J. Colloid Interface Sci. *52*, 251–259 (1975b)
Stevens, R. E., Carron, M. K.: Simple field test for distinguishing minerals by abrasion pH. Am. Mineral. *33*, 31–49 (1948)
Swaine, D. J., Schneider, J. L.: The geochemistry of underground water. In: Salinity and water use. Talsma, T., Philip, J. R. (eds.). New York: Macmillan Press, Ltd. 1971, pp. 3–23
Takahashi, H.: Effect of dry grinding on kaolin minerals. I. Kaolinite. Bull. Chem. Soc. Jpn. *32*, 235–245 (1959)
Townsend, F. C., Reed, L. W.: Effect of amorphous constituents on some mineralogical and chemical properties of Panamanian latosol. Clays Clay Miner. *19*, 303–310 (1971)
van Olphen, H.: An introduction to clay colloid chemistry. New York: Interscience Pub. 1963.
van Raij, B., Peech, M.: Electrochemical properties of some oxisols and alfisols of the tropics. Soil Sci. Soc. Am. Proc. *36*, 587–593 (1972)
Webb, T. L., Krüger, J. E.: Carbonates. In: Differential thermal analysis. Mackenzie, R. C. (ed.) London: Academic Press, 1971, Vol. 1, pp. 343–361
West, R. R.: Ceramics. In: Differential thermal analysis. Mackenzie, R. C. (ed.) London: Academic Press, 1972, Vol. 2, pp. 149–179
Weyl, P. K.: The solution behavior of carbonate materials in sea water. Stud. Trop. Oceanogr. *5*, 178–228 (1967)

Yaalon, D. H., Brenner, J., Koyumdjisky, H.: Weathering and mobility sequence in minor elements on a basaltic pedomorphic surface, Galilee, Israel. Geoderma *12,* 233–244 (1974)
Yaalon, D. H., Jungreis, C., Koyumdjisky, H.: Distribution and reorganization of manganese in three catenas of Mediterranean soils. Geoderma *7,* 71–78 (1972)
Yariv, S.: Infrared study of grinding kaolinite with alkali metal chlorides. Powder Technol. *12,* 131–138 (1975)
Yariv, S., Birnie, A. C., Farmer, V. C., Mitchell, B. D.: Interactions between organic substances and inorganic diluents in D. T. A. Chem. Ind. 1744–1745 (1967)

Chapter 5
Kinetic Properties of Colloid Solutions

1. Kinetic Properties of Particles Dispersed in Still Fluids

1.1 Brownian Movement

Particles dispersed in a colloid solution are engaged in ceaseless irregular movement that can be observed in the microscope. Adjacent particles are not moving in the same direction and not at the same velocity, which may indicate that this movement does not result from microscopic convection currents in the system. This phenomenon, called Brownian movement, was first noticed with pollen grains by the English botanist Robert Brown in 1827.

Einstein in 1905 and Smoluchevski in 1906 developed a quantitative theory of Brownian movement, according to which the movement of microscopic bodies is a visible manifestation of the random thermal movement of molecules of the dispersing medium. Brownian movement is caused by the difference in the sum of impacts received from molecules that hit the particles from opposite sides. It is obvious that the smaller the particle the less the probability of the molecular bombardment being exactly belanced and therefore the more intensive will be the movement.

The thermal movement of fluid molecules and of particles is the basis of all properties such as diffusion and osmosis.

1.2 Diffusion

Diffusion of colloid particles is one of the consequences of Brownian movement. The diffusion of colloids differs only in degree from that of molecules and electrolytes, and the rates of diffusion are often of similar magnitudes.

When a certain phase is dispersed along the x axis in the absence of external force fields and convections (e.g., solid particles in still water), the mass, dm, transferred across an area, S, during an interval, dt, as a result of a *concentration gradient*, dc/dx, is given by Fick's first law

$$dm = -DS(dc/dx)dt, \tag{5.1}$$

where D is the diffusion coefficient. Both concentration and concentration gradients are functions of the two components, time, t, and position, x. Since diffusion always proceeds in the direction of decreasing concentration, the sign of dm should be opposite to that of the gradient dc/dx.

A diffusion flux, i_d, is defined as the amount of substance transferred across a unit area per a unit time. In accordance with Eq. (5.1) the diffusion flux can be expressed as

$$i_d = dm/S\,dt = -D\,(dc/dx). \tag{5.2}$$

The diffusion flux, being a function of the concentration gradient, is dependent on time and position, and the derivatives with respect to t and x are partial derivatives.

The relationship between c, x, and t can be determined as follows. Let two equal parallel sections S be disposed at x_1 and x_2 at right angles to the direction of diffusion. According to Eq. (5.1), the amount $dm_1 = -DS(\delta c/\delta x)_{x=x_1}\,dt$ will enter across the section at x_1 during the time dt, whereas the amount $dm_2 = -DS(\delta c/\delta x)_{x=x_2}\,dt$ will leave across the section at x_2. During the time $\Delta t = t_2 - t_1$, the increase of the amount of the disperse phase between x_1 and x_2 will be

$$\Delta m = \int_{t_1}^{t_2} (dm_1 - dm_2) = DS\int_{t_1}^{t_2} [(\delta c/\delta x)_{x=x_2} - (\delta c/\delta x)_{x=x_1}]\,dt. \tag{5.3}$$

The amount of substance inside the space limited between the two sections is

$$m = \bar{c}\,S\,x, \tag{5.4}$$

where $\bar{c}$ is the average concentration of the disperse phase in this space and x is the distance between the two sections (the thickness of the space). If this distance is infinitely small and is equal to dx, c may be regarded to be homogeneous inside the space, and the amount of the disperse substance in the space is

$$m = c\,S\,dx. \tag{5.5}$$

As a result of diffusion, the concentration of the disperse phase may change. It is c_{t1} at the time t_1 and becomes c_{t2} at the time t_2. The change in the amount of disperse phase during the time Δt in an infinitely thin layer is $dm = c_{t2}Sdx - c_{t1}Sdx$. Integrating between x_1 and x_2 gives

$$\Delta m = S\int_{x_1}^{x_2} (c_{t2} - c_{t1})\,dx. \tag{5.6}$$

Both equations (5.3) and (5.6) describe Δm. Equating these expressions for Δm gives

$$DS\int_{t_1}^{t_2} [(\delta c/\delta x)_{x=x_2} - (\delta c/\delta x)_{x=x_1}]\,dt = S\int_{x_1}^{x_2} (c_{t2} - c_{t1})\,dx. \tag{5.7}$$

S can be canceled from both sides of Eq. (5.7) and the equation can be solved by means of the theorem of mean values. According to this theorem, if the function $f(u)$ is continuous in the interval (a, b) and keeps the same sign, a point e can be found between a and b for which the relation (5.8) holds

$$\int_a^b f(u)\,du = f(e)\,(b - a). \tag{5.8}$$

This theorem can be proved with the help of Figure 5.1, in which the abscissa and the ordinate describe u and $f(u)$, respectively. The integral $\int_a^b f(u)du$ is shown by the area AA'B'B. A point E' can be found on the curve A'B' so that the area A'E'A'' will be equal to the area B'E'B''. It follows that the area AA'B'B is equal to the area AA''B''B. From the Figure it is obvious that the area AA''B''B is equal to $f(e)\,(b-a)$. The mean value theorem is thus proved.

Considering the left side of Eq. (5.7) it may be assumed that

$$f(u) = f(t) = (\delta c/\delta x)_{x=x_2} - (\delta c/\delta x)_{x=x_1},$$

$du = dt$, $a = t_1$, $b = t_2$, $e = t'$ where $t_1 < t' < t_2$ and $\Delta t = t_2 - t_1$. Considering the right side of Eq. (5.7) it may be assumed that $f(u) = f(x) = c_{t2} - c_{t1}$, $du = dx$, $a = x_1$, $b = x_2$, $e = x'$, where $x_1 < x' < x_2$ and $\Delta x = x_2 - x_1$. Equation (5.7) becomes

$$D\, f(t')\,(t_2 - t_1) = f(x')\,(x_2 - x_1)$$

or

$$D[(\delta c/\delta x)_{x=x_2} - (\delta c/\delta x)_{x=x_1}]_{t=t'}\,\Delta t = (c_{t2} - c_{t1})_{x=x'}\,\Delta x. \tag{5.9}$$

This equation can be solved by means of the finite increment theorem of Lagrange. According to this theorem, if the function $f(u)$ is continuous in the interval (a, b) and differentiable at every interior point of this interval, a point k can be found in this interval for which the relation (5.10) holds:

$$f'(k) = [f(b) - f(a)]/(b - a). \tag{5.10}$$

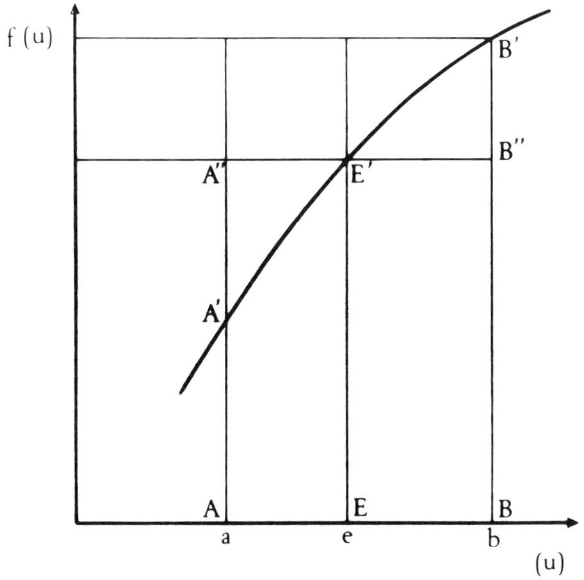

Fig. 5.1. A representative curve for $f(u)$ versus u

This theorem can be proved with the help of Figure 5.2, where the abscissa and the ordinate describe u and $f'(u)$ respectively. The integral $\int_a^b f'(u)du$ is equal to $f(b) - f(a)$ and is shown by the area AA′B′B. A point K′ can be found on the curve A′B′ so that the area A′K′A″ will be equal to the area B′K′B″. It follows that the area AA′B′B is equal to the area AA″B″B or to $f'(k)(b - a)$. Hence

$$f'(k)\,(b-a) = \int_a^b f'(u)du = f(b) - f(a)$$

or

$$f'(k) = [f(b) - f(a)]/(b-a).$$

Multiplying and dividing the left side of Eq. (5.9) by $\Delta x = x_2 - x_1$ and the right side by $\Delta t = t_2 - t_1$ gives

$$\frac{D\,[(\delta c/\delta x)_{x=x_2} - (\delta c/\delta x)_{x=x_1}]_{t=t'}\ \Delta t \Delta x}{x_2 - x_1} = \frac{(c_{t2} - c_{t1})_{x=x'}\ \Delta x \Delta t}{t_2 - t_1}. \tag{5.11}$$

Concerning the left side of this equation, it may be assumed that $f(u) = (\delta c/\delta x)_{t=t'} = f(x)$, $a = x_1$, $b = x_2$, $k = x''$, where $x_1 < x'' < x_2$ and $f'(k) = f'(x'') = (\delta^2 c/\delta x^2)_{\substack{x=x'' \\ t=t'}}$.

Concerning the right side of (5.11), it may be assumed that $f(u) = f(t) = c_{x=x'}$, $a = t_1$, $b = t_2$, $k = t''$ where $t_1 < t'' < t_2$ and $f'(k) = f'(t'') = (\delta c/\delta t)_{\substack{x=x' \\ t=t''}}$. Equation 5.11 becomes

$$D(\delta^2 c/\delta x^2)_{\substack{x=x'' \\ t=t'}}\ \Delta t \Delta x = (\delta c/\delta t)_{\substack{x=x' \\ t=t''}}\ \Delta x \Delta t. \tag{5.12}$$

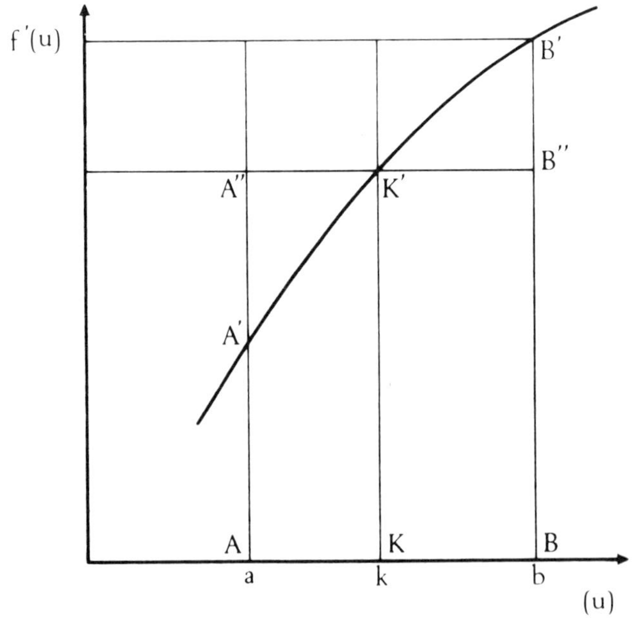

Fig. 5.2. A representative curve for $f'(u)$ versus u

On reaching the limit with respect to Δt and Δx, the values x' and x'' as well as t' and t'', become identical, respectively, with x and t, and the following equation is obtained

$$\delta c/\delta t = D(\delta^2 c/\delta x^2), \tag{5.13}$$

which is known as Fick's second law. By solving this differential equation the concentration of the suspended matter can be expressed as a function of time and position. The formulation of this law for the three-dimensional diffusion is $\delta c/\delta t = D\nabla^2 c$, where ∇^2 is the *Laplace operator* and $\nabla^2 = \delta^2/\delta x^2 + \delta^2/\delta y^2 + \delta^2/\delta z^2$.

The diffusion coefficient, D, depends on the properties of the disperse phase and the dispersing medium. For colloid particles D is related to the friction coefficient, B, between the particles and the dispersing medium by the expression

$$D = k\,T/B, \tag{5.14}$$

where k is a constant and T is the absolute temperature. The magnitudes of B and D depend on the dimensions and shapes of the particles. For spherical particles having a radius r, when the viscosity of the dispersing medium is η, B equals $6\,\Pi\eta r$ and Eq. (5.14) becomes $D = kT/6\,\Pi\eta r$.

1.3 Sedimentation

Particles are subjected to the terrestrial gravitational field and, owing to their weight, settle at an appreciable rate. The force of gravity, G, on a particle with an effective mass m can be expressed as

$$G = m\,g, \tag{5.15}$$

where g is the acceleration due to gravity. The effective mass of the particle is equal to its mass in a vacuum less the mass of the displaced dispersing medium, and for m we obtain

$$m = \rho\,v - \rho_0 v = v(\rho - \rho_0) = m_0(1 - \rho_0/\rho), \tag{5.16}$$

where ρ and ρ_0 are the densities of the disperse phase and dispersing medium, respectively, m_0 is the particle mass in a vacuum, and v is the volume of the particle.

During the first stage of sedimentation the movement of the particles along a vertical axis h is accelerated. As the rate of sedimentation increases the friction also increases and becomes so high that after a few seconds the particle settles at a uniform rate, $u = dh/dt$, which may be expressed by

$$u = m\,g/B, \tag{5.17}$$

where B is the friction coefficient. For spherical particles, where $m = (4/3)\,\Pi r^3(\rho - \rho_0)$ and $B = 6\,\Pi\eta r$, we obtain

$$u = 2\,r^2 g(\rho - \rho_0)/9\,\eta. \tag{5.18}$$

The sedimentation velocity of spherical particles is directly proportional to the square of particle radius and inversely proportional to the viscosity of the dispersing medium. This is *Stokes law of settling.*

In most geologic systems sedimentation occurs with accompanying processes, such as flocculation, dissolution, sorption, and desorption of solutes, lateral transport, and diffusion. If the settling velocities in disperse systems are to be correctly described, these processes must be taken into consideration (see, e.g., Lerman et al., 1974; Hamilton et al., 1974). A reduction in the rate of deposition occurs when the fine, suspended sediment is entrapped in viscous layers, caused for example by occasional erosion, bottom roughness, and orogenic resuspension, and was calculated by McCave and Swift (1976).

A simple case will now be analyzed in which sedimentation and diffusion in still water along the vertical axis h are taken into consideration. The sedimentation flux i_s is defined by

$$i_s = u\,c \tag{5.19}$$

and with Eq. (5.17) it becomes

$$i_s = (mg/B)\,c. \tag{5.20}$$

Comparison of the sedimentation flux with the diffusion flux (Eq. 5.2) provides a convenient method of determining whether sedimentation or upward diffusion is the predominating process in the given case. When $|i_s/i_d| \gg 1$, sedimentation predominates, whereas in the case of $|i_s/i_d| \approx 1$, both processes take place simultaneously. This ratio is obtained from Eqs. (5.2), (5.14), and (5.20) as follows:

$$\frac{i_s}{i_d} = -\frac{m\,g}{k\,T}\cdot\frac{c}{dc/dh} = -\frac{\nu(\rho-\rho_0)\,g}{k\,T}\cdot\frac{c}{dc/dh}\,. \tag{5.21}$$

From this equation it follows that sedimentation plays a greater role in the case of systems with coarser particles. Sedimentation is always the major process during the first stage, when $dc/dh = 0$. After some time, as the sedimentation process progresses, the uniform distribution of the substance is disturbed, dc/dh becomes finite, and increases with time, and i_d increases until it becomes equal to i_s. At this point no further mass transfer takes place, equilibrium is established, and $i_s/i_d = -1$.

It is possible to calculate the equilibrium distribution of concentration in the terrestrial gravitational field as follows

$$-(m\,g/k\,T)\cdot[c/(dc/dh)] = 1.$$

The variables may be equated as follows

$$-(dc/c) = (m\,g\,dh/k\,T).$$

Integrating between 0 and h and correspondingly between c_0 at $h = 0$ and c_h at h, we obtain

$$\ln(c_0/c_h) = m\,g\,h/k\,T. \tag{5.22}$$

This equation describes an equilibrium distribution. The time necessary for the establishment of this equilibrium does not appear in this equation. This equation does not include the coefficient B, which represents the mass transfer and the accompanying friction.

2. Kinetic Properties of Particles in Flowing Fluids

2.1 Laminar and Turbulent Flow

There are two modes of fluid flow, *laminar* and *turbulent.* It is possible to differentiate between these two modes with the help of "streamlines". A streamline is an imaginary line to which at any instant the velocity vector of all the fluid elements lying on it, is tangent. When the flow velocity is below a certain critical value, the flow obtained is laminar, characterized by the smooth linearity of the streamlines representing it. The streamlines of this uniform flow are parallel to each other.

When the flow velocity exceeds a certain critical value, the flow obtained is turbulent, and the streamlines for the movement are highly distorted and nonparallel. The magnitude of the velocity vector is not constant. It increases as the streamlines converge and decreases when the streamlines diverge. The pattern of streamlines persists as long as the flow field remains steady. In an unsteady and nonuniform flow, on the other hand, a given pattern of streamlines can only have a temporary existence.

In water moving in a channel in laminar flow, parallel layers of the fluid shear one above the other. The shearing stress is proportional to the resistance of the fluid layers to movement, and is determined by the viscosity, η, and by the change in velocity from one layer to the next. At any point at a distance y above the channel bed, where the layer of water is flowing at velocity v, the shearing stress, τ, (see Ch. 9) can be expressed as

$$\tau = \eta\,(dv/dy) \qquad (5.23)$$

The layer of maximum velocity lies below the water surface. In the proximity of the bed or the channel walls, the velocity is zero and increases with the distance from the boundary layer. The vertical profile of velocity has the shape of a parabola since there is no mixing of the moving layers.

In turbulent flow the magnitude and direction of the velocity vector measured at a point must vary from instant to instant as different portions of the randomly eddying fluid pass the point. There are mass and momentum exchanges upward, downward, and lateral to the direction of the propagation of the fluid and the system is well mixed. A turbulent flow resists distortion to a much greater degree than a laminar flow of the same fluid. Because of the relative movement of sizeable portions of fluid in turbulent flow, this flow would appear to have a very high viscosity compared to laminar flow. This is, however, an apparent viscosity that varies with the character of the turbulence. Equation (5.23) for the turbulent flow becomes

$$\tau = (\mu + \eta)\,(dv/dy) \qquad (5.24)$$

where μ is the "eddy viscosity".

The factors that affect the critical velocity where laminar flow becomes turbulent are viscosity and density, ρ, of the fluid, depth of water, and roughness of the channel surface. *Reynolds number* is most commonly used to distinguish between laminar and turbulent flow and is expressed by

$$N_R = \rho(VR/\eta) \tag{5.25}$$

where V is the mean velocity and R is the hydraulic radius. Flow is laminar for small values of the Reynolds number (less than 500) and turbulent for higher ones. The Reynolds number for flows in streams is generally over 500.

2.2 Fluid Drag

If an object is dropped in water the downward force will accelerate it so that it will drop at an ever-faster rate. As the object moves through the water, the water must move around the object to permit it to sink, at the same time exerting an upward force on the object. This is the drag of the water, which impedes the downward motion of the object. The drag increases as the relative motion between the object and the water is increased. When a fluid and a solid body are in relative motion, forces always are exerted to oppose the motion and enforce equilibrium.

There are three types of drag. *Viscous* (or *laminar*) *drag* occurs at very small Reynolds numbers, when viscous forces predominate over inertial ones, because of deformation of the fluid not merely close to the surface of the solid with which the fluid is in relative motion, but also at a considerable distance away from it. As the Reynolds number becomes high, the eddy viscosity of the fluid becomes high and the layer of fluid that becomes deformed is very thin and is pressed more and more closely against the surface of the solid body. This type of drag is known as *surface* (or *turbulent*) *drag.* The third kind of drag is known as *form drag.* It is associated with flow separation at discontinuities in the shape of the body. Usually form drag will accompany surface drag and may be the greater of the two.

The drag force, F_D, is given by the basic equation

$$F_D = C_D A \rho(V^2/2) \tag{5.26}$$

in which A is the projected area of the solid body on a plane normal to the direction of movement, ρ is the fluid density, and V is the relative velocity of fluid and body. V is a vectorial quantity that introduces a vectorial property into Eq. (5.26). C_D is a variable dimensionless coefficient of drag and is a function of the Reynolds number and body geometry. For a spherical particle with radius r, $A = \Pi r^2$ and Eq. (5.26) becomes

$$F_D = C_D \Pi r^2 \rho_0(V^2/2). \tag{5.27}$$

The equation for the relative velocity of fluid and body is obtained from the following equation of motion

$$ma = m(dV/dt) = F_W - F_B - F_D \tag{5.28}$$

where m is the particle mass, F_D is the drag force of the liquid, F_W is the weight of the particle, F_B is the upward resisting forces of buoyancy, and a is the acceleration. For a spherical particle with radius r, $F_W = (4/3)\,\Pi r^3 \rho\, g$, $F_B = (4/3)\,\Pi r^3 \rho_0 g$, and the effective weight of the particle is $F_W - F_B = (4/3)\,\Pi r^3\,(\rho - \rho_0) g$. As the particle begins to sink, it accelerates until the force of the drag is equal to the downward force and $F_W - F_B - F_D = 0$. Once it reaches this velocity, the two forces are equal and opposite, the particle will continue to fall at a uniform rate, and $dV/dt = 0$. For a spherical particle, Eqs. (5.27) and (5.28) now become

$$4/3\,\Pi r^3\,(\rho - \rho_0)\,g = C_D\,\Pi r^2\,\rho_0\,V^2/2$$

and

$$V = \left[\frac{4}{3}\,\frac{2\,\Pi r^3\,(\rho - \rho_0)}{C_D \Pi r^2\,\rho_0}\,g\right]^{1/2} = \left[\frac{8}{3}\,\frac{r(\rho - \rho_0)}{C_D\,\rho_0}\,g\right]^{1/2}. \tag{5.29}$$

Similar equations for V are obtained for nonspherical particles, but empirically determined shape factors must be used.

2.2.1 Sedimentation Rate of Small Particles in a Laminar Flow

Since the drag coefficient, C_D, depends on the Reynolds number, it varies with the mode of flow. If the flow is turbulent, C_D for spherical particles has a value of approximately 0.5. If the fluid is motionless, or if the flow is slow laminar, and its rate is negligible compared to the rate of movement of the particle, u, then C_D is found to be equal to $24/N_R$ (Rouse, 1961). It can be demonstrated that for small particles in the latter case sedimentation obeys Stokes' law. If the rate of flow of the fluid approaches zero, the rate of settling of the particle approaches the relative velocity of fluid and particle. Under these conditions the hydraulic radius, expressed in the Reynolds number, equals the particle diameter. According to (5.25), $C_D = 24/N_R = 24\,\eta/\rho_0 u\,2r$. Substituting the value of C_D in (5.29) gives

$$u^2 = \frac{8}{3}\,r\,\frac{(\rho - \rho_0)}{\rho_0}\cdot\frac{\rho_0 u\,2r}{24\,\eta}\,g$$

or

$$u = \frac{2}{9}\frac{r^2}{\eta}\,(\rho - \rho_0)\,g = k'r^2$$

which is Stokes' law, given by equation (5.18).

From the foregoing discussion it follows that for Stokes' law to apply the fluid must be completely nonturbulent and free of boundary effects. In aqueous media it works well for particles up to 0.18 mm that have viscous (laminar) settling but not for larger ones that have turbulent (surface) settling.

2.2.2 Sedimentation Rate of Large Particles in a Still Fluid

Large particles move more rapidly through the fluid than small particles do. For small, slowly moving grains, the fluid moves around each grain as a smooth stream in a laminar flow. If the grain moves more rapidly, the smooth laminar flow of the fluid around the grain becomes disturbed and the flow becomes turbulent. Under these conditions the drag increases markedly, and for spherical particles $C_D = 1/2$. Since the fluid is still, the net velocity of the fluid is zero, and u, the rate of settling of the particle, is equal to V, the relative velocity of fluid and particle. Substituting these values in Eq. (5.29) gives

$$u = \left[\frac{8}{3}\,\frac{r(\rho - \rho_0)}{(1/2)\,\rho_0}\,g\right]^{1/2} = k''\,r^{1/2}, \qquad (5.30)$$

where k'' is a constant. Thus, while the settling rate for small particles is proportional to the square of the radius, for large particles the settling rate increases only as the square root of the radius.

Between these two extremes there is a range in size wherein the flow of the fluid around the grain gradually changes from laminar to turbulent. A plot of the variation in settling rate in water for grains of a density of 2.65 (quartz) is shown in Figure 5.3.

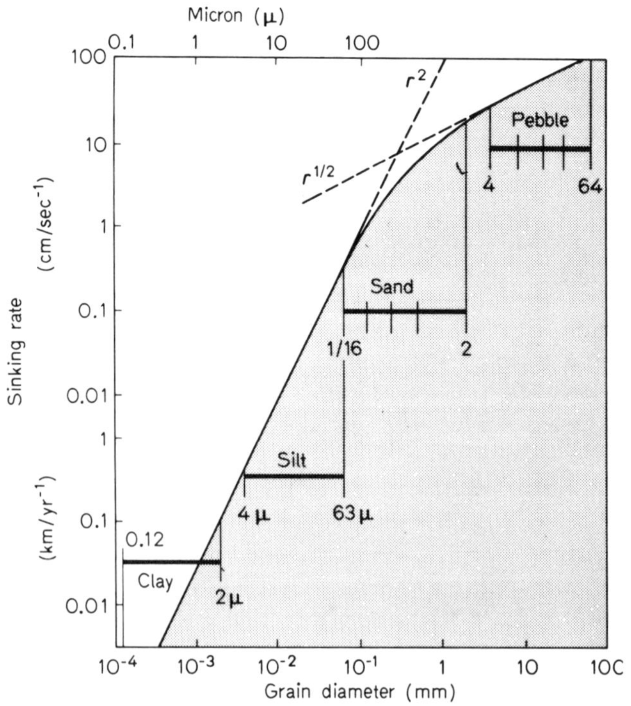

Fig. 5.3. Sinking rate of average-shape quartz grains in pure water (After Bagnold, 1962). (From Weyl, 1970; used with permission of the publisher)

The transition between settling associated with laminar and turbulent flow occurs between 0.1 and 2.0 mm, grain sizes characteristic for sand particles.

2.3 Dispersion of Particles in a Turbulent System

A turbulent flow is associated with a mass exchange up, down, and parallel to the system bed. If a fluid element with a density higher than the average density of the fluid reaches a spot below the particle, or if a fluid element with a density lower than the average reaches a spot above the disperse particle, the particle will move upward. By the same mechanism fluctuations in the density of the fluid may cause the particle to move horizontally. This irregular motion prevents the particles from settling out because the vertical stirring produced by the irregular flow counteracts the settling of the particles to the bottom.

The degree of sedimentation of particles depends on the efficiency of the vertical stirring relative to the rate of sedimentation. The rate of stirring depends on the Reynolds number and increases with the velocity of the flow, whereas the rate of sedimentation depends on the grain size. For a given flow velocity, particles larger than a particular size will settle out while smaller particles will be dispersed in the system. A critical particle-size limit for a certain flow velocity of stream water cannot be defined exactly, for it depends on the shapes of the particles and on the parameters of the stream flow. A rough boundary between transportation and deposition as a function of grain size and stream velocity is given in Table 5.1.

The stream velocity at the boundary between transportation and deposition corresponds roughly to the sedimentation rate of the grain in the water. If the particle sinks at a velocity greater than the mean flow velocity of the stream, it will settle out and be deposited, while grains having a slower sedimentation rate will be carried along by the stream. Water bodies in the hydrosphere are often brown in color because they contain silt and clay particles in suspension. If the Reynolds number of the water body becomes small, the fine particles gradually settle out, and the water becomes clear.

When suspended particles and hydrophilic colloids reach the oceans, seas, or lakes, with time they spread out and become diluted by mixing with additional water. When,

Table 5.1. Rate of sedimentation of silicate and quartz grains having a density of 2.65 (after Bagnold, 1962) and the relationship between the critical particle size and current velocity in the sedimentation process from stream water (after Hulström, 1939)

Sediment	Diameter of particle in nm	Sedimentation rate		Current velocity of stream leading to sedimentation
Clay	$< 2.0 \times 10^3$	< 0.1	km/year	
Clay	$2.0 \times 10^3 – 3.9 \times 10^3$	0.1–0.5	km/year	< 0.1 cm/s
Silt	$3.9 \times 10^3 – 6.2 \times 10^4$	0.5–124.4 (0.01–4.0	km/year mm/s)	< 0.4 cm/s
Sand	$6.2 \times 10^4 – 2.0 \times 10^6$	0.4–20.0	cm/s	< 20.0 cm/s
Gravel	$2.0 \times 10^6 – 4.0 \times 10^6$	20.0–35.0	cm/s	< 35.0 cm/s
Pebble	$4.0 \times 10^6 – 6.4 \times 10^7$	35.0–120.0	cm/s	< 120.0 cm/s

for example, the water body is maintained at a high Reynolds number, sedimentation will be prevented, and, after thousands of years, the colloids will have been uniformly distributed throughout the ocean. In the previous section of the present chapter, molecular diffusion, which takes place owing to the random molecular motion of the fluid, was shown to give rise to a uniform distribution of the suspended material. On the time-space scale of the ocean, molecular diffusion is so slow as to be negligible.

Turbulent eddies play an important role in the dispersion of particles and solutes in bodies of water. This type of dispersion, which may be regarded as diffusion by stirring, is called *eddy diffusion.* Large, uniform motions, such as those of waves and ocean currents, are accompanied by turbulent motions that produce mixing. Eddies play an important part in ocean circulation and in the energy transfer processes related to ocean circulation. Eddies have been observed in many parts of the oceans. Many of these eddies are related to topographic features like seamounts and islands, but they also occur in the open ocean. According to Wyrtki et al. (1976), eddy motion in the ocean is generated in areas of strong mean shear flow and is subsequently spread over the whole ocean.

2.4 Entrainment of Sediment

As the velocity of fluid flow over a bed of sediment is increased a stage is reached wherein the intensity of the applied force is large enough to cause sediment to move from the bed into the flow. This critical stage of erosion or entrainment, known as the *threshold of movement,* can be defined in terms of a value of the boundary shear stress.

When dealing with critical boundary shear stress, the differences in behavior between *cohesionless* and *cohesive* (or *sticky*) sediments must be considered. Cohesionless sediments, such as clear sands and gravels, are formed of loose grains that are bound together by very weak interparticle interactions. When they are eroded, particles leave the bed singly. Sticky sediments, on the other hand, consist of silt or clay-sized particles, which have high hydration energies and in which water molecules serve as bridges joining solid particles. The material entrained from such beds generally consists of variously sized aggregates of clay or silt particles. In the present section the entrainment of particles from cohesionless sediment will be analytically treated. An analytical treatment of the entrainment of "sticky" material requires a knowledge of the types and strengths of the interparticle interactions (see soil detachment in Ch. 3).

The forces that control the entrainment of a spherical particle of radius r and density ρ from a cohesionless bed are demonstrated in Figure 5.4 (Allen, 1970). The Figure shows:

1. The effective weight of the particle is given by $F_g = (4/3)\,\Pi\, r^3\,(\rho - \rho_0)\,g$, and the force acting downward on the particle is equal to $F_g \sin \alpha$.

2. The fluid stream exerts a tangential drag force per unit area of τ_0 on the bed, whence the fluid force acting on the single grain becomes

$$F_D = (\tau_0/N) \tag{5.31}$$

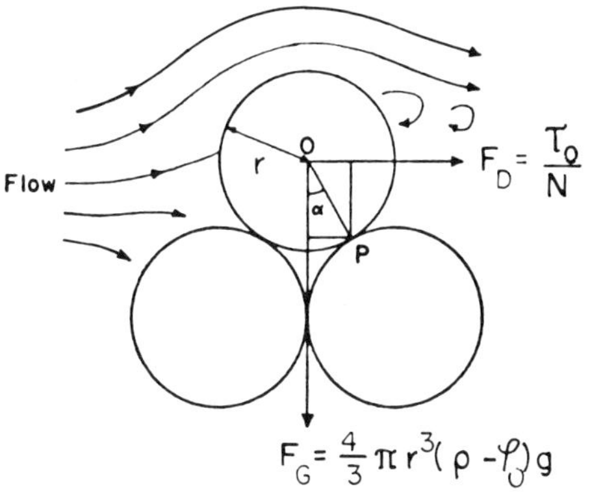

Fig. 5.4. Equilibrium of a spherical particle on a bed of similar particles beneath a fluid stream. (After Allen, 1970)

in which N is the number of grains exposed to the drag on a unit area of the bed, and the force acting to transfer the particle into the fluid phase is equal to $F_D \cos \alpha$.

The grain will pivot about a point P and the threshold of movement will be reached when the moment of the drag force about the pivot equals the moment of the body force about the pivot. In other word, the stress at the threshold will be given by equating the field forces that just cause movement of the grain with the body force holding the grain in place. When these forces are equal, $F_g \sin \alpha = F_D \cos \alpha$ or

$$F_D = F_g \frac{\sin \alpha}{\cos \alpha} = F_g \tan \alpha. \tag{5.32}$$

From Eqs. (5.31) and (5.32) we obtain

$$\tau_{\text{crit}} = F_g \tan \alpha N = (4/3)\,\Pi\, r^2\, N \tan \alpha\, (\rho - \rho_0)\, r\, g = k'''r \tag{5.33}$$

in which τ_{crit} is the critical stress, the angle α is related to the packing of the grains, and the quantities to the left of the bracket can be regarded as forming a complex constant (r^2 and N have a reciprocal relationship). The factor k''' is a variable factor taking into account the character of the flow and the bed surface. Equation (5.33) indicates that the critical stress is directly proportional to the particle radius.

The relationship between the threshold stress and particle size of cohesionless material is not as simple as suggested by the preceding analysis because of the variation of the character of the flow around the grains with the grain size and with the velocity of the flow. Quartz-density grains in water obey Eq. (5.33) only when the particles have a diameter larger than 6 mm. A rough estimation of stream velocity that results in entrainment of particles of various sizes is given in Figure 5.5. From the Figure it is obvious that it is more difficult to initiate motion than to maintain it. To erode the ground, that is, to initiate a downslope motion, requires higher flow velocities than to maintain this motion by suspension.

There are random impingements of turbulent eddies on the bottom of a stream. The instantaneous velocity fluctuations associated with these eddies cause the grains to be entrained, starting with the smallest and lightest. As the flow intensity at the bottom of the stream increases, frequency and magnitude of the eddies increase, and

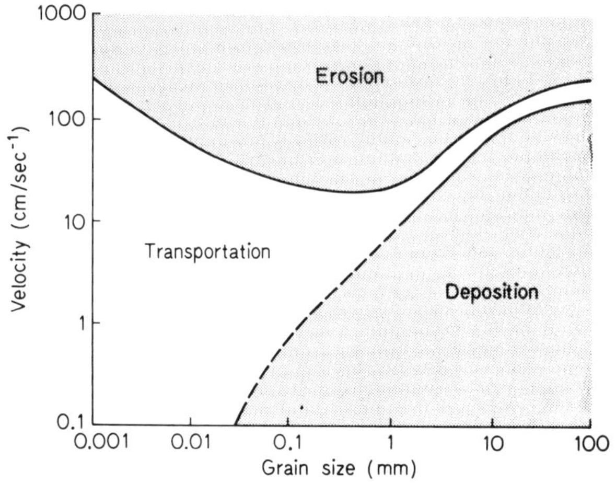

Fig. 5.5. The relationship between current velocity and the fate of sediment particles. (After Hjulström, 1939). (From Weyl, 1970: Used with permission of the publisher)

movement becomes more general. Finally all the grains, including the largest, are in motion everywhere on the bed (Sutherland, 1967). The flow intensity at the bottom may increase as a result of the following two factors: (1) an increase in the velocity of the water flowing in the stream and (2) an increase in the roughness of the bottom of the stream. As the size of the particles becomes smaller, the bed becomes more nearly regular and the flow becomes laminar. In the presence of bigger particles the flow at the bottom may become sufficiently turbulent to entrain the big particles, but at higher levels of the stream the velocity is not sufficient to keep the big particle in suspension, and it sinks. Such particles move mostly along the bottom as "bed load", being suspended for only short periods. These "short jumps" are referred to as *saltation.* In streams the finest material with a diameter less than 0.02 or 0.03 mm is transported entirely in suspension and moves approximately at the velocity of the water, while sand is transported mainly by saltation (Pettijohn et al., 1972). The relationship between current velocity and the fate of sediment particles is shown in Figure 5.5.

References

Allen, J. R. L.: Physical processes of sedimentation. An introduction. London: George Allen and Unwin, Ltd. 1970

Bagnold, R. A.: Mechanics of marine sedimentation. In: The sea. New York: Interscience Publishers, 1962, Vol. 3, pp. 507–528

Einstein, A.: Über die von der molekularkinetischen Theorie der Wärme geforderte Bewegung von in ruhenden Flüssigkeiten suspendierten Teilchen. Ann. Physik *17,* 549 (1905)

Hamilton, E. L., Moore, D. G., Buffington, E. C., Sherrer, P. L.: Sediment velocities from sonobuoys: Bay of Bengal, Bering Sea, Japan Sea and North Pacific. J. Geophys. Res. *79,* 2653–2668 (1974)

Hjulström, F.: Transportation of detritus by moving water. In: Recent marine sediments. Trask, P. D. (ed.), Am. Assoc. Petrol. Geol. Tulsa, 1939, pp. 5–31

Lerman, A., Lal, D., Dacey, M. F.: Stokes' settling and chemical reactivity of suspended particles in natural waters. In: Suspended solids in water. Gibbs, R. J. (ed.), New York: Plenum Press, 1974, pp. 17–47

McCave, I. N., Swift, S. A.: A physical model for the rate of deposition of fine grained sediments in the deep sea. Geol. Soc. Am. Bull. *87,* 541–546 (1976)

Pettijohn, F. J., Potter, P. E., Siever, R.: Sand and sandstone. Berlin: Springer-Verlag 1972

Rouse, H.: Elementary mechanics of fluids. New York: John Wiley 1961

Smoluchowski, M. von: Zur kinetischen Theorie der Brownschen Molekularbewegung und der Suspensionen. Ann. Physik *21,* 756–780 (1906)

Sutherland, A. J.: Proposed mechanism for sediment entrainment by turbulent flow. J. Geophys. Res. *72,* 6183–6194 (1967)

Weyl, P. K.: Oceanography, an introduction to the marine environment. New York: John Wiley & Sons 1970

Wyrtki, K., Magaard, L., Hager, J.: Eddy energy in the oceans. J. Geophys. Res. *81,* 2641–2646 (1976)

Chapter 6
Colloid Geochemistry of Silica

Silica, SiO_2, is common in nature and occurs as seven distinct minerals, of which five show crystalline structures (quartz, tridimite, cristobalite, coesite, and stishovite) and two are amorphous (opal-A and lechatelierite). Lechatelierite is a silica glass and is very rare. Coesite and stishovite are also very rare and are found only in meteoritic craters, where they have been formed from quartz because of the high pressure resulting from the meteoritic impact. Opal is obtained by deposition from aqueous silica solutions at low temperatures. It is deposited by thermal waters associated with igneous activity and is also secreted by sponges, radiolaria, and diatoms. The most common form of silica is quartz.

Amorphous silica is an important component of soils, coating particles of quartz, feldspar, and hornblende, as well as clay minerals (McKyes et al., 1974). It also plays an important role as a cementation agent in many sediments. Monosilicic acid is an important constituent of natural aqueous solutions. In the pH range of normal natural water the monosilicic acid occurs predominantly as the uncharged $Si(OH)_4$ species. However, suspended species of silicates are important and, according to Garrels and Mackenzie (1972), the dominant transfer of silicon in the sedimentary cycle occurs in the form of suspended quartz or suspended clay minerals. In their model the ratio of silica in the suspended flux to silica in the dissolved flux is about 4 to 1. From the Cambrian period (some $500–600 \times 10^6$ years ago) and in the present period most of the dissolved silica is removed from the oceans in the tests of organisms as opaline silica. Eventually, it is transformed to quartz as chert and/or overgrowth on detrital quartz grains or reincorporated into silicate minerals. According to Wise and Weaver (1974) chert formation is a "maturation" process, i.e., the aging of colloid silica system occurs as follows: The amorphous precipitate of biogenous silica is transformed into opal-CT, a poorly ordered crystalline form of silica that contains cristobalite and tridymite, and on further aging, into quartz.

During the Precambrian period, prior to the advent of siliceous organisms, amorphous silica probably precipitated from marine waters by non-biogenous mechanisms. According to Oehler (1976) the diagenetic reactions culminating in biogenous and nonbiogenous chert formation may differ. Some nonbiogenous cherts were formed through an intermediate opal-CT stage and some crystallized directly from silica gel without the intervention of an intermediate crystalline phase such as opal-CT. The quartz crystallized primarily in the form of radially fibrous, chalcedonic spherulites, which later recrystallized to form anhedral grains, producing the interlocking mosaic texture typical of most cherts. The following diagenetic sequences were suggested by Oehler:

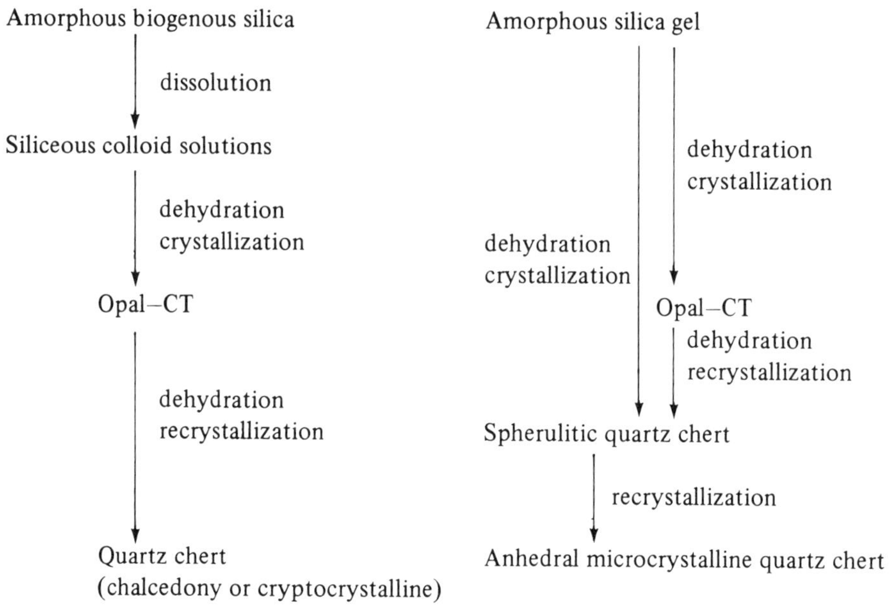

Silicic acid is adsorbed from soil solutions by living organisms. In plant tissue the silicic acid polymerizes and the opal formed accumulates in the form of incrustations or completely fills the cells in small bodies of cellular shape, which are known as *phytoliths*. The structure of the phytolith depends on the type of plant and on the geologic and climatic environments of its formation and can be used as a tool for the identification of these environments (Bartoli and Beaucire, 1976). The biogenous silica returns to soil together with plant debris, and phytoliths are gradually transformed to secondary quartz through the intermediate state of chalcedony (Kutuzova, 1968).

1. Surface Chemistry of Silica

1.1 Functional Groups on Silica

Most studies on silica surfaces have been performed with amorphous silicas having colloid dimensions, partly because of their large surface area and partly because of their technical importance. Amorphous silica is similar to the crystalline modifications in the close ordering of the atoms but differs in that the three-dimensional array of the SiO_4 tetrahedra is not as regular.

The surface chemistry of silica is somewhat less complex than that of other solid materials. Two types of surface groups cover the silica particle whether it is amorphous or crystalline: the silanol (I) and siloxane (II) groups:

(I) –SiOH (II) –Si–O–Si–

The structure of a particle of silica can be visualized as a network of interlinked SiO_4 tetrahedra. Hydroxyl groups are attached to the surface due to the tendency of silicon to complete tetrahedral coordination, and each particle of silica can be considered as a macromolecule of polysilicic acid. Surfaces covered with silanol groups are hydrophilic in nature while those covered with siloxane groups are hydrophobic. Usually both groups may be present in the same crystal.

The ratio between the silanol groups and the SiO_4 tetrahedra is very high for amorphous silica and decreases with increasing degree of crystallinity. On both hydrophilic and hydrophobic silica three types of silanol groups can occur. If two hydroxyls are attached to silicon atoms that are separated by at least one other silicon atom, one obtains "isolated silanols" (I-a). When the silanol groups are direct neighbors, under which circumstances they show hydrogen bonding, they are called paired or "vicinal" (I-b). In the third type, two hydroxyls are present on a given silicon atom, a configuration known as "geminal" (I-c).

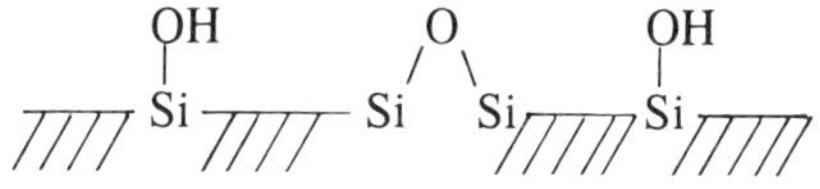

Ia

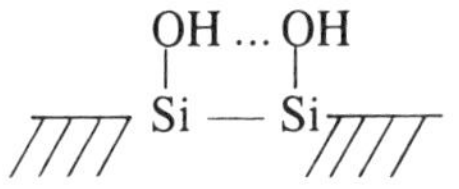

Ib

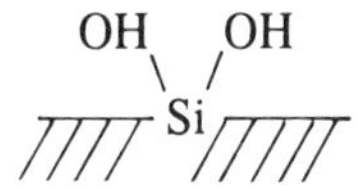

Ic

Surface silanol groups are weakly acidic, with pK_as in the range of 9–10 (West et al., 1973), and therefore provide sites that may interact with molecules or functional groups possessing electron densities locally concentrated on the periphery, forming H-bonding, whose strength depends very much on the basic properties of the adsorbed molecules. Many investigators have shown that it is not possible to describe the surface acidity only in terms of the three types of silanol described above but that it is necessary to postulate the presence of sites of other types. West et al. (1973) suggest that surface impurities such as Al^{3+} or the surface sulfonic groups, $-Si-O-SO_3H$, which are present in negligible amounts, are responsible for sites with a rather strong acidity. Clar-Monks and Ellis (1973) suggest that anomalous adsorption sites are associated with the ease of dehydration of vicinal silanol groups and the formation of a siloxane group. A strong H-bond between the silanol and a neighbor siloxane group also decreases the acidity of such sites. Steric hindrance, topography, kinks, and population of hydroxyls on silica surface are also responsible for anomalies in adsorption sites. For example, two vicinal hydroxyls attached to two silicons with an optimum separation are highly reactive. There is a correlation between the pore sizes of various silica gels and the concentrations of these reactive hydroxyls. Combinations of the different surfaces functional groups cause a "heterogeneous" surface to be formed.

Siloxanes are unreactive. The strong bond between the Si and O atoms and the partial π interaction cause the oxygen to lose much of its basicity and to show extremely little tendency of participating in hydrogen bonds.

All the surface hydroxyl groups are lost when silica is heated in air between 180 °C and 1050 °C. Water is evolved and each water molecule desorbed is the condensation product of two silanol groups, resulting in the formation of siloxane groups according to Eq. (6.I)

$$2\,(\gtrdot SiOH) \rightarrow \gtrdot Si-O-Si \lessdot + HOH. \qquad (6.I)$$

The surface hydroxyl concentration may be measured in this way. The dehydration of silanol groups is reversible and rehydroxylation of the siloxane bonds will take place instantaneously if the dehydration is carried out below 450 °C. No rehydration will occur via adsorption of water vapor if the silica has been heated to higher temperatures. Under liquid water a slow incomplete rehydration may occur. The difference in behavior between a siloxane group that is obtained at low temperature and that obtained at high temperature has been ascribed to strained siloxane bonds, which are obtained during the dehydroxylation process at low temperatures. At this stage only the oxygen moves to a position between the two Si atoms. Thus the tetrahedral coordination of each of the two Si atoms is disturbed. At temperatures above 450 °C the thermal energy is high enough to cause other atoms in the solid particle to move to such positions that the strain in the siloxane group is relieved, thus causing the change in rehydration characteristics. Hydrolysis of a stable siloxane group would result in stressed silanol bonds. According to Hair (1967) heating above 400 °C causes a drastic irreversible elimination of adjacent hydroxyls and at 800 °C only isolated silanols remain. Siloxane bonds are readily opened by strong bases according to the equation

$$\gtrdot\text{Si–O–Si–} + 2\,\text{NaOH} - 2\,(\gtrdot\text{SiONa}) + \text{HOH}. \qquad (6.\text{II})$$

From structural considerations it is to be expected that the maximal surface concentration of silanols on silica will be 5 per 1 nm^2. One rarely finds a packing density higher than 3.3 per nm^2. This indicates that in some positions on the surface of amorphous silica the siloxane group is more stable than the two silanol groups from which it is derived (Snoeyink and Weber, 1972). Siloxane is the stable surface group found on particles of crystalline silica.

1.1.1 Sorption of Water by Silica Gel

The silica gel surface has a significant affinity for water. Water molecules can be sorbed either physically or chemically. The weight loss in air between room temperature and 180 °C is due to loss of water commonly regarded as physically adsorbed. Physically adsorbed water can also be removed under vacuum at temperatures less than 100 °C. The water released by heating to temperatures above 180 °C due to the condensation of the silanol groups is regarded as chemically sorbed.

The amount of water physically adsorbed on silica depends on the partial pressure of the water vapor and is directly related to the number of hydroxyls present on the surface. Hydrogen bonds are formed by surface silanols with single water molecules; as can be seen from perturbation of infrared bands characteristic of silanol groups (Takamura et al., 1964). The first adsorbed water monolayer is strongly held by the silica gel surface but the following adsorbed layers are loosely bound (Fripiat, 1965).

With low water contents in the silica gel, free silanols donate protons to water oxygens. With higher water contents hydrogen-bond containing clusters are formed before all the free silanols are taken up by monomeric water molecules. From heat of reaction measurements it appears that clustering of water molecules is favored over bonding between monomeric water and single silanols. On silica gel surfaces water is stabilized in small clusters containing five or six molecules, with heat of clustering similar to the heat of liquefaction of water. The clustered molecules, however, are more restricted translationally and rotationally in their motion than monomers weakly bonded to the silanols and therefore have a lower entropy. Thus, at low coverage, water adsorption on silanols is favored by entropy, whereas at higher coverages clustering is favored by energy stabilization in the cluster (Klier and Zettlemoyer, 1977).

Since the hydration energy of silica gel is less than the heat of liquefaction of water, silica gel is a hydrophobic colloid. Some aluminosilicates have higher surface hydroxyl concentrations than silica gel and closer spacing of the hydroxyls. During water adsorption, multiple water-hydroxyl bonds are formed, accompanied by higher heats of sorption, and the surface is occupied to a higher degree before clustering occurs. Water clusters are more easily formed on silica surfaces with the lowering of surface densities of silanol groups because the increasing distances between hydroxyls prevent multiple water-hydroxyl bonds from being formed. Quartz, having almost no surface hydroxyls, allows the formation of water clusters at a very low water coverage.

Physically adsorbed water may prevent the formation of hydrogen bonds between immediately adjacent surface silanol groups. A drying temperature of 100–120 °C, even with thorough outgassing of the silica, may not be sufficient to drive off strongly bound water, and the possibility also exists that adsorbed water may be held at higher temperatures in very small pores and interfere with the estimation of the quantity of silanol groups.

Silica gel is rigid, that is, the swelling and shrinking occur only to a limited extent. With such a gel, the vapor pressure curve on dehydration (ABCD of Fig. 6.1) differs over part of its course from the curve on rehydration (Alexander and Johnson, 1949). It appears that over the part AB water is being removed from between large particles, over BC from the capillaries in these particles, and over CD from the polar boundary layer. The volume shrinkage on drying practically ceases at point B. From the observed vapor pressure (p_R) over the BC region, as compared with the normal vapor pressure (p_0) for water, the radius of the capillaries can be calculated from the Kelvin Eq. (2.33). Values of the order 2.5–3.0 nm were obtained for silica gel. The hysteresis region can be explained as follows if the gel consists of a network of rigid capillaries, as idealized according to the structures shown in Fig. 6.2a and b. In Fig. 6.2a the gel is seen to consist of a series of upright and inverted cones, filled with liquid when the gel has been completely wetted, which behave independently of one another if the walls are perfectly rigid. Since $R_1 \gg R_2$ it follows from the Kelvin Equation (2.33) that $p_{R1} \gg p_{R2}$

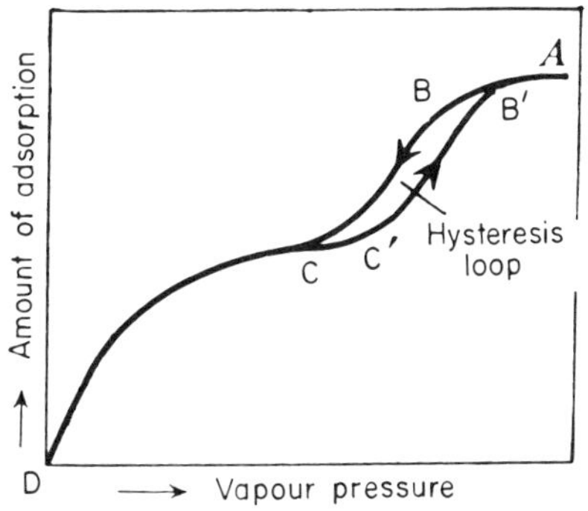

Fig. 6.1. Adsorption–desorption/vapor pressure for silica gels, showing hysteresis loop. For explanation see text

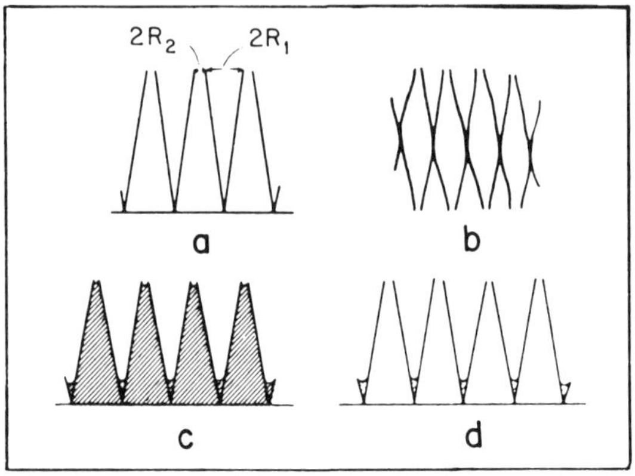

Fig. 6.2a–d. Idealized pore structure of silica gels. (After Alexander and Johnson, 1949)

where R_2 and R_1 are the radii of the circles inscribed by the water–air interfaces on the walls of the upright cones and inverted cones, respectively, and p_{R1} and p_{R2} are the respective vapor pressures. On reducing the total pressure (*AB* in Fig. 6.1) the inverted cones will begin to empty first, and the equilibrium vapor pressure will decrease as the liquid is removed from the inverted cone, since R_1 is steadily decreasing. When p_R falls to p_{R2} (Fig. 6.2c) the cones will commence to empty, and having once started, they will empty rapidly since R increases as dehydration proceeds (*BC* in Fig. 6.1). On rehydration of the completely dry gel (*DC′* in Fig. 6.1), since R_1 is initially smaller than R_2, $p_{R1} \ll p_{R2}$. Consequently the inverted cones will commence to fill first. The cones will commence to fill at a stage described in Fig. 6.2d, when $p_R \geqslant p_{R2}$. From this stage the inverted cones will not continue to fill until the cones will be completely filled (Fig. 6.2c). When allowance is made for the differences between the assumed model and an actual rigid gel, the two curves shown in Figure 6.1 may readily be explained. If the capillaries are not completely rigid, as assumed above, it is possible that collapse of the cones may occur as dehydration proceeds and this may also lead to hysteresis.

Morariu and Mills (1973) studied water adsorption on various silicas and demonstrated that the more acidic silica has certain sites with a higher heat of adsorption and a more heterogeneous surface. They also demonstrated that the adsorption does not take place initially at sites with highest energy and then progressively on the remaining sites in decreasing order of energy, as would be expected, but that adsorption occurs equally at all sites.

1.1.2 Sorption of Metal Ions

The silanol group on the surface of silica reacts with bases as a weak acid. In the reaction with NaOH the H^+ ions are replaced by Na^+ ions, which, unlike protons, cannot form a nondissociated group by entering the electron shell of the oxygen. Weaker bases, such as $Ca(OH)_2$, can react in a similar way. The neutralization reaction can be described by the chemical equations:

$$\geqslant SiOH + NaOH \rightarrow \geqslant SiO^- + Na^+ + HOH \qquad (6.III)$$

$$\equiv SiOH + Ca(OH)_2 \rightarrow \equiv SiO\text{–}(CaOH) + HOH \quad (6.IVa)$$

$$2\,(\equiv SiOH) + Ca(OH)_2 \rightarrow 2\,(\equiv SiO)^- + Ca^{2+} + 2\,HOH \quad (6.IVb)$$

Silanol groups can be determined by titration with strong bases, but only dilute solutions can be used for this purpose to avoid dissolution of the silica. There is a difference between the amount of $Ca(OH)_2$ and NaOH adsorbed by the silica, since $Ca(OH)_2$ may react partly according to (6.IVa) and partly according to (6.IVb) (Iler, 1975).

When silica gel is titrated with NaOH, all surface silanol groups are neutralized at pH 9.0. At higher pH values, siloxane bonds in the surface are opened according to (6.II) and a maximum in the adsorption of the Na^+ occurs usually at pH 10.5 to 10.6, which corresponds to a packing density of approximately 5 OH groups per nm^2. On further addition of alkali the silica begins to dissolve. The isoelectric point of silica is near pH 2, that is, the pH at which the particle of silica in solutions has no electric charge. At pH 10 a negative charge of 1.8 per nm^2 was determined (Snoeyink and Weber, 1972).

Biogenous silica controls the hydrogeochemistry of certain elements in interstitial water in deep-sea sediments. According to Donnelly and Merrill (1977) considerable quantities of Mg are adsorbed onto biogenous opal. The adsorption is sensitive to pH. Na, Fe, and Mn are also adsorbed by opal, but K is not adsorbed.

According to Malati et al. (1974) the cations adsorbed by silica are hydrated and the adsorption of cations may be envisaged as an *ion exchange* between a hydrated cation and a hydrated proton held in the outer Helmholtz layer. The adsorption of the alkali metal ions is exothermic. On the other hand, the adsorption of the alkaline-earth cations, which exhibit a higher adsorption affinity for quartz and silica, is endothermic.

Hydration energy of metallic cation plays an important role in determining sorption affinity of the ions. Dalton et al. (1962) studied the ion exclusion effect with silica gel and found that ion exclusion can be correlated with the ability of ions to react as structure makers or breakers of aqueous solutions. Structure-making ions were excluded, most of them not being able to attain as high a concentration near the silica surface as in the bulk solution. Cesium, which is the largest of all the alkali metal cations and is a structure breaker of liquid water, has the highest absorption affinity toward silica and quartz among the alkali metals (Malati et al., 1974). It has the smallest hydrated size and is able to penetrate the small pores of silica gel. Structure-making ions are excluded from the silica surface because of the inability of large hydrated cations to penetrate the smaller pores.

Entropy changes also play an important role in determining sorption affinity. The exchange of two monovalent cations for one divalent cation is accompanied by an increase in entropy because of the increase of the number of ions in the solution, thus increasing the degree of disorder in the system. Such entropy contribution can explain why the alkaline-earth cations have a greater affinity for silica than the alkali metal cations.

Many metal ions can also be adsorbed from nearly neutral or weakly acidic solutions. On mixing a silica sol suspension of pH 3.2 and an $AlCl_3$ solution of pH 4, a lower pH of 2.7 results. This is caused by hydrolytic adsorption as represented by the schematic formula

$$AlCl_3 + 2\,H_2O \underset{(aq)}{\longrightarrow} Al(OH)_2Cl + 2\,HCl$$

$$\geq Si\text{–}OH + Al(OH)_2Cl + H_2O \rightarrow \geq Si\text{–}OH \cdot Al(OH)_3 + HCl \qquad (6.V)$$

Hydrolytic adsorption occurs similarly with other polyvalent cations such as ferric ions. Hydrolytic adsorption seems to play an important role in the geochemical mechanism of accumulation of many polyvalent trace metals and on the behavior of silica in natural waters (Hurd, 1972a). (See also pages 164, 211–219).

The presence of amorphous silica has a great effect on the concentration of aluminum in ground water, rivers, and lakes. This ion is quantitatively adsorbed onto silica in the pH range of 3.8 to 4.2. At high concentrations of $AlCl_3$ polynuclear complexes of aluminum occur in solution and more than the stoichiometric amount of the cation is adsorbed from solution together with small amounts of chloride ions, which are easily removed by washing. When silica is freshly coated with aluminum ions, it acquires a positive charge. This is a result of the high positive charge of the adsorbed polymeric ions being in the inner Helmholtz layer and the low degree of dissociation of the surface group $\geq$Si–O–Al (the aluminum in this group belongs to the inner Helmholtz layer). After prolonged standing in slightly alkaline solution a recrystallization of the amorphous silica surface in the presence of the sorbed aluminum ions results in tetrahedra in which Al is substituted for Si atom. The net charge of the amorphous silica becomes more negative as a result of this substitution. In the laboratory the chemisorbed aluminum is removed by mineral acids and in nature by polyprotic organic acids (pages 162–163).

1.1.3 Sorption of Organic Cations

The adsorption of cationic surfactants onto silica was studied by Bijsterbosch (1974). At low surfactant concentrations the positive surfactant ions are attracted by the negatively charged surface sites. In the adsorbed state they are oriented with the apolar chain parallel to the surface. At higher concentrations, depending on the ratio between the surface density of silanol groups and the size of the organic ion, two different mechanisms are postulated (see Fig. 2.8a, b page 124, for monolayer structures):

1. With very long organic chains the ions would gradually be oriented perpendicular to the surface thus causing the adsorption to increase from a first to a second plateau in an adsorption isotherm. When the surfactant cations adsorbed in the first step change their orientation, space is provided for additional adsorption of identical ions. In the first step all dissociated silanol groups are neutralized by the surfactant cation. In the second step, the additional ions will interact with the siloxane bonds. Consequently, the attractive forces will be of a type other than electrostatic, probably van der Waals interaction. A reverse orientation occurs, with the apolar chain toward the surface of the silica and the ionic moiety in the solution.

2. With shorter organic chains a bilayer is built up at higher concentrations. The first step implies local coverage of the surface with hydrophobic groups and the next step would then be adsorption of a second layer of surfactant ions at these hydrophobic sites, with van der Waals forces operating between the apolar hydrocarbon chains. As a result, the second layer will be oriented with the ionic moieties toward the solution.

1.1.4 Proton Donor Properties of Silanol Groups

Ammonia is a weak base with $pK_b = 5$. The adsorption of ammonia onto Cabosil silica, which is an extremely pure form of silica and does not contain any alumina or other oxides, was studied by infrared spectroscopy (Little, 1966). A hydrogen bond is formed between the proton of the silanol group and the nitrogen lone pair electrons of the ammonia. During this interaction the OH radical of the silanol group does not dissociate and the proton stays chemically bonded to the oxygen, but this bond is weakened. The process can be described by the following chemical equation

$$\geq Si{-}OH + NH_3 \rightarrow {-}Si{-}O{-}H \ldots NH_3 \qquad (6.VI)$$

In this system the basic strength of the oxygen atom is higher than that of the nitrogen atom.

Small amounts of ammonia are chemisorbed on a very dry silica. The amount sorbed is very small in comparison with that sorbed by a nondry silica. During the reaction the chemisorbed ammonia dissociates and reacts with strained surface siloxane linkages, giving silanol and silamine according to the equations;

$$NH_3 \rightleftharpoons NH_2^- + H^+$$

$$-Si{-}O{-}Si- + H^+ \rightarrow -Si{-}OH + -Si^+$$

$$-Si^+ + NH_2^- \rightarrow -Si{-}NH_2 \qquad (6.VII)$$

The maximum number of these strained siloxane sites is 1.4 per nm^2 (Hair, 1967).

Pyridine, C_5H_5N, an organic base, weaker than ammonia, with $pK_b = 8.8$, was adsorbed on silica, forming hydrogen bonds with the surface hydroxyl groups, as follows $-SiOH \ldots NC_5H_5$ (Little, 1966). Organic amines react similarly.

There are many aromatic derivatives among the organic compounds that are found in geologic systems. Such molecules possess π electron densities locally concentrated on the aromatic ring. After elimination of adsorbed water the surface of silica is covered by hydroxyl groups that may interact with aromatic compounds. The aromatic ring is the site responsible for hydrogen bondings (Semples and Rouxhet, 1975). Such interactions considerably affect the adsorption properties of the solid. For example, the heat of adsorption of benzene on the hydroxylate surface is much higher than that of hexane while they become similar after the silica has been dehydroxylated. During the migration of organic compounds through soils and ocean sediments of silica, these interactions are responsible for the marked retention of aromatic compounds compared with aliphatic hydrocarbons.

The adsorption of benzene, the simplest aromatic compound, serves as an example of interaction between π electrons and silanol groups. The infrared spectrum of adsorbed benzene reveals that the adsorbed state is somewhere between that of a gas and a liquid. In the liquid state there are some interactions between the benzene molecules while there are no interactions in the gaseous state. If the surface proton interacts specifically with the aromatic ring, then it would be expected that those vibrations of the benzene molecule that are most perturbed would be those associated with the aromaticity of the compound. The greatest perturbations are observed at low coverage and with samples of silica that have a majority of the hydroxyl groups on the surface. This interaction decreases with increasing number of sorbed benzene molecules or decreasing number of surface hydroxyl groups (Hair, 1967).

Since both aromatic and lone-pair electron compounds interact specifically with surface hydroxyl groups during the adsorption process, the mechanism of the adsorption of compounds such as aniline, $C_6H_5NH_2$, and phenol, C_6H_5OH, on a hydroxylated silica surface is of interest. Because of the aromatic ring, aniline is less basic than the parent ammonia or aliphatic amines and phenol is more acidic than either water or aliphatic alcohols. Nevertheless, the silanol group is more acidic than both these functional groups and in the adsorbed state the aniline and the phenol molecules are proton acceptors, being hydrogen bonded to the silanol group via the lone pairs of electrons on the aniline nitrogen and phenol oxygen atoms respectively, the silanol group being the proton donor (Kisselev and Lygin, 1964).

The mechanism of adsorption of nitrobenzene, $C_6H_5NO_2$, differs from that of aniline or phenol, but is similar to that of benzene. The nitro group is too weak a base to accept a proton from the silanol group and the aromatic ring rather than the nitro group accepts hydrogen from the silanol group (Hair, 1967).

The adsorption of alcohols onto silica also occurs at room temperature. Alcoholic OH groups are hydrogen-bonded with the surface hydroxyl groups, the silanol being the proton donor. When a silica is heated with an alcohol, esterification of the silanol groups may occur, which can be described as follows

$$\geq Si{-}OH + C_nH_{2n+1}OH \rightarrow \geq Si{-}O{-}C_nH_{2n+1} + HOH \qquad (6.VIII)$$

The chemisorbed reaction products, called estersils, are extremely hydrophobic. To obtain such groups a temperature of 190 °C is necessary for primary alcohols and 275 °C for secondary alcohols.

1.1.5 Adsorption of Nonpolar Gases

Adsorption of nonpolar gases takes place at very low temperatures, such as −170 °C or −190 °C and at very high pressures. It is therefore to be expected that such reactions will be only of minor importance in geologic systems. However, they seem to be of greater importance in other systems of the cosmos. Many of the methods used in the laboratory for the determination of the surface area of minerals, rocks, and soil samples are based on the adsorption of these gases.

The infrared spectra of silica treated with argon, krypton, xenon, nitrogen, oxygen, and methane at temperatures between −170 and −190 °C has been studied (Little, 1966; Hair, 1967). The intensity of the free hydroxyl band of the silanol group at 3750 cm^{-1} decreases progressively with increasing adsorption, while a new band of increasing intensity appears at lower frequencies. The position of the band maximum depends on the degree of saturation of the silica. Condensation of bulk adsorbate onto the silica surface produces a more perturbed hydroxyl band of greater intensity at lower frequencies. Dry air at 30 °C and atmospheric pressure does not lead to any perturbation of the OH band, indicating that there is no interaction between air and silanol groups under normal atmospheric conditions. The frequency displacements for the silanol hydroxyl band resulting from the adsorption of several nonpolar gases are presented in Table 6.1, together with the values obtained for silica treated with long-chain aliphatic hydrocarbons, alkanes and alkenes, aromatic compounds, and polar compounds of low molecular weights, measured at 30 °C. It is obvious from the Table that the perturbation of the OH band caused by the adsorption of the nonpolar molecules is very small compared to the perturbation caused by the proton acceptors. This means that the interaction between the silanol proton and the nonpolar molecule is very slight. Adsorption takes place predominantly with the formation of van der Waals bonds between the silica surface and the small molecule. Heat of adsorption of nonpolar molecules is essentially independent of the surface density of the silanol groups while that of the polar molecule is dependent on this surface density (Table 6.2).

1.1.6 Surface Area Measurements of Silica

The low temperature adsorption isotherm of nonpolar gases, such as nitrogen, is commonly used to determine the surface area of silica. From the adsorption isotherm

Table 6.1. Surface hydroxyl frequency displacements for adsorption of organic compounds onto amorphous silica. (After Little, 1966)

Adsorbate	Temp. °C	OH frequency (cm^{-1})	Frequency displacement (cm^{-1})
None		3750	–
Argon	−170	3742	8
Krypton	−170	3734	16
Xenon	−170	3731	19
Nitrogen	−170	3726	24
Oxygen	−170	3738	12
Methane	−170	3718	32
Sulfur dioxide	24	3635	115
Methyl chloride	24	3640	110
Acetone	25	3420	330
	75	3445	305
	135	3480	270
Ammonia	25	2930	820
	75	3000	750
	100	3040	710
	150	3110	640
n-Hexane	25	3705	45
Benzene	25	3640	110
Toluene	25	3620	130
p-Xylene	25	3605	145
Mesitylene	25	3590	160
Nitrobenzene	25	3610	140
Phenol	25	3400	350
Aniline	25	3200	550
Diethyl ether	25	3300	450
Pyridine	25	2900	850
Diethylamine	25	2760	990
Trimethylamine	25	2760	990

Table 6.2. Comparison of heats of adsorption on hydroxylated and strongly dehydroxylated silica surfaces for various pairs of molecules similar in size and in total polarizability (or in certain cases the number of carbon atoms or the sum of carbon and oxygen atoms), but differing greatly in local electron density distribution (some contain π bonds and lone electron pairs). (After Little, 1966)

Adsorbate	Heats of adsorption in kcal/mol adsorbate	
	Hydroxylated silica	Dehydroxylated silica
Nitrogen	2.8	2.2
Argon	2.1	2.1
Ethylene	5.2	3.8
Ethane	4.4	4.2
Benzene	10.2	8.6
n-Hexane	8.8	8.8
Diethyl ether	15.0	8.5
n-Pentane	7.3	7.3

the amount of gas adsorbed to form a monolayer is determined. Knowing the surface area of one single molecule, the surface area of the mineral can be computed. The method, named BET (after Brunaur, Emmet, and Teller, see Ch. 2), is commonly used to determine the surface area of clay minerals, rocks, and soils.

Another method commonly used for surface area determination of minerals is based on the sorption of organic molecules that have a high affinity for the surface of the mineral examined. For example, the surface area of silica gel can be estimated by adsorption of methyl-red dye from benzene solution. This dyestuff is adsorbed on surfaces coated with hydroxyl groups. Pore size is very important relative to these measurements, however, because of the relatively large size of the methyl-red molecule. This molecule cannot penetrate into very small pores and surface area determinations by this method are sometimes lower than those obtained by the BET method.

p-Nitrophenol is also used to determine the surface area of silica. Only 20% of the nitrogen BET surface area is covered with *p*-nitrophenol when the silica has a pore size of 1.4–2.3 nm. Nitrogen molecules are smaller than those of *p*-nitrophenol and may penetrate into those pores that the latter cannot penetrate.

1.1.7 Surface Functional Groups on Crystalline Silica

In principle, there is no difference in surface groups on quartz and on amorphous silica (Snoeyink and Weber, 1972). There is a disturbed layer of amorphous character present on the quartz surface, which is readily dissolved by water or by hydrofluoric acid. The formation of this layer can result either from (1) slow partial dissolution of surface Si atoms, leaving a thin vitreous layer of silica; (2) adsorption of a monolayer of silicic acid on the crystalline mineral; (3) adherence of very small particles on the larger quartz particles, or (4) the grinding of quartz particles, producing a disturbed surface layer (Henderson et al., 1970). No essential difference in reaction behavior and in the packing density of surface groups between quartz and silica gel was observed.

1.1.8 Oxidation of Organic Compounds on Silica Surface

Silica possesses active properties. The shaking of siliceous dust with various organic compounds in air effect their oxidation, and in the case of certain aromatic compounds, causes hydroxylation of the compounds. According to Schofield et al. (1964) oxidation and hydroxylation of aromatic compounds occur on the surface of silica, both in dry and moist states. In the moist state the predominant reaction is the formation of a complex between the organic compound and iron impurities in the silica. Ferric ions oxidize the organic molecules. The ferrous ions obtained may be catalytically oxidized by the air oxygen on the silica surface.

1.2 Functional Groups on Silica–Alumina

Aluminum is one of the major elements in silicate rocks and it is therefore to be expected that most of the amorphous silica in geologic systems contains this element on the surface as well as in the bulk (Paces, 1973). When the concentration of Al in amorphous silica is high and the Al/Si ratio is 0.85–1.72, the mineral is known as allophane or imogolite. X-ray study shows that both these minerals are amorphous, and their infrared spectra have very broad bands that are difficult to interpret (Wada, 1967). Imogolite is somewhat more crystalline than allophane (Russell et al., 1969). These amorphous aluminum-silicate minerals are the principal clay constituents in volcanic ash soils (Henmi and Wada, 1976).

Amorphous silica–alumina layers appear also on the surface of crystalline minerals, such as the feldspars and zeolites, and on the edges of most clay minerals.

1.2.1 Synthetic Silica–Alumina Gels

Dried co-gels of silica and alumina are of immense importance in the petroleum industry where these materials form the basis of the process used for the cracking of crude oil and the preparation of high-quality fuels. The chemistry of the surfaces of silica–alumina has been thoroughly studied (Little, 1966; Hair, 1967; Cloos et al., 1969).

The acid sites on silica–alumina are considered to be of two types: (1) *Lewis acid sites* such as Al^{3+} ion in the crystal edge, which are capable of accepting electrons from the adsorbate molecules and (2) *Brønsted acid sites* such as Si–OH or Al–OH and H_3O^+, which can donate a proton to the adsorbate molecule. The nature of the acidic sites can be explained in a number of ways:

1. The isomorphous substitution of a trivalent Al for a tetravalent Si ion in a tetrahedral silica structure will result in an excess negative charge in the oxide anion lattice. To maintain electrical neutrality the silica–alumina particle must have a number n_i of exchangeable cations of charge e_i, where $\Sigma n_i e_i$ is equal to the number of Al ions. If the exchangeable ions are protons (or H_3O^+ ions) a Brønsted acidic site is obtained. The proton is only weakly attracted by the huge negative oxide anion lattice, as shown by

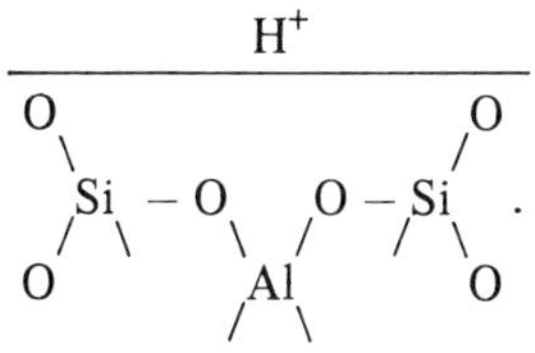

Since the electrostatic attraction forces between the protons and the huge negative double-oxide anion species are weak, such a Brønsted acid is stronger than a silanol group.

2. The isomorphous substitution of Al for Si in tetrahedra located at a crystal edge results in each Al atom being linked by means of electron pair bonds to three oxygens in the silica structure. Such an Al atom will require an additional electron pair to complete its p orbitals. Thus, it provides an electron acceptor site (Lewis acid) at the surface and may be coordinated by molecules having lone pair electrons (Lewis bases), as shown by:

$$O-\underset{\underset{O}{|}}{\overset{\overset{OH}{|}}{Si}}-O-\underset{\underset{O}{|}}{\overset{}{Al}}-O-\underset{\underset{O}{|}}{\overset{\overset{OH}{|}}{Si}}-O.$$

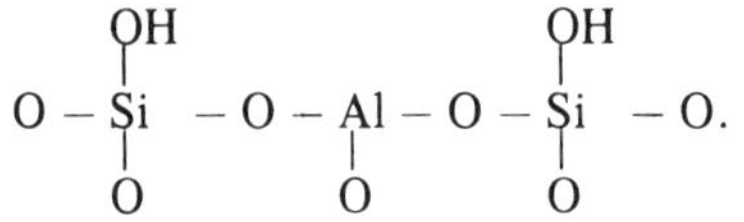

3. Water molecules may be coordinated at the Lewis acid site to provide two independent sites of Brønsted acid of different strengths. One site, being a free proton or H_3O^+ ion, is of a higher acid strength than that of the silanol group. The second Brønsted site is a proton belonging to an Al–OH group and is a weaker acid than the silanol group, because of the smaller inductive effect of the trivalent Al compared with that of the tetravalent Si ion, as shown by

$$O-\underset{\underset{O}{|}}{\overset{\overset{OH}{|}}{Si}}-O-\underset{\underset{O}{|}}{\overset{\overset{\overset{H^+}{\cdots}}{\overset{OH}{|}}}{Al}}-O-\underset{\underset{O}{|}}{\overset{\overset{OH}{|}}{Si}}-O.$$

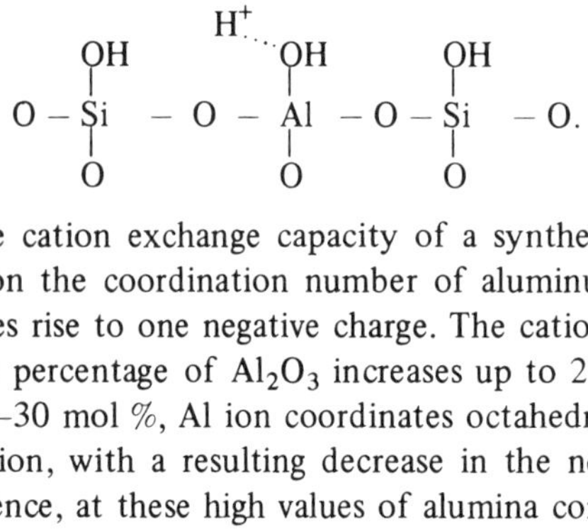

The cation exchange capacity of a synthetic silica–alumina gel is largely dependent upon the coordination number of aluminum. Each Al ion in fourfold coordination gives rise to one negative charge. The cation exchange capacity of the gel increases as the percentage of Al_2O_3 increases up to 20–30 mol %. When Al_2O_3 increases above 20–30 mol %, Al ion coordinates octahedrally, in addition to the tetrahedral coordination, with a resulting decrease in the net negative charge of the gel. As a consequence, at these high values of alumina concentrations, the cation exchange capacity of the gel decreases (DeKimpe et al., 1961). The cation exchange capacity of the gel is highly dependent upon pH, and upon type and concentration of salt used for the reaction.

Many organic reactions take place on surfaces of freshly synthesized silica–alumina gels but not on those of silica gels because of the lower acidity of the latter. For example, the mechanism of polymerization of unsaturated hydrocarbons on silica–alumina gels can be shown by the following equation for the polymerization of *iso*butene

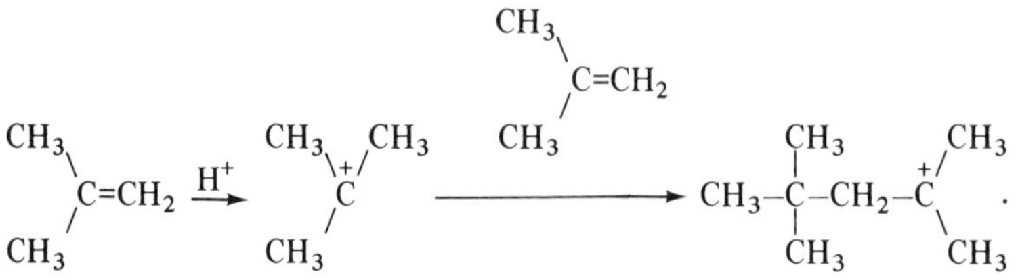

$$\underset{CH_3}{\overset{CH_3}{\diagdown}}C=CH_2 \xrightarrow{H^+} \underset{CH_3}{\overset{CH_3\diagdown\ _+/CH_3}{C}} \xrightarrow{\overset{CH_3}{\diagdown}C=CH_2 \atop CH_3 \diagup} CH_3-\underset{CH_3}{\overset{CH_3}{\underset{|}{C}}}-CH_2-\underset{CH_3}{\overset{CH_3}{\diagup^+}}C \diagdown \qquad . \quad (6.IX)$$

The positive ion in this equation is designated as *carbonium* and is stabilized on silica–alumina surfaces.

The catalytic activity of the surface depends on its humidity. Water competes with the organic molecules on the sorption sites. In the presence of excess water there is no adsorption and the catalyzed reactions that should follow adsorption do not take place. When the silica–alumina is dehydrated some of the Brønsted acid sites are converted into Lewis acid sites. Organic compounds having a high affinity for interaction with Al atoms will now be adsorbed, but those reactions that require the presence of protons may or may not occur. It is to be concluded that the presence of a minimum amount of water is required for carbonium ion to be formed.

Rouxhet and Sempels (1974) distinguished between two types of hydroxyl groups on silica–alumina surfaces. The first type was identical with hydroxyls present on the surface of silica. The second type, the proportion of which increased with the Al content of the gel, was of a higher acid strength, with a pK_a between −4 and −8. Some proton acceptors such as ammonia and pyridine, which form hydrogen bonds with hydroxyls of the first type, may become protonated during the interaction with hydroxyls of the second type.

1.2.2 Surface Properties of Allophane and Imogolite

The surface properties of allophane and imogolite differ in some respects from those of freshly prepared synthetic silica–alumina gels and may be regarded as aged gels. These minerals have large surface areas, large water-holding capacities, high phosphorus adsorption coefficients, and a strong tendency to adsorb humic substances (Wada and Inoue, 1967). The particles of these minerals have a permanent negative charge resulting from substitution of Si by Al ions in the structure. The negative charge increases with increased pH (Fieldes and Schofield, 1960). At pH 10.5 only about 20–30% of the Si–OH groups are ionized (Wada, 1967).

There are marked variations in cation exchange capacity values reported in the literature for allophane and imogolite clays. The allophanic clay cation exchange capacity values show no systematic variation with the mole percent of Al_2O_3 while the cation exchange capacity and surface charge density of imogolite decrease systematically as the mole percent of Al_2O_3 increases. According to Jenne (1972) this behavior of allophane results from the fact that at higher aluminum contents, Al ions go into sixfold rather than fourfold coordination.

Cation exchange reactions occur with these minerals according to two different mechanisms (Eckstein et al., 1970; Wada and Harada, 1969). Exchange capacities (with Na, K, Ca, and Mg) in the range of 20–26 meq/100 g, were found to be due to the presence of the permanent negative charge of the particles resulting from the substitution of Si by Al. Besides the exchange reaction with the metallic cations, a pH-dependent cation exchange reaction also occurs with some of the hydroxyl groups on the edges of the particles, releasing protons. For example, an exchange reaction with Li ions can be described by the following equation

$$\equiv Si{-}OH + Li^+ \rightarrow \equiv Si{-}OLi + H^+ \qquad (6.X)$$

The adsorbed Li is easily replaced or hydrolyzed when the Li concentration in the external solution is reduced. The number of effective broken edge exchange sites is highly dependent on the acidity and on the concentration of the solution, generally increasing with both concentration and pH (Table 6.3).

Yariv (1969) studied the catalytic effect of various clay minerals and sands on the reduction of ferric iron in the ferric-ferricyanide complex. This reduction reaction is catalyzed by protons and is, therefore, dependent on the surface acidity of the minerals. The surface of silica gel and sand, being more acidic than that of alumina, catalyzes the reduction of iron. The surface acidity of allophane, like that of alumina, is not sufficiently high to catalyze the reaction.

1.3 Opals

Opal is a disordered form of silica. There are large natural deposits of "common opal", an opaque, milk-white amorphous silica of no commercial value. In a few scattered deposits around the world, usually in association with weathered volcanic rocks, "precious opal" is to be found, a mineral that can exhibit sparkling colors ranging across the entire visible spectrum, from violet to deep red, and that is valued as a gemstone. Jones and Segnit (1971) proposed a classification for opals based on normal X-ray diffraction and on water content. According to this classification the group of opal minerals is subdivided into opal-A (diffuse bands in the X-ray pattern, absence of long-range order, noncrystalline), opal-CT (disordered cristobalite-tridymite stacking sequences, poorly crystalline), and opal-C (normal cristobalite pattern, crystalline). The term "opal A", thus defined, includes various forms of opals. Langer and Flörke (1974), on the basis of low-angle X-ray-scattering, separate hyalites from other opal varieties within this group. Hyalite has a glass-like continuous three-dimensional network structure producing no remarkable low-angle X-ray scattering, while other varieties of opal-A, due to electron density discontinuities, do show low-angle scatter-

Table 6.3. The effect of lithium and ammonium on cation exchange of allophane and imogolite. (After Eckstein et al., 1970)

Sample	Saturating treatment			Displaced cations in meq/100 g					Li released by excess Ca
	Cation	Normality	pH	Na	K	Ca	Mg	Sum of cations	
Allophane	Li	1.0	4.5	11.4	4.1	6.0	0.5	22.0	43.0
		1.0	8.2	14.5	5.7	4.6	1.4	26.2	86.5
		0.5	8.2	8.1	3.8	4.2	0.9	17.0	31.5
		0.05	8.2	2.6	0.8	3.9	0.6	7.9	31.4
	NH_4	1.0	7.0	2.5	0.5	4.7	1.2	8.9	–
Imogolite	Li	1.0	4.5	9.7	3.6	6.1	0.5	19.9	89.4
		1.0	8.2	12.3	5.5	3.6	0.7	22.1	124.0
		0.5	8.2	7.8	3.3	3.0	0.6	14.7	29.0
		0.05	8.2	2.5	0.8	2.3	0.4	6.0	31.0
	NH_4	1.0	7.0	2.3	0.6	4.7	1.3	8.9	–

ing typical for gel-like structures of noncrystalline silica. They subdivide the opal-A group into opal-AG (gel-like) and opal-AN (network, glass-like).

The content of silanol groups differs significantly in different opal types, reflecting their different genesis and microstructure. Opal-AN (hyalite) is deposited by quenching of high-temperature supercritical SiO_2–HOH fluid phase on cold substrates (Flörke, 1972). This class shows the highest SiOH content in accordance with the origin of the minerals from a fluid containing monomeric H_4SiO_4, dimeric $H_6Si_2O_7$, or higher associated species $Si_nO_{n-1}(OH)_{2n+2}$. Though the nature of silica species and the equilibria between them in supercritical fluids, from which opal-AN is deposited, are not yet clear, these fluid phases contain hydrous silica species with low values for n at high concentrations, and many OH groups are retained during the rapid solidification on a cold substrate. Opal-CT shows a very low silanol fraction. This type of mineral is formed slowly from highly diluted true solutions at relatively low temperatures, where few crystal nuclei are formed, and crystal growth is slow. Opal-AG is formed by rapid precipitation from supersaturated true solutions or from sols in sediments some 10–100 m below the earth's surface (Jones and Segnit, 1966). This rapid reaction at low temperatures leads to the formation of relatively large amounts of silanol groups (Langer and Flörke, 1974).

In all opals, molecular physically adsorbed water is present in two different modifications. Type A corresponds to single water molecules trapped in small cages within the Si–O matrix. Water molecules in these cages are free from hydrogen bondings. Type B occurs in relatively large voids as liquid water or as multimolecular films on inner surfaces. In this type of water, the HOH molecules are hydrogen bonded. The amount of Type B water differs in the different opal types. Opal-CT, with its large inner surface and mostly irregular microstructure, contains the greatest amount, and opal-AN, with its dense network, has the lowest water content (Langer and Flörke, 1974).

Opal gem stone consists of transparent spherical particles of amorphous silica that are tightly packed together in good optical contact. The material is transparent and has the refractive index of amorphous silica, which lies between 1.435 and 1.455, depending on the degree of hydration. At the interfaces of the voids there is a reduction in the refractive index because the voids contain only air, water vapor, or liquid water. These tiny, regularly arranged discontinuities act as light-scattering points, forming a three-dimensional diffraction grating from which beams of a single wavelength are reflected at specific angles of incidence and reflection when the spacing between the voids is of the order of the wavelength. Specimens containing only very small spheres, and therefore showing small values of spacings between voids, exhibit no visible colors. Violet light, with a wavelength of 400 nm, first appears when the diameter of the sphere is 138 nm. Opals with larger diameters show a greater variety of colors ranging from violet and blue (at low viewing angles) to green, yellow, or orange (at high angles). Red first appears when the diameter of the sphere is 241 nm and the material is viewed with the light source very near the eye. Opals that show flashes of deep red can exhibit all the other colors of the spectrum when they are viewed at suitable angles. The light diffraction effects are strongest in opals with distinct voids. The colors are less intense when the voids are smaller and when the amount of silica deposited between the spheres is larger. Irregular arrays of voids give rise to a general scattering of white light, causing an opal to have the milky appearance described by the term opalescent. Fine particles of light-absorbing material, such as carbon or oxides of iron and titanium, darken the stone. The most valuable form of opal, the black opal, contains these particles in combination with spheres of uniform size packed in highly regular arrays (Darragh et al., 1976).

2. Silica in Aqueous Solutions

The aqueous chemistry of silica is dominated by the properties of the silicic acid monomer $Si(OH)_4$ and by its tendency to join or polymerize with other $Si(OH)_4$ units to form hydrophilic polysilicic acids $Si_nO_{2n-m}(OH)_{2m}$ or hydrophobic amorphous silica $SiO_2 \cdot xH_2O$. The solubility of silica in water is dependent on many factors, including the state of combination of the silica, whether it is in the form of a crystalline or an amorphous mineral, the temperature, pressure, pH of solution, and the presence of certain metallic cations, such as Al^{3+} and Fe^{3+}, organic ions, and the presence of F^-, with which silicon forms the soluble coordination ion SiF_6^{2-}. The solubility of crystalline minerals depends on the type of mineral and on the particle size.

2.1 Polymerization and Depolymerization of Silicic Acid

Silicic acid is a very weak acid, the first dissociation constant being

$$K = [H^+] \times [H_3SiO_4^-]/[H_4SiO_4] = 10^{-9.9}.$$

Hence, the solubility of silica is not affected by pH at values below about 9, but greatly increases at higher pH values. Silicate ions exist in appreciable amounts only at a pH above 9. In a more acid solution, the silicic acid is essentially un-ionized.

In aqueous solutions silicic acid establishes in time an equlibrium concentration of monomeric molecules. However, if a "considerably" polymerized silica is dissolved in water, some time may be needed to establish equilibrium conditions. As long as the total concentration of silica in water is less than 100–140 mg SiO_2/liter at 25 °C, a true solution of monosilicic acid is obtained.

If the total amount of silica in water is greater than 140 mg/liter at 25 °C and at pH below 9, the excess silica will in time polymerize and may form colloid suspension or even precipitate. The condensation process is reversible. At pH above 9 the tendency of aqueous silica solution, even at concentrations greater than 140 mg/l, is to form units containing one Si atom per unit. The dimer $H_6Si_2O_7$ is stable only at pH below 10.5.

The condensation of two units of monomeric silicic acid is catalyzed by a proton. The reaction mechanism can be described as follows:

1. Protonation of a silanol group and the formation of a transition complex

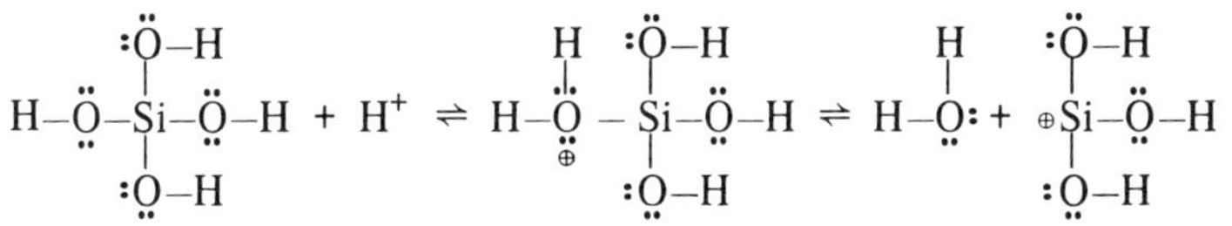

$$\begin{array}{c}\ddot{\text{O}}-\text{H}\\|\\\text{H}-\ddot{\text{O}}-\text{Si}-\ddot{\text{O}}-\text{H}\\|\\\ddot{\text{O}}-\text{H}\end{array} + \text{H}^+ \rightleftharpoons \begin{array}{c}\text{H} \quad \ddot{\text{O}}-\text{H}\\\ddots \quad |\\\text{H}-\ddot{\text{O}}-\text{Si}-\ddot{\text{O}}-\text{H}\\\oplus \quad |\\\ddot{\text{O}}-\text{H}\end{array} \rightleftharpoons \text{H}-\ddot{\text{O}}: + \begin{array}{c}\text{H} \quad \ddot{\text{O}}-\text{H}\\|\\\oplus\text{Si}-\ddot{\text{O}}-\text{H}\\|\\\ddot{\text{O}}-\text{H}\end{array} \quad (6.\text{XI})$$

$$\text{Si(OH)}_4 + \text{H}^+ \rightleftharpoons [\text{Si(OH)}_3(\text{OH}_2)]^+ \rightleftharpoons \text{HOH} + [\text{Si(OH)}_3]^+.$$

The positively charged Si atom in the transition complex is a very strong Lewis acid (an electron pair acceptor), since it shares only three electron pairs directed to oxygens located at three (out of four) vertices of a tetrahedron. Similar transition complexes of the general formula $[(\equiv\text{SiO})_p^+ \text{Si(OH)}_{3-p}]$ (where p varies between 0 and 3) can be obtained from silanol groups.

2. Interaction between the transition complex and a silanol group, the release of a proton and the formation of a siloxane group is as follows:

$$\begin{array}{c}\text{H}-\ddot{\text{O}}:\\|\\\text{H}-\ddot{\text{O}}-\text{Si}^\oplus\\|\\\text{H}-\ddot{\text{O}}:\end{array} + \begin{array}{c}\text{H} \ddot{\text{O}}-\text{H}\\\ddots \quad |\\:\ddot{\text{O}}-\text{Si}-\ddot{\text{O}}-\text{H}\\|\\:\ddot{\text{O}}-\text{H}\end{array} \rightleftharpoons \begin{array}{c}\text{H}-\ddot{\text{O}}: \quad \text{H} \quad \ddot{\text{O}}-\text{H}\\|\\\text{H}-\ddot{\text{O}}-\text{Si} - \text{O} - \text{Si} -\ddot{\text{O}}-\text{H}\\|\quad\quad\oplus\quad |\\\text{H}-\ddot{\text{O}}: \quad\quad\quad :\ddot{\text{O}}-\text{H}\end{array} \rightleftharpoons \begin{array}{c}\text{H}-\ddot{\text{O}}: \quad \ddot{\text{O}}-\text{H}\\|\quad\quad|\\\text{H}-\ddot{\text{O}}-\text{Si}-\text{O}-\text{Si}-\ddot{\text{O}}-\text{H}\\|\quad\quad|\\\text{H}-\ddot{\text{O}}: \quad :\ddot{\text{O}}-\text{H}\end{array} + \text{H}^+$$

(6.XII)

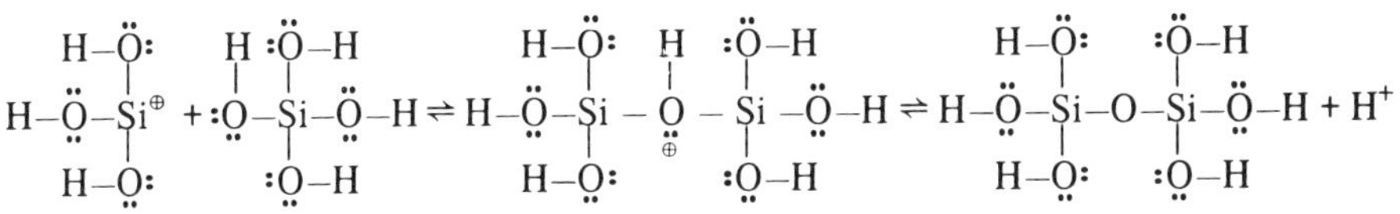

$$[\text{Si(OH)}_3]^+ + \text{Si(OH)}_4 \rightleftharpoons [\text{Si}_2(\text{OH})_7]^+ \rightleftharpoons \text{H}_6\text{Si}_2\text{O}_7 + \text{H}^+.$$

The proton is involved in this reaction only as a catalyst and the number of protons at the end of the condensation process is equal to their number at the beginning of the process.

The condensation can proceed further according to the same mechanism resulting in a polymeric silicic acid of the general formula $\text{Si}_n\text{O}_{2n-m}(\text{OH})_{2m}$. When $n \leq 6$ the acid behaves as a true soluble species, but at higher values of n the solution becomes colloidal. Hydrophilicity of the colloid decreases and the hydrophobicity increases when the value of $2m/(2n-m)$ decreases, the latter being the ratio between silanol and siloxane groups in the polymer. This ratio usually decreases with increasing value of n, either by drying the solution or by aging. The last stage of the polymerization is achieved when every Si atom is part of four siloxane groups whereby a hydrophobic colloid of amorphous silica is obtained.

Protons also catalyze the depolymerization of silica, and the same equations, from right to left, may describe the sequence of the mechanism of depolymerization. The "transition complex" $[(-\text{SiO})_p\text{Si(OH)}_{3-p}]^+$ can either react with a silanol group, giving rise to a condensation reaction, or it can react with an OH^- ion to give a new silanol group. With $[\text{Si(OH)}_3]^+$ this results in a monomeric silicic acid, according to the equation

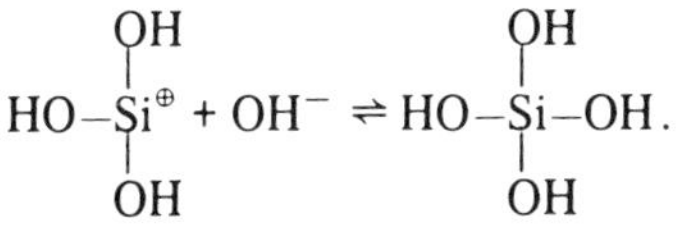

$$\begin{array}{c}\text{OH}\\|\\\text{HO}-\text{Si}^\oplus\\|\\\text{OH}\end{array} + \text{OH}^- \rightleftharpoons \begin{array}{c}\text{OH}\\|\\\text{HO}-\text{Si}-\text{OH}.\\|\\\text{OH}\end{array} \quad (6.\text{XIII})$$

With increasing pH values the concentration of OH^- ions becomes high, compared to that of H_4SiO_4, and the probability of the "transition complex" interacting with OH^- or with a silanol group increases and decreases, respectively. However, the rate of depolymerization at high pH values is slow because of the low concentration of the proton catalyst. At pH between 5 and 6 the rate of polymerization and depolymerization is at a maximum. At lower pH values the concentrations of the transition complexes $\geq Si-OH_2^+$ and $\geq Si^+$ become very high but at the same time the concentration of the $\geq SiOH$ groups, necessary to react with the transition complexes, decreases. The reaction rate therefore decreases.

The surface area and properties of the hydrophobic silica gel obtained by the condensation of silicic acid depend on the pH of the solution in which the silica polymer grows. In solutions with pH below 5 the growth of the polymer results from the condensation of silicic acid polymers whereas in solutions with pH above 5 the particles grow by the condensation between polymeric and monomeric species. Silica gel achieves the highest specific surface area (~ 800 m^2/g SiO_2) when the condensation is carried out from solutions of pH 5 and both growth mechanisms take place simultaneously (Coudurier et al., 1971).

2.2 Solubility of Silica

The results reported in the literature for the solubility of amorphous silica at 25 °C for pH below 9 range between 100 and 180 mg SiO_2/l (1.7–3.0 mmol/l). The variations in these results are undoubtedly due to the differences in solubility characteristics of the various forms of silica (Shell, 1962). It appears that the "true" values for the solubility of amorphous silica in water range from 100 to 140 mg/l at 25 °C. The solubility of amorphous silica in water as affected by pH was determined by Alexander et al. (1954) and the values are given in Table 6.4. This solubility increases rapidly with increasing pressure (Jones and Pytkowicz, 1973). Krauskopf (1956) studied dissolution and precipitation of silica at low temperatures and came to the following conclusions:

1. The solubility of amorphous silica increases with increasing temperature from 60–80 mg/l at 0 °C, 100–140 mg/l at 25 °C to 300–380 mg/l at 90 °C.

2. The solubility of amorphous silica barely changes in the pH range 0–9, but increases rapidly above pH 9 (Fig. 6.3).

Table 6.4. Solubility of amorphous silica at 25 °C. (From Alexander et al., 1954)

pH	mg SiO_2/l
1.0	140
2.0	150
3.0	150
4.2	130
5.7	110
7.7	100
10.26	490
10.60	1120

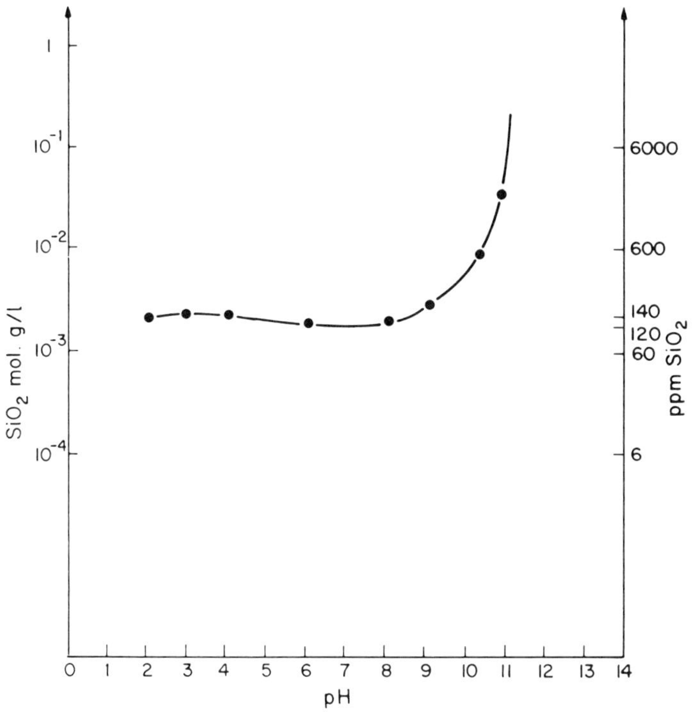

Fig. 6.3. Solubility of amorphous silica as a function of pH. (After Krauskopf, 1956)

3. An unsaturated solution of silica gradually dissolves any colloid amorphous silica present, until equilibrium is reached. The rate of approach to equilibrium increases with increasing temperature.

4. Amorphous silica has about the same solubility in seawater as in fresh water. Hence the silica of most stream water, which is in true solution as H_4SiO_4, should not form a new solid phase on contact with seawater. The concentrations of dissolved silica in seawaters are lowered due to uptake by organisms.

5. A freshly prepared supersaturated solution of silica usually does not precipitate the silica immediately, but in time a new colloid phase will be formed. In dilute solutions very small hydrophilic polymers are obtained that form a colorless, transparent colloidal solution. The colloidal solution is remarkably stable with respect to time, temperature changes, and mechanical disturbances. The polymers obtained from concentrated solutions are large and hydrophobic. In weakly basic solutions, where the condensation rate is low and the aggregation rate is high, flocculent masses are precipitated. In weakly acid solutions, where the condensation rate is high, a gel is precipitated.

6. Colloidal silica is precipitated by evaporation, or by coprecipitation with other colloids, or by means of fairly concentrated solutions of electrolytes. Silica in true solution is not affected by electrolytes.

The solubility of the crystalline forms of silica is lower. Measurements are extremely difficult because of the slowness with which equilibrium is established and

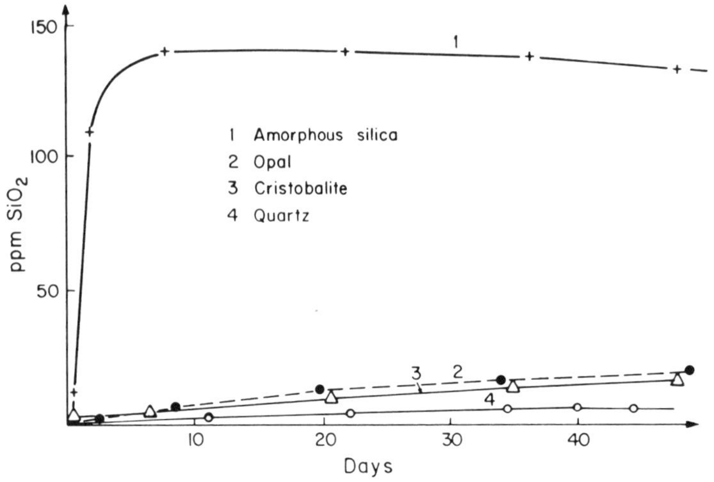

Fig. 6.4. The rate of solubility of amorphous silica, opal, cristobalite, and quartz. (After Wey and Siffert, 1961)

because of the solubility effects due to impurities, especially aluminum. The commonest form of silica, quartz, has a solubility of about 6–14 mg SiO_2/l (0.1–0.23 mmol/l) at ordinary temperatures. Other forms of crystalline silica show similar or somewhat higher values than those of quartz (Fig. 6.4). Increase in chloride or sulfate ion concentration, increase in pH value, and rise of temperature accelerate the dissolution rates of the crystalline forms of the silica (Kamiya et al., 1974).

The presence of certain organic hydroxy acids causes a slight increase in the solubility of amorphous and crystalline silica. Sodium citrate, for example, dissolves 170 mg SiO_2/l at pH 7.5. The solubility of silica in the presence of other sodium salts of organic acids, although less than that with sodium citrate, is still greater than that in pure water. The highest solubility is obtained at about pH 7 and decreases with decreasing pH value of the solution.

Silicon forms soluble complexes with catechols (derivatives of 1,2 dihydroxy benzene), which are common in humic substances. According to Barnum et al. (1973) the structure of the complex of Si with pyrocatechol is as follows:

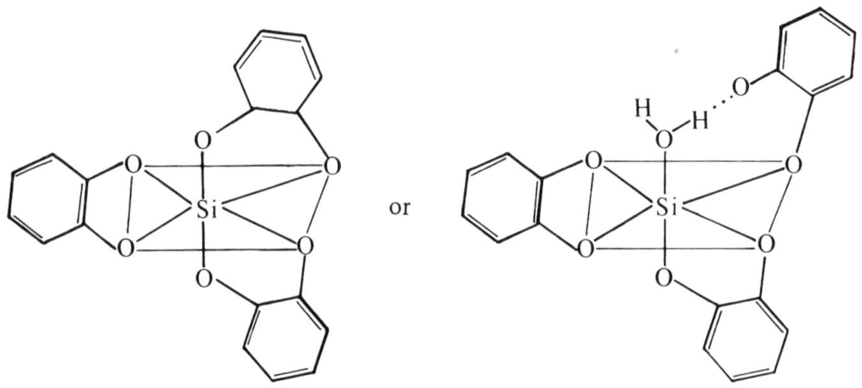

At pH above 9 the formation of Si–catechol complexes is assumed to be complete (Bartels, 1964), and at pH 9.6 the dissolution rate of quartz at 25 °C in pyrocatechol solution is of the same order as in 0.1*N* NaOH (Bach and Sticher, 1966).

2.2.1 The Effect of Ions Present in Solution on the Solubility of Silica

The solubility of monosilicic acid in water is not affected by the presence of monovalent metallic cations. This is also true for seawater. Multivalent metallic cations may drastically change the solubility of silicic acid. Okamoto et al. (1957) pointed out that aluminum is able to reduce the solubility of silica in water. An amount of 20 mg Al/l reduced the solubility of silica to 15mg/l at pH values between 8 and 9.

The effect of aluminum and magnesium ions on the solubility of silica at various pH values is shown in Figures 6.5 and 6.6, after Wey and Siffert (1961). At pH values

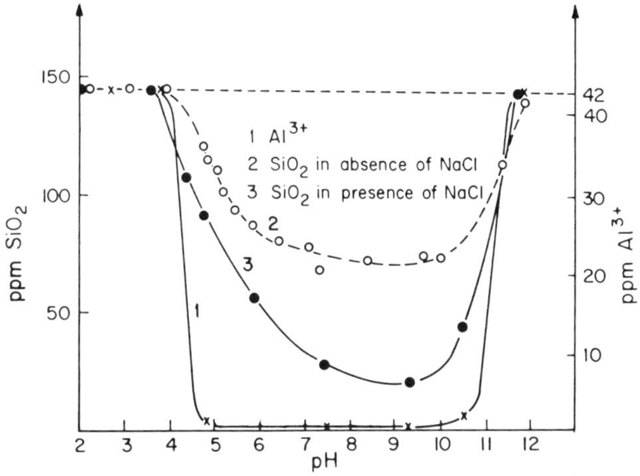

Fig. 6.5. Solubility of amorphous silica in the presence of Al^{3+} and solubility of Al^{3+} in the presence of silica as a function of pH. (After Wey and Siffert, 1961)

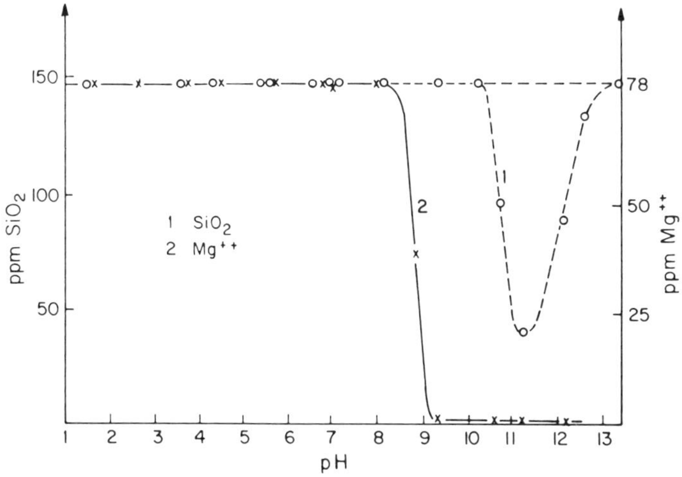

Fig. 6.6. Solubility of amorphous silica in the presence of Mg^{2+} and solubility of Mg^{2+} in the presence of silica as a function of pH. (After Wey and Siffert, 1961)

between 5 and 10.5, at which aluminum is hardly soluble, the solubility of silica decreases drastically. In the presence of Mg the solubility of silica decreases rapidly in the narrow pH range between 10 and 12. For very high concentrations of magnesium the solubility of silica decreases even at pH values lower than 10. In the preceding systems, the dissolved silica precipitates as aluminosilicate or as magnesiumsilicate.

Sorption of multivalent cations and organic materials by amorphous silica and the interaction of the sorbates with the surface functional groups on the silica affects the solubility of the amorphous material. Boehm and Schneider (1962) and Liefländer and Stöber (1960) showed that amorphous silica is remarkably protected against dissolution by chemisorbed aluminum. After 3 weeks, only 6 mg SiO_2/l were dissolved at pH 8.2.

2.2.2 Dissolution of Opal in Water

The amount of dissolved silica and the rate of dissolution of opal depend on the water content of the mineral sample. Solubility increases with increasing water content of the opal sample. After 24 days of contact with water, solutions of 12.0 and 0.8 mg SiO_2/l (0.200 and 0.012 mmol/l) were obtained from opals having 13.7% and 5.25% of water, respectively. This phenomenon is interpreted as follows. Opal is obtained by the dehydration of silica gel, a process that is accompanied by the growth of relatively large colloid particles at the expense of the smaller ones. Thus, as the water content is decreased, particles would become larger, surface and surface energy would be reduced, and solubility and dissolution rate become less (Huang and Vogler, 1972).

Hurd (1972b) studied several factors that affect the dissolution rate of biogenic opal. The rate of dissolution of a single biogenic opal test is a function of solution temperature, degree of saturation, available surface area of the test, and any protective coating layer the test may have. There appears to be a tendency toward slower dissolution rates with increasing weight percent of soluble fraction in solution. Lowering the temperature decreases the dissolution rate constants. Unstirred solutions dissolve more slowly than stirred ones. Variation of specific surface area of tests from one species to the next and from different portions of the same test could produce differing overall rates of dissolution. Delicate tests dissolve more rapidly than robust ones. The crushing of tests and breakdown of cell protoplasm during digestion creates greater specific surface area and partially removes protective organic coatings from the test, increasing rate of dissolution. Tests dissolve more slowly if pretreated with various metals. Metal silicates, formed by the organisms by concentration from the surrounding environment, may be incorporated in the test, retarding the dissolution rate. Biogenic opal has a solubility lower than amorphous silica because of trace impurities obtained either by secondary adsorption from seawater by the test surface or incorporated into the test by the organism itself.

The slight solubility of the soil opal phytolite, compared to that of amorphous silica, is also due to the presence of protective organic coatings and the adsorption of aluminum. According to Bartoli and Selmi (1977) the rate of dissolution of phytolites and the extent of the biodegradation of the organic matter in soils are parallel. In soils having low biogenous activity the phytolites are more stable.

2.2.3 Dissolution of Quartz

Because quartz is one of the most abundant minerals, it is a potential source of silica for the formation of secondary minerals. The extent to which the silica of quartz is involved in the formation of authigenic minerals is determined by the solubility characteristics of this mineral. Results for the solubility of quartz at 25 °C, as stated in the literature, range from about zero to 30 mg SiO_2/l (0–0.5 mmol/l). Many factors influence the rate and extent of dissolution of quartz. Such factors include the presence or absence of a disturbed surface layer (produced by grinding), particle size (surface area), pH and ionic composition of the aqueous solution, and equilibrium time. Mechanical grinding produces a disturbed layer composed of amorphous silica on the surface of quartz particles. This disturbed surface layer markedly increases the solubility of silica. Quartz particles pulverized by glacial grinding are more soluble than normal quartz. The amount of silica released from quartz ground in nature through movements by ice, water, and wind (i.e., quartz with a disturbed surface layer) may be sufficient to support the genesis of montmorillonite and other layer silicates at temperatures existing in the soils, while that amount released from quartz not having a disturbed layer would be insufficient (Henderson et al., 1970).

The rate of dissolution of quartz in water at a fixed temperature is constant and is dependent upon the volume of solution and the surface area of quartz. The rate of dissolution increases with increasing temperature.

The solubility of quartz has been calculated from theoretical considerations to be independent of pH below pH 8. However, Henderson et al. (1970) showed that the silica dissolved from quartz increases in the pH range above 4, from 0.6 mg SiO_2/l at pH 4 to 19 mg/l at pH 10 and that the dissolution versus pH curve is apparently sigmoidal.

Quartz particles adsorb SiO_2 from solutions supersaturated in silica with respect to quartz. Monosilicic acid rather than silicate anion is the species of silica adsorbed. According to Stöber (1967) the layer of adsorbed silica controls the final concentration of silica in solution. The soluble silicic acid is only partly in equilibrium with solid silica (either quartz or vitreous or any other form) and partly with amorphous silica, which is sorbed on the crystalline particle. Stöber reported that the amount of silica removed from a pH 8.4 buffer solution was directly related to the surface area of the suspended quartz particles, indicating adsorption of SiO_2 rather than quartz crystal growth. After the adsorption of two or three monolayers of monosilicic acid from solution onto the quartz surfaces, the concentration of silica remaining in solution would level off. According to Stöber, after several washings of quartz crystals from amorphous sorbed silica, a solubility of 2.9 mg SiO_2/l (0.05 mmol/l) is obtained within 24 h in the presence of NaCl as a catalyst and $NaHCO_3$ as a buffer.

2.2.4 Diagenesis of Siliceous Oozes

The transformation of amorphous silica to intermediate opalline phases and to the stable phase quartz and the epitaxial growth of quartz processes that occur during the diagenesis of siliceous oozes, occur mainly through the mechanism of dissolution–reprecipitation. All factors that give rise to enhancement of the dissolution also give rise to enhancement of the degree of diagenesis. The solubility of silica and the rate of each step of the transformation of amorphous silica to quartz increases with alkalinity. For example, the activation energy of opal-CT to quartz transformation is lower in KOH solutions (14.4 kcal/mol, Mizutami, 1966) than in H_2O (23.2 kcal/mol, Ernst and Calvert, 1969). Reaction mechanisms and reaction rates depend on temperature (Campbell and Fyfe, 1960; Fyfe and McKay, 1962). The presence of Mg, Al, Fe, and Mn is an important requirement in the conversion process of amorphous silica to the crystalline variations (Harder and Flehming, 1970).

Amorphous silica can crystallize directly into quartz only when the silica concentration in the solution is below the inferred equilibrium solubility of opal-CT. When silica in solution reaches concentrations above the equilibrium solubility of opal-CT, as in interstitial waters of most siliceous oozes, a disordered metastable silica phase, such as opal-CT, is likely to crystallize instead of the ordered stable, poorly soluble quartz. Under hydrothermal conditions (3 kb, 100 to 300 °C, 25 to 5200 h), whereby the equilibrium solubility of quartz is increased, Oehler (1976) crystallized quartz from silica gel with no intermediate crystalline phase. The crystallization sequence involves three stages. In the initial stage the rate of quartz formation is extremely slow and is controlled by the sluggish nucleation rate of primary quartz crystallites in the gelatinous matrix. In the second stage the rate of quartz formation increases markedly and is controlled by the crystal growth from the solution supersaturated with respect to quartz. In the third stage the formation of new quartz is very slow and the rate of formation is controlled by the diffusion of atoms from the amorphous phase to the crystalline phase through the aqueous phase. Quartz crystallites nucleate on the surfaces of organic matter, which provides chemically reactive sites conducive to crystallite nucleation. According to Oehler and Schopf (1971) small quartz grains predominate in organic-rich chert areas, where numerous reactive sites for crystallite nucleation exist, while larger quartz grains predominate in organic-poor chert areas, where the density of potential nucleation centers is comparatively lower. Mackenzie and Gees (1971) crystallized quartz directly from sea water at 20 °C, but the solubility never exceeded 0.073 ± 0.005 mmol SiO_2/l.

According to Kastner et al. (1977), the presence of nuclei containing magnesium and OH^- with a Mg/OH ratio of 1.2, enhances the rate of formation of opal-CT. The nuclei appear to serve as sites upon which opal-CT precipitation takes place. These nuclei form rapidly at 150 °C. At lower alkalinity and at lower temperatures, the rate of nuclei formation of critical size and the subsequent formation of opal-CT are reduced. In the presence of clay minerals the transformation of opal-A into opal-CT slows down due to the recrystallization of the clay minerals to more Mg-rich clays, at the same time lowering the concentration of Mg and the alkalinity of the solution. Consequently, insufficient OH and Mg are available for the formation of the nuclei and the subsequent crystallization of opal-CT.

The diagenesis of siliceous oozes is strongly dependent on the nature of the host sediments. In deep-sea sediments most cherts occurring in clayey layers are predominantly disordered cristobalite, while cherts in carbonate sediments are predominantly quartz (Lancelot, 1973). In deep-sea sediments opal-CT is widespread in both carbonate and clay-rich sediments, but in the latter the degree of disorder of the opal-CT crystals is generally higher than in carbonate sediments of the same geologic age

(Keene, 1976). In addition, the quartz-to-opal-CT ratio is generally much higher in younger carbonate sediments than in older clay-rich sediments. According to Kastner et al. (1977) most quartz in cherts, even in carbonate sediments, had an opal-CT precursor. The degree of disorder of the opal-CT phase depends on the Mg, OH^-, and silica concentrations in solution during the formation of the silica phase. Opal-CT that crystallizes in the presence of clay minerals and a great excess of soluble silica, grows very fast around the small number of nuclei and has a high degree of disorder, whereas opal-CT that crystallizes from solutions poor in silica but relatively rich in Mg and OH, arrives very quickly at the stage where the crystal growth is slow and is essentially controlled by the diffusion of atoms from the solution to the crystalline, less-soluble phase, and the phase obtained has a low degree of disorder. The less disordered opal-CT in carbonate sediments needs to undergo fewer steps of progressive ordering than the more disordered opal in clay-rich sediments (Murate and Larson, 1976). (See pages 158–160, on crystal growth by the mechanism of orientation).

2.3 Silica Sorption by Minerals

Silicic acid, H_4SiO_4, or its anionic species $[H_3SiO_4]^-$, can be sorbed onto the surfaces of stable oxides (such as silica or one of the modifications of Al_2O_3 or Fe_2O_3) or onto the edges of clay particles (Mott, 1970). This process is one of the basic geochemical reactions occurring in the hydrosphere, in soils, and in sediments. The interaction of these minerals with silica seems to be of geochemical importance only in aqueous environments, under which circumstances the surfaces of the sorbent are hydroxylated. Reversible protonation or deprotonation of chemisorbed water can then be expected in one of two ways depending upon pH, giving rise to either a negative or a positive surface site as follows:

$$\left[\mathrm{M}\begin{matrix}/\mathrm{OH_2^{1-\delta}}\\ \backslash\mathrm{OH_2^{1-\delta}}\end{matrix}\right]^{2-2\delta} \xleftarrow{H^+} \left[\mathrm{M}\begin{matrix}/\mathrm{OH_2^{1-\delta}}\\ \backslash\mathrm{OH^{-\delta}}\end{matrix}\right]^{1-2\delta} \xrightarrow{OH^-} \left[\mathrm{M}\begin{matrix}/\mathrm{OH^{-\delta}}\\ \backslash\mathrm{OH^{-\delta}}\end{matrix}\right]^{-2\delta} \qquad (6.XIV)$$

where M represents the cation at the edge of the crystal network, OH and OH_2 are chemisorbed surface ligands, and δ is the negative charge gained by the oxygen atom from M–OH bond polarization and may have values between 0 and 1. The presence of protonated or deprotonated ligands on the surface gives rise to a surface charge density that is equal to the sum of charges of the individual species, $\Sigma[OH^{-\delta}] - \Sigma[OH_2^{1-\delta}]$ per unit surface area (Ch. 4).

2.3.1 Surface Charge and Specific Anion Sorption

There are two ways in which an anion added to the system can be sorbed: (1) It can be electrostatically attracted if the net surface charge is positive, i.e., at a pH below the zero point of charge. The anion does not exchange with ligands in the inner Helmholtz layer. This is similar to cation exchange on constant charge surfaces and is referred to as *nonspecific adsorption*. Examples are NO_3^- and Cl^-. These anions, derived from very strong acids, tend not to form covalent bonds and do not affect the surface potential. (2) The anion can be exchanged with surface ligands and form partly covalent bonds with lattice cations. This can occur whether the surface has a net positive or negative charge, and the amount adsorbed is far greater than would be expected for nonspecifically adsorbed species, which are adsorbed according to their relative abundance. This kind of sorption is referred to as *specific sorption.* Most anions fall into the specifically sorbed category. They are generally derived from weak acids.

The extent to which an anion is specifically adsorbed onto a hydrous oxide or a clay mineral surface depends on pH, largely because of the dissociation of the conjugate acid. The maximum sorption of silicate is greatest at pH 9.2, which is close to the pK_1 of monosilicic acid. Hingston et al. (1967) measured the maximum sorption, termed by them *sorption envelopes,* for a series of other anions on geothite and found a convincing correlation between the pK of the acid and pH of maximum sorption. The isoelectric point of the oxide has no influence on the process, but the adsorption of the anion has itself shifted this point to a lower pH, showing that the surface must have become more negative through specific anion sorption. (page 217).

The sorption of silicic acid by either pure oxide or soil at constant pH and temperature proceeds by the Langmuir equation up to a maximum, which may be calculated. This implies a uniform unimolecular sorption on a surface without interactions between the sorbed molecules. However, the Langmuir plot is only applicable if the undissociated species, H_4SiO_4, and the anion, $[H_3SiO_4]^-$, are considered to be involved in the sorption mechanism. Hence it may be concluded that both species must be present in the solution for the sorption to take place.

It is possible to show mathematically why the maximum sorption occurs at a pH value close to the pK_a value. For silicic acid

$$K_a = \frac{[H_3SiO_4^-][H^+]}{[H_4SiO_4]} \tag{6.1}$$

and

$$[H_4SiO_4] + [H_3SiO_4^-] = [Si] \tag{6.2}$$

Solving Eq. (6.1) and (6.2) simultaneously, we obtain

$$[H_3SiO_4^-] = \frac{K_a}{K_a + [H^+]}[Si] \tag{6.3}$$

$$[H_4SiO_4] = \frac{[H^+]}{K_a + [H^+]}[Si]. \tag{6.4}$$

At any given pH, there will be a mixture of anions and undissociated acid, and the probability of either being sorbed is proportional to its fractional concentration in solution. The probability of both being sorbed together will be proportional to $[H_3SiO_4^-] \times [H_4SiO_4]/\{[H_3SiO_4^-] + [H_4SiO_4]\}^2$. Hence the total silica sorbed will be proportional to the square root of this expression. Substituting back for $[H_4SiO_4]$ and $[H_3SiO_4^-]$ from Eq. (6.3) and (6.4), the sorption of silica is proportional to

$\sqrt{K_a[H^+]/\{K_a + [H^+]^2\}}$. If V is the actual amount of silica sorbed at peak adsorption, then the amount of silica adsorbed at any other pH will be

$$A = 2\,V\,\sqrt{\frac{K_a[H^+]}{K_a + [H^+]^2}}\,. \tag{6.5}$$

From Eq. (6.5) A can be calculated in terms of V at any pH, or if V is known, a theoretical sorption envelope can be constructed. By differentiation it is seen that this expression has a minimum value when $K_a = [H^+]$, or when pK_a = pH.

2.3.2 Mechanism of Specific Sorption of Silicic Acid

Since the specific sorption of anions results from exchange with surface ligands followed by the formation of localized, partly covalent bonds between silicate oxygens and metal cations at the crystal network edges, it is clear that (1) the presence of proton acceptors in the system is essential, and (2) that the process must allow the surface to become more negative. Four regions of different pH values with respect to the point of zero charge of the sorbent and the pK_a of the sorbate will be considered here.

1. At a pH Below the Point of Zero Charge. At such a pH the surface groups $[-M(OH_2)_2]^{2-2\delta}$ predominate and H_4SiO_4 is almost entirely in solution. It would be expected that sorption would be very low since both species are proton rich. A ligand exchange, followed by proton loss, would not be favored at low pH, since the concentration of proton acceptors is very low, as can be shown by

$$[-M(OH_2)]^{2-2\delta} + H_4SiO_4 \rightarrow [-M(OH_2)(OSiO_3H_3) + H^+ + H_2O. \tag{6.XV}$$

There still remain some $[-M(OH)]^{-\delta}$ groups at low pH values, and the mechanism of sorption for these groups is similar to that described next.

2. At the Point of Zero Charge of the Sorbent. As the pH rises there is an increasing number of $[-M(OH)]^{-\delta}$ groups on the surface. These sites induce proton dissociation of the silicic acid, whereupon the silicate anion is able to displace a coordinated water molecule from the surface:

$$[-M(OH_2)(OH)]^{1-2\delta} + H_4SiO_4 \rightleftharpoons [-M(OH_2)_2]^{2-2\delta} + H_3SiO_4^- \rightleftharpoons$$

$$[-M(OH_2)(OSiO_3H_3)]^{1-2\delta} + H_2O. \tag{6.XVI}$$

The surface H_2O is a stronger proton donor than liquid water and will dissociate as follows

$$[-M(OH_2)\,(OSiO_3H_3)]^{1-2\delta} \rightleftharpoons [-M(OH)\,(OSiO_3H_3)]^{-2\delta} + H^+. \tag{6.XVII}$$

The dissociation of surface H_2O is encouraged by the inductive effect of the electrophilic $H_3SiO_4^-$ group bonded to M. The process renders the surface more negative than it was.

3. At pH Between the Point of Zero Charge and pH = 9.6, the pK_a *of* H_4SiO_4. There is now an abundance of neutral surface sites and also an increasing number of dissociated anions. These can react by straightforward ligand exchange with water molecules on the surface. The negative charge change is one unit per one M–OH surface group.

$$[-M(OH_2)(OH)]^{1-2\delta} + H_3SiO_4^- \rightleftharpoons [-M(OH)(OSiO_3H_3)]^{-2\delta} + H_2O. \quad (6.XVIII)$$

4. At pH above 9.6. As the pH rises still further there remain fewer and fewer proton donors, but it might be possible for the anion to replace a surface hydroxyl according to

$$[-M(OH)_2]^{-2\delta} + H_3SiO_4^- \rightleftharpoons [-M(OH)(OSiO_3H_3)]^{-2\delta} + OH^-. \quad (6.XIX)$$

This anion exchange process does not permit the surface to become more negative, and it may occur only as a secondary mechanism and to a small extent. However, at high pH when OH^- concentration becomes very high, the reaction described in (6.XIX) proceeds from right to left.

This mechanism sequence demonstrates why both proton acceptor and donor species must be present, and why the maximum sorption must occur at pH = pK_a, for here the amount of $H_3SiO_4^-$ equals the amount of un-ionized acid.

2.4 Silica in Natural Waters

Studies of dissolved silica concentrations and gradients in natural waters reveal that its concentrations are well below the equilibrium solubility for amorphous silica. This is true for soil and ground waters, rivers, lakes, and seas and also for the interstitial waters of deep-sea sediments. The concentration of silica in the surface waters of the ocean is between 0.1 and 2.0 mg SiO_2/l and rises to 5–10 mg/l in the deep layers. In the interstitial waters of sea sediments this concentration may sometimes be between 10 and 80 mg/l. In stream water and ground water the concentration of silica depends on climatic and weathering factors and is between 7 and 43 mg/l (Davis, 1964), but in a few cases rises up to 60 mg/l. These figures indicate that most of the silica in natural waters is in "true" solution of a monosilicic acid and not in a polymeric state. Near hot springs a small portion of the silica may be, at least temporarily, in the form of hydrophylic colloids because the concentration in the original hot water is high. Atmospheric precipitation may contain up to 0.1 mg dissolved SiO_2/l.

Silica in ground water is primarily dependent on the rocks and minerals contacting the water. There is almost no influence of pH, salinity, climatic region, vegetation, or temperature on silica concentrations. According to Garrels and Christ (1965), with allowance being made for the prevailing pH and the Al concentrations, the silica concentrations in ground waters in many instances lie in the range that would be expected when kaolinite is in equilibrium with the solution.

When water percolates through silicate rocks or through soils, silicic acid and some of the metallic ions are leached (Ch. 3). The extent to which an ion is leached depends on the solubility of its oxide. In general, the higher the solubility of the oxide the higher will be the rate of its leaching. At the same time new amorphous phases are formed on the surfaces of the particles that do not go into solution. These amorphous phases are slowly recrystallized into clay minerals. According to Briner and Jackson (1970) the following weathering sequence appears to be applicable to the transformation of feldspar into clay minerals:

$$\begin{array}{ccccc} \text{Feldspar} & \rightleftharpoons & \text{Amorphous material} & \rightleftharpoons & \text{1:1 clay minerals.} \\ & & \updownarrow & & \\ & & \text{2:1 clay minerals} & & \end{array}$$

The transformation of potassium feldspar into the 1:1 clay mineral kaolinite occurs by an intense washing of the feldspar and leaching of the K and SiO_2 according to the equation

$$2\,KAlSi_3O_8 + 3\,H_2O \rightarrow Al_2Si_2O_5(OH)_4 + 4\,SiO_2 + 2\,KOH. \qquad (6.XX)$$

With less intense washing of the amorphous phase, and if the percolating water is rich in potassium, only a small fraction of this ion will be leached together with silica and the resulting amorphous phase will recrystallize into the 2:1 clay mineral illite according to the equation

$$3\,KAlSi_3O_8 + 2\,H_2O \rightarrow KAl_2(Al, Si_3)O_{10}(OH)_2 + 6\,SiO_2 + 2\,KOH. \qquad (6.XXI)$$

Further washing of the sediment results in further leaching of the silica and weathering of the illite and the formation of kaolinite. The leaching of silica from kaolinite leaves gibbsite as a residue.

The weathering sequence developed on basalts, as occurs in parts of the New England area of New South Wales, demonstrates the importance of leaching of silica from rock material as a source for silica in ground water (Fig. 6.7) (Loughnan, 1969). Montmorillonite is the initial weathering product but, as leaching increases, this mineral is rendered unstable through the loss of soluble constituents and is replaced by kaolinite. Ultimately, destruction of the kaolinite through desilicification results in a residue of bauxite minerals. The cation exchange capacity of the soil rises to a maximum when montmorillonite is the dominant mineral but, with the increasing development of kaolinite and bauxite, it decreases gradually toward zero. The pH of ground water at the beginning of the weathering process is slightly alkaline through leaching of the bases. It becomes progressively more acid because of the higher fraction of silicic acid in the soluble solids. The acidity reaches a maximum at the point of conversion of montmorillonite to kaolinite. From then on the leaching of silicic acid is reduced causing the return of the pH toward neutrality.

The diagenesis of buried smectites during compaction and conversion of argillaceous sediment to lithified shale results in the leaching of silica by percolating water and in the conversion of minerals rich in silica into minerals poorer in silica. The conversion of smectite into illite is associated with an increase of the Al/Si ratio in the tetrahedral unit of the 2:1 clay mineral together with the sorption of K ions. Al and K supplied by feldspar are trapped by the clay fraction while silica is leached by percolating water (see Ch. 10).

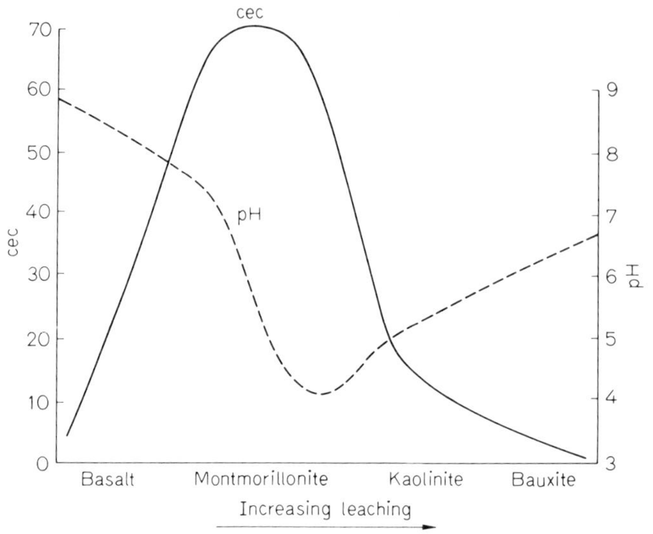

Fig. 6.7. Changes in cation exchange capacity and pH with increased silica leaching from basalt from New England area of New South Wales. (From Loughnan, 1969: Used with permission of American Elsevier Publishing Company, Inc.)

Compared with other major dissolved constituents, the concentration of silica in the large rivers is generally within certain limits. The mean silica contents for various river waters from various parts of the world are given in Table 6.5 (Livingstone, 1963). Many major constituents in river water show an inverse relationship between concentration and discharge. This is normally thought to be caused by dilution of baseflow by storm run-off. In the case of silicon the concentration in solution is insensitive to change in discharge compared with other major components. Toward the mouths of the large rivers, silica tends to be remarkably uniform on a world-wide basis. According to Edwards and Liss (1973) dissolution and precipitation of silicate minerals, although the ultimate source of most silica in solution, is unlikely to be sufficiently rapid to maintain buffering of dissolved silica in river water. Also the biologic mechanism does not seem adequate to maintain near constant concentration through the rapidly changing discharge of a storm hydrograph or to explain the small global interriver

Table 6.5. Mean SiO_2 content of river waters of the world in mg/liter. (After Livingstone, 1963)

North America	9.0 mg SiO_2/l
South America	11.9 mg SiO_2/l
Europe	7.5 mg SiO_2/l
Asia	11.7 mg SiO_2/l
Africa	23.2 mg SiO_2/l
Australia	3.9 mg SiO_2/l
World	13.1 mg SiO_2/l

variability. These investigators believe that the specific adsorption of silicic acid by soil particles that are suspended in rivers, and the reverse desorption reaction are responsible for buffering of dissolved silica. A similar sorption process may also occur in lakes. The silicon cycle in some lakes seems to be intimately linked to the formation and dissolution of solid phase iron and manganese oxides, in response to changes in redox potential. Reactions of this type may explain the large-scale removal of dissolved silica, above that accounted for by biologic processes, found in Lake Biwa, Japan (Kato, 1969).

The concentration of soluble silica in river water is generally 10 to 15 times that of surface water in the open sea. In addition to dissolved silica reaching the oceans with the river waters, significant fluxes of dissolved silica come out of the sea floor. Schink (1968) estimated that the annual flux was about 10% of the dissolved silica in the mixed layer of the ocean. Fanning and Pilson (1974) calculated a flux equal to about 14% of the riverine input. Dissolution of silica from solid particles within the sediment provides much of the silica that diffuses out. Deposited biogenic amorphous silica is probably the main source of this flux. Silica is also provided by anion and ligand exchange between seawater and exposed rocks on the sea floor and by sea weathering of suspended river and air-borne particulate material entering the ocean. This influx seems to be small in comparison with the inflow in river water (Burton and Liss, 1973).

The low concentrations of silica in seawaters indicate that the removal of dissolved silica from seawater is fast. It occurs either by biologic uptake due to the activity of organisms such as diatoms and radiolaria that use silica for shell material, or by inorganic precipitation and sorption.

Bien et al. (1958) provided evidence for an almost complete silicon removal through nonbiogenous means at the mouth of the Mississippi River. Nonbiogenous removal increases with riverine silica concentrations, suspended solid load, and salinity. It seems that nonbiogenous removal is predominantly the result of sorption of silicic acid onto the surfaces of clay minerals and amorphous hydroxides of Fe, Al, and Mn. According to Burton and Liss (1973) such a mechanism is important for the removal of silica in some estuaries where changes in pH lead to the formation of hydrous iron oxide phases, which are active sorbents for silicic acid. Dissolved silica in the estuary is in an environment where the ionic strength changes rapidly, and turbulent mixing produces a higher concentration of suspended sediment than is found in the sea or in many rivers. Moreover, the solid minerals, subjected to the turbulent mixing, possess very active surfaces. Siever (1968) observed that ground kaolinite, montmorillonite, and illite tend to increase their reactivity toward aqueous silica. The effects of mechanical disintegration of minerals on the surface activity were discussed in Chapter 4.

According to Heath and Dymond (1973) the solubility of silica in pore waters of deep-sea sediments containing opaline silica depends on whether the deposits are oxidized or reduced. Pore waters of oxidized deposits contain between 22 and 35 mg SiO_2/l whereas reduced sediments contain only between 10 and 20 mg SiO_2/l. Under oxidizing conditions the original organic coatings of the frustrules and the tests are partly oxidized and removed, making the silica more susceptible to dissolution.

The co-precipitation of multivalent metallic cations with silica has some effect on the concentration of these elements in natural waters. For example, Gac et al. (1977)

observed that during the evaporation of water in Lake Chad, magnesium silicate precipitates as an amorphous material, giving rise to Mg depletion of the water. The amorphous material is aged to form smectite minerals.

2.4.1 Solubility of Silica in Soil Solutions

Water traveling through the soil profile may contain appreciable silica leached from the soil. Surface run-off from rain can acquire 1–3 mg SiO_2/l during the first few minutes after rain contacts the soil. Silica concentration in soil waters is controlled by the following four factors: (1) rate of dissolution of unstable silicates (Mainly amorphous silica), (2) rate of precipitation of stable silicates, (3) rate of movement of silica-bearing solutions out of the system, and (4) rate of plant uptake (Kittrick, 1969). Soil-leaching studies indicate that only a few hours to a few days are needed for the silica concentration to achieve a constant value in water recycled through a soil column. The same silica concentration is attained starting with water containing either more or less silica than that at "equilibrium". However, some silica is released from soil rather rapidly. McKeague and Cline (1963a, b) showed that in saturated soil–water mixtures, the silica in solution after 10 days was only twice that after the first 5 min. After the first day the rate of silica dissolution became very low. They also showed that pH has a marked effect on silica concentrations in soil solutions. They concluded that a pH-dependent reaction involving the adsorption of dissolved silica on soil particle surface plays a major role in controlling the solubility of silica in soil solutions.

Beckwith and Reeve (1963) reported that after shaking soils with solutions containing silicic acid, the concentration of the acid was controlled by a pH-dependent sorption. A sorption maximum occurs at about pH 9.5 with decreasing sorption at either higher or lower pH values (see pages 273–276).

Jones and Handreck (1963) established that silica in soil solutions is entirely monosilicic acid and is commonly present in the range of 0.5–0.7 mmol/l. Elgawhary and Lindsay (1972) studied the solubility of silica in various soils. The silica dissolved from the soil samples (45–60 mg SiO_2/l) was only half of that expected from the solubility of amorphous silica. On addition of excess amorphous silica the concentration of the dissolved fraction increased and reached a level that is usually obtained with amorphous silica. In a separate experiment the investigators added a solution of sodium silicate to the soil system. The concentration of silicon in the aqueous solution reached equilibrium when the silica level in the solution was the same as that found for the untreated soil in the absence of amorphous silica or sodium silicate. Equilibrium was achieved within 30–50 days. They concluded that the solid phase silica in the soils maintained the quantity of dissolved Si below the solubility product of amorphous silica but above that of quartz. They suggested that silica in soil is an unstable intermediate that accumulates because it precipitates more readily than quartz. Eventually the amorphous form slowly converts to more stable forms. Rather than a discrete solid phase of amorphous silica per se in soils, there may be present silicate matrices that range from freshly precipitated amorphous silica to crystalline quartz.

According to Yariv (1974) the low solubility of amorphous silica in soils is not surprising, as neither sorption of silicic acid nor partial crystallization of silica controls this concentration. The amorphous silica in soils sorbs Al, Fe, and organic ions. The sorption of polyvalent cations and organic materials by amorphous silica greatly reduces the solubility of the solid phase and controls the concentration of silica in the soil solutions (page 270).

2.4.2 Silica in Stream Water

The concentration of most dissolved constituents in stream water decreases with increasing discharge, but the silica content is less variable than that of any of the other major dissolved constituents. This means that the proportion of silica in the dissolved solids becomes greater with increasing discharge and implies that the rate of silica release from soils increases more rapidly than that of the other dissolved solids during storm run-off. Silica concentration in subsurface run-off waters is related to the length of time of contact with the soil, the soil–water ratio, and the rainfall history prior to the time of sampling. Storm run-off appears to acquire most of its silica within a few days. Silica concentrations in stream water remain relatively constant during periods of high discharge, despite the decrease in dissolved solids.

Kennedy (1971) studied silica variation in the Mattole River of Northern California with time and discharge. His model is useful for describing the behavior of silica in many other streams. Suspended sediment is an unlikely source for most of the silica found in the Mattole during high flow, and the only other possible source is the soil itself. Silica release from the soil must occur rapidly because the time for the rain to fall, pick up silica, and travel an average of 20 miles to the sampling site may take 5–6 h. He concluded that the concentration of the dissolved silica varies in a consistent manner during the storm run-off as water from various sources enters the stream. Silica is more concentrated in water that seeps through the soil (subsurface flow) than in overland flow or in ground water.

When rain begins after a dry period, the water easily enters the pore spaces in the soil. This water dissolves readily soluble materials. The electrolytes are released more rapidly than the silica on the first contact of dry soil with water. At this stage the ratio between dissolved SiO_2 and electrolytes is as low as, or lower than that present in the stream. As rain continues, the initial rainfall will be carried downward into the soil along with the readily soluble salts. If rainfall increases in intensity or if the soils have low permeability, the pores of the surface soil become saturated, and water flows over the land surface to produce *overland flow.* A part of this overland flow infiltrates the soil further downslope, and a part continues on to join rivulets and becomes part of a larger stream. At this stage an appreciable amount of both silica and electrolytes is derived from surface soils as water passes over or through them. Since the highly soluble salts have already been leached from surface soil in the first stages, the ratio between silica and electrolytes increases with time. Nevertheless, the absolute concentrations of both silica and electrolytes decrease.

Some of the water that has entered the soil moves laterally down the gradient through the upper soil horizons and emerges to join the overland flow. This *subsurface run-off* transports some of the readily soluble soil materials initially and, as time of contact increases, some of the less readily soluble material will also go into solution and be carried along. With continuing rainfall, an increasing amount of subsurface run-off will join the overland flow to form the *storm run-off.* Subsurface run-off that has moved laterally through only a few inches of soil may rejoin surface flow within one hour or less, and cause the initial increase in silica concentration observed during a stream rise. An increase in silica concentration while discharge is still rising, is attributed to an increasing proportion of subsurface run-off in the stream flow. Water in subsurface flow has a greater contact time with soil than water in surface flow, resulting in a higher concentration of silica. In many streams, the proportion of ground water in stream flow is very small during periods of storm run-off, so that the stream chemistry is largely controlled at such times by chemical reactions at the land surface.

As rainfall decreases, overland flow also decreases, but subsurface run-off continues for several days after the storm, the water gradually draining from the soil. At this stage silica concentration continues to increase to a leveling-off point, while concentrations of dissolved electrolytes still continue to decrease. The leveling off of the silica concentration indicates the end of overland flow. A "steady-state" concentration of dissolved silica in soil water is reached shortly after the storm. With decreasing discharge, the proportion of ground water in the stream flow increases and silica concentration slowly decreases.

Suspended-sediment concentrations increase sharply with the discharge rate and reach a maximum at or, commonly, shortly before peak discharge. Peak sediment concentrations consistently occur at times of minima in silica concentrations, since both minimum silica and maximum sediment concentrations coincide with maximum contribution of overland run-off to stream flow, for at that time maximum dilution of silica-rich subsurface flow by direct run-off would occur and maximum erosive capability of direct run-off would exist. Also, under these conditions, the sorption of silicic acid by the suspended sediments removes some of the silica.

References

Alexander, A. E., Johnson, P.: Colloid science. Oxford: Clarendon Press, 1949, Vol. 2

Alexander, G. B., Heston, W. M., Iler, R. K.: The solubility of amorphous silica in water. J. Phys. Chem. *58*, 453–455 (1954)

Bach, R., Sticher, H.: Abbau von Quarz mit Brenzcatechin. Experientia *22*, 515–516 (1966)

Barnum, D. W., Kelley, J. M., Poocharoen, B.: Magnesium and guadinium salts of a silicon-catechol coordination complex. Inorg. Chem. *12*, 497–498 (1973)

Bartels, H.: Stability of Si complexes. Helv. Chim. Acta *47*, 1605–1609 (1964)

Bartoli, F., Beaucire, F.: Accumulation du silicium dans les plantes vivantes en melieux pédogénétiques tempérés aérés. C. R. Acad. Sci. Ser. D, *282*, 1947–1950 (1976)

Bartoli, F., Selmi, M.: Sur l'evolution du silicium végétal en milieu pédogénétique aérés acides. C. R. Acad. Sci. Ser. D, *284*, 279–282 (1977)

Beckwith, R. S., Reeve, R.: Studies on soluble silica in soils. I. The sorption of silicic acid by soils and minerals. Aust. J. Soil Res. *1*, 157–168 (1963)

Bien, G. S., Contois, D. E., Thomas, W. H.: The removal of soluble silica from fresh water entering the sea. Geochim. Cosmochim. Acta *14*, 35–54 (1958)

Bijsterbosch, B. H.: Characterization of silica surfaces by adsorption from solution. Investigation into the mechanism of adsorption of cationic surfactants. J. Colloid. Interface Sci. *47*, 186–198 (1974)

Boehm, H. P., Schneider, M.: Über die Bindung von Aluminum aus Aluminochlorid-Lösungen an Siliciumdioxyd-Oberflächen. Z. Anorg. Allg. Chem. *316*, 128–133 (1962)

Briner, P., Jackson, M. L.: Mineralogical analysis of clays in soils developed from basalts in Australia. Isr. J. Chem. *8*, 487–500 (1970)

Burton, J. D., Liss, P. S.: Processes of supply and removal of dissolved silicon in the oceans. Geochim. Cosmochim. Acta *37*, 1761–1773 (1973)

Campbell, A. S., Fyfe, W. S.: Hydroxyl ion catalysis of the hydrothermal crystallization of amorphous silica. Am. Mineral. *45*, 464–468 (1960)

Clark-Monks, C., Ellis, B.: The characterization of anomalous adsorption sites on silica surfaces. J. Colloid Interface Sci. *44*, 37–49 (1973)
Cloos, P., Leonard, A. J., Moreau, J. P., Herbillon, A., Fripiat, J. J.: Structural organization in amorphous silicoaluminas. Clays Clay Miner. *17*, 270–287 (1969)
Coudurier, M., Baudru, B., Donnet, J. B.: Etude de la polycondensation de l'acide disilicique. II. Influence du pH sur la cinétique et le mécanism de la réaction des polycondensations de disilicique. Texture des produits formés. Bull. Soc. Chim. Fr. 3154–3160 (1971)
Dalton, R. W., McClanaham, J. L., Maatman, R. W.: The partial exclusion-electrolytes from the pores of silica gel. J. Colloid. Sci. *17*, 207–219 (1962)
Darragh, P. J., Gaskin, A. J., Sanders, J. V.: Opals. Sci. Am. *234*, (4), 84–95 (1976)
Davis, S. N.: Silica in streams and ground water. Am. J. Sci. *262*, 870–891 (1964)
DeKimpe, C., Gastuche, M. C., Brindley, G. W.: Ionic coordination in aluminosilicic gels in relation to clay mineral formation. Am. Mineral. *46*, 1370–1381 (1961)
Donnelly, T. W., Merrill, L.: The scavenging of magnesium and other chemical species by biogenic opal in deep sea sediments. Chem. Geol. *19*, 167–186 (1977)
Eckstein, Y., Yaalon, D. H., Yariv, S.: The effect of lithium on the cation exchange behavior of crystalline and amorphous clays. Isr. J. Chem. *8*, 335–342 (1970)
Edwards, A. M. C., Liss, P. S.: Evidence for buffering of dissolved silicon in fresh waters. Nature (London) *243*, 341–342 (1973)
Elgawhary, S. M., Lindsay, W. L.: Solubility of silica in soils. Soil Sci. Soc. Am. Proc. *36*, 439–442 (1972)
Ernst, W. G., Calvert, S. E.: An experimental study of the recrystallization of porcelanite and its bearing on the origin of some bedded cherts. Am. J. Sci. *267-A*, 114–133 (1969)
Fanning, K. A., Pilson, M. E. Q.: The diffusion of dissolved silica out of deep sea sediments. J. Geophys. Res. *79*, 1293–1297 (1974)
Fieldes, M., Schofield, R. K.: Mechanisms of ion adsorption by inorganic soil colloids. N. Z. J. Sci. *3*, 563–579 (1960)
Flörke, O. W.: Transport and deposition of SiO_2 with H_2O under supercritical conditions. Krist. Tech. *7*, 159–166 (1972)
Fripiat, J. J.: Surface chemistry and soil science. In: Experimental pedology. Hallsworthy, E. G., Crawford, D. V. (eds.). London: Butterworths, 1965, pp. 3–13
Fyfe, W. S., McKay, D. S.: Hydroxyl ion catalysis of the crystallization of amorphous silica at 330 °C. Am. Mineral. *47*, 83–89 (1962)
Gac, J., Droubi, A., Fritz, B., Tardy, Y.: Geochemical behavior of silica and magnesium during the evaporation of waters in Chad. Chem. Geol. *19*, 215–228 (1977)
Garrels, R. M., Christ, C. E.: Solutions, minerals and equilibria. New York: Harper and Row, 1965
Garrels, R. M., Mackenzie, F. T.: A quantitative model for the sedimentary rock cycle. Mar. Chem. *1*, 27–41 (1972)
Hair, M. L.: Infrared spectroscopy in surface chemistry. New York: Marcel Dekker, Inc. 1967
Harder, H., Flehming, W.: Quarz Synthese bei tiefen Temperaturen. Geochim. Cosmochim. Acta *34*, 295–305 (1970)
Heath, G. R., Dymond, J.: Interstitial silica in deep-sea sediments from the North Pacific. Geology *1*, 181–184 (1973)
Henderson, J. H., Syers, J. K., Jackson, M. L.: Quartz dissolution as influenced by pH and the presence of a disturbed surface layer. Isr. J. Chem. *8*, 357–372 (1970)
Henmi, T., Wada, K.: Morphology and composition of allophane. Am. Mineral. *61*, 379–390 (1976)
Hingston, F. J., Atkinson, R. J., Posner, A. M., Quirk, J. P.: Specific sorption of anions. Nature (London) *215*, 1459–1461 (1967)
Huang, W. H., Vogler, D. L.: Dissolution of opal in water and its water content. Nature (London) Phys. Sci. *235*, 157–158 (1972)
Hurd, D.: Interactions of biogenic opal, sediment and seawater in the central Equatorial Pacific. Prepared for the Office of Naval Research of the United States Government, Hawaii Institute of Geophysics, University of Hawaii, 1972a

Hurd, D.: Factors affecting solution rate of biogenic opal in seawater. Earth Planet. Sci. Lett. *15*, 411–417 (1972b)

Iler, R. K.: Coagulation of colloidal silica by calcium ions, mechanism and effect of particle size. J. Colloid Interface Sci. *53*, 476–488 (1975)

Jenne, E. A.: Surface charge density dependency on Al_2O_3 content in imogolite. Clays Clay Miner. *20*, 101–103 (1972)

Jones, J. B., Segnit, R. E.: The occurrence and formation of opal at Coober Pedy and Andamooka. Aust. J. Sci. *29*, 129–133 (1966)

Jones, J. B., Segnit, R. E.: The nature of opal, Part I. J. Geol. Soc. Aust. *18*, 419–422 (1971)

Jones, L. H., Handreck, K. A.: Effects of iron and aluminum oxides on silica in solution in soils. Nature (London) *198*, 852–853 (1963)

Jones, M. M., Pytkowicz, R. M.: The solubility of silica in seawater at high pressure. Bull. Soc. R. Sci. Liege *42* C, 118–120 (1973)

Kamiya, H., Ozaki, A., Imahashi, M.: Dissolution rate of powdered quartz in acid solution. Geochem. J. *8*, 21–26 (1974)

Kastner, M., Keene, J. B., Gieskes, J. M.: Diagenesis of siliceous oozes–I. Chemical controls on the rate of opal-A to opal-CT transformation–an experimental study. Geochim. Cosmochim. Acta, *41*, 1041–1059 (1977)

Kato, K.: Behavior of dissolved silica in connection with oxidation-reduction cycle in lake water. Geochem. J. *3*, 87–97 (1969)

Keene, J. B.: Distribution, mineralogy and petrography of biogenic and authigenic silica in the Pacific Basin. Ph. D. Thesis, University of California, San Diego, 1976

Kennedy, V. C.: Silica variation in stream water with time and discharge. In: nonequilibrium systems in natural water chemistry. Adv. Chem. Ser. *106*, Am. Chem. Soc., Washington, D.C.: 94–130 (1971)

Kisselev, A. V., Lygin, V. I.: Infrared spectroscopy of solid surface and adsorbed molecules. Surf. Sci. *2*, 236–246 (1964)

Kittrick, J. A.: Soil minerals in the Al_2O_3-SiO_2-H_2O system and a theory of their formation. Clays Clay Miner. *17*, 157–167 (1969)

Klier, K., Zettlemoyer, A. C.: Water at interface: Molecular structure and dynamics. J. Colloid Interface Sci. *58*, 216–229 (1977)

Krauskopf, K. B.: Dissolution and precipitation of silica at low temperatures. Geochim. Cosmochim. Acta *10*, 1–26 (1956)

Kutuzova, R. S.: Silica transformation during the mineralization of plant residues, Soviet Soil Sci. (English trans.) *7*, 970–978 (1968)

Lancelot, Y.: Chert and silica diagenesis in sediments from the Central Pacific. In: Inital reports on the deep sea drilling project. Winterer, E. L., Ewing, J. I. (eds.). U.S. Govern. Printing Office, 1973, Vol. *17*, pp. 377–405

Langer, K., Flörke, O. W.: Near infrared absorption spectra of opals and the role of water in these minerals. Fortschr. Mineral. *52*, 17–51 (1974)

Liefländer, M., Stöber, W.: Topochemische Reaktionen von Quarz, Amorphen Siliciumdioxyd und Titandioxyd mit Aluminiumtriisobutyl. Z. Naturforsch. *15B*, 411–413 (1960)

Little, L. H.: Infrared spectra of adsorbed species. London: Academic Press 1966

Livingstone, D. A.: Chemical composition of rivers and lakes. Geol. Survey Prof. Paper 440-G, Washington, D. C.: U. S. Govern. Printing Office 1963

Loughnan, F. C.: Chemical weathering of silicate minerals. New York: American Elsevier Publ. Co. 1969

Mackenzie, F. T. Gees, R.: Quartz synthesis at earth surface conditions. Science *173*, 533–534 (1971)

Malati, M. A., Mazza, R. J., Sheren, A. J., Tomkins, D. R.: The mechanism of adsorption of alkali metal ions on silica. Powder Technol. *9*, 107–110 (1974)

McKeague, J. A., Cline, M. G.: Silica in soil solution. I. The forms and concentration of dissolved silica in aqueous extracts of some soils. Can. J. Soil Sci. *43*, 70–82 (1963a)

McKeague, J. A., Cline, M. G.: Silica in soil solution. II. The adsorption of monosilicic acid by soil and by other substances. Can. J. Soil. Sci. *43*, 83–96 (1963b)

McKyes, E., Sethi, A., Young, R. N.: Amorphous coatings on particles of sensitive clay soils. Clays Clay Miner. *22,* 427–433 (1974)
Mizutami, S.: Transformation of silica under hydrothermal conditions. J. Earth Sci., Nagoya Univ. *14,* 56–89 (1966)
Morariu, V. V., Mills, R.: A study of water adsorbed on silica. The effect of surface nature, temperature and coverage on the spin-spin relaxation time. II. Z. Phys. Chem. Neue Folge *83,* 41–53 (1973)
Mott, C. J. B.: Sorption of anions by soils. S. C. I. Monograph No. *37,* Sorption and transport processes in soils. 1970. pp. 40–53
Murata, K. J., Larson, R. R.: Diagenesis of Miocene siliceous shales, Temblor Range, California. J. Res. U. S. Geol. Surv. *3,* 553–566 (1975)
Oehler, J. H.: Hydrothermal crystallization of silica gel. Geol. Soc. Am. Bull. *87,* 1143–1152 (1976)
Oehler, J. H., Schopf, J. W.: Artificial microfossils: Experimental studies of permineralization of blue-green algae in silica. Science *174,* 1229–1231 (1971)
Okamoto, G., Okura, T., Goto, K.: Properties of silica in water. Geochim. Cosmochim. Acta *12,* 123–132 (1957)
Paces, T.: Steady-state kinetics and equilibrium between ground water and granitic rock. Geochim. Cosmochim. Acta *37,* 2641–2663 (1973)
Russell, J. D., McHardy, W. J., Fraser, A. R.: Imogolite, a unique aluminosilicate. Clay Mineral. *8,* 87–99 (1973)
Rouxhet, P. G., Sempels, R. E.: Hydrogen bond strengths and acidities of hydroxyl groups on silica-aluminum surfaces and in molecules in solution. J. Chem. Soc. Faraday Trans. I, *70,* 2021–2032 (1974)
Schink, D. R.: Observations relating to the flux of silica across the sea floor interface (abstract). Eos. Trans. AGU *49,* 335 (1968) (Quat., Fanning and Pilson, 1974)
Schofield, P. J., Ralph, B. J., Green, J. H.: Mechanism of hydroxylation of aromatics on silica surfaces. J. Phys. Chem. *68,* 472–476 (1964)
Sempels, R. E., Rouxhet, P. G.: H Bonding properties of hydroxyl held on the surface of silica. Bull. Soc. Chim. Belg. *84,* 361–370 (1975)
Shell, H. R.: Silicon. In: Treatise on analytical chemistry. Kolthoff, I. M., Elving, P. J. (eds.) New York: Interscience Pub. 1962, Part II, Vol. 2, pp. 107–206
Siever, R.: Establishment of equilibrium between clays and seawater. Earth Planet. Sci. Lett. *5,* 105–110 (1968)
Snoeyink, V. L., Weber, W. J.: Surface funtional groups on carbon and silica. Prog. Surf. Membr. Sci. *5,* 63–119 (1972)
Stöber, W.: Formation of silicic acid in aqueous suspensions of different silica modification. In: Equilibrium concepts in natural water systems. Adv. Chem. Ser. *67,* 161–182 (1967)
Takamura, T., Yoshida, H., Inazuka, K.: Infrared characteristic bands of highly dispersed silica. Kolloid Z. Z. Polym. *195,* 12–16 (1964)
Wada, K.: A structural scheme of soil allophane. Am. Mineral. *52,* 690–708 (1967)
Wada, K., Harada, Y.: Effects of salt concentration and cation species on the measurements of cation exchange capacities of soils and clays. Proc. Int. Clay Conf., Tokyo *1,* Heller, L. (ed.). Jerusalem: Israel Univ. Press, 1969, pp. 561–571
Wada, K., Inoue, T.: Retention of humic substances derived from rotted clover leaves in soils containing montmorillonite and allophane. Soil Sci. Plant Nutr. *13,* 9–16 (1967)
West, P. B., Haller, G. L., Burwell, R. J., Jr.: The catalytic activity of silica gel. J. Catalysis *29,* 486–493 (1973)
Wey, R. Siffert, B.: Reaction de la silice monomoleculaire en solution avec les ions Al, Mg. Genèse et synthèse des argiles. Colloq. Int. C. N. R. S. *105,* 11–23 (1961)
Wise, S. W., Jr., Weaver, F. M.: Chertification of oceanic sediments. In: Pelagic sediments on land and under sea. Spec. Pub. Int. Ass. Sediment *1,* 301–326 (1974)
Yariv, S.: Reactions of some clays with ferric-ferricyanide (Prussian brown), Isr. J. Chem. *7,* 453–461 (1969)
Yariv, S.: Solubility of silica in soils. Soil Sci. Soc. Am. Proc. *38,* 693 (1974)

Chapter 7
Colloid Geochemistry of Clay Minerals

In considering surface charge density of clay minerals on a microscale, it can be seen that the distribution of the electric charge is not homogeneous. Although electroneutrality in the bulk is maintained, these minerals have localized point-sites of high and low electric charges, either positive or negative. On a macroscale, the silicate unit layers are almost always negatively charged and the regions where the metallic exchangeable cations are situated are positively charged. The positive regions are located at the liquid–solid or gas–solid interfaces of all natural clay minerals and also at the interlayer spaces of smectites, vermiculites, and chlorites.

The negative surface charge of the silicate layer may be accounted for by: (1) Isomorphous substitution, that is, the presence of "foreign" ions in the crystal structure that have a valence lower than those of the ions proper to the lattice; (2) broken bonds at the edges of the 1:1 and 2:1 units, on noncleavage surfaces; (3) proton dissociation–association of exposed OH groups and (4) specific sorption of anions such as silicate and aluminate. The negative charge originating from (1) is the "permanent charge" and does not depend on the pH of the environment, whereas the charges derived from (2), (3), and (4) are variable and are pH dependent. The contribution of (2) and (3) to the net negative charge of the unit increases with increasing pH value of the environment whereas the contribution of (4) increases only up to a certain pH, at which the specific sorption of the anion is maximal and then decreases. In most naturally occurring cases the contribution of the specifically sorbed anion to the total net negative charge is only of minor importance.

1. Functional Groups on Clay Minerals and Ion Exchange Reactions

1.1 "Broken-bond" Surfaces

When a 1:1 or a 2:1 type unit layer is disrupted, exposing a surface not parallel to the "flat" unit layer, the valences of crystal atoms that are exposed at the surface are not completely compensated, as they are in the interior of the crystal. These surfaces are called "broken-bond" surfaces or "edge" surfaces. The unsatisfied charges may be balanced by the sorption of cations and anions, either specifically by chemisorption or nonspecifically by electrostatic attraction, the latter ions being easily exchangeable.

The surface properties depend greatly on the exposed atoms. That part of the surface at which the octahedral sheet is broken may be compared to the surface of an alumina or magnesia particle, whereas that part of the surface at which the tetrahedral sheet is broken may be compared to that of silica or silica-alumina. The surface charge is determined by sorbed "potential-determining ions".

Protons are important potential-determining ions. In acid solution the net charge of the surface is positive and the positive charge further increases with decreasing pH. Protons occupy positions near oxygens. Depending on the basic strength of the oxygen, the $-Mg-O^-$ functional group is the first site to be protonated, followed by octahedral$-Al-O^-$, tetrahedral$-Al-O^-$ and $-Si-O^-$ groups, respectively. The $-Mg-OH$ and $-Al-OH$ functional groups are further protonated only at a very low pH.

The broken-bond surface has a high affinity for Al ions. These ions are preferentially sorbed on the broken bond surface even from very dilute solutions, contributing a net positive charge to this surface. These ions can be leached by an acid, and thereby the clay surface becomes less positive (Buchanan and Oppenheim, 1968).

The anions OH^- and $Al(OH)_4^-$ are also important potential-determining ions. They are responsible for the broken bond surface becoming negatively charged in an alkaline solution. The reactions occurring in alkaline and acid solutions are summarized in Scheme 7.1. Practically, it is difficult to estimate the point of zero charge of the broken-bond surfaces of clay minerals for the following two reasons:

1. The point of zero charge changes as a result of the high specific sorption of cations and anions. When the surface becomes more negative through specific anion adsorption the point of zero charge shifts to higher pH values whereas specific sorption of cations shifts this point to lower pH values (page 217).

2. The contribution of the broken-bond surface to the total net charge of most clay minerals is only minor in comparison with that of the large flat surface, the latter being always negative.

A number of studies indicate that the edges of clay particles are positively charged at pH below 7 or 8 although some data suggest that the edges are already neutralized at pH 6 (Swartzen-Allen and Matijević, 1974). Kaolinite shows a point of zero charge at a pH of about 5 (Ferris and Jepson, 1975).

Broken-bond surface occupies an appreciable proportion (10–20%) of the total crystal area of kaolinite [$(15–40) \times 10^3$ m^2/kg]. On the other hand in expanding-type smectite minerals, less than 10% and more often only 2–3% of the total (external and interlayer) area of $700–800 \times 10^3$ m^2/kg is occupied by the crystal edge. Hence the influence of the crystal edges on the pH dependent-charge and sorption of anions is much more in evidence with kaolinite than with expanding-type minerals (Grim, 1968).

At a pH below the point of zero charge, nonspecific (Coulombic) adsorption of anions can occur, whereas at higher pH values nonspecific adsorption of cations can occur. In kaolin-serpentine group minerals, broken bonds are the major cause of exchange capacity. Their contribution to the total exchange capacity of illites and chlorites is smaller. This contribution is even smaller in the case of the expanding smectites and vermiculites, where the broken bonds are responsible for only 20% of the total cation exchange capacity. The range of the total cation exchange capacity of the common clay minerals is given in Table 7.1.

1.1.1 Sorption of Cations

It should be expected that the broken-bond surface exchange capacity will depend on the surface area. Ormsby et al. (1962) examined this suggestion by studying the exchange capacity of kaolinites. They obtained a linear relation between surface areas and

Scheme 7.1. A schematic representation of the interaction of the broken-bond surface with some potential-determining ions

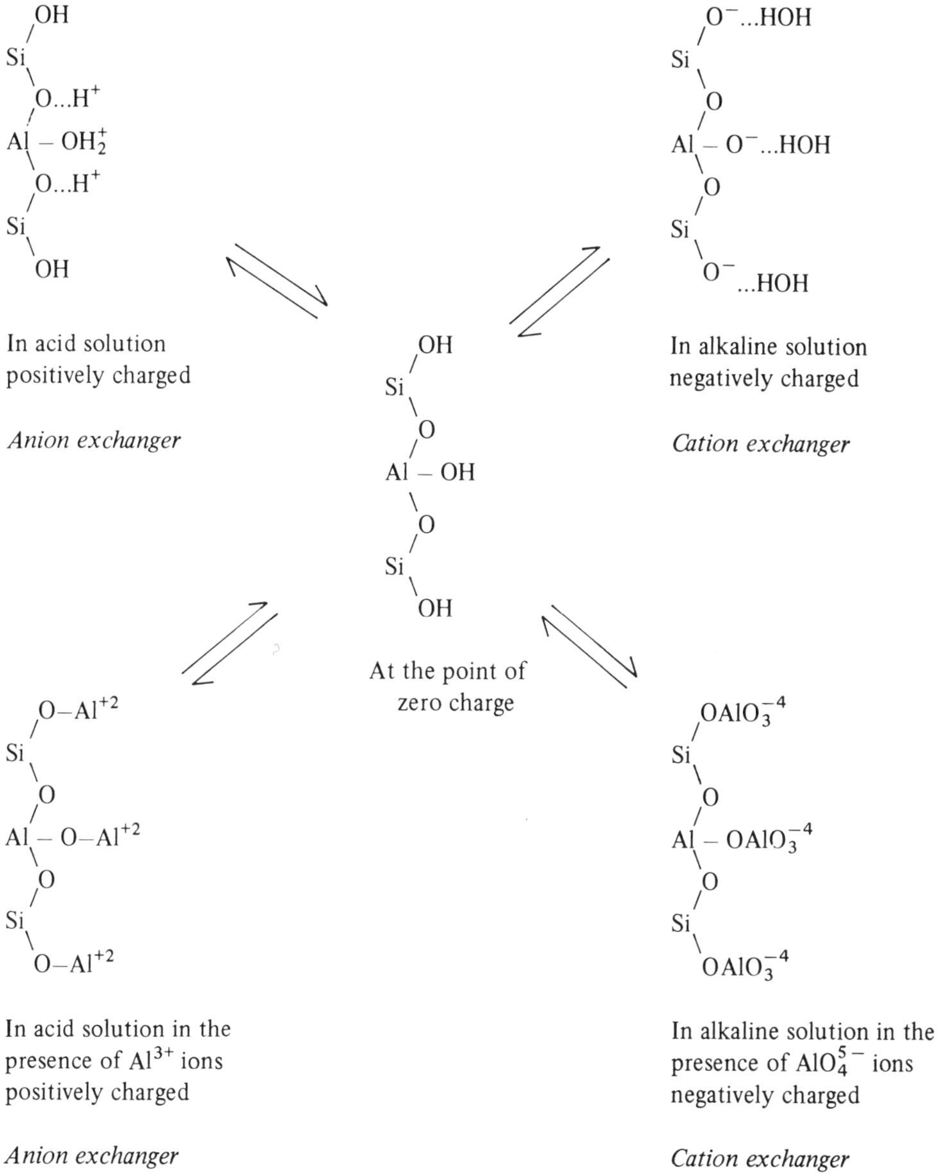

Table 7.1. Cation exchange capacity of common clay minerals in meq/100 g. (After Grim, 1968)

Mineral	Exchange capacity
Kaolinite	3–15
Halloysite · 2 HOH	5–10
Halloysite · 4 HOH	40–50
Smectite	60–120
Illite	10–40
Vermiculite	100–160
Chlorite	10–40

exchange capacities when areas were increased by decreasing the particle size or by changing from well-crystallized to poorly crystallized kaolinites.

Broken-bond surfaces show a strong affinity for Li. In addition to exchange reactions occurring with the exchangeable metallic cations, lithium replaces protons in some of the hydroxyls on the crystal edges as follows

$$\rangle Al{-}OH + Li^+ \rightleftharpoons \rangle Al{-}OLi + H^+. \qquad (7.I)$$

The adsorbed Li is easily replaced when Li concentration in the external solution is reduced during leaching with solutions of polyvalent cations. The adsorption increases with both pH and concentration of Li, in agreement with (7.I) (Eckstein et al., 1970).

The adsorption of various chlorides by kaolinite was studied by Ferris and Jepson (1975). Adsorption of Cl^- falls from about 1.0 meq/100 g at pH 1 to zero at about pH 8. Chloride adsorption at pH 6.8 is about 0.1 meq/100 g. Adsorption of Na^+ increases from zero at pH 2 to over 3 meq/100 g at pH 12. Experiments with different chloride solutions and at a common pH of about 7.0 show that the cation adsorption decreases in the order $Ca^{+2} > Cs^+ > Na^+ > Li^+$.

It is difficult to determine the structure of ionic species of the polyvalent metals adsorbed onto the broken-bond surface. These ions tend to hydrolyze and to polymerize in aqueous solution. The mechanism of these reactions was discussed in previous chapters. Some of the surface functional groups, being proton donors, are catalysts for these reactions. Polymerization and formation of hydroxy complexes are minor with the divalent cations, but become significant with cations of higher valency. In natural systems this is important, especially in the geochemical behavior of iron and aluminum. Polymeric hydroxy cationic complexes may bridge between the clay particles and organic or inorganic anions, resulting in adsorption of these anions by the clay. Smith and Emerson (1976) identified several polymeric species of exchangeable aluminum on kaolinite. When prepared from H-kaolinite, the charge per Al ion was 3, indicating that there was no polymerization or hydroxy complex formation, whereas when prepared from K-kaolinite or when the pH of Al-kaolinite was raised, the average charge per Al ion was 1.4 and 0.5, respectively, as a result of hydroxy–polymeric complex formation. The adsorption of hydroxy complexes of polyvalent ions is termed "hydrolytic adsorption" (see Ch. 4). As a result of polymerization of exchangeable Al, kaolinites with divalent cations are better catalysts than Al-kaolinite for acid-catalyzed reactions (Saltzman et al., 1974).

1.1.2 Sorption of Anions

There seem to be four mechanisms for the specific sorption of anions: (1) replacement of exposed OH groups, a mechanism similar to that described in Chapter 6 for the sorption of monosilicic acid by hydrous oxides; (2) sorption of tetrahedral anions of weak acids such as phosphate, arsenate, borate, etc., which have about the same size and geometry as the silica tetrahedra and which fit onto the edge of silica tetrahedral sheets and grow as extensions of these sheets; (3) coordination with nonhydrated polyvalent cations, such as Al, which are Lewis acid sites, and (4) sorption on polymeric hydroxy cations sorbed earlier on the broken-bond surface.

As shown in the previous chapter, maximum specific sorption of anions from aqueous solution occurs when the pH is equal to the pK_a of the conjugate acid. For silicic acid maximum sorption by various clay minerals occurs at about pH 9.6, which is also the pK of silicic acid. Adsorption isotherms of phosphate by kaolinite reveal that adsorption decreases as the pH is increased. The difference in adsorption between silicic and phosphoric acids (Fig. 7.1) lies in the differences in the behavior of the acids. Silicic acid behaves as a monoprotic acid whereas phosphoric acid behaves as a triprotic acid. In the case of phosphate, both HPO_4^{2-} and $H_2PO_4^-$ anions can accept or donate protons. Both these anions can be sorbed by the clay, but since HPO_4^{2-} contributes a greater negative charge to the surface than does the $H_2PO_4^-$ ion, the equivalent amounts of P adsorbed decrease as the pH and the proportion of HPO_4^{2-} increase. When the pH of the system is so high above the point of zero charge that the surface cannot contribute protons, a fully charged ion like PO_4^{3-} in solution will not adhere to the surface (Kafkafi and Bar-Yosef, 1969).

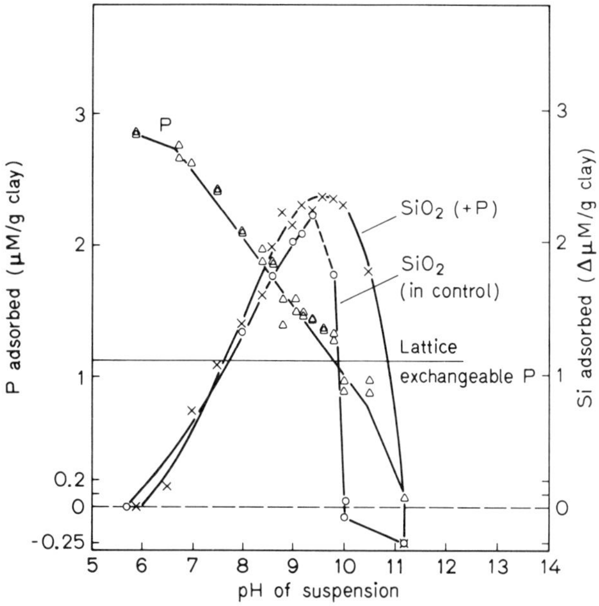

Fig. 7.1. The effect of pH on the amounts of adsorbed P and SiO_2 on kaolinite. (From Kafkafi and Bar-Yosef, 1969: Used with permission of Keter Publishing House, Ltd.)

Clay minerals play an important role in phosphate fixation in soils and in sediments. Kafkafi et al. (1967) differentiated between *fixed* (7.II) and *exchangeable* (7.III) phosphates. They observed that not all the phosphate adsorbed by kaolinite is isotopically exchangeable. The stability of the nonexchangeable phosphate variant is due to the formation of the six-membered ring structure (7.II).

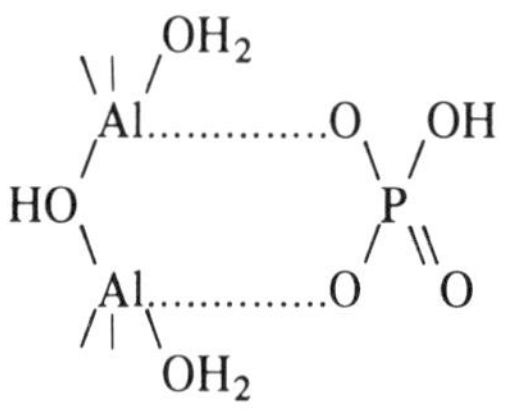

(7.II)

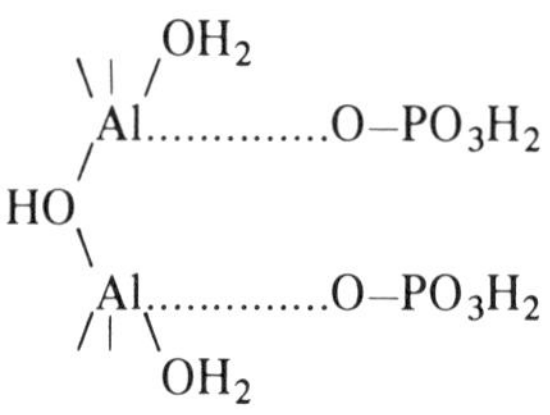

(7.III)

1.1.3 Water at the Broken-bond Surface

Water sorption onto this surface takes place via two mechanisms: (1) hydrogen bonding between the water molecules and exposed hydroxyls or oxygen atoms and (2) hydration of exchangeable ions. The first mechanism results in a hydrophilic hydration interface, similar to that described for silica and silica-alumina. Hydration numbers for cations sorbed on kaolinite are given in Table 7.2. According to Fripiat (1964), when physical adsorption of water occurs on outgassed kaolinite, the film consists first of molecules that form clusters around partially or fully dehydrated cations.

Water at the broken-bond surface can be divided into three regions: the hydrophilic hydration interface region, the cation hydration region, both having a high degree of order, and the third, having a low degree of order, which serves as a bridge between the first two regions.

The cation hydration region has the nature of Brønsted acid. As the water content of the clay drops below that required to saturate the solvation shells of the exchangeable cations, the residual water dissociates more readily and hence is more acidic. Also, the removal of water exposes Al atoms located at edge sites, creating Lewis acid sites (Solomon et al., 1971; Conley and Althoff, 1971).

Table 7.2. Cation hydration number on kaolinite surface. (After Fripiat, 1964)

Sample	Hydration number
Li kaolinite	2.1–2.7
Na kaolinite	1.6–2.1
K kaolinite	1.7–2.0
Mg kaolinite	4.9–10.0
Ca kaolinite	4.3–7.8

1.2 Interlayer Space of 2 : 1-Type Clay Minerals

An interlayer space of a 2 : 1-type clay mineral can be regarded as the product of the overlapping of the diffuse double layers of two parallel flat surfaces of unit layers. The electric field in the interlayer is the result of the presence of exchangeable cations. This model is in agreement with the following characteristics of the interlayer space:

1. Cation adsorption and anion exclusion occur in the interlayer space (the latter known as the *Donnan exclusion*). In general, cations of higher charge are preferentially sorbed onto the interlayers by ion exchange.

2. The size of the diffuse double layer and the associated swelling of the interlayer space decrease with increasing surface charge density of the flat layer, and with the charge and concentration of the adsorbed ions. In Table 7.3, however, it can be seen that a certain surface charge is required for swelling to occur. Talc and prophyllite, the two 2 : 1 minerals with a zero surface charge, are not swollen or wetted by water. Illite can be regarded as if the anhydrous K ions are specifically adsorbed onto the inner Helmholtz layer of the flat oxygen sheet, resulting in a zero surface charge density. This mineral behaves like talc and pyrophyllite and does not expand. Potassium vermiculite shrinks, and K montmorillonite, depending on the circumstances, may exhibit, both limited and extensive swelling. Na vermiculite gives a 1.48-nm spacing even in dilute solutions and Na montmorillonite swells extensively. Lithium vermiculite and montmorillonite exhibit extensive swelling (Quirk, 1968). The repulsion between two overlapping double layers is discussed further in Chapter 8.

3. Suspensions of montmorillonites treated with Na and salts of other monovalent ions give interlayer spacings proportional to $C^{-1/2}$, where C is the salt concentration in the suspension (Norrish, 1954).

4. Since the oxygen plane is composed predominantly of oxygen atoms belonging to siloxane groups that do not form stable hydrogen bonds with water molecules, the

Table 7.3. Charge density and crystalline swelling of 2 : 1 clay minerals. (After Quirk, 1968)

Minteral type	Charge per unit cell	nm² per charge	Interlamellar cation	Interlayer spacing in nm in dilute suspension
Talc	0	0		0.93 nm
Prophyllite	0	0		0.91
Illite	1.3	0.7	K	1.00
Vermiculite	1.3	0.7	Ca	1.4–1.5
			Mg	1.4–1.5
			Li	≫ 4.00
			Na	1.4–1.5
			K	1.16
			Cs	1.20
Montmorillonite	0.67	1.4	Mg	1.90
			Ca	1.90
			Li	≫ 4.00
			Na	≫ 4.00
			Ka	1.5 and ≫ 4.00
			Cs	1.20

hydration of this surface is hydrophobic in character. Swelling and the resulting structure of the interlayer water are the outcome of the attraction forces between water molecules and the exchangeable ionic species and the thermal motion of water molecules in the environment of the mineral. For example, the exchange of Cu by an organic hydrophobic tetraalkyl-ammonium ion such as tetrapropylammonium, leads to the contraction of smectites from 1.9–2.0 nm to about 1.45 nm (McBride and Mortland, 1975).

5. The basic strength of the oxygen plane and the ability to form hydrogen bonds with water molecules depend on whether or not there is an isomorphous tetrahedral substitution in the clay of Al for Si. Since the charge of Al is lower than that of Si it induces a weaker effect on the electrons of the O atom, and the basic strength of the oxygen plane increases. Thus the Si–O–Al group forms stronger hydrogen bonds with water molecules than the Si –O–Si group. The characteristics of the hydration atmosphere of the oxygen plane depend on whether the charge in the clay mineral originates from a tetrahedral or an octahedral substitution. Hydrophilicity of this hydration, which increases with tetrahedral substitution, may account for many of the differences between these two subgroups. In montmorillonite only a small fraction of the interlayer water forms hydrogen bonds with the oxygen sheet. This water fraction becomes higher in saponite and still higher in vermiculite (Yariv and Heller, 1970; Farmer and Russell, 1971). Consequently, the interlayer of montmorillonite is more easily expanded than that of vermiculite and more free water can be sorbed.

A simple model for nonfixed exchangeable cations in the expanded interlayer space of an electrolyte free suspension of a mineral presupposes that they are randomly distributed. In accordance with the discussion on water structure in the Introduction, six zones having differing water structures may be distinguished in the system. The zones that contain the water molecules comprising the hydration atmosphere of the ions (Fornes and Chaussidon, 1975) and the solid–liquid boundary layer at the flat oxygen plane are distinguished as zones A_m and A_o, respectively (the subscripts m and o signify metallic ion and oxygen plane, respectively). The bulk water is zone C. Zones B_{om}, B_{oc}, and B_{mc} are disordered zones separating the ordered zones A_o and A_m, A_o and C, and zones A_m and C, respectively. By the removal of liquid water new water structures are obtained that are closely related to those of zones A_m, A_o, and B_{om}. In zone A_o the clay/water interface exerts an ordering influence on the water structure, reducing the thermal amplitudes of the intermolecular vibrations and consequently reducing the density of the water layers in the immediate neighborhood of the clay surface. In zone A_m the attenuation of molecular motion is obtained from the strong polarizing field of the exchangeable cation.

Ion exchange is a stoichiometric process. As a consequence of the electroneutrality requirement, any counterions that leave the interlayer are replaced by an equivalent amount of other counterions. Deviations occur as, for example, when electrolyte is adsorbed during exchange. Although Donnan exclusion normally keeps the ionic content at a very low level, concentrated salt solutions weaken the exclusion effect. Salt intercalation, that is, the entry of cations and anions into the interlayer region, has been observed in vermiculites and smectites. With a 6 M NaBr solution, an amount equivalent to as much as one-third of the exchange capacity may be intercalated (Walker, 1963).

Another example of deviation in the stoichiometric process of ion exchange is hydrolytic adsorption. This occurs when a hydroxy complex or a polymeric oxycation is formed in the interlayer during an ion exchange with polyvalent cations.

The counterion within the interlayer must migrate into the external solution, while a concurrent migration of the exchanging ion occurs in the opposite direction. In the bulk solution any concentration differences are balanced by convection and agitation, but in the interlayer and the immediate vicinity of the crystals concentration gradients develop and migration is a function of diffusion. This interdiffusion of ions is the rate-determining step in ion exchange processes (Walker, 1963).

1.2.1 Water in the Interlayer Space

When clay is exposed to water vapor the exchangeable cations hydrate first (Low, 1961), forming zone A_m. Depending upon atmospheric humidity and hydration ability of the cation, two different types of hydrated cations appear. When the hydration number of the metal is $\leqslant 4$ and it has a square coordination, the hydrated cation occupies a flat orientation parallel to the layers. This is the case with most alkali metal cations (monovalent) and with Cu^{2+} in smectites at a relatively high humidity of 40%. When the hydration number of the cation is above 4 it occupies the center of an octahedron, two triangular surfaces lying parallel to the oxygen planes. This is the case with most di- and trivalent cations even at a relatively low humidity of less than 10%. The basal spacings obtained for minerals with planar and octahedral hydrated ions are ~1.2 and ~1.43 nm, respectively (Walker, 1956). On further exposure to water vapor the remainder of the surface hydrates, and zones A_o and B_{om} are formed. At this stage water molecules penetrate into vacancies in the interlayer space. A monolayer is obtained in the first case, whereas a dilayer results in the second case, with a slight expansion of the clay to values above 1.25 and 1.48 nm respectively. When the clay is in contact with liquid water, osmotic forces caused by the relatively high ionic concentration between the layers may lead to a continuous swelling. The interlamellar water is slightly more densely packed than liquid water (Hougardy et al., 1970).

The interlayer space reveals basic and acidic sites, the oxygen planes being electron donors and water molecules of zone A_m being proton donors. Hydration zones and structures of various hectorites at various hydration states are shown in Figures 7.2 and 7.3 (Prost, 1975).

At the hydrophobic surface water polymers are obtained, forming zone A_o. The OH groups of water are not directed toward the hydrophobic surface siloxane-type oxygens, and the organization of the water molecules is such that they are closely linked to one another by hydrogen bonds. To understand the formation of this hydrophobic water region, one should consider the interlayer space as a cage wherein the thermal amplitude of the water molecules is reduced. When the water molecules enter into the cage they undergo a loss in entropy, while at the same time a new structure is formed. Li and Na ions do not break the water structure of zone A_o of smectite minerals, since, because of their small sizes, these ions can fit into the interstitial cavities. The intense ionic fields of multivalent cations cause local destruction of the water structure of this region and the formation of zone A_m. Whereas hydrogen bonds between water molecules of zone A_m and the oxygen plane restrict the swelling of the clay above a certain limit, the presence of zone A_o does not limit osmotic swelling to

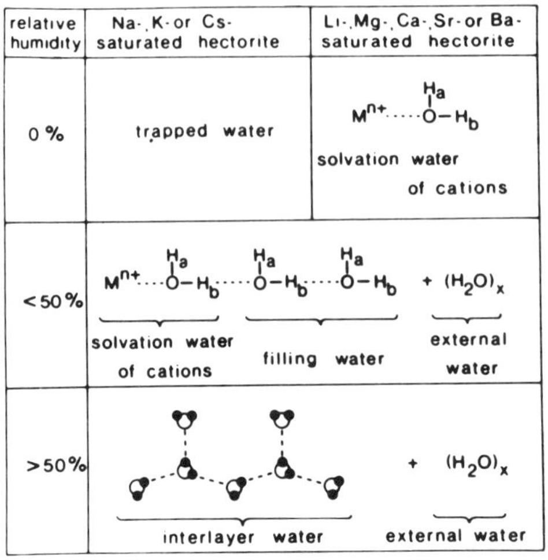

Fig. 7.2. Schematic representation of adsorbed water states on hectorite. (From Prost, 1975: Used with permission of Applied Publishing Ltd.)

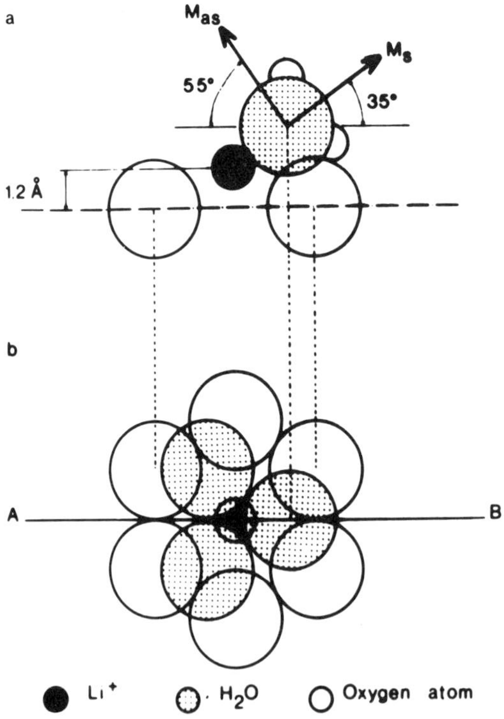

Fig. 7.3 a and b. Orientation of the water molecules with respect to Li^+ cation and oxygen atoms of the oxygen plane in Li-hectorite. (From Prost, 1975: Used with permission of Applied Publishing Ltd.)

any degree. It is possible to produce a large water uptake in vermiculite by breaking down the natural magnesium–water complexes. This can be achieved by the exchange of Mg for Li whereby water of zone A_m is exchanged for water of zone A_o (Forslind and Jacobsson, 1975; Walker and Garrett, 1967).

Water of zone A_m dissociates under the polarizing effect of the metallic cation as follows

$$M(OH_2)_x \rightarrow M(OH)(OH_2)_{x-1} + H^+. \qquad (7.IV)$$

A similar polarizing effect occurs in aqueous salt solutions resulting in an increasing number of hydrogen bonds between water molecules and extension of the ordered zone A. Because of steric hindrance in the interlayer space, zone A_m cannot extend as much as zone A in liquid water and is therefore a stronger proton donor. Moreover, water structure in the interlayer space contains more lattice vacancies than normal water, and the dielectric constant of this water is lower than that of liquid water; consequently, protons are more mobile than in liquid. According to Touillaux et al. (1968) the degree of dissociation of water is 10^7 times higher in the adsorbed state than in the liquid. The proton-donating tendency increases as the water content of the interlayer space decreases.

The polarization of the interlayer water is strongly affected by the exchangeable cation. Greater polarizing ability of the interlayer cations increases both the strength and number of acid sites per surface unit area (Frenkel, 1974). Acid strength of hydrates of some common ions decreases in the following sequence: Al, Fe, Mg, Ca, Li, Na, K, and Cs. Mortland and Raman (1968) showed that the degree of protonation of sorbed ammonia depended upon the hydrolysis constant of the interlayer hydrate and that the degree of protonation may be regarded as a measure of the acidity of the interlayer space. In base-saturated montmorillonite and vermiculite, ammonia forms hydrates $M{-}OH_2 \ldots NH_3$, or reacts with protons originating from water dissociation, forming NH_4^+. In montmorillonite saturated with Al, NH_3 reacts with protons to form ammonium ion.

The acid strength of water of region A_m depends on whether the charge of the clay mineral originates from tetrahedral or octahedral substitution. From nuclear magnetic resonance studies on Li hectorite, in which the charge originates from octahedral substitution, Conrad (1976) suggested that the trihydrated Li cation is fixed above the six oxygens of the hexagonal cavity by delocalized bonds in which three of the protons of the hydrated Li can be equally distributed among twelve equivalent positions with respect to the oxygens forming the hexagonal cavity. In this symmetric structure the equal density of protons throughout the ring facilitates the removal of one proton, which accounts for the acidic properties of the clay.

Oxygen sheets of layers, wherein the charge originates from the tetrahedral substitution, form strong hydrogen bonds with polarized water molecules having the structure shown in (7.V):

$$\begin{array}{l} \qquad\qquad\quad\ \ /\text{Si, Al} \\ \text{M}\ldots\underset{\displaystyle\text{H}}{\underset{|}{\text{O}}}{-}\text{H}\ldots\text{O} \\ \qquad\qquad\quad\ \ \backslash\text{Si, Al} \end{array} \qquad (7.V)$$

As a result of this hydrogen bond formation, the proton-donating ability of this water molecule is diminished, and the water oxygen, acting as an electron pair donor, becomes a basic site. The strength of these hydrogen bonds and hence the number of basic sites of this type, increase with increasing polarizing power of the exchangeable cations. The bonds are detected mainly in the presence of exchangeable Al (Yariv et al., 1969). Protonation, which involves breaking of the H bonds with the oxygen planes, is therefore more readily effected with octahedral charged than with tetrahedral charged minerals (Yariv and Heller, 1970).

Water in zone B_{om}, not being structured, may interact with negative charge sites on the oxygen planes. Some water molecules will be oriented with the positive ends of the dipoles toward the oxygen planes, thus imparting a basic character to the interlayer water. The area of A_o decreases as that of B_{om} increases. Zone B_{om} is present in Cs-montmorillonite (Yariv et al., 1969). This exchangeable cation is large enough to break the structure of zone A_o but too large to form zone A_m. The size of zone B_{om} depends on the polarizing power of the exchangeable cation. The greater the polarizing power of the cation, the more acidic the associated hydration shell and the greater is the extension of zone B_{om}. Zone B_{om} forms an important fraction of interlayer water in Al montmorillonite in which every discrete zone A_m spreads over large areas. However, since the total number of Al ions in the interlayers is only one-third of the number of monovalent ions, the total area occupied by A_m is comparatively small, leaving a considerable fraction of the interlayer for adsorption of water through the formation of zones B_{om} and A_o. In most cases "acidic" water predominates over "basic" water and obscures the presence of the latter.

1.2.2 Selective Sorption and Fixation of Cations by Clay Minerals

1. Fixation by Sorption Onto Frayed Edges. Certain cations are sorbed more selectively than others and are fixed, that is, they resist replacement by other cations. Cations with low hydration energy, such as K^+, NH_4^+, Rb^+, and Cs^+, allow interlayer dehydration and collapse to take place. This subject has been reviewed by Sawhney (1972). These ions may become fixed in interlayer positions. The layer charge of the mineral also affects the interlayer collapse and hence the degree of cation fixation. The greater the layer charge within the smectites, the greater is the K fixation. K ions are held more tightly in dioctahedral than in trioctahedral minerals. The following two factors may explain the stronger fixation of K in dioctahedral minerals: (1) the hydroxyl of the octahedral sheet in trioctahedral minerals is vertically oriented, whereas in dioctahedral minerals it is inclined; the positive pole of the OH group is nearer the K ion in the first case, resulting in a greater repulsion; (2) K–O distances are shorter in dioctahedral minerals than in trioctahedral minerals.

Weathering in mica proceeds from the edges inward. Thus, partially weathered mica or illite in soils should consist of a collapsed 1.0-nm central core and expanded frayed edges (Fig. 7.4). Cations that produce interlayer collapse would be selectively sorbed at the frayed edges to produce a more stable collapsed structure, similar to that of the central core. This chemical reaction is the reverse of the reaction that occurs during the first steps of the weathering of the mica. Bolt et al. (1963) found that the selectivity of the expanded frayed edges of illite for the sorption of K was about 500-fold greater than that of Ca. Because of the similarity of K and NH_4^+ ions and the small diffusion distance from solution to these sites, ammonium ions are effective in replacing K. Conversely, hydrated ions, Ca, and Mg are not selectively sorbed by these sites, and hence

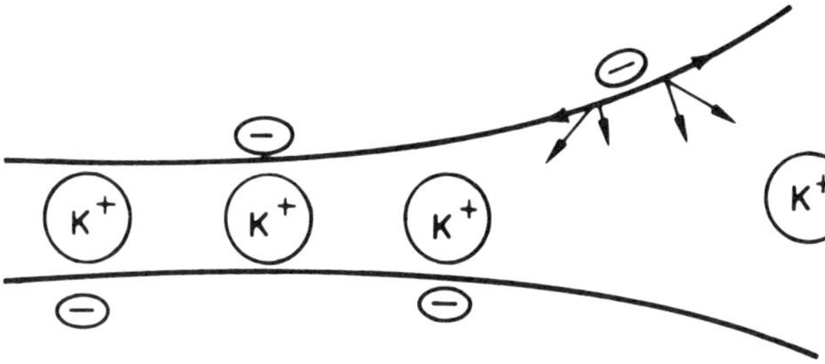

Fig. 7.4. Selective sorption of K^+ ion in a frayed edge of a weathered mica sheet. (From Sawhney, 1972: Used with permission of the author and the publisher)

they are not as effective in replacing K. The selective sorption of K by the clay fraction in soils and in the oceans is due to the presence of frayed edges resulting from weathering in micas.

Sorption of K or Cs ions by montmorillonite produces a monolayer of water in the interlayers, giving a 1.2-nm interlayer spacing, while vermiculite collapses to give a 1.0-nm spacing and a 1.08-nm spacing on saturation with K and Cs, respectively. In vermiculite, progressive sorption of K or Cs causes collapse in alternate layers, producing regularly interstratified 1.0- and 1.5-nm layer sequences. When micas and K vermiculite are exhaustively leached with solutions of cations, such as Ca and Mg, that produce interlayer expansion, K is replaced by these cations. Conversely, K is not replaced effectively by cations that produce interlayer collapse in minerals.

Fixation of Cs ions by layer silicates appears to be similar to that of K at the frayed-edge sites of partially weathered micas but unlike K in interlayers of vermiculite or montmorillonite. K and ammonium ions replace Cs fixed at the frayed edges as well as in the interlayer space more readily than Ca and Mg ions. Cs ions having a smaller hydration energy than K ions hold the layers together more strongly than K ions, so that hydrated Ca and Mg, which can expand layers held by K ions, are ineffective in producing layer expansion in Cs saturated minerals. Consequently, Cs ions are not readily replaced by these ions as are the K ions. Anhydrous K ions can diffuse into the interlayers and replace the Cs ions.

Kaolonite, peat, and sandy minerals do not fix Cs. The highest Cs concentrations in soil samples and fresh-water sediments are found in the micaceous mineral fractions and are associated with illites and muscovite micas (Francis and Brinkley, 1976).

Nonexchangeable ammonium ions are found in clays and shales. Some clays in Israel in which illite is the dominant mineral were found to contain over 2 meq NH_3/100 g clay, which comprised some 75% of the total nitrogen of the sample. Montmorillonite-rich clays in marine sediments with only a small amount of illite, contained only 0.5–1.5 meq NH_3/100 g clay comprising about 60% of the total nitrogen in the inorganic form. Montmorillonitic red beds and kaolinitic clays, both of residual weathering and of hydrothermal origin, contained only trace amounts of inorganic ammonium ion (Yaalon and Feigin, 1970).

2. Fixation by Thermal Sinking Into Crystal Framework Vacancies. Other cations may also be fixed by clay minerals. When minerals saturated with exchangeable cations of sizes up to and including Ca are dehydrated at high temperatures, small unhydrated ions sink into crystal framework vacancies becoming very difficult to hydrate or exchange. Depending on the size of the cation, the type of the mineral, and the temperature of the reaction, penetration into the hexagonal holes in the tetrahedral sheet and still deeper penetration into the octahedral sheet may occur (Russell and Farmer, 1964).

3. Fixation by Hydrogen Bonding Between Hydrated Cations and Oxygen Plane. Fixation of Li occurs on smectites that have a large degree of substitution in the tetrahedral sheet. This type of fixation occurs at low temperatures. It does not occur with minerals whose charges originate from octahedral substitution. Li has a strong tendency to form hydrates, the water shell being highly acidic. The hydrogen bond of the hydrated ion with the more basic oxygen plane of tetrahedral substituted smectites is stronger than that with octahedral substituted smectites (Eckstein et al., 1970).

4. Fixation by Polymerization of Multivalent Cations. In smectites, protonation of sorbed alkaline materials may lead to precipitation of metal hydroxides or hydroxy cationic complexes in the clay interlayers, according to the reaction

$$M^{n+} + nHOH + nB \rightarrow M(OH)_n + nBH^+ \qquad (7.VI)$$

where M is a metal cation of charge n and B is an organic or an inorganic proton acceptor. As a result of this reaction the hydroxylated metal becomes nonexchangeable or fixed. This type of fixation occurs in nature with Al, Fe, and to a smaller extent with divalent metals. The product formed in (7.VI) polymerizes with some hydrated cations, and the entire polymer is fixed. Furthermore the fixed material blocks the exit of other hydrated interlayer cations. In this way monomeric cations also become fixed (Heller-Kallai et al., 1973 b, 1973 c).

The proportion of the interlayer hydroxy-aluminum polymer in the total interlayer aluminum increases with the alkanity of the system. The cation exchange capacity reduction of the clay is apparently due to the formation and entrapment of hydroxy Al polymer in the interlayers. This capacity is therefore reduced with increasing alkalinity of the environment. According to Keren et al. (1977) the exchange capacity of smectites is reduced only if the hydrolysis of the aluminum takes place in the interlayer space but not if smectites interact with $Al(OH)_3$.

The fixation of polyvalent cations is further increased upon drying. According to Weismiller et al. (1967), dehydration of a hydroxy Al polymer–smectite complex results in a closer association of the hydroxyls with the oxygen plane of the clay.

1.3 The Flat Oxygen and Hydroxyl Planes

The surface characteristics of oxygen and hydroxyl planes can be inferred from the properties of the intercalation complexes of kaolinite and halloysite. Under certain conditions kaolin-type minerals can intercalate certain inorganic salts and a variety of organic compounds (Wada, 1961; Weiss et al., 1963). The salts or organic compounds penetrate the interlayer space of kaolin-like layers and so expand the crystal from a basal spacing of ~0.72 nm to about 1.00–1.42 nm. The penetrating species that break the strong electrostatic and van der Waals types of interactions between the layers may form hydrogen bonds with surface hydroxyls as was proved from infrared spectroscopy by Ledoux and White (1966). Basal hydroxyls are very poor proton donors and may form hydrogen bonds only with very strong bases such as hydrazine. Strong proton donors may form hydrogen bonds with oxygens located on the tetrahedral sheet surface. Since the oxygen plane is a very poor electron donor, these hydrogen bonds are very weak.

The flat oxygen plane, even in kaolinite, exhibits a constant negative charge, arising from occasional isomorphous replacement of silicon by aluminum in the tetrahedral sheet (Range et al., 1969). Exchangeable cations that compensate this charge are distributed over the planar surface of the silicate. This was observed by McBride (1976) during electron spin resonance study of exchangeable Cu and Mn on kaolinite. Divalent exchange ions are about 1.1–1.2 nm apart on kaolinite surfaces. Exchangeable planarly hydrated ions, such as $Cu(H_2O)_4^{2+}$, are oriented parallel to the flat oxygen planes. The mobility of cations located in the vicinity of the flat oxygen plane, compensating the permanent charge of the mineral, is greater than that of ions located at the broken-bond surface, compensating the pH-dependent charge.

Hydroxides and hydrous oxides of polyvalent cations such as Al, Fe, and Mn are often found in nature covering clay minerals. The conditions determining the association between ferric hydroxides and kaolinite surfaces were discussed by Greenland (1975). Iron hydroxide precipitated in the presence of kaolinite behaves as an amphoteric hydroxide with a point of zero charge at about pH 7. It precipitates onto kaolinite surfaces only under acid conditions. At higher pH values, which are necessary for the formation of a well-crystallized goethite, the products are essentially a mixture of goethite and kaolinite crystals. The main factor determining whether the iron hydroxide is attached to the clay surface is the charge on the hydroxide. Precipitation at pH below the point of zero charge leads to positively charged gel particles that are adsorbed on the negative basal oxygen planes of the kaolinite. Precipitation at higher pH values leads to the formation of negatively charged ferric hydroxide and the development of a separate phase.

The interaction occuring when water is adsorbed on the flat planes can be visualized from the work of Yariv and Shoval (1976) on the interaction between alkali halides and hydrated halloysite. The hydration of both types of basal sheets results in a hydrophobic hydration boundary layer of water, with no hydrogen bonds between water molecules and the oxygen or hydroxyl sheets. The negative oxygens of the water molecules in the boundary layer are oriented toward the hydroxyl plane whereas the positive hydrogens are oriented toward the oxygen plane, but there is no localized interaction between water molecules and the basal sheets. Hydrogen bonds occur between the water molecules. The presence of large ions such as Cs, Rb, and K in the boundary layer results in the disruption of the hydrophobic hydration structure of the water. Water molecules may now coordinate with the alkali cations, and at the same time they may interact with basal hydroxyls and basal oxygens of kaolin-like layers, forming hydrogen bonds as follows:

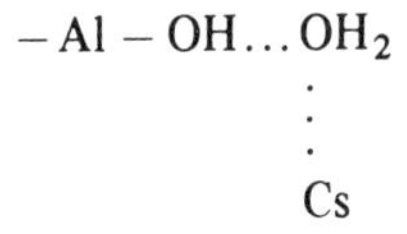

(7.VII)

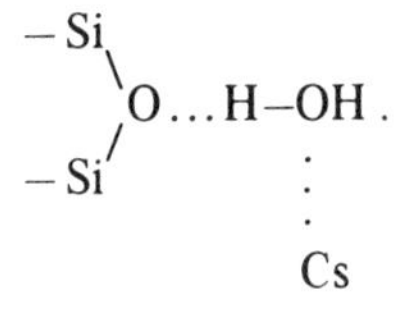

(7.VIII)

The degree of structuring and the size of water clusters in the adsorbed water layer are affected by the cation size and increase from Cs to K. The ability of the adsorbed water layer to donate protons to the oxygen plane or to accept protons from the hydroxyl plane, in both cases forming localized H bonds, increases with decreasing degree of structuring of this water layer. Hydrogen bond formation either by proton-accepting from hydroxyls (7.VII) or by proton-donating to oxygens (7.VIII) is affected by the cations in the order Cs > Rb > K. No hydrogen bonds are formed in the presence of Na ions because the size of this ion is such that it can apparently fit into the interstitial cavities of water with minimum disruption of the water structure. Kaolinite forms hydrogen bonds only with Cs hydrate.

2. Interaction Between Clay Minerals and Organic Compounds

The presence of nucleophilic (electron donor) sites and electrophilic (electron acceptor) sites on the surfaces of clay minerals and the electric charges of the unit layers enable these minerals to sorb polar organic compounds and ions. Organic compounds may be sorbed within the interlayer spaces of 1 : 1 and 2 : 1-type minerals, on the broken-bond surface, and on the outer flat oxygen and hydroxyl planes. The interaction may be of an electrostatic or van der Waals type and of a short-range type, which comprises H and chemical bonds and π interactions and is strongly dependent upon distance. The subject has been reviewed by Mortland (1970) and by Theng (1974) and the present section is based mainly on these reviews, supplemented by results reported in more recent studies.

2.1 Sorption of Organic Ions by Clay Minerals

2.1.1 A Model for the Structure of the Double Layer in the Presence of Organic Ions

The present model is based on a model suggested by Yariv (1976). The attraction of organic ions with a charge opposite to that of the solid phase and the repulsion of ions with a charge of the same sign, and the formation of the diffuse double layer, are predominantly due to "long-range" electrostatic forces. For simplicity the organic ion will be considered as consisting of two distinct moieties, a hydrophilic head, where the electric charge is concentrated, and a hydrophobic tail, a straight or branched saturated hydrocarbon chain, which is organophilic. This is an amphipathic or "soap" ion, which forms micelles in aqueous solutions. The clay surface will be considered to be flat with a uniform charge density. In the first stage the possibility of "short-range" forces or "keying" will be ignored and the interaction between water molecules and clay surface will be considered to be purely electrostatic.

The aqueous boundary phase is divided into regions of distinct dielectric behavior. Region A (dielectric constant ϵ_1) consists of a layer of preferentially oriented water molecules in contact with the clay surface. This is the "inner Helmholtz" layer where the adsorbed "bare" ions are without their hydration shells and the charged functional groups are in the immediate vicinity of the interface. Region B (dielectric constant ϵ_2) is a region of both free water molecules and molecules attached to hydrated ions. This is the "outer Helmholtz" layer, defined by the closest approach of a fully hydrated charge group to the boundary. Region C (dielectric constant ϵ_3) is the diffuse layer. ϵ_4 is the dielectric constant of the bulk water.

In the double layer, organic ions are arranged so as to accommodate their amphipathic character (Fig. 7.5). Because of electrostatic attraction, and since $\epsilon_1 < \epsilon_2 < \epsilon_3 < \epsilon_4$, the hydrophilic head of the organic ion will almost always be at a distance equal to or less than the distance between the hydrophobic moiety and the clay surface.

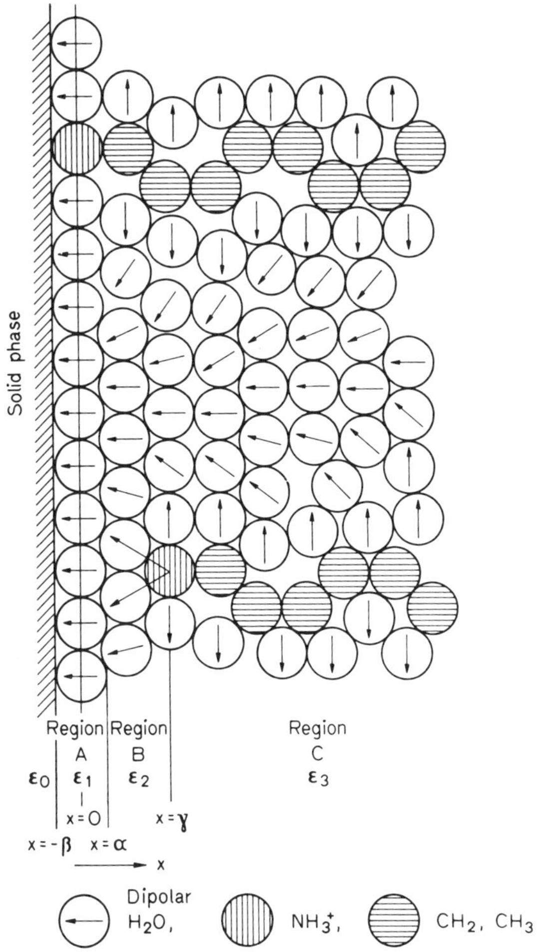

Fig. 7.5. Schematic cross section through an electric double layer of a flat surface, in the presence of alkylammonium ions. Induced dipoles of H_2O are shown by arrows. (From Yariv, 1976: Used with permission of the publisher)

The long-range electrostatic attraction is minimized by the following reactions:

1. When the organic ion is in the bulk solution, the hydrophobic moieties are associated to form dimers or micelles. They must dissociate before they can enter into the double layer. The free energy change during this dissociation process, ΔG_d, depends on the ratio of the surface area of the monomeric ion to that of the polymeric form. For aliphatic tails it increases with increasing number of carbons.

2. When the organic ion is in the bulk aqueous solution the hydrophilic head is hydrated. It must be dehydrated before it may penetrate into the inner Helmholtz layer. If it reaches regions B or C only, it need not dehydrate, but the symmetry of the hydration shell is destroyed. The free energy change during this process, ΔG_h, depends mainly on the polarizability and charge of the hydrophilic group, and to some extent also on the structure and composition of the noncharged hydrophobic moiety of the ion. Bodenheimer et al. (1966a) showed that in the clay interlayers amines display their true basicity, unaffected by solvating molecules, whereas in the bulk aqueous solution hydration shells have great influence on the basicity of the amine.

3. Polarized water molecules are repelled from the double layer to the bulk of the solution by the penetrating organic ions. The change in free energy is given by Eq. (7.1)

$$\Delta G_p = \frac{1}{8\Pi}\left(1 - \frac{1}{\epsilon}\right)x^2 v, \tag{7.1}$$

where x is the intensity of the electric field at the location of the center of the ion in the double layer and v is the volume of the organic ion together with its hydrophobic and hydrophilic hydration shells.

Water in the vicinity of the noncharged moiety of the organic ion will have the "hydrophobic hydration" structure. There may be repulsion forces resulting from the self atmosphere potentials of the organic and inorganic ions present in the double layer, and also from the electric field induced by the clay surface (arrows in Fig. 7.5). Due to these repulsion forces in the diffuse and the Helmholtz layers, the hydrophobic alkyl tail will point away as far as possible from the clay surface toward regions with low intensities of the electric field. The most stable configuration is obtained when the tail is perpendicular to the clay surface.

At low concentrations the organic cations are adsorbed as individual counterions, but at higher concentrations they associate through interaction of the hydrophobic moieties of those ions adsorbed in the Helmholtz layer, even if this association requires that the angle that develops between the oxygen plane and the long axis of the tail be nonperpendicular. For any fixed ionic concentration, this angle decreases with decreasing length of alkyl chain. For a definite length of an alkyl chain, the angle decreases with decreasing concentration. The angle also depends on the charge density of the clay surface, and it will increase with increasing charge density. The change in free energy that results from this association is defined as the double layer association energy, ΔG_a. The associated species formed at the solid–liquid interface are named *hemimicelles*.

The system is stabilized by hemimicelle formation and there is a rapid rise in the fraction of organic ions penetrating into the double layer as the bulk concentration increases above a certain value. It is also to be expected that the affinity of the double layer for the organic ions will further increase with the increased occupancy of the exchange sites by organic cations. Adsorption studies of organic ammonium ions by Na-montmorillonite (Theng, 1971) reflect the importance of hemimicelle formation in the adsorption mechanism. When the sodium ions are only partly exchanged by the organic ions, some of the interlayers are occupied mainly by the organic ions and the remainder by Na ions.

Because of the low polarizability of the hydrophobic chain, the electric field induced by the clay surface may become screened by the hydrophobic moiety of the organic ion. This increases the ease of dehydration of organo-clay complexes. Repulsive forces that occur between two similar double layers are thereby decreased, and the saturated organo-clay complexes tend to coagulate.

The increase in the number of carbon atoms in the alkyl group is paralleled by an increase in the sorption energy, approximately 400 cal/mol per CH_2 group (Cowan and White, 1958). In the earlier literature the increase of free energy with the chain length of the alkylammonium ions was attributed to the effect of increasing van der Waals forces. Vansant and Uytterhoeven (1972) showed that for small organic ammonium ions the van der Waals interactions are negligible and that hydration and coulombic interaction between the organic cation and the clay mineral are more important.

Short-range forces begin to operate when organic ions penetrate into the inner Helmholtz layer. For example, hydrogen bonding between polarized water molecules coordinated to cations and the oxygen plane of a mineral with tetrahedral substitution of Al for Si may occur. Hydrated ions, inorganic as well as organic, may thus penetrate into the Helmholtz layer. For broken-bond surfaces the effect of short-range forces is more critical than that of long-range forces.

The probability of the penetration of organic ions into the inner Helmholtz layer is given by

$$v = g \exp - (Z_i\varphi + \Delta G_a + \Delta G_s - \Delta G_d - \Delta G_h - \Delta G_p)/kT, \qquad (7.2)$$

where Z_i is the charge of the ion, φ the inner Helmholtz electric potential, ΔG_s the short-range energy, k the Boltzmann Constant, and g a statistical factor that depends on the concentration of the organic ion. Since ΔG_a greatly increases with increasing concentration, it follows from Eq. (7.2) that the probability of penetration of organic ions into the inner Helmholtz layer also increases with increasing concentration.

2.1.2 Arrangement of Alkylammonium Ions in the Interlayer Space of 2:1 Clay Minerals

The c spacing measurements of tactoids of mica-type expanding layer silicates saturated with alkylammonium ions can provide considerable information on the structure of the interlayer space. Much work in this field has been done and reviewed by Weiss and co-workers (Weiss, 1963; Lagaly et al., 1970). The interlayer space can be regarded as being derived from the overlapping of two parallel diffuse double layers. A correlation indeed exists between the model of the structure of an isolated double layer and the experimental results of the sorption of ammonium ions by clay minerals, as shown here.

1. Surface with a Low Charge Density. The interaction between water molecules and a clay surface with a low charge density is weak; consequently, water is easily desorbed from the inner Helmholtz layer, and a hydrophobic alkyl chain may be able to contact the oxygen plane, giving rise to weak van der Waals interactions (Fripiat et al., 1969). The long axis of the hydrophobic moiety of the organic species is parallel to the oxygen plane. As a result of the overlapping of the double layers belonging to parallel aluminosilicate layers, the ammonium group rotates so that the plane containing the three protons of the NH_3 group is no longer parallel to the oxygen plane and seems to be hydrated (Yariv and Heller, 1970). Diffusivity of alkylammonium ions increases with water content as a result of the "unkeying" of the cation from the silicate surface (Gast and Mortland, 1971).

The screening of the clay surface electric field by the hydrophobic moiety, discussed previously, leads to a decrease in the electrostatic repulsion of these surfaces. This has a great effect on the parallel aggregation and formation of tactoids and on the dehydration of the interlayer. Consequently, Na-rich layers expand much more upon contact with water than do the organic-rich layers (Theng, 1971).

Depending on the size of the alkyl chain and on the degree of saturation of the clay, the interlayer organic ions can lie in the interlayer space as a single ionic, a bi-ionic, or a pseudo tri-ionic layer, parallel to two opposing silicate layers. With long alkyl chains the "double layer association energy" may become predominant and exceed the van der Waals attraction. The long axis of the hydrophobic moiety will no longer be parallel to the oxygen plane.

2. Surface with a High Charge Density. The electrostatic field in region A is too strong for an exchange of polar water molecules with alkyl chains to take place. The long axis of the hydrophobic moiety will be tilted with respect to the oxygen plane of the silicate layer. The angle of tilting increases with increasing charge density of the unit layer.

Walker (1967) studied the interaction of n-alkylammonium ions with vermiculite. There is a "keying in" of the ammonium group in a hexagonal hole. Infrared evidence indicates that N–H...O interactions exist but are weak. With the N–C bond normal to the silicate, and the alkyl chain in the ideal trans-trans configuration, the chain makes an angle of $54^\circ 44'$ with the silicate layer plane. Bi-ionic layer complexes of vermiculite have been observed at high solute concentrations. In the bilayer complex the two layers are accommodated back to back, so that their hydrophobic tails are buried in the interlayer space interior and their hydrophilic heads constitute the top and bottom inner Helmholtz layer (Fig. 7.6 a–c).

2.1.3 Sorption of Organic Anions by Clay Minerals

Under certain conditions anions are sorbed on the broken-bond surface (see Sect. 1.1.2 of this chapter and Ch. 8, Sect. 3). From Eq. (7.2) it is to be expected that anionic species will be negatively sorbed within the interlayer space, which has a negative potential. Only if the short-range energy ΔG_s exceeds the sum of hydration energies ΔG_h and ΔG_p, micelle dissociation energy ΔG_d and coulombic repulsion $Z_i\varphi$, will anions be positively adsorbed. Negative adsorption of anionic herbicides by montmorillonite was observed by Frissel and Bolt (1962). In the presence of polyvalent metallic cations, benzoate anions were adsorbed due to short-range interaction of benzoate and the metallic cations (Yariv et al., 1966). Anions may be sorbed from aqueous solutions by clay minerals if they are able to form stable coordination species that are positively

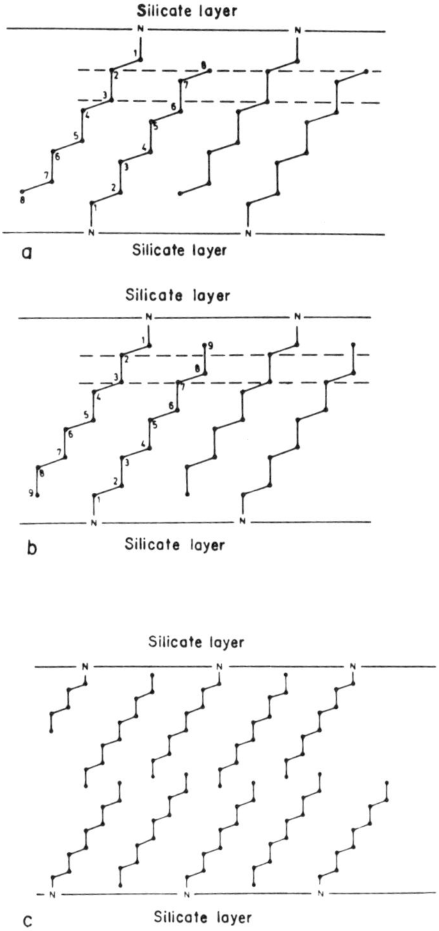

Fig. 7.6 a–c. Arrangement of the vermiculite and mica complexes with interpenetrating alkylammonium ions attached to opposing surfaces. NH_3^+ groups ionically bonded to the silicate surfaces are indicated by N, **a** monolayer of even-C alkylammonium complex; **b** monolayer of odd-C alkylammonium complex; and **c** double layers of alkylammonium complex. (After Walker, 1967)

charged (Yariv and Bodenheimer, 1964 a; Yariv et al., 1964 b). It seems probable that organic cations having more than one functional group may act as bridges between the organic anions and the clay surface, this enabling the sorption of the anions.

2.2 Sorption of Organic Polar Molecules by Clay Minerals

The process of adsorption of organic polar molecules and the distribution of the organic matter in the double layer depend on the nature of the exchangeable inorganic cation and on the hydration state of the clay. From the preceding section and from the discussion on the energy of surface dehydration which was given in section 1.1.1 this chapter, it appears that when the clay surface has a low charge density, water in the inner Helmholtz layer may be easily displaced by hydrated cations. The kind of exchangeable cation present, with its associated water molecules, determines the acidity of the clay surface. Within the double layer, the self atmosphere of a metallic cation is nonsymmetric about the given ion center, the water molecules lying closer to the clay surface being more polarized than those more remote. Nevertheless, sorbed hydrated cations are better proton donors than cations in the bulk solution, since the dielectric constant of the solvent in the double layers is lower than that in the bulk (Mortland, 1968).

2.2.1 Sorption Via the Mechanism of Proton Transfer

On the basis of their interaction in the double layer, organic molecules may be divided into those that are proton acceptors and those that are proton donors.

1. Proton Acceptors: Strong bases are protonated during sorption, yielding positive ions. The extent of this reaction depends on the basicity of the organic compounds and the polarizability of the metallic cation. The organic cation thus obtained may reach region A. The sorption of aliphatic amines within smectite interlayers exemplifies this process as follows (Laura and Cloos, 1975).

$$C_nH_{2n+1}NH_2 + H_2O \ldots M - Clay \rightarrow C_nH_{2n+1}NH_3 \cdot (OH)M - Clay \qquad (7.IX)$$

where M is a metallic exchangeable cation. The protonation of nitrogen in heterocyclic aromatic compounds is as follows (Cruz et al., 1968):

(7.IXa)

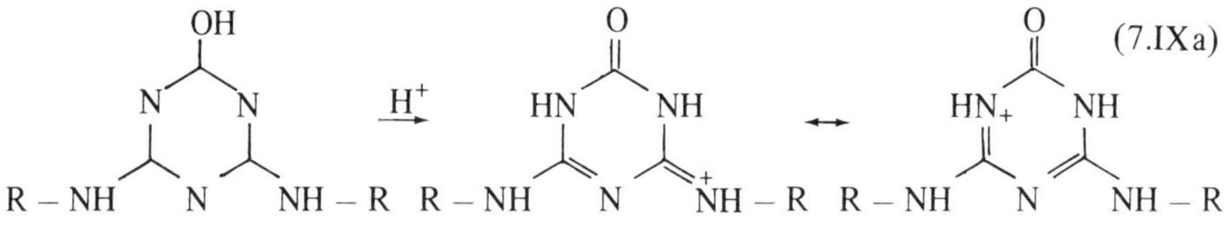

With decreasing basicity of the organic compounds or acidity of exchangeable metallic cations, association species are obtained with hydrogen bonding, wherein a water molecule acts as a proton donor. This can be illustrated by the sorption of aromatic amines such as aniline, $C_6H_5NH_2$ (Yariv et al., 1968), which are weak bases, as follows

$$C_6H_5NH_2 + H_2O \ldots M - Clay \rightarrow C_6H_5\overset{H}{\underset{H}{N}} \ldots H\overset{H}{O} \ldots M - Clay \qquad (7.X)$$

in which, the nitrogen atom serves as the nucleophilic site (Fig. 7.7c).

The conclusions drawn in paragraph 2.1.1 concerning the tilting of the hydrophobic chains of organic ions and the formation of hemimicelles are also valid for polar molecules. From the same analysis it follows that their relative concentration at the interface increases with an increase in their total concentration, or increasing length of alkyl chain. Their distribution in the double layer also depends on the charge of the inorganic cations with which they are associated. It follows from Gouy-Chapman calculations (Ch. 2, Sect. 4.1) that with increasing cation charge more molecules will approach the clay surface.

2. Proton Donors: These molecules may interact with two different surface sites: (1) basic sites in the oxygen plane and (2) negative poles of water molecules in the hydration spheres of cations.

A direct interaction of protons of the organic compounds with the oxygen plane (formation of hydrogen bonds) requires a low surface charge density of the clay surface and a low hydration energy of the inorganic exchangeable cations. This can be illustrated by the sorption of indoles, which are very weak acids, onto Cs-montmorillonite (Sofer et al., 1969), as follows:

$$C_8H_6NH + O(-Si\leqslant)_2 \rightarrow C_8H_6NH \ldots O(-Si\leqslant)_2 \qquad (7.XI)$$

This type of hydrogen bond may also occur in the presence of other cations if they are low in concentration at the surface and their hydration shells (zone A_m) occupy only a small fraction of the interlayer space.

Due to hydrogen bonding between water coordinated to polyvalent cations and oxygen planes (see structure 7.V) water molecules may react as bases and accept protons from acids through hydrogen bonding (Heller and Yariv, 1969). This can be illustrated by the sorption of phenol onto Al-montmorillonite (Saltzman and Yariv, 1975), as follows

$$C_6H_5OH + Al \ldots \underset{\displaystyle H}{\underset{|}{O}}{-}H \ldots O(-Si, Al\leqslant)_2 \rightarrow C_6H_5O{-}H \ldots \overset{\displaystyle Al}{\overset{|}{\underset{\displaystyle H}{\underset{|}{O}}}}{-}H \ldots O(-Si, Al\leqslant)_2. \qquad (7.XII)$$

Aniline and its derivatives form a wide range of organo-clay complexes with smectites in which the organic compound reacts either as a proton donor or acceptor. The various possible assemblages that are formed in the interlayer space are shown in Figure 7.7.

2.2.2 Sorption Via the Formation of Coordination Compounds

Transition metals in the double layers may form stable coordination complexes with amines that are good electron donors (Bodenheimer et al., 1962, 1963 a, b, 1966 a, b; Laura and Cloos, 1970). With di- and polyamines, which may coordinate to form 5- or 6-member chelates (7.XIII a, b), complexes are obtained in the interlayer space that are stable to hydrolysis. Coordination compounds of monoamines are easily hydrolyzed by water and may be stable only under dry conditions, or when the amines are in great excess.

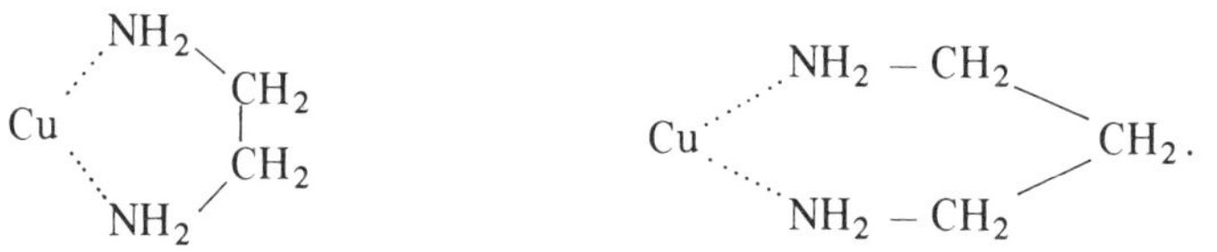

(a) (b) (7.XIII)

Stable coordination complexes of amino acids may form at the clay surface if strongly chelating cations such as copper, cobalt, and zinc are present. Bodenheimer and Heller

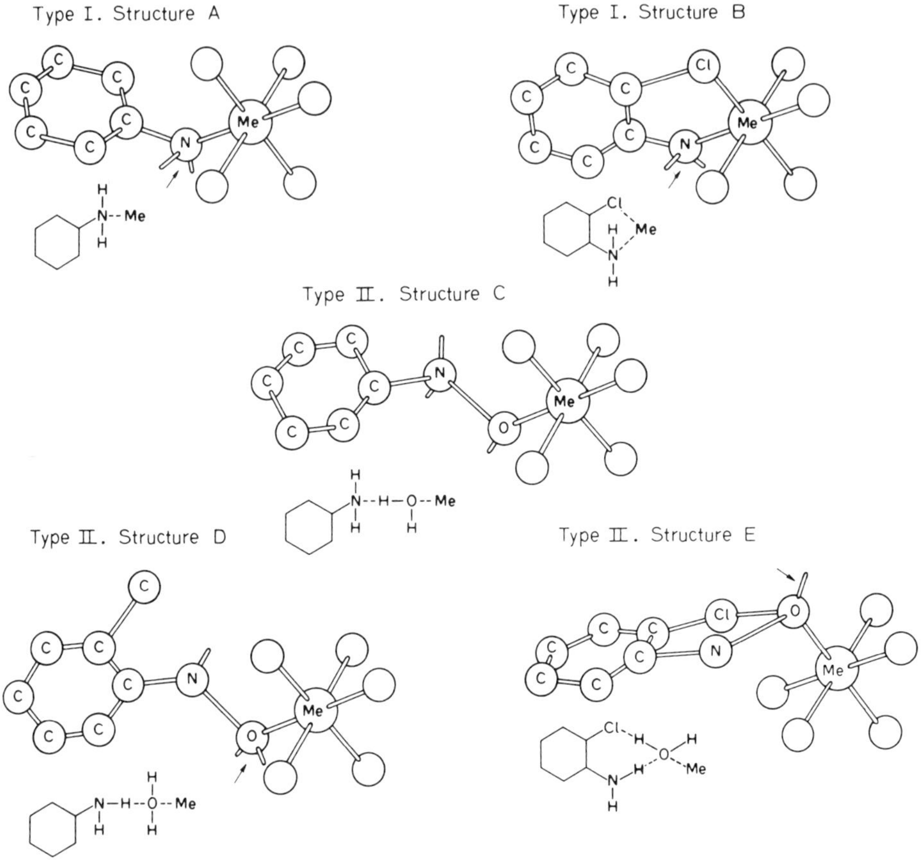

Fig. 7.7. Configuration of complexes of major and transition metal cations with aromatic amines and water in clay interlayers. (From Heller and Yariv, 1969: Used with permission of Keter Publishing House, Ltd.)

(1967) showed that the sorption onto montmorillonite of amphoteric and basic amino acids from dilute solutions at pH above 5 was enhanced in the presence of copper ions. Under similar conditions most amino acids are not sorbed by montmorillonite in the absence of copper ions. During the formation of the complex the amino acid is deprotonated and the anionic species forms a chelate in which the nitrogen and one of the oxygen atoms of the groups NH_2 and CO_2 are the nucleophilic sites (Heller-Kallai et al., 1972) as follows:

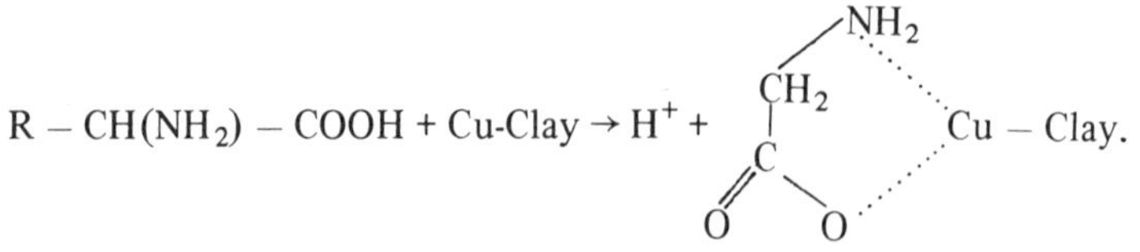

(7.XIV)

2.2.3 Sorption Via π Interactions

Aromatic molecules are sorbed onto the interlayers of Cu smectites via complex formation with the metal ion. Benzene forms two distinct types of complexes: (1) a type I species in which the Cu is edge-bonded to the benzene and aromaticity is retained, and (2) a type II species in which the benzene is oxidized by the metal cation to form an organic radical cation and the ring is distorted with some localization of the C=C bonds (Doner and Mortland, 1969; Mortland and Pinnavaia, 1971). The color of the type I complex is green and that of type II is red. Type II complex formation is not unique to benzene, and analogous species can be formed with aromatic molecules containing two or more benzene rings (Rupert, 1973). Phenol and alkyl-substituted benzenes can also complex with Cu via donation of π electrons, although only type I analogs are observed (Pinnavaia and Mortland, 1971; Fenn and Mortland, 1972). Thiophene forms with Cu smectite a type II complex (Cloos et al., 1973).

Ag smectites also form complexes via π interactions, but in all systems investigated, only type I complexes were observed (Clementz and Mortland, 1972). Fe(III) smectites and VO_2(II) smectites, in which the metallic cations are oxidizing agents, form with aromatic compounds only type II complexes (Pinnavaia et al., 1974).

2.2.4 Sorption of Fatty Acids by Clay Minerals.

It has been suggested that fatty acids are transformed into paraffins during burial of sediments (Kvenvolden and Weiser, 1967) and that this conversion is catalyzed by clay surfaces (Shimoyama and Johns, 1971; Waples, 1972). Sorption of organic acids onto the double layer of the flat unit layer surface is affected by the exchangeable inorganic cations. The carboxylic group is bound to the cation through a water bridge (Yariv et al., 1966).

Sorption of fatty acids from aqueous solution is predominantly due to their interaction with the broken-bond surfaces. This sorption depends on the pH of the environment. Since these acids are weak, their dissociation and the concentration of negatively charged organic species depend on the pH of the aqueous environment. The potential of the double layer also depends on the acidity of the system. Adsorption takes place by the mechanism of specific sorption of anions. There is an optimal pH value at which the surface charge is still positive and the concentration of the carboxylate anion is sufficiently high for maximum sorption to occur. At higher pH values, at which the positive surface charge decreases and may even become negative, sorption decreases, although the concentration of the anionic species increases. At lower pH values a decrease in the concentration of the negative species leads to a decrease in sorption of the acid. The maximum sorption of these acids from aqueous solutions occurs when the pH is equal to the pK_a of the acid (see pages 273–276).

Meyers and Quinn (1971, 1973) studied the sorption of long-chain fatty acids with various clay minerals from aqueous solutions. Their results are in good agreement with the preceding model of sorption mechanism and the structure of the double layer. The pK_a of the acids they studied was about 5. For systems of $pH \geqslant 5$ they found that fatty acid sorption by clay decreased 6 to 9% per pH unit as the pH became more basic. They found that the sorption increased with increasing concentration and chain length or decreasing temperature. These phenomena are in agreement with Eq. (7.2). These investigators studied the effect of NaCl salinity on heptadecanoic acid adsorption by montmorillonite and found it to increase with increasing salt concentration. This is a result of the increase of the "double layer association energy". As will be shown in the next chapter, an increase of the salt concentration leads to a compression of the diffuse double layer. It is then to be expected that the probability of organic chains associating and forming hemimicelles will be enhanced.

2.3 Organic Reactions Catalyzed by Clay Minerals

Adsorbed organic molecules may undergo chemical reactions on the clay surface. These reactions can be either positively or negatively catalyzed by the clay mineral. These reactions are divided into three groups: (1) oxidation–reduction reactions; (2) polymerization–depolymerization reactions, (3) transformation, synthesis, and decomposition reactions.

Simple colorless organic molecules convert to their colored derivatives when they are brought into contact with clay minerals. A well-known example of this type of reactions is the coloration of smectites by benzidine. The colorless neutral molecule of benzidine is converted by oxidation to its blue derivative, monovalent semiquinone, and by further oxidation to the yellow derivative, divalent quinone, according to

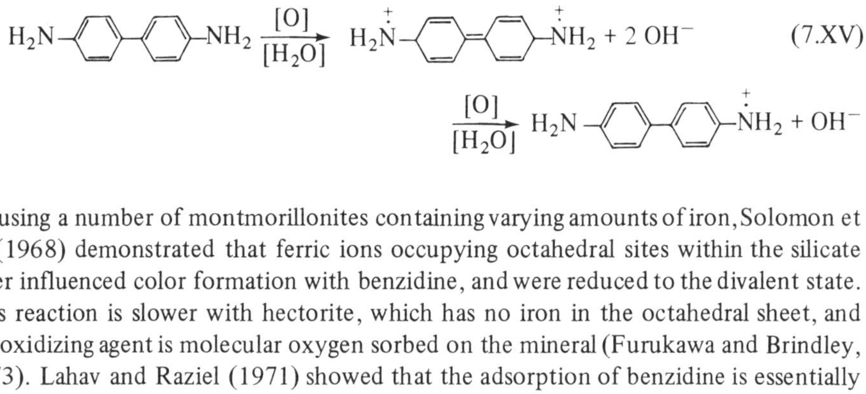

(7.XV)

By using a number of montmorillonites containing varying amounts of iron, Solomon et al. (1968) demonstrated that ferric ions occupying octahedral sites within the silicate layer influenced color formation with benzidine, and were reduced to the divalent state. This reaction is slower with hectorite, which has no iron in the octahedral sheet, and the oxidizing agent is molecular oxygen sorbed on the mineral (Furukawa and Brindley, 1973). Lahav and Raziel (1971) showed that the adsorption of benzidine is essentially one of cation exchange, the exchange reaction, however, being apparently irreversible. Besides influencing the amount adsorbed, the pH of the system also determines the color of the resultant complex. The divalent yellow quinone is formed mainly in acidic solutions, when the pH is below 2, whereas the blue semiquinone is obtained in more alkaline solutions. Radical complexes of this type seem to be responsible for the dark color of the organic–clay associations in soils and sediments.

Clays are able to influence and interfere with polymerization reactions. The initiation of polymerization of organic monomers by clays involves the conversion of the appropriate monomer to a reactive intermediate. The use of clays as a support for polypeptide synthesis has received much attention see, e.g., Degens and Mathéja, 1968; Fripiat et al., 1966; Paecht-Horowitz et al., 1970; Heller-Kallai et al., 1972). Polypeptide formation appears to increase with increasing surface acidity of the clay. It is considerably enhanced by prolonged standing in contact with excess amino acid solution and by heating. The formation of macromolecules by polymerization of adsorbed monomers in the smectite interlayers is not always possible. The conformation of the intercalated monomers is clearly an important factor in determining whether or not polymerization occurs (Theng, 1974).

Yoshino et al., (1971) synthesized a range of amino acids together with purines, pyrimidines, and some hydrocarbons, by reacting CO, H_2, and NH_3 at 473 to 973 K in the presence of montmorillonite. Paraffins can be cracked catalytically. Catalytic cracking requires the presence of acidic sites, capable of supplying available protons that initiate the reaction. By losing an hydride, the paraffin molecule becomes a carbonium ion. The carbonium ion produced may undergo a number of rearrangements (Theng, 1974).

The water contents of clay minerals affect the catalyzed reactions. Those reactions that are catalyzed by Lewis acid sites are best performed on dry clays whereas the reactions that make use of protons require slightly humid clay. Saltzman et al. (1976) showed that oven-dried kaolinite has a catalytic effect upon hydrolysis of certain insecticides. However, the addition of water to the dry kaolinite, to the extent required for ion hydrate formation, affects the degradation kinetics of the insecticides and increases the degradation rate. Further increase in the moisture content results in a sharp decrease in the degradation rate.

It has been suggested that simple organic molecules such as amino acids, purine and pyrimidine bases, and pentose sugars could have been readily formed on the primitive earth and thus made available for further transformation in the chemical evolution processes. Polypeptides and polynucleotides of significant lengths were obtained from monomers under abiotic, primordial conditions (Calvin, 1969). Clays provide the most likely surface on which small molecules could have been adsorbed and thus concentrated as a first step leading to polymerization. According to Lahav and Chang (1976) a high surface concentration is essential in the condensation reactions during chemical evolution. These are more likely to occur in dehydrated and frozen systems. Condensation rates and the size distribution of the oligomers formed during dehydration will be enhanced in changing environments, which represent systems where processes such as rainstorms, tidal variations, flooding, dehydration, and freezing take place in cyclic manners, one after another.

3. Solubility of Clay Minerals

Dissolution of clay minerals in water may proceed by the three following mechanisms: (1) the replacement of adsorbed metallic cations by H from water; (2) the depolymerization of the tetrahedral and octahedral layers; and (3) diffusion of atoms from the framework to the solid–liquid interface and from there to the bulk solution. These processes result in the formation of neutral or anionic species of monosilicic acid and hydrated species of the exchangeable and lattice metallic cations such as Na, K, Ca, Mg, Al, and Fe.

The replacement of metallic cations by H from water can be regarded as a step in the hydrolysis of the clay mineral. For example,

$$\text{Na-clay} + \text{HOH} \rightarrow \text{H-clay} + \text{Na}^+ + \text{OH}^-. \qquad (7.\text{XVI})$$

Barshad (1960) showed that a small fraction of the adsorbed protons, which had accumulated on the clay surface, entered the interior of the crystal to displace Al, Mg, and Fe whenever these were present. The metallic ions that were released from the crystal interior were able to exchange the residual adsorbed protons on the clay surface.

The depolymerization step is catalyzed by protons. The mechanism of the reaction is similar to that described in Chapter 6 for the depolymerization of silica. Protons are attached to nucleophilic sites on the surface or in the framework of the mineral followed by breaking Si–O, Mg–O, Fe–O, or Al–O bonds. The process begins at either the broken-bond surfaces or the flat oxygen planes. The latter proceeds via the formation of a transition complex of the type

$$\{[\gtrdot Si - \overset{\oplus}{O}(H) -]_3 Si - \overset{\oplus}{O}(H) [-Al, Mg(-OH)_2(-O-)_3]_2\}. \qquad (7.XVII)$$

The activation energy for the formation of this transition complex is very high, since it requires the penetration of protons into the unit layer and their attachment to oxygen atoms already coordinated to three or four atoms (Al, Mg, and Si atoms). In trioctahedral minerals the protons must exchange structural Mg atoms whereas in dioctahedral minerals the central oxygen atom must rotate so that a proton may be attached to the free pair of electrons. However, since clay minerals exhibit defects of the type of lattice vacancies, which lower the activation energy of lattice diffusion, formation of the transition complex may result. The negative charge on montmorillonite results in part from substitution of Mg for Al in the octahedral sheet. The nucleophilic site around the isomorphous substituent will attract penetrating protons more strongly. Thus, this is the preferred site for the formation of the transition complex (7.XVII). The diffusion of divalent Mg through a negative oxygen plane requires a lower activation energy than the diffusion of trivalent Al in di- and trioctahedral minerals. Octahedral substitution causes montmorillonite to be less stable and more susceptible to dissolution through diffusion of atoms in the solid phase.

Clay minerals are sparingly soluble compounds. The aqueous solubility of certain clay minerals is shown in Table 7.4. Most rapid dissolution occurs within the first 24 h and then begins to slow down. However, a very slow dissolution continues after 100 days and lasts for 2–2.5 years (Lerman et al., 1975).

The composition of the residue is determined by several of the following consecutive processes:

1. In the first stage a dissolution process of the surface proceeds via mechanisms (1) and (2) and is congruent, that is, the equivalent ratio between the elements other than oxygen and hydrogen in solution is equal to the ratio in the solid state. The congruent dissolution of several clay minerals can be described by the following chemical equations: for kaolinite

$$Al_2Si_2O_5(OH)_4 \rightarrow 2\,Al^{3+} + 2\,SiO_4^{4-} + 2\,H^+ + HOH \qquad (7.XVIII)$$

for montmorillonite

$$NaAl_7MgSi_{16}O_{40}(OH)_8 + 16\,HOH \rightarrow Na^+ + 7\,Al^{3+} + Mg^{2+} + 40\,H^+ + 16\,SiO_4^{4-} \qquad (7.XIX)$$

Table 7.4. Analytical data on dissolved species in the dissolution of clay minerals in deionized water, at room temperature. Samples were equilibrated for 102 days. (After Huang and Keller, 1973)

Species	Kaolinite	Montmorillonite	Illite
	Moles/liter	Moles/liter	Moles/liter
Na^+		0.170×10^{-2}	0.870×10^{-5}
K^+		0.169×10^{-4}	0.361×10^{-4}
Mg^{+2}	0.206×10^{-6}	0.393×10^{-4}	0.219×10^{-4}
Ca^{+2}		0.823×10^{-5}	0.389×10^{-4}
Al^{+3}			0.251×10^{-5}
$Al(OH)^{+2}$	0.185×10^{-7}		0.530×10^{-6}
$Al(OH)_2^+$	0.862×10^{-6}		0.178×10^{-5}
$Al(OH)_4^-$	0.464×10^{-7}	0.371×10^{-5}	
$Fe(OH)^{+2}$			0.323×10^{-5}
$Fe(OH)_2^+$		0.287×10^{-4}	0.108×10^{-5}
H_4SiO_4	0.160×10^{-4}	0.161×10^{-3}	0.798×10^{-4}

Kaolinite from Georgia with the specific formula $(Al_{1.98}Mg_{0.02})\,Si_2O_5\,(OH)_4$

Montmorillonite from Wyoming (Clay spur) with the specific formula $(Na_{0.27}Ca_{0.10}K_{0.02})\,(Al_{1.52}Fe'''_{0.19}Mg_{0.22})\,(Si_{3.94}Al_{0.06})\,O_{10}\,(OH)_2$

Fithian illite with the specific formula $(K_{0.59}Na_{0.02}Ca_{0.01})\,(Al_{1.54}Fe'''_{0.29}Mg_{0.23})\,(Si_{3.47}Al_{0.47}Al_{0.53})\,O_{10}\,(OH)_2$

and for illite

$$KAl_2Si_3AlO_{10}(OH)_2 \rightarrow K^+ + 3\,Al^{3+} + 2\,H^+ + 3\,SiO_4^{4-}. \qquad (7.XX)$$

2. The dissolution follows mechanisms (3) and becomes incongruent, and the rate of dissolution of any ion depends on its diffusion rate in the solid phase.

3. The species obtained from the dissolution of Si, Al, and Fe are pH dependent. The process taking place in water yields soluble species such as $H_3SiO_4^{1-}$ and H_4SiO_4, Al^{3+}, $[Al(OH)]^{2+}$, $[Al(OH)_4]^-$, Fe^{3+} and $[Fe(OH)]^{2+}$ and insoluble species such as $SiO_2.xHOH$, $Al(OH)_3$ and $Fe_2O_3.yHOH$. The effect of acidity on the solubility mechanism of the clay mineral is shown by means of the reactions of kaolinite, in which the alkalinity of the system increases from Eq. (7.XXI) to Eq. (7.XXV) as follows:

$$Al_2Si_2O_5(OH)_4 + 6\,H^+ \rightarrow 2\,Al^{3+} + 2\,H_4SiO_4 + HOH \qquad (7.XXI)$$

$$Al_2Si_2O_5(OH)_4 + 4\,H^+ + HOH \rightarrow 2\,[Al(OH)]^{2+} + 2\,H_4SiO_4 \qquad (7.XXII)$$

$$Al_2Si_2O_5(OH)_4 + 5\,HOH \rightarrow 2\,Al(OH)_3 + 2\,H_4SiO_4 \qquad (7.XXIII)$$

$$Al_2Si_2O_5(OH)_4 + 2\,OH^- + 5\,HOH \rightarrow 2\,[Al(OH)_4]^- + 2\,H_4SiO_4 \qquad (7.XXIV)$$

$$Al_2Si_2O_5(OH)_4 + 4\,OH^- + 3\,HOH \rightarrow 2\,[Al(OH)_4]^- + 2\,H_3SiO_4^-. \qquad (7.XXV)$$

In natural waters in which the pH ranges between 4 and 9, aluminum and ferric iron occur as insoluble hydroxides.

4. The soluble species obtained in process (3) may be sorbed onto the surfaces of the nonsoluble fraction of the mineral particle. Silica is sorbed by the mechanism of specific anion sorption whereas the metals are sorbed either by cation exchange or by the hydrolytic adsorption mechanism. In this process clays in aqueous suspensions evolve toward aluminum-saturated clays.

5. Polyvalent cations react with silicic acid in various ways. They form sparingly soluble silicates and catalyze the formation of polysilicic acid. They are then sorbed onto the polymer, resulting in the flocculation and settling out of the silica.

From processes 3–5 it follows that certain amounts of soluble species revert from aqueous solution to the solid phase. These amounts depend on the chemical properties of the element and on the chemical environment.

From this model a few important conclusions on the solubility of clay minerals can be drawn. The solubility is incongruent and depends on the acidity of the solution. In general acids remove alkali, alkaline earth, Fe and Al metals from the solid phase whereas silica dissolves in basic solutions. Mg-rich minerals are much more soluble than those rich in Al, with Fe-rich minerals somewhere in between. In dioctahedral clay minerals Mg is released in preference to Al. Solubility increases with increasing specific surface area and decreasing particle size of the mineral.

Barshad and Foscolos (1970) studied the factors affecting the rate of conversion of acidic vermiculites, montmorillonites, and illites into Al clays. They found that the interchange reaction is a first-order reaction and that the rate of the reaction is directly proportional to the surface charge density and to the MgO content in the interior of the clay crystal.

Dissolution by process (1) predominates with those clay minerals in which the charge originates from tetrahedral substitution, whereas dissolution by process (2) predominates with minerals having octahedral substitution. The dissolution of illite, a mineral with a tetrahedral substitution, was shown by Feigenbaum and Shainberg (1975) to proceed mainly via process (1). At pH above 3, the rate of Al release was similar to the rate of K release, whereas the rate of Fe and Mg release was about three times that of K and Al release. Shainberg et al. (1974), comparing the chemical stability of four Na-saturated montmorillonites differing in their octahedral isomorphous substitution, found that the dissolution proceeded via process (2) and was proportional to the square root of the time, and that the proportionality was linearly related to the degree of octahedral substitution.

In a study of the dissolution of montmorillonite in acid, Churchman and Jackson (1976) showed that the mineral particles were not in a unique state of equilibrium with all solutions during the process. A secondary solid phase, enriched in silica, appeared to control the activities of ions and other solutes during the initial stages of the reaction of montmorillonite with acid aqueous solutions. Later, a secondary solid phase was formed, which was enriched in alumina relative to montmorillonite. Churchman and Jackson questioned the validity of deriving thermodynamic properties of montmorillonite by the application of a single-step dissolution equation to results of solubility studies in mildly acidic solutions.

Bar-On and Shainberg (1970) leached Na-montmorillonite with distilled water and found that exchangeable Na was replaced predominantly by Mg and to a smaller extent by Al (process 4), which were dissolved from the crystal by a selective diffusion mechanism. The concentration of Na in the effluent was constant at a value of 1.1×10^{-4} mol/l until all exchangeable Na was released from the clay. At this stage 15% of the clay had dissolved and the exchange capacity decreased markedly (Fig. 7.8).

The presence of organic compounds will change the solubility characteristics of clay minerals as follows:

1. Carboxylic Acids. These acids compete with silicic acid for the sites at which specific sorption of anions occurs. They are stronger acids than H_4SiO_4 and will replace the adsorbed silica in acidic solutions. Consequently, the solubility of silica increases in the presence of these acids. This competition and the solubility of the mineral depend on the sorption affinity of the acid for the clay surface.

2. Chelating Agents. $\alpha\beta$ Dicarboxylic acids and α hydroxy acids, which may form stable soluble coordination compounds with Al, Fe, and Mg, in which the complexed species is negatively charged, increase the solubility of these elements. A chelate formation between iron and fulvic acid was suggested by Kodama and Schnitzer (1973) to be responsible for the very high dissolution of iron-rich chlorites in fulvic acid.

3. Reducing Agents. Since the solubility of ferric oxide is much lower than that of ferrous hydroxide, the presence of organic reducing agents greatly increases the solubility of this element.

4. Flocculating Agents. Some types of organic compounds may serve as flocculating agents. The mechanism of this process will be described in the next chapter. After the agents have been sorbed and the clay mineral flocculated, the specific surface area of the solid phase decreases and consequently the total solubility of the mineral decreases.

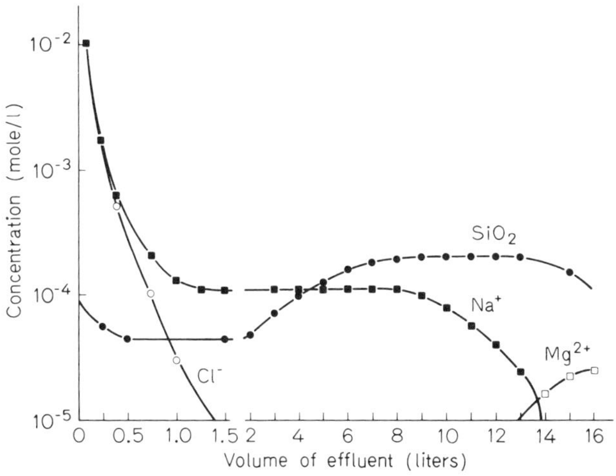

Fig. 7.8. The concentration of Na, Mg, Cl, and SiO_2 during the leaching of Na-montmorillonite. (From Bar-On and Shainberg, 1970: Used with permission of the publisher and of the authors)

Huang and Keller (1971) showed that the total weight of clay minerals dissolved by strongly complexing organic acids (representative of components of humic acids) exceeds that dissolved by distilled water by factors of 5 to 75. Dissolved Si in acid solution is 2–35 times greater than its concentration in distilled water whereas dissolved Al in acid solution exceeds its distilled water solubility by a factor of 3–500. The dissolution behavior of Fe parallels that of Al. The explanation for the relatively high dissolution of Al and Fe in organic acids lies in their chelation with the acids. Ca, Na, and K may be released from the clay minerals by ion exchange reactions rather than by destruction of the framework. These ions typically are so highly soluble in pure water that chelation processes are not detectable. However, aspartic acid, which is an amino acid, is especially active in dissolving Ca and Mg.

The process of leaching silica from clay minerals by natural waters was discussed in the previous chapter. According to Huang and Keller there is a difference between leaching in an environment where humus is abundant and in one which is poor in organic constituents. In the first case, Al and Fe are dissolved in excess of congruence with Si, and the residue, known as "podsol", is rich in silica, whereas in the second, leaching of silica occurs in preference to Fe and Al and the weathering end product, which is known as "laterite", is rich in alumina and ferric oxide. The processes are called *podsolization* and *lateritization*, respectively.

4. Environmental Effects on Clay Mineralogy

This subject has been reviewed by Keller (1970) and by Millot (1970) and the present section is based mainly on these reviews, supplemented by results reported in more recent studies.

Parent material of clay minerals includes (1) solid nonclayey minerals, such as tectosilicates, inosilicates, glasses, and so on; (2) ions in aqueous solution; (3) colloid phases, including gels; and (4) previously existing clay minerals or other layer silicates. Two categories of clay minerals can develop from these parent materials: (a) newly formed, i.e., those which are clay minerals for the first time and (b) those resulting from a change, either chemical or mineralogic, of previously existing clay minerals. The first category, which can be entitled "the primary stage" mineral, is produced from parent materials (1), (2), and probably (3). The second category, which can be entitled "the N+1 stage" clay mineral, is produced from parent material (4), which may interact with (2) and (3).

The surroundings of the reacting material, alternatively called the *environment*, determines the energy imposed upon the materials during the time of formation and throughout the changes of the mineral. Geologic, chemical, and physical characterizations of energy are given in Table 7.5. The clay mineral formed in any geologic environment may be a stable or equilibrium product of the reaction, or alternatively, it may be a metastable form controlled by the reaction rate. To relate geology and physical chemistry most effectively, it may be necessary to define the environment geochemically, outline the chemical reaction, then determine the latitude of the range over which the reaction is operative, and translate this range into geologic conditions. While the physicochemical reaction defines the possible directions of the process, the geologic observation shows what are the actual and real products.

The geochemical cycle of clay minerals goes through three zones: (1) the weathering zone, in which clay minerals are neoformed from nonclayey minerals; (2) the erosion and sedimentation zone, in which clay minerals are transformed; and (3) the zone of burial, which is the zone of diagenesis and metamorphism of the minerals.

A primary-stage clay mineral or rock carries the imprint of its genesis, and the environment of its formation may be inferred from the nature and characteristics of the clay. When a primary-stage clay mineral is transported to a new environment, three types of behavior are possible. The first type, in which there is no visible change in the clay mineral, is found when the reaction rate of the mineral with the new environment is too slow or the activation energy is too high for any chemical reaction to occur. The second possibility is the condition in which the clay mineral, transferred to the new environment, shows characteristics of two specific environments. The combination of characteristics enables an accurate interpretation to be made of the several argillic events that have occurred. A third alternative is that the new environment into which the clay mineral has been introduced reacted with the clay mineral and the product reflects only this new environment. Should this happen, all traces of the argillic past will have been obliterated.

The modification of the clay mineral crystal under the influence of erosion or sedimentation environments is called *transformation. Diagenesis* defines changes in the clay mineral structure that take place in the crust zone located between sedimentation and metamorphism zones (Millot, 1972).

Table 7.5. Geologic, chemical, and physical characterizations of energies or environments, imposed on constituent material. (After Keller, 1970)

1. Geologic characterizations of environments:
(a) Marine, fluviatile, lacustrine, aeolian, etc.
(b) A or B-zone of soil, hydrothermal, below ground-water table etc.
(c) Climatic, humid, tropical, arid, etc.
(d) Oxidizing, acidic, alkaline, etc.
(e) Organic, inorganic, living, dead
(f) Uniform vs. fluctuating conditions in time
(g) Uniformity of conditions in widespread space or megaenvironment vs. uniformity over a few nm units distance, i.e., microenvironment

2. Chemical characterizations of environments:
(a) Concentration of protons, pH
(b) Concentration of electrons, Eh, or oxidation potential
(c) Concentration of various metallic cations
(d) Concentration of monomeric silicic acid and soluble polymeric silica
(e) Polymorphs of SiO_2 present
(f) Concentration of soluble aluminum
(g) Concentration of organic and inorganic compounds that form soluble complexes with silicon and/or aluminum

3. Physical characterizations of environments:
(a) Temperature
(b) Pressure
(c) Time–duration of the interaction of a single geochemical event
(d) An interval separating two (or more) distinguishable events in a particular argillic sequence

Soil clay minerals may be either inherited directly from the parent material, more or less epigenetically altered, or produced by the process of chemical weathering in the soil. Marine clay minerals may also be either unchanged or modified detrital terrigeneous and submarine volcanogenic material or they may be autigenic minerals precipitated from marine solutions. According to Chukhrov (1966) the mineral composition of marine clays derived from humid zones is primarily unchanged terrigeneous material and authigenesis is relatively unimportant. In contrast, marine clay minerals derived from arid regions, where the effect of chemical weathering is small, are formed during the process of sedimentation and in subsequent transformation. Detrital illites and chlorites from weathering zones in polar regions do not undergo transformation in marine environments and are found in recent marine sediments in polar regions whereas detrital kaolinites formed in tropical lateritic soils are found in the equatorial areas (Biscaye, 1965).

Not only do environments influence the clay minerals, but these minerals influence the environment around them (Wedepohl, 1970). Sillén (1967) considered the effects of silicates on the composition of ocean water. He concluded that clay minerals constitute the major factor determining the pH of ocean water because of their ability to sorb protons. Weaver (1967), after extensively studying ion exchange by clay minerals in the ocean, concluded that clays extract and fix K ions from the ocean and contribute easily desorbable Na ions to the ocean (pages 298–299).

The concentrations of transition and heavy metals in natural waters and in soil solutions are controlled by clay minerals. The processes proposed for this control mechanism include surface sorption and penetration of ions into the interior of the crystal (Jenne, 1968). Surface sorption takes place by one of the following chemical reactions: (1) ion exchange, (2) surface ion complex formation between the transition or heavy metallic ion and an organic compound, and (3) hydrolysis of the metallic cation and the sorption of an anionic hydroxy complex or hydrous oxide. Jenne concluded that the hydrous oxides of Mn and Fe furnish the principal control for the fixation of transition and heavy metals. These metals form hydrous oxides, which coat particles of clay minerals. The common occurrence of these oxides as coating allows the oxides to exert chemical activity far out of proportion to their total concentrations.

Rare earth elements in nature accumulate by adsorption on clay minerals. The content and distribution of these elements in clays are strongly influenced by environmental factors. Under neutral and alkaline conditions the rare earth elements are accumulated by ion exchange, occupying the interlayer space of 2 : 1-type minerals. By increasing the hydrogen ion concentration, the adsorbed ions are readily removed. During illitization and fixation they occupy positions in the interlayers similar to those occupied by K in micas (Roaldest, 1973).

The behavior and stability of organic compounds in nature are strongly dependent on clay minerals. These induce chemical transformation of organic matter, which is usually explained on the basis of the catalytic action of the clay surface, through either hydrolysis, oxidation, reduction, or decomposition of the organic matter adsorbed. They also affect the migration of organic matter. This will be discussed further in Chapters 8 and 10.

4.1 Origin of Primary-stage (Neoformation) Clay Minerals and Their Related Environment

Processes by which primary-stage clay minerals are formed include: (1) weathering of nonlayered silicate minerals by leaching; (2) genesis of clay minerals by interactions between oxides or nonlayered silicate minerals and aqueous solutions; (3) direct crystallization from solutions; (4) crystallization from colloid gels; and (5) alteration of nonclayey layered silicates, such as biotite or muscovite.

4.1.1 Genesis of Clay Minerals by Weathering of Nonlayered Silicates

Weathering is the most important clay-forming process. The physicochemical reactions that occur during this geologic process were discussed in Chapter 3. Any of the clay mineral families can be produced in the weathering zone. The genesis of each of the clay minerals can be related to certain distinguishing materials or energies. Geologic environments for kaolinitization and smectitization of aluminosilicates are summarized in Table 7.6.

Kaolin Family Minerals: Any aluminum silicate parent material can yield kaolin by weathering provided that K, Na, Ca, Mg, and Fe are leached away and substituted by H. In aqueous weathering systems H is readily available, so that the external action is confined to the removal of the metallic cations and silica from the system. From the

Table 7.6. Geologic characteristics of environments of primary kaolinitization and smectitization. (After Keller, 1970)

Kaolinitization	Smectitization
A. Removal of Ca, Mg, Fe, Na, K 1. Precipitation exceeds evaporation 2. Leaching Permeable rocks Percolating water 3. Oxidation of Fe^{2+} to Fe_2O_3 or precipitation of FeS_2	A. Retention of Mg, Ca, Fe, Na 1. Evaporation exceeds precipitation, as in semi-arid climate 2. Ineffective leaching Stagnant water and water logging Lakes and marine basins Low effective permeability of rocks 3. Fe not combined with O and S 4. Silicates characterized by high specific surface area and high surface energy, adsorbing Mg, Ca, and Fe ions
B. Addition of H^+ 1. Fresh water 2. Acids, Sulfur compounds, carbonic and organic acids. Organisms	B. Alkalinity Solutions rich in alkali and alkaline earth metal ions in weathering zones
C. High Al : Si ratio 1. Leaching 2. Peptization of silica with Na and K and organic peptizing agents 3. High concentration of Al	C. Retention of silica 1. Ineffective leaching 2. Flocculation of silica with Ca, Mg, and Fe and certain organic flocculating agents 3. Clay-size cristobalite

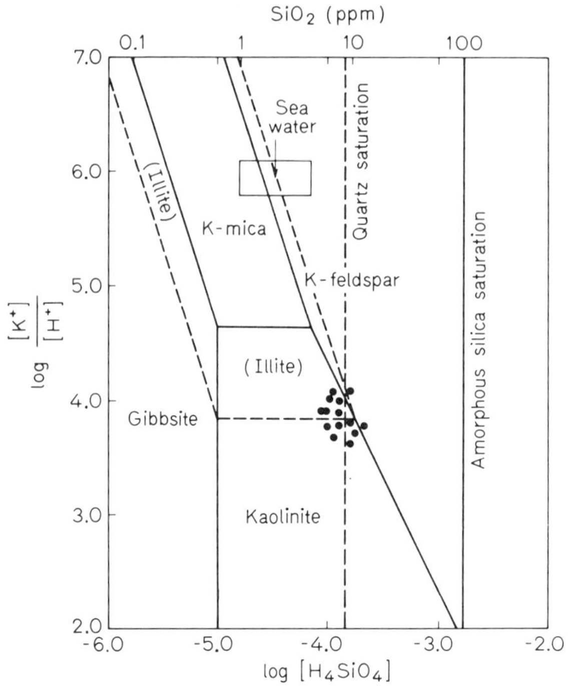

Fig. 7.9. Stability diagram of the system $K_2O-Al_2O_3-SiO_2-H_2O$. (From Keller, 1970: Used with the permission of the publisher)

stability diagram of Figure 7.9 it is obvious that at any concentration of H_4SiO_4 occurring in geologic solutions, such as at saturation with respect to quartz, kaolinite is stable when the ratio of $[K^+]/[H^+]$ is below a certain value. Acidity of ground-water weathering solutions need not be high for kaolinite to be formed. Even at the pH of ocean water, if the concentration of K is low, kaolinite may persist.

Although kaolinite may result from hydrogenation of any aluminum silicate, silicates rich in K and Na apparently yield kaolinite more readily than those rich in Ca and Mg. Granites are weathered directly to kaolinite, whereas gabbros are commonly converted first into smectites. K and Na silicates formed during the hydrolysis process of weathering are highly soluble and are leached away, enriching the residue in Al. This takes place in the weathering of granites. In gabbros, where Ca, Mg, and Fe are abundant, silicates of low solubility are obtained. Moreover, the amorphous silica gel that is obtained adsorbs these elements and flocculates. Hence, a high concentration of SiO_2 is maintained in the argillizing system and results in the formation of smectite minerals.

Smectite Family Minerals: Analyses of smectites obtained from various origins indicate that this group is characterized by a wide variety of elements that may occupy positions in the octahedral sheet, by variable distribution of charge between octahedral and tetrahedral sheets, and by a variety of exchangeable ions in the interlayers. It is therefore logical to suggest that the conditions under which these minerals are formed are the broadest of all the clay mineral families. Stability relationships of some phases in the system $Na_2O-Al_2O_3-SiO_2-H_2O$ at 25 °C and 1 atmosphere are shown in Figure 7.10. A complete stability diagram for smectite minerals requires additional parameters to describe activity of elements such as Mg, Ca, and Fe relative to H.

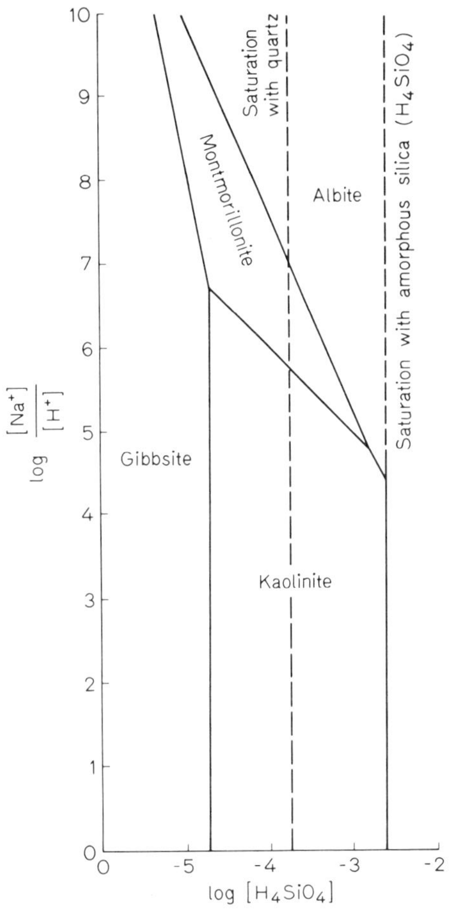

Fig. 7.10. Stability diagram of the system. $Na_2O-Al_2O_3-SiO_2-H_2O$. (From Keller, 1970: Used with the permission of the publisher, The Society of Economic Paleontologists and Mineralogists, The American Association of Petroleum Geologists)

Formation of smectite in preference to kaolinite occurs in regions where the quantity of water is sufficient to dissolve the parent rock, but which then evaporates, leaving the soluble salt in the weathering system, with almost no leaching. Such a situation obtains, for example, in a semi-arid region. Volcanic ash is a common parent material for montmorillonite. Alteration of the ash may have occurred under nonmarine conditions such as in lakes and in percolating water or in marine basins. Volcanic ash is composed of phases of a very low degree of crystallinity and a very high surface energy; consequently, it is easily leached and recrystallizes with even small amounts of percolating water. However, cations such as Ca, Mg and Fe, and silicic acid are adsorbed by the remaining surface-active solid fraction, leading to the formation of smectite (mainly montmorillonite). These elements can be leached from the argillizing system under mild climatic conditions, that is, with a high acidity of the weathering zone and in the presence of large quantities of water, giving rise to the formation of allophane and imogolite, or halloysite. Mixed-layer kaolinite-montmorillonite has been described as being developed through the weathering of volcanic ash under acidic conditions (see, e.g., Schultz et al., 1971).

Illites: From Figure 7.9 it is obvious that a high ratio of $[K^+/H^+]$ versus H_4SiO_4 is the factor responsible for illite formation in argillizing systems of aluminum silicates. The weathering of K-feldspar to illite occurs at high potassium and silica concentrations and at a low acidity of the reacting solution. At low concentrations of K and SiO_2 and at a high acidity, kaolinite will be the weathering product.

Weathering Products of Ultrabasic Rocks: Minerals of ultrabasic rocks are very susceptible to weathering. Under conditions of weathering these rocks alter to serpentinites. The alteration of phlogopites follows serpentinization of the olivines and alteration of the spinel. Serpentinization of the olivine depends on the nature and circulation of aqueous solutions. Olivine alters to serpentine mainly by losing Mg and Fe, whereas phlogopite alters to vermiculite and chlorite by losing alkalis and alumina. Phlogopite also forms talc, which is later generated into serpentine and vermiculite by loss of H_2O and Fe. The ions removed from altered primary minerals either precipitate in the parent rock or can be removed from the source rock by aqueous solutions and precipitate elsewhere to form secondary Mg, Fe, and Ni deposits (see, e.g., Rimsaite, 1972).

4.1.2 Genesis of Clay Minerals by the Interaction Between Oxides or Nonlayered Silicate Minerals and Solutes

Apart from weathering processes that produce clay minerals by leaching ions from the weathered rock material, the interaction of nonclayey minerals with various solutions and the addition of solutes, such as SiO_2, also may result in genesis of clay minerals. The replacement of gibbsite by chlorite in ocean sediments in Waimea Bay, Hawaii, can be used to demonstrate the phenomenon (Swindale and Fan, 1967). When transported to the sea, the gibbsite was deposited in a solution containing silicic acid, Mg, K, and H ions, in which it is unstable. Part of the deposited gibbsite interacts with the environmental solution to give chlorite. Replacement is evidence that the guest material (chlorite in the present case) was less soluble in the prevalent environment than the host substance (gibbsite). The chemical reaction that occurs during the replacement is spontaneous.

Neoformation of montmorillonite in the Bekaa Plain (Lebanon) is another example of this mechanism (Paquet and Millot, 1972). Under a mean annual rainfall of some 300–500 mm, in poorly drained lowlands where humid seasons alternate with dry seasons, the ions liberated into solutions on highlands and brought to the plain during humid seasons become concentrated during dry seasons and authigenesis of montmorillonite takes place by their interaction with minerals of the plain.

Kaolinitization in some localities is the result of a process of addition of silica onto an aluminous support (Delvigne, 1965). During hydrothermal alteration of silicates, kaolinite may result from the addition of silica to an aluminic phase (Tchoubar, 1965; Oberlin and Couty, 1970).

4.1.3 Genesis of Clay Minerals by Direct Precipitation from Solutions

Clay minerals deposited from solutions show loose aggregates of euhedral crystals, with a chemical composition approximating that of the ideal structural formula and showing the presence of nearly all theoretically possible X-ray diffractions. Neoformation takes place to a considerable extent in basin areas where the neighboring land area has undergone intense weathering. The material is largely in solution with little or no detrital material. Smectite, palygorskite and sepiolite were precipitated under such conditions, for example in various areas in West Africa (Millot, 1970).

Well-formed kaolinite crystals occur in small geodes in illitic shales near Keokuk, Iowa (Keller et al., 1966), in cavities in the Mississippian Chouteau, Missouri, and in the Joplin lead-zinc district in Southwest Missouri (Tarr and Keller; 1937). The characteristic texture of kaolinite that has crystallized from solution within a cavity is made up of euhedral plates, 5–15 μm diameter, which occur individually or face to face, aggregated in packets in loosely expanded books up to 20 μm in thickness. Porosity of the clay mass is high (Keller, 1976). The geochemistry of deposition of these kaolinites is obscure. A typical Keokuk geode may contain 15 g of kaolinite. From conventional solubility data Keller showed that the volume of solution required to bring in the alumina would be over 2 million liters. He suggested that complexed ions of Al with organic matter, which result in more concentrated Al solutions, may have provided Al transport. Another possibility is that solutions of hydrophilic colloids of polymeric aluminum hydroxide supplied the Al for the kaolinite formation in the cavities. The colloid solution is stable as long as the aqueous system is turbulent, but when it comes to rest in a cavity, alumina species grow by the mechanism of condensation and in the presence of silica kaolinite precipitates (see Ch. 3).

Rex (1966) prepared a stability diagram for illite, kaolinite, feldspar, and their aqueous solutions in a system of $K_2O–Al_2O_3–SiO_2–H_2O$, which is shown in Figure 7.9. The Figure indicates that the concentration of K must have been very low for kaolinite to crystallize at pH 7.

Dickite crystallizes from warm solutions in preference to kaolinite. It maintains euhedrism to a higher degree than kaolinite in packed lumps of clay. Keller (1976) suggested that Hg, Pb, and Zn are important catalysts for the formation of dickite and may be more important for the formation of this mineral than hydrothermal temperature alone.

4.1.4 Genesis of Clay Minerals by Aging and Recrystallization of Amorphous Gels

This process can be demonstrated by the formation of the well-crystallized kaolinite found in flint clays of North America (Keller, 1968). The characteristic properties of this kaolinite, which distinguish it from other kaolins, are its high resistance to slaking in water, lack of plasticity when wetted, and high uniformity and homogeneity in texture and composition. It is clearly different in properties from common types of kaolinites. In flint clay the kaolinite tactoids are randomly oriented and interlocked. The kaolinite had been crystallized from a cold aluminosilicate amorphous gel and from very small particles of a colloid size, by the mechanism of orientation, forming a nearly monomineralic homogeneous rock of interlocking crystals. The parent material was composed of hydrated alumina and hydrated silica that accumulated in low lying, marshy basins. Some of the parent material may have been fine-grained kaolin minerals from weathering zones washed into the basins. These materials underwent diagenesis, being transformed from quasi-illitic and kaolinitic material into an essential monomineralic kaolinite by recrystallization. Ions dissolved from these weathering parent-material sources and from the recrystallized materials during aging were transported by feeble currents into vegetation-ringed basins.

4.1.5 Genesis of Clay Minerals by Weathering of Layered Silicates

The weathering of dioctahedral micas gives illites (Jørgensen and Rosenqvist, 1963). The weathering of trioctahedral micas is an extremely fast process even in glacial climates (Reynolds, 1971). They are weathered to vermiculite with a mixed layered mineral as an intermediate stage (Mitchell, 1961; Norrish, 1972; Englund and Jørgensen, 1973). (See Ch. 3).

4.2 Environmental Relationships of N + 1 Stage Clay Minerals

The properties and characteristics of a clay mineral may change when the environment of the mineral is changed. For example, kaolinite transported by fresh-water streams to the ocean is subjected to dissolution and flocculation. Dissolution of kaolinite in ocean water may be expected because of the differences in composition between ocean water and an aqueous solution of this mineral. Kaolinite dissolves slowly in ocean water. If other clay minerals are associated they yield equilibrium amounts of dissolved SiO_2 sooner, and thus possibly inhibit further dissolution of kaolinite. If the amount of kaolinite brought to the ocean exceeds that which can be dissolved, it can either flocculate and settle out or be changed by transformation under marine conditions. Some investigators suggest that kaolinite reacts with silica and K ions to form illite. Such a reaction requires an extremely high activation energy, since it involves the transformation of a 1:1-type mineral into a 2:1-type mineral. Siever and Woodford (1973) suggest that the sorption of silica and K by kaolinite may result in two products: (1) a siliceous kaolinite, or "anauxite", which forms at a low K concentration, and (2) an illite-like phase that occurs at higher K values. Maynard (1975) suggests that in a first step a monolayer of adsorbed silica is formed around the kaolinite. In a second step amorphous versions of K-montmorillonite and illite are likely products. Evidence of recent marine sediments suggest that little, if any, transformation of kaolinite to a 2:1-type or other clay mineral occurs while the kaolinite is at or near the depositional interface with ocean water.

After deposition and burial, kaolinite may be converted to illite if the temperature is raised to 175–200 °C with burial pressure, and if the availability of K is greater than that which occurs in ocean water. This will be discussed in Chapter 10.

Smectites of primary-stage weathering on land are often carried into the oceans by rivers. They may be lost by dissolution, and in some localities it has been demonstrated that they were converted into other clay minerals of a 2:1-type, such as chlorite-vermiculite mixed-layers before deep burial (see, e.g., Huang et al., 1975). The rate of sedimentation of smectites is slower than that of kaolinites. Smectites are relatively stable in a semi-arid climate, in water-logged soil, or at the depositional interface of an open circulation marine environment. If pore water percolating through smectites has a low $[K^+]/[H^+]$ ratio, silica, magnesium, and alkali metals are leached and smectites will change to kaolinites. On further leaching bauxite will be formed. This transformation is associated with a drop in the cation exchange capacity of the clay fraction (Loughnan, 1969). (See also pages 277–278).

Relatively little illitization of smectites takes place at or near the depositional interface of marine sedimentation. After deep burial of marine and nonmarine deposits, large amounts of illite are formed from smectites by diagenesis. Elevated temperatures and pressures accompanying deep burial of sediments are important for formation of illite from smectites. Sufficient K and Al must be available. The diagenesis process will be discussed in Chapter 10.

Podzolic weathering of illites is the reverse reaction and will frequently result in the formation of smectite and interstratified illite-smectite (see, e.g., Weir and Rayner, 1974).

Degraded illites differ from montmorillonites in having a much higher content of tetrahedral aluminum. Like smectite minerals they may be expanded with organic compounds such as ethylene glycol, but unlike montmorillonite, they become nonexpanding after being saturated with K ions. Degraded illites are converted into illites in seawater and in recent marine sediments. Only the interlayer cations are replaced, and no substantial change is undergone by the silicate portion of the 2 : 1 clay mineral (Jonas, 1975).

Stephen (1952) suggests that the weathering of chlorite at low pH is an ion exchange reaction between interlayer Mg and H. This process results in a vermiculite-like mineral. In regions of low chemical weathering, chlorite may pass unaltered from rock to soil and to the oceans. In general, chlorites persist during transport from weathering regions on the land to the ocean and during settling out, and their presence in marine sediments serves as an indicator of provenance.

After deep burial, chlorite is formed from smectites, provided that sufficient Mg is available. The diagenesis will be discussed in Chapter 10. According to Heller-Kallai et al. (1973 a) chloritization of marine clays in a semiclosed basin is the result of prolonged contact of the clays during early diagenesis with water, which may have been more concentrated than normal water. Depositional interruptions resulting in the formation of hard impervious beds over either shallow or deep ocean bottoms could provide favorable regions for such a diagenetic process. Chlorite is formed diagenetically by prolonged percolation of seawater through previous layers overlying the impervious ones.

Vermiculites and smectites interlayered by hydroxy–cation associations and polymers, particularly those of Al, have frequently been found in soils. According to Rich (1968), moderately active weathering is required to provide the Al ions, while the pH should be slightly acid, there should be frequent wetting and drying of the clay, and the environment should be poor in organic matter, which may cause the Al ions to become soluble. The transformation of vermiculites and smectites into interlayered hydroxy complexes is known as *chloritization* (this term in its broadest sense includes any hydroxy interlayering). This transformation is associated with a drop in the cation exchange capacity of the clay fraction, since interlayer cations become nonexchangeable. Sorption and subsequent desorption of organic proton acceptors may cause partial chloritization of smectite minerals (Heller-Kallai et al., 1973 b).

References

Bar-On, P., Shainberg, I.: Hydrolysis and decomposition of Na-montmorillonite in distilled water. Soil Sci.. *109*, 241–246 (1970)

Barshad, I.: The effect of the total chemical composition and crystal structure of soil minerals on the nature of exchangeable cations in acidified clays and in naturally occuring acid soils. Trans. 7th Int. Conf. Soil Sci. *2*, 435–444 (1960)

Barshad, I., Foscolos, A. E.: Factors affecting the rate of interchange reaction of adsorbed H on the 2:1 clay minerals. Soil Sci. *110*, 52–61 (1970)

Bodenheimer, W., Heller, L.: Sorption of α-amino acids by copper montmorillonite. Clay Mineral. *7*, 167–176 (1967)

Bodenheimer, W., Heller, L., Kirson, B., Yariv, S.: Organometallic clay complexes, Part II. Clay Miner. Bull. *5*, 145–154 (1962)

Bodenheimer, W., Heller, L., Kirson, B., Yariv, S.: Organometallic clay complexes, Part III. Proc. Int. Clay Conf., Stockholm. Rosenqvist, I. Th. (ed.) Oxford: Pergamon Press, 1963b, vol. II. pp. 351–360

Bodenheimer, W., Heller, L., Kirson, B., Yariv, S.: Organometallic clay complexes, Part IV. Isr. J. Chem. *1*, 391–403 (1963c)

Bodenheimer, W., Heller, L., Yariv, S.: Organometallic clay complexes, Part VI, Copper montmorillonite-alkylamines. Proc. Int. Clay Conf., Jerusalem. Heller, L., Weiss, A. (eds.) Jerusalem: Israel Prog. Sci. Trans., 1966a, vol. I, pp. 251–261

Bodenheimer, W., Heller, L., Yariv, S.: Infrared study of copper montmorillonite-alkylamine. Proc. Int. Clay Conf., Jerusalem. Heller, L., Weiss, A. (eds.) Jerusalem: Israel Prog. Sci. Trans., 1966b, vol. II, pp. 171–174

Bodenheimer, W., Kirson, B., Yariv, S.: Organometallic clay complexes. Part I. Isr. J. Chem. *1*, 69–78 (1963a)

Biscaye, P. E.: Mineralogy and sedimentation of recent deep sea clay in the Atlantic Ocean and adjacent seas and oceans. Bull. Geol. Soc. Am. *76*, 803–832 (1965)

Bolt, G. H., Sumner, M. E., Kamphorst, A.: A study of the equilibria between three categories of potassium in an illitic soil. Soil Sci. Soc. Am. Proc. *27*, 294–299 (1963)

Buchanan, A. S., Oppenheim, R. C.: The surface chemistry of kaolinite. I. Aust. J. Chem. *21*, 2367–2371 (1968)

Calvin, M.: Chemical evolution. New York and Oxford: Oxford Univ. Press 1969

Chukhrov, F. V.: Origin and geochemistry of clays. Introduction to section 2, Proc. Int. Clay Conf., Jerusalem. Heller, L., Weiss, A. (eds.) Jerusalem: Israel Prog. Sci. Transl., 1966, vol. II, pp. 71–74

Churchman, G. J., Jackson, M. L.: Reaction of montmorillonite with acid aqueous solutions, solute activity control by a secondary phase. Geochim. Cosmochim. Acta *40*, 1251–1259 (1976)

Clementz, D. M., Mortland, M. M.: Interlamellar metal complexes in layer silicates-III. Silver(I)-arene complexes in smectities. Clays Clay Miner. *20*, 181–186 (1972)

Cloos, P., Vande Poel, D., Camerlynck, J. P.: Thiophene complexes on montmorillonite saturated with different cations. Nature (London) Phys. Sci. *243*, 54–55 (1973)

Conard, J.: Structure of water and hydrogen bonding on clays studied by ^{7}Li and ^{1}H NMR. In: Magnetic Resonance. San Francisco: A. C. S. Pub. 1976, pp. 85–93

Conley, R. F., Althoff, A. C.: Surface acidity in kaolinite. J. Colloid Interface Sci. *37*, 186–195 (1971)

Cowan, C. T., White, D.: The mechanism of exchange reaction occurring between sodium montmorillonite and various n-primary aliphatic amine salts. Trans. Faraday Soc. *54*, 691–697 (1958)

Cruz, M., White, J. L., Russell, J. D.: Montmorillonite-s-triazine interactions. Isr. J. Chem. *6*, 315–323 (1968)

Degens, E. T., Mathéja, J.: Origin, development and diagenesis of biogeochemical compounds. J. Brit. Interplanet. Soc. *21*, 52–82 (1968)

Delvigne, J.: Pédogenese en zone tropicale. La formation des minéraux secondaires en milieu ferralitique. Ph. D. Thesis, University of Louvain 1965

Doner, H. E., Mortland, M. M.: Benzene complexes with copper (II) montmorillonite. Science *166*, 1406–1407 (1969)

Eckstein, Y., Yaalon, D. H., Yariv, S.: The effect of Li on the cation exchange behavior of crystalline and amorphous clays. Isr. J. Chem. *8*, 335–342 (1970)

Englund, J. O., Jørgensen, P.: A chemical classification system for argillaceous sediments and factors affecting their composition. Geol. Foeren. Stockholm Foerh. *95*, 87–97 (1973)

Farmer, V. C., Russell, J. D.: Interlayer complexes in layer silicates, the structure of water in lamellar ionic solutions. Trans. Faraday Soc., *67*, 2737–2749 (1971)

Feigenbaum, S., Shainberg, I.: Dissolution of illite. A possible mechanism of potassium release. Soil Sci. Soc. Am. Proc. *39,* 986–990 (1975)

Fenn, D. B., Mortland, M. M.: Interlamellar metal complexes in layer silicates II. Phenol complexes in smectites. Proc. Int. Clay Conf., Madrid. Serratosa, J. M. (ed.) Madrid: Div. Ciencias C. S. I. C., 1973, pp. 591–603

Ferris, A. P., Jepson, W. B.: The exchange capacities of kaolinite and the preparation of homoionic clays. J. Colloid Interface Sci. *51,* 245–259 (1975)

Fornes, V., Chaussidon, J.: Near infrared spectroscopic studies of clay-water systems. Proc. Int. Clay Conf., Mexico. Bailey, S. W. (ed.) Wilmette, IL: Applied Publishing Ltd., 1975, pp. 383–390

Forslind, E., Jacobsson, A.: Clay-water system. In: Water, a comprehensive treatise, Franks, F. New York: Plenum Press, 1975, Vol. 5, pp. 173–248

Francis, C. W., Brinkley, F. S.: Preferential adsorption of ^{137}Cs to micaceous minerals in contaminated freshwater sediments. Nature (London) *260,* 511–513 (1976)

Frenkel, M.: Surface acidity of montmorillonite. Clays Clay Miner. *22,* 435–441 (1974)

Fripiat, J. J.: Surface properties of alumino-silicates. Clays Clay Miner. *12,* 327–358 (1964)

Fripiat, J. J., Cloos, P., Calicis, B., Makay, K.: Adsorption of amino acids and peptides by montmorillonite. II. Identification of adsorbed species and decay products by infrared spectroscopy. Proc. Int. Clay Conf., Jerusalem. Heller, L., Weiss, A. (eds.) Jerusalem: Israel Prog. Sci. Trans., 1966, vol. I, pp. 233–245

Fripiat, J. J., Pennequin, M., Poncelet, G., Cloos, P.: Influence of the van der Waals force on the infrared spectra of short aliphatic alkylammonium cations held on montmorillonite. Clay Miner. *8,* 119–134 (1969)

Frissel, M. J., Bolt, G. H.: Interaction between certain ionizable organic compounds (herbicides) and clay minerals. Soil Sci. *94,* 284–291 (1962)

Furukawa, T., Brindley, G. W.: Adsorption and oxidation of benzidine and aniline by montmorillonite and hectorite. Clays Clay Miner. *21,* 279–288 (1973)

Gast, R. G., Mortland, M. M.: Self-diffusion of alkylammonium ions in montmorillonite. J. Colloid Interface Sci. *37,* 80–92 (1971)

Greenland, D. J.: Charge characteristics of some kaolinite-iron hydroxide complexes. Clay Miner. *10,* 407–416 (1975)

Grim, R. E.: Clay Mineralogy. New York: McGraw-Hill 1968

Heller, L., Yariv, S.: Sorption of some anlines by Mn, Co, Ni, Zn and Cd montmorillonite. Proc. Int. Clay Conf., Tokyo. Heller, L. (ed.) Jerusalem: Israel Univ. Press, 1969, vol. I, pp. 741–755

Heller-Kallai, L., Nathan, Y., Zak, I.: Clay mineralogy of triasic sediments in Southern Israel and Sinai. Sedimentology *20,* 513–521 (1973a)

Heller-Kallai, L., Yariv, S., Riemer, M.: Effect of acidity on the sorption of histidine by montmorillonite. Proc. Int. Clay Conf., Madrid. Serratosa, J. M. (ed.) Madrid: Div. Ciencias C. S. I. C., 1973b, pp. 651–662

Heller-Kallai, L., Yariv, S., Riemer, M.: The formation of hydroxy interlayers in smectites under the influence of organic bases. Clay Miner. *10,* 35–40 (1973c)

Hougardy, J., Serratosa, J. M., Stone, W., van Olphen, H.: Interlayer water in vermiculite: thermodynamic properties, packing density, nuclear pulse resonance and infrared absorption. Spec. Discuss. Faraday Soc. *1,* 187–193 (1970)

Huang, W. H., Doyle, L., Chiou, W. A.: Clay mineral studies of surface sediments from the shelf of the Northeastern and Eastern Gulf of Mexico. Proc. Int. Clay Conf., Mexico. Bailey, S. W. (ed.) Wilmette, IL: Applied Publishing Ltd, 1975, pp. 55–70

Huang, W. H., Keller, W. D.: Dissolution of clay minerals in dilute organic acids at room temperature. Am. Mineral. *56,* 1082–1095 (1971)

Huang, W. H., Keller, W. D.: Gibbs free energy of formation calculated from dissolution data using specific mineral analyses. III-Clay minerals. Am. Mineral. *58,* 1023–1028 (1973)

Jenne, E. A.: Controls on Mn, Fe, Co, Ni, Cu and Zn concentrations in soils and water: the significant role of hydrous Mn and Fe oxides. In Trace Inorganics in Water. Adv. Chem. Ser., *73,* 337–387 (1968)

Jonas, E. C.: Crystal chemistry of diagenesis in 2:1 clay minerals. Proc. Int. Clay Conf., Mexico. Bailey, S. W. (ed.) Wilmette, IL: Applied Publishing Ltd., 1975, pp. 3–13

Jørgensen, P., Rosenqvist, I. Th.: Replacement and bonding conditions for alkali ions and hydrogen in dioctahedral and trioctahedral micas. Nor. Geol. Tidsskr. *43,* 497–536 (1963)

Kafkafi, U., Bar-Yosef, B.: The effect of pH on the adsorption and desorption of silica and phosphate on and from kaolinite. Proc. Int. Clay Conf., Tokyo. Heller, L. (ed.) Jerusalem: Israel Univ. Press, 1969, vol. I, pp. 691–696

Kafkafi, U., Posner, A. M., Quirk, J. P.: Desorption of phosphate from kaolinite. Soil Sci. Soc. Am. Proc. *31,* 348–353 (1967)

Keller, W. D.: Flint clay and flint clay facies. Clays Clay Miner. *16,* 113–128 (1968)

Keller, W. D.: Environmental aspects of clay minerals. J. Sediment. Petrol. *40,* 788–813 (1970)

Keller, W. D.: Scan electron micrographs of kaolins collected from diverse environments of origin. I. Clays Clay Miner. *24,* 107–113 (1976)

Keller, W. D., Pickett, E. E., Reesman, A. L.: Elevated dehydroxylated temperature of the Keokuk geode kaolinite–a possible reference mineral. Proc. Int. Clay Conf., Jerusalem. Heller, L., Weiss, A. (eds.) Jerusalem: Israel Prog. Sci. Trans. 1966, vol. I, pp. 75–85

Keren, R., Gast, R. G., Barnhisel, R. I.: Ion exchange reactions in non-dried Chambers montmorillonite hydroxy-Al-complexes. Soil Sci. Soc. Am. Proc. *41,* 34–39 (1977)

Kodama, H., Schnitzer, M.: Dissolution of chlorite minerals by fulvic acid, Can. J. Soil Sci. *53,* 240–243 (1973)

Kvenvolden, K. A., Weiser, D.: A mathematical model of geochemical process: Normal paraffin formation from normal fatty acids. Geochim. Cosmochim. Acta *31,* 1281–1309 (1967)

Lagaly, G., Stange, H., Taramasso, M., Weiss, A.: N-alkylpyridinium derivatives of mica-type layer silicates. Isr. J. Chem. *8,* 399–408 (1970)

Lahav, N., Chang, S.: The possible role of solid surface area in condensation reactions during chemical evolution: Reevaluation, J. Mol. Evol. *8,* 357–380 (1976)

Lahav, N., Raziel, S.: Interaction between montmorillonite and benzidine in aqueous solutions. I. Adsorption of benzidine on montmorillonite. Isr. J. Chem. *9,* 683–689 (1971)

Laura, R. D., Cloos, P.: Adsorption of ethylenediamine on montmorillonite saturated with different cations–I. Copper montmorillonite coordination. Reunión Hispano-Belga de Minerales de la Arcilla, Madrid, 76–86 (1970)

Laura, R. D., Cloos, P.: Adsorption of ethylenediamine on montmorillonite saturated with different cations–III. Na-, K- and Li-montmorillonite and hydrogen bonding. Clays Clay Miner. *23,* 61–69 (1975)

Ledoux, R. L., White, J. L.: Infrared study of hydrogen bonding of organic compounds on oxygen and hydroxyl surfaces of layer lattice silicates. Proc. Int. Clay Conf., Jerusalem. Heller, L., Weiss, A. (eds.) Jerusalem: Israel Prog. Sci. Transl., 1966, vol. I, pp. 361–374

Lerman, A., Mackenzie, F. T., Brickner, O. P.: Rates of dissolution of alumino silicates in seawater. Earth Planet. Sci. Lett. *25,* 82–88 (1975)

Loughnan, F. C.: Chemical weathering of the silicate minerals. New York: American Elsevier Pub. Co. 1969

Low, P. F.: Physical chemistry of clay water interaction. Adv. Agron. *13,* 269–327 (1961)

Maynard, J. B.: Kinetics of silica sorption by kaolinite with application to seawater chemistry. Am. J. Sci. *275,* 1028–1048 (1975)

McBride, M. B.: Origin and position of exchange sites in kaolinite: an ESR study. Clays Clay Miner. *24,* 88–92 (1976)

McBride, M. B., Mortland, M. M.: Surface properties of mixed Cu(II)-tetraalkylammonium montmorillonites. Clay Miner. *10,* 357–368 (1975)

Meyers, P. A., Quinn, J. G.: Fatty acid clay minerals association in artificial and natural seawater solutions. Geochim. Cosmochim. Acta *35,* 628–632 (1971)

Meyers, P. A., Quinn, J. G.: Factors affecting the association of fatty acids with mineral particles in seawater. Geochim. Cosmochim. Acta *37,* 1745–1759 (1973)

Millot, G.: Geology of Clays. Farrand, W. R., Paquet, H., (transl.). New-York-Heidelberg-Berlin: Springer-Verlag 1970

Millot, G.: Data and tendencies of recent years in the field genesis and synthesis of clays and clay minerals. Proc. Int. Clay Conf., Madrid. Serratosa, I. M. (ed.) Madrid: Div. Ciencias C. S. I. C., 1973 pp. 151–157

Mitchell, B. D.: The influence of soil forming factors on clay genesis. Colloq. Int. C. N. R. S., *105*, 139–147 (1961)

Mortland, M. M.: Protonation of compounds at clay mineral surfaces. 9th Int. Cong. Soil Sci. Adelaide, Trans. *1*, 691–699 (1968)

Mortland, M. M.: Clay organic complexes and interactions. Adv. Agron. *22*, 75–117 (1970)

Mortland, M. M., Pinnavaia, T. J.: Formation of copper(II) arene complexes on the interlamellar surfaces of montmorillonite. Nature (London) Phys. Sci. *229*, 75–77 (1971)

Mortland, M. M., Raman, K. V.: Surface acidity of smectites in relation to hydration, exchangeable cations and structure. Clays Clay Miner. *16*, 393–398 (1968)

Norrish, K.: The swelling of montmorillonite. Discuss. Faraday Soc. *18*, 120–134 (1954)

Norrish, K.: Factors in the weathering of mica to vermiculite. Proc. Int. Clay Conf., Madrid. Serratosa, J. M. (ed.) Madrid: Div. Ciencias, C. S. I. C., 1973, pp. 717–732

Oberlin, A., Couty, R.: Conditions of kaolinite formation during alteration of some silicates by water at 200 °C. Clays Clay Miner. *18*, 1161–1173 (1970)

Ormsby, W. C., Shartsis, J. M., Woodside, K. H.: Exchange behavior of kaolins of varying degrees of crystallinity. J. Am. Ceram. Soc. *45*, 361–366 (1962)

Paecht-Horowitz, M., Berger, J., Katchalsky, A.: Prebiotic synthesis of polypeptides by heterogeneous polycondensation of amino acid adenylates. Nature (London) *228*, 636–639 (1970)

Paquet, H., Millot, G.: Geochemical evolution of clay minerals in the weathered products and soils of Mediterranean climates. Proc. Int. Clay Conf., Madrid. Serratosa, I. M. (ed.) Madrid: Div. Ciencias, C. S. I. C., 1973, pp. 199–206

Pinnavaia, T. J., Hall, P. L., Cady, S. S., Mortland, M. M.: Aromatic radical cation formation on the intracrystal surfaces of transition metal layer lattice silicate. J. Phys. Chem. *78*, 994–999 (1974)

Pinnavaia, T. J., Mortland, M. M.: Interlamellar metal complexes on layer silicates. I. Copper(II)-arene complexes on montmorillonite. J. Phys. Chem. *75*, 3957–3962 (1971)

Prost, R.: Interaction between adsorbed water molecules and the structure of clay minerals. Hydration mechanism of smectites. Proc. Int. Clay Conf. Mexico. Bailey, S. W. (ed.) Wilmette, IL: Applied Publishing Ltd., 1975, pp. 351–359

Quirk, J. P.: Particle interaction and soil swelling. Isr. J. Chem. *6*, 213–234 (1968)

Range, K. J., Range, A., Weiss, A.: Fire-clay type kaolinite or fire-clay mineral? Experimental classification of kaolinite-halloysite minerals. Proc. Int. Clay Conf., Tokyo. Heller, L. (ed.) Jerusalem: Israel Univ. Press, 1969, vol. I, pp. 3–13

Rex, R. W.: Authigenic kaolinite and mica as evidence for phase equilibria at low temperature. Clays Clay Miner. *13*, 95–104 (1966)

Reynolds, R. C.: Clay mineral formation in an alpine environment. Clays Clay Miner. *19*, 361–374 (1971)

Rich, C. I.: Hydroxy interlayers in expansible layer silicates. Clays Clay Miner. *16*, 15–30 (1968)

Rimsaite, J.: Genesis of chlorite, vermiculite, serpentine, talc and secondary oxides in ultrabasic rocks. Proc. Int. Clay Conf., Madrid. Serratosa, J. M. (ed.) Madrid: Div. Ciencias, C. S. I. C., 1973, pp. 291–302

Roaldest, E.: Rare earth elements in Quaternary clays of the Numedal area, southern Norway. Lithos *6*, 349–372 (1973)

Rupert, J. P.: Electron spin resonance spectra of interlamellar Cu(II)-arene complexes on montmorillonite. J. Phys. Chem. *77*, 784–790 (1973)

Russell, J. D., Farmer, V. C.: Infrared spectroscopic study of the dehydration of montmorillonite and saponite. Clay Miner. Bull. *5*, 443–464 (1964)

Saltzman, S., Mingelgrin, U., Yaron, B.: Role of water hydrolysis on parathion and methylparathion on kaolinite. J. Agric. Food Chem. *24*, 739–743 (1976)

Saltzman, S., Yariv, S.: Infrared study of the sorption of phenol and p-nitrophenol by montmorillonite. Soil Sci. Soc. Am. Proc. *39*, 474–479 (1975)

Saltzman, S., Yaron, B. Mingelgrin, U.: The surface catalyzed hydrolysis of parathion on kaolinite. Soil Sci. Soc. Am. Proc. *38*, 231–234 (1974)

Sawhney, B. L.: Selective sorption and fixation of cations by clay minerals: a review. Clays Clay Miner. *20*, 93–100 (1972)

Schultz, L. G., Shepard, A. O., Blackmon, P. D., Starky, H. C.: Mixed-layer kaolinite-montmorillonite from the Yucatan peninsula, Mexico. Clays Clay Miner. *19*, 137–150 (1971)

Shainberg, I., Low, P., Kafkafi, U.: Electrochemistry of Na-montmorillonite suspensions. I. Chemical stability of the suspensions. Soil Sci. Soc. Am. Proc. *38*, 751–756 (1974)

Shimoyama, A., Johns, W. D.: Catalytic conversion of fatty acids to petroleum-like paraffins and their maturation. Nature (London) Phys. Sci. *232*. 140–144 (1971)

Siever, R., Woodford, N.: Sorption of silica by clay minerals. Geochim. Cosmochim. Acta *37*, 1851–1880 (1973)

Sillén, L. G.: The ocean as a chemical system. Science *156*, 1189–1197 (1967)

Smith, B. H., Emerson, W. W.: Exchangeable aluminum on kaolinite. Aust. J. Soil Res. *14*, 43–53 (1976)

Sofer, Z., Heller, L., Yariv, S.: Sorption of indoles by montmorillonite. Isr. J. Chem. *7*, 697–712 (1969)

Solomon, D. H., Loft, B. C., Swift, J. D.: Reactions catalyzed by minerals. IV. The mechanism of the benzidine blue reaction on silicate minerals. Clay Miner. *7*, 389–397 (1968)

Solomon, D. H., Swift, J. D., Murphy, A. J.: The acidity of clay minerals in polymerizations and related reactions. J. Macromol. Sci. Chem. *A 5*, 587–601 (1971)

Stephen, I.: A study of weathering with reference to the soils of the Malvern Hills. J. Soil Sci. *3*, 20–33 (1952)

Swartzen-Allen, S. L., Matijevic, E.: Surface and colloid chemistry of clays. Chem. Rev. *74*, 385–400 (1974)

Swindale, L. D., Fan, P. F.: Transformation of gibbsite to chlorite in ocean bottom sediments. Science *157*, 799–800 (1967)

Tarr, W. A., Keller, W.D.: Some occurrences of kaolinite deposited from solution. Am. Mineral. *22*, 933–935 (1937)

Tchoubar, C.: Formation de la kaolinite à partir d'albite altérée par l'eau à 200 °C. Etudes en microscopie et diffraction électroniques. Bull. Soc. Fr. Mineral. Cristallogr. 483–518 (1965)

Theng, B. K. G.: Adsorption of alkylammonium cations by porous crystals. N. Z. J. Sci. *14*, 1026–1039 (1971)

Theng, B. K. G.: Formation, properties and practical applications of clay organic complexes. J. R. Soc., N. Z. *2*, 437–457 (1972)

Theng, B. K. G.: The chemistry of clay-organic reactions. London: Adam Hilger 1974

Touillaux, R., Salvador, P., Vandermeersche, C., Fripiat, J. J.: Study of water layers adsorbed on Na and Ca montmorillonites by the pulsed nuclear magnetic resonance technique. Isr. J. Chem. *6*, 337–348 (1968)

Vansant, E. F., Uytterhoeven, J. B.: Thermodynamics of the exchange of n-alkylammonium ions on Na-montmorillonite. Clays Clay Miner. *20*, 47–54 (1972)

Wada, K.: Lattice expansion of kaolin minerals by treatment with potassium acetate. Am. Mineral. *46*, 78–91 (1961)

Walker, G. F.: Mechanism of dehydration of Mg-vermiculite. N. A. S. N. R. C., Publ. *456*, 101–115 (1956)

Walker, G. F.: Ion exchange on clay minerals. Proc. Int. Clay Conf., Stockholm. Rosenqvist, I. Th. (ed.) Oxford: Pergamon Press, 1963, vol. II, pp. 259–263

Walker, G. F.: Interaction of n-alkylammonium ions with mica-type layer silicates. Clay Miner. *7*, 129–143 (1967)

Walker, G. F., Garrett, W. G.: Chemical exfoliation of vermiculite and the production of colloidal dispersions. Science *156*, 385–387 (1967)

Waples, D. W.: Catalytic formation of hydrocarbons from fatty acids. Nature (London) *237*, 63–64 (1972)

Weaver, C. E.: Potassium, illite and the ocean. Geochim. Cosmochim. Acta *31*, 2181–2196 (1967)

Wedepohl, K. H.: Environmental influences on the chemical composition of shales and clays. In: Physics and chemistry of the earth. Ahrens, L. H., Press, F., Runcorn, S. K., Urey, H. C. (ed.), Oxford: Pergamon Press, 1970, Vol. 8, pp. 307–333

Weir, A. H., Rayner, J. H.: An interstratified illite-smectite from Denchworth series soil in weathered Oxford clay. Clay Miner. *10*, 173–187 (1974)

Weismiller, R. A., Ahlrichs, J. L., White, J. L.: Infrared studies of hydroxy-aluminum interlayer material. Soil. Sci. Soc. Am. Proc. *31*, 459–463 (1967)

Weiss, A.: Organic derivatives of mica-type layer silicates. Angew. Chem. *2*, 134–143 (1963)

Weiss, A., Thielepape, W., Göring, G., Ritter, W., Schäfer, H.: Kaolinit-Einlagerungs-Verbindungen. Proc. Int. Clay Conf., Stockholm. Rosenqvist, I. Th. (ed.) Oxford: Pergamon Press, 1963, vol. I, pp. 287–305

Yaalon, D. H., Feigin, A.: Non-exchangeable ammonium ions in some clays and shales of Israel. Isr. J. Chem. *8*, 425–434 (1970)

Yariv, S.: Organophilic pores as proposed primary migration media for hydrocarbons in argillaceous rocks. Clay Sci. *5*, 19–29 (1976)

Yariv, S., Bodenheimer, W.: Specific and sensitive reactions with the aid of montmorillonite. Pyrocatechol and its derivatives. Isr. J. Chem. *2*, 197–200 (1964a)

Yariv, S., Bodenheimer, W., Heller, L.: Organometallic clay complexes, Part V. Isr. J. Chem. *2*, 201–208 (1964b)

Yariv, S., Heller, L.: Sorption of cyclohexylamine by montmorillonite. Isr. J. Chem. 935–945 (1970)

Yariv, S., Heller, L., Kaufherr, N.: Effect of acidity in montmorillonite interlayers on the adsorption of aniline derivatives. Clays Clay Miner. *17*, 301–308 (1969)

Yariv, S., Heller, L., Sofer, Z., Bodenheimer, W.: Sorption of aniline by montmorillonite. Isr. J. Chem. *6*, 741–756 (1968)

Yariv, S., Russell, J. D., Farmer, V. C.: Infrared study of the adsorption of benzoic acid and nitrobenzene in montmorillonite. Isr. J. Chem. *4*, 201–213 (1966)

Yariv, S., Shoval, S.: Infrared study of the interaction between alkali-halide and hydrated halloysite. Clays Clay Miner. *24*, 253–261 (1976)

Yoshino, D., Hayatsu, R., Anders, E.: Origin of organic matter in early solar systems. III. Amino acids, catalytic synthesis. Geochim. Cosmochim. Acta *35*, 927–938 (1971)

Chapter 8
Interaction between Solid Particles Dispersed in Colloid Systems

1. Interaction Between Solid Particles Dispersed in a Gaseous Phase

The adhesion between two particles is the result of factors such as (a) chemical and non-chemical bond formation, which may be at or away from the particle–particle interface, and (b) creation and release of mechanical strains and stresses at and around that interface. The area of contact between the two particles is called the *adhesive area.*

The attractive forces between different media include all those types of interactions that contribute to the cohesion of solids, such as primary chemical bonds as well as secondary van der Waals interactions. In addition, there are interaction forces such as hydrogen bonds, electronic charge-transfer bonds, and electrostatic double layer forces. The interactions between particles can be classified as follows:

1. Class I includes the *long-range attractive interactions* that result from van der Waals and electrostatic forces. These forces act not only at the immediate adhesive area, but also contribute considerably to the overall adhesive forces as a result of their appreciable magnitudes extending outside the actual interface.

2. Class II includes the *short-range attractive interactions* resulting from various types of chemical bonds as well as hydrogen bonds and π interactions. It is thought that the broken bonds formed during the exposure of new surfaces have been reacted with molecules such as H_2O, giving rise to chemically saturated surface coatings (Ch. 4). Since the adhering surfaces are usually chemically saturated, class II bond formation across the interface occurs only to a small extent in the atmosphere.

3. Class III includes *interfacial reactions* such as sintering effects, mutual diffusion of mineral materials into one another, and condensation and mutual dissolution followed by crystal growth. The interactions of class III cannot be strictly separated. Sintering between two solid bodies is a growth of their common adhesive area by one or more of the following mechanisms: (1) recrystallization; (2) surface or bulk diffusion; (3) evaporation and recondensation; and (4) creep. This generally requires elevated temperatures and pressures, conditions that are achieved during metamorphism.

Classes I and III determine the size of the adhesive area, whereas contributions from class II to the overall adhesion can only result if an adhesive area of finite size has already been established by interactions of class I or III.

Adhesion of one particle to another particle involves the following steps: (1) the particles come into contact at one point through a contact area of atomic dimensions; (2) by long-range attraction forces between them, the particles are subjected to a moment of force so that several contacts are formed; and (3) by interaction forces the contact area increases until the attractive forces and the forces resisting the further deformation at the interface are in equilibrium. An adhesive area of finite size is formed between the adhering surfaces.

Association of particles may result from active collisions between the suspended particles. The association process is known as *coagulation*. Calculations in which the kinetics of coagulation of aerosols is related to the Brownian movement of the particles can be found in the literature (Fuchs, 1964; Fuchs and Sutugin, 1970).

1.1 Van Der Waals–London Forces Between Disperse Particles

The van der Waals forces may comprise three components: dipole orientation (Keesom), induction (Debye), and dispersion (London) forces. The first two forces require a permanent dipole moment for at least one of the two particles. On the other hand, only dispersion forces act between nonpolar particles. Each solid body contains local electric fields of a wide range of frequencies that originate from spontaneous polarization of the constituent atoms and molecules, that is, the fluctuations of the electronic atmospheres of atoms relative to their nuclei. These fluctuations produce a time-dependent dipole moment. The intensity of these fields increases with increasing dielectric constant of the body. A phase difference in the fluctuating dipoles leads to mutual attraction between the particles.

Because of the additivity of the atomic forces, the attraction between macroscopic particles is much stronger than that between gaseous atoms and molecules, and it increases with increasing size of particle. The forces decrease with the distance between the particles (Table 8.1). Calculation of the van der Waals interaction energy between particles requires an allowance to be made for particle size and shape. Such calculations can be found in the literature (Sonntag and Strenge, 1972; Krupp, 1967). For the simple case of clay mineral platelets it is described by the following equations (Low, 1968):

$$F'_A = A/h^3 \tag{8.1}$$

and

$$F''_A = B/h^4 \tag{8.2}$$

where F_A is the force (usually per cm^2), h is the distance between the particles, and A and B are constants having values that depend on the mineralogic composition of the particles. The first equation is appropriate when $h < 10$ nm and the second when $h > 2000$ nm.

Table 8.1. Van der Waals interaction at small interparticle separations (h). (After van den Tempel, 1972)

	Force	Energy
2 Thick plates per cm^2	$A/6\Pi h^3$	$A/12\Pi h^2$
Plate-sphere	$AR/6h^2$	$AR/6h$
2 Spheres	$AR/12h^2$	$AR/12h$

h-distance; R-radius of sphere; A-Hamaker constant

1.2 Electrostatic Forces Between Disperse Particles

Two solids in contact with one another, in general, charge each other electrostatically. Such charges can be observed upon rubbing two solids against each other. The presence of moisture may give rise to ionic effects at the interface. Attraction occurs between the particle charge and its image charge in another particle. With a high surface charge density, the effect of the space charge inside the particle can normally be neglected. In particles immersed in a liquid, electrostatic forces between them tend to become small because of the enhanced dielectric constant of the liquid compared with a gaseous environment and because of the sorption phenomena involved, whereby the charges of the particles are shielded.

1.3 Effect of Adsorbed Water Monolayer on Desert Varnish

The effect of adsorbed layers on the force of adhesion has been found to be appreciable in certain systems. It is to be expected that these effects are due to more specific interaction forces in which the adsorbed monolayer serves as a bridge between the adhering particles.

A very low humidity seems to be responsible for the postdepositional patination staining of cherts and carbonate rocks in deserts by detrital dark metal oxides. Hydrated oxides of metals such as Al, Fe, and Mn are suspended in the atmosphere, and are kept moving by winds. These atmospheric particles may become positively charged when they collide with a rock surface, whereas carbonates and cherts become negative by the mechanism of proton transfer. Only those particles that can become positively charged are electrostatically attracted by the negative field of the rock surface and are retained. Electrostatic and van der Waals interactions are not sufficiently great to be responsible for the strong adhesion between the oxide pigments and the rock surface. Moreover, this phenomenon is characteristic for deserts. As a possible explanation, it may be assumed that chemisorbed water molecules at both surfaces form hydrogen bonds if the surfaces are sufficiently close together, and adsorbed water molecules form a layer with a conformation favorable for the formation of strong hydrogen bonds. It seems plausible that adsorbed water dipoles first give rise to repulsion. In the desert, excess water is easily desorbed and a stable water bridge is formed after a long period of contact between the surfaces of the adherents due to slow reorientation of the residual water molecules in the surface layer. The colored, thin outer layer produced on the surface of a rock after long exposure is called *patina.*

2. Aggregation of Particulate Matter in the Hydrosphere

2.1 Interactions Between Solid Particles Dispersed in a Liquid Medium

The stability of disperse systems and their readiness to coagulate depend on the sign and the magnitude of the potential energy of interaction, V, which is determined by the superposition of the electrostatic repulsive energy and the van der Waals atractive energy

$$V = V_{el} + V_{vw}. \tag{8.3}$$

Van der Waals forces are the means of attraction between like and unlike disperse particles. The electrostatic repulsion between like particles is the result of their like charge. However, repulsive forces cannot be calculated by the introduction of the quantities of the surface charge in the simple expression of Coulomb's law since the particles are surrounded by counterions and together with their diffuse double layer they are electrically neutral entities.

When two particles approach each other their diffuse double layers interpenetrate and deform. This deformation is the basis for the quantitative description of the electrostatic repulsion. Two types of double layer interpenetration can be discerned, both resulting in repulsion between the particles. The first type of double layer interpenetration is characteristic for surfaces whose potential is determined by the specific sorption of certain ions whereas the second type is characteristic of surfaces with a constant surface charge determined by the dissociation of certain functional groups located on the particle surface or by diadochic substitution of certain elements for other elements of a different charge in the interior of the crystal.

In the first type a thermodynamic equilibrium is maintained during the overlapping of the diffuse double layers, since the migration of counter-, co-, and potential-determining ions into and out of the interface is fast (Fig. 8.1 a and b). The potential on the interface does not change while the charge of the double layer decreases. In the limiting case of two interfaces in contact, the double layer charge becomes zero. The free energy of the double layer increases with decreasing surface charge, and the process is responsible for the repulsion between the particles (Verwey and Overbeek, 1948).

In the second type the rearrangement of charge carriers is a slower process than the approach of the particles to each other as a result of thermal motion. In this case the interpenetration of double layers occurs at constant charge but increasing potential. The force driving the two particles apart per unit area is given by the difference between the values of the electrostatic forces at the surface before and after the overlapping of the two double layers, when the surface potential is equal to φ_0 and φ_i respectively, and ϵ is the dielectric constant of the medium, namely,

$$\frac{\epsilon}{8\Pi}\left[\left(\frac{d\varphi_0}{dx}\right)^2_{x=0}-\left(\frac{d\varphi_i}{dx}\right)^2_{x=0}\right].$$

The concentration of the ions in the interparticle separation region increases as the separation decreases, resulting in an increase of the osmotic pressure. The difference

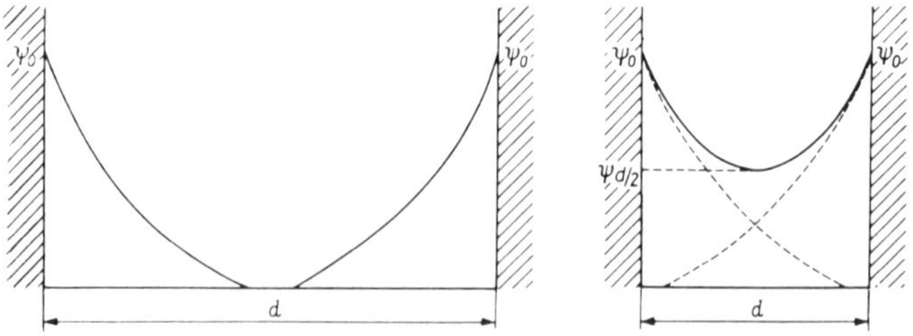

Fig. 8.1. Potential vs. distance curves for two undisturbed double layers **a** and for two overlapping double layers **b**. (From Sonntag and Strenge, 1972: Used with the permission of Keter Publishing House)

between the values of the osmotic pressure in the interparticle region before and after the overlapping of the two double layers is also a quantitative means to evaluate the force driving the two particles apart and is known as *osmotic repulsion* (Ohshima, 1974).

The energies of repulsion for particles at relatively great distances from each other are almost the same for both types, but are greater for constant charge type than for constant potential type at small distances (Fig. 8.2).

Calculation of the interaction energy between the particles as a sum of electrostatic (V_{el}) and van der Waals (V_{vw}) components involves an allowance for particle size and shape. Such calculations can be found in the literature (Verwey and Overbeek, 1948). A simple case of two identical plate-like particles separated by gaps that are very narrow in comparison with the linear dimensions of these plates, will be considered here. This simple case is characteristic of the behavior of suspensions of most clay minerals.

When two suspended particles approach each other as a result of their Brownian movement, until their diffuse double layers overlap, the free energy of the system increases. Work must be performed, resulting in deformation of the double layer structure. It is possible to calculate the work done by infinitely distant particles approaching until their double layers overlap. It equals the repulsion energy or the repulsion potential at a certain interparticle distance and is defined as follows:

$$V_{el} = (2G_h - G_\omega) \tag{8.4}$$

where G_h and G_ω are the free energies of the diffuse electric double layers per cm^2 when the distance between the plate particles is $h = 2d$ and when the distance is such that there is no overlapping of the two double layers respectively. In a symmetric electrolyte solution, where the cation and anion have the same charge of opposite signs, the repulsion force per cm^2 of the layer, F_{el}, which arises from the deformation of the dif-

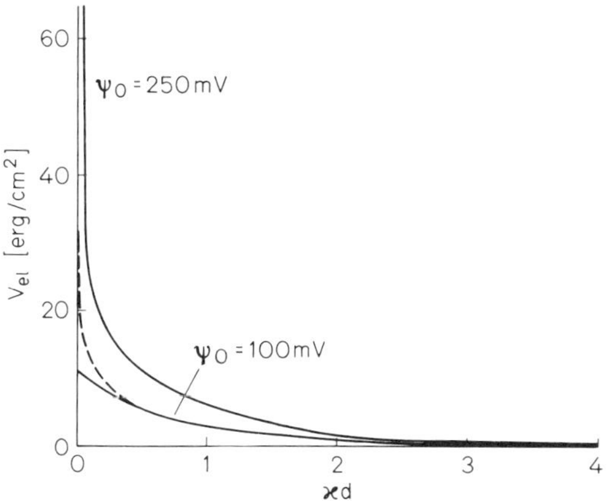

Fig. 8.2. Repulsion energy vs. distance for constant potential (*solid curves*) and for constant diffuse double layer charge (*broken curve*). (From Sonntag and Strenge, 1972: Used with the permission of Keter Publishing House)

fuse layer, is given by

$$F_{el} = 64\, cRT\gamma^2 e^{-\kappa h} \tag{8.5}$$

where h is the width of the gap between the two layers, the expression $1/\kappa$ is defined as the thickness of the diffuse double layer (see Ch. 2, Eq. 2.80) and γ is defined as follows:

$$\gamma = \frac{[\exp(zF\varphi_0/2RT)] - 1}{[\exp(zF\varphi_0/2RT)] + 1} \tag{8.6}$$

wherein φ_0 is the electric potential at the solid surface, z is the charge of the counterion, and F is the Faraday constant.

The potential energy of repulsion is obtained by integration of Eq. (8.5) in the form

$$V_{el} = -\int_0^h F_{el} dh = \frac{64\, cRT\gamma^2}{\kappa} e^{-\kappa h}. \tag{8.7}$$

From (8.7) it is obvious that the repulsion potential decreases roughly exponentially with increasing particle separation. In Figure 8.3, three potential curves are shown that are valid for the same particles for different electrolyte concentrations. Due to the compression of the double layer at increasing electrolyte concentrations, the range of the repulsion is considerably reduced.

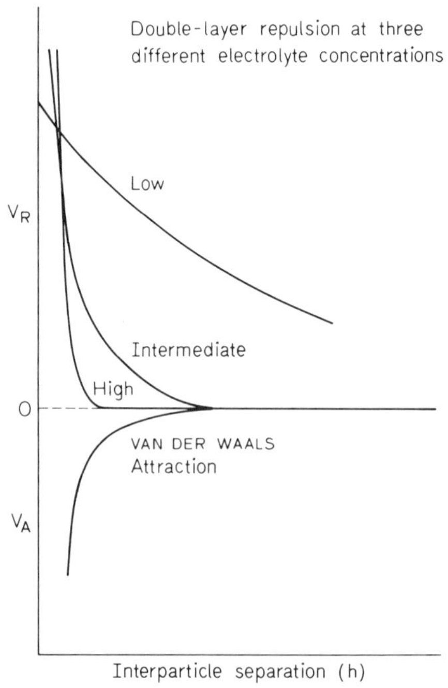

Fig. 8.3. Repulsive and attractive energy as a function of interparticle separation at three electrolyte concentrations

The van der Waals attraction energy between two unit layers of a 2:1 clay mineral, such as montmorillonite, is given by

$$V_{vw} = -\frac{A}{48F}\left[\frac{1}{0.5h^2} + \frac{1}{(0.5h+\Delta)^2} - \frac{2}{(0.5h+0.5\Delta)^2}\right]. \tag{8.8}$$

Here h is the distance between the plates, measured between the planes of the centers of the oxygens of the tetrahedral sheet, and Δ is the thickness of the unit layer measured between the same planes. For 2:1 minerals $\Delta = 0.66$ nm. The constant A is approximately 10^{-12} (van Olphen, 1963).

For very thick plates, e.g., clay tactoids, the van der Waals force of attraction per cm^2 of the layer is simplified and is given by Eq. (8.1), $F_{vw} = -K/h^3$, where K is a constant depending on the nature of the dispersing medium, the thickness of the plate, and its chemical composition. The van der Waals attraction energy can be calculated by integrating this equation as follows:

$$V_{vw} = -\int_{\infty}^{h} F_{vw}\,dh = -(K/2h^2). \tag{8.9}$$

The total interaction energy is given by the following equation:

$$V = V_{el} + V_{vw} = (64\,cRT\gamma^2/\kappa)\,(e^{-\kappa h}) - (K/2h^2). \tag{8.10}$$

This equation describes the dependence of V on h. Figure 8.4 shows this dependence for a certain ratio of constants as obtained by summing the attraction and repulsion

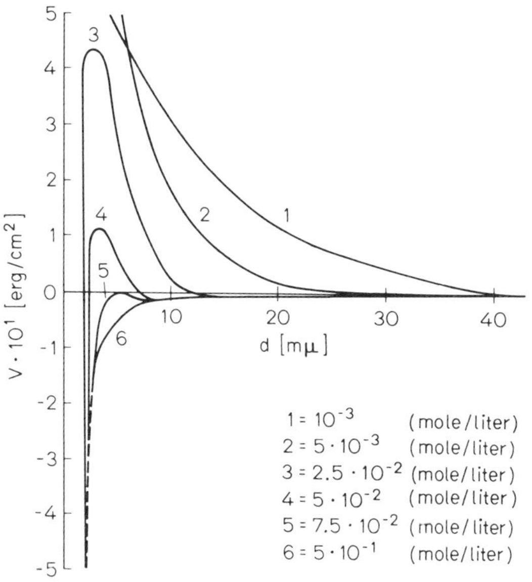

Fig. 8.4. Interaction energy vs. distance for $A = 5 \times 10^{-13}$ erg, $\varphi_\delta = 100$ mV and variable ionic strength (*1*) 10^{-3} mol/l; (*2*) 5×10^{-3} mol/l (*3*) 2.5×10^{-2} mol/l; (*4*) 5×10^{-2} mol/l; (*5*) 7.5×10^{-2} mol/l; (*6*) 5.0×10^{-1} mol/l. (From Sonntag and Strenge, 1972: Used with the permission of Keter Publishing House)

potentials for each particle distance, at various electrolyte concentrations, referred to as *interaction energy curves.*

At very short distances from the solid surface, a short-range repulsion occurs to which there are two possible contributors. One, which is the resistance to the interpenetration of the crystal lattices, known as the *Born repulsion,* becomes effective as soon as extruding lattice points or regions come into contact. A second short-range repulsion is the result of high adsorption of the liquid molecules onto the solid–liquid interface. Usually one or two monomolecular layers of the solvent are held rather tightly by the particle surface, even if it forms a hydrophobic colloid in an aqueous dispersion. For the distance between the two particles to become less than the thickness of the adsorbed water layers on both particles, the adsorbed water must be desorbed. The superposition of the Born repulsion on the van der Waals attraction forms a *potential well* at a very short interparticle distance, called the *primary minimum* (Fig. 8.5).

A knowledge of the interaction forces between unlike particles in aqueous suspensions is very important in understanding certain geochemical phenomena, for example, the flotation of mineral grains in the oceans by interaction with gas bubbles, the coating of grains of one mineral by very small grains of another mineral or by amorphous material, and the interaction of humic substances with mineral particles. The interaction between two dissimilar particles is determined by the same laws that determine interaction between like particles (Usui, 1972). It is caused by the superposition of dispersion on electrostatic forces due to interpenetration of double layers.

Both van der Waals (dispersion) and electrostatic forces for two unlike particles have been calculated. For large distances F_{vw} is a monotonic function of distance. For smaller distances this relationship changes because of charge reversals resulting from the variations in the dielectric constants. The electrostatic force depends on the particle charges. If the particle charges have dissimilar signs, attraction forces exist. If the particle charges have the same sign but the potentials of the individual particles are different, the magnitude of the repulsion energy is determined by the particles with the lower potentials. The higher potential changes the radius but not the strength of the interacting

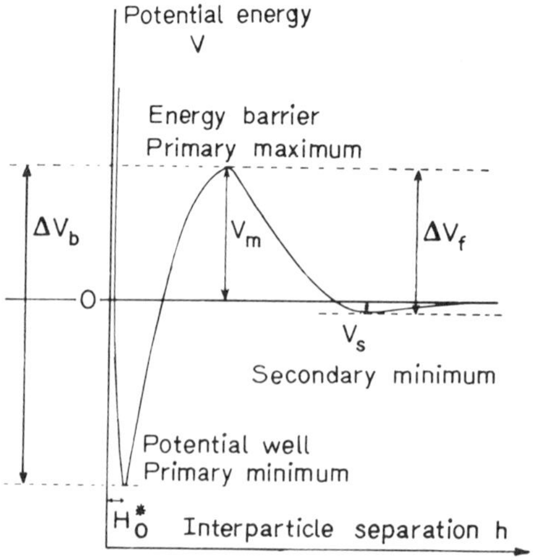

Fig. 8.5. General form of the curve of potential energy as a function of interparticle separation for interaction between two particles

force (Fig. 8.6) and the interacting force reverses its sign at very small distances. Despite the like charges, the particles will attract each other at small distances if their potentials are different (Sonntag and Strenge, 1972).

2.2 Stability of Aqueous Hydrophobic Colloid Solutions and Suspensions

The stability of a geologic colloid system depends on the ability of the system to maintain its disperse state. Unlike molecularly disperse systems lyophobic dispersions have only a limited stability. Coarse suspensions are unstable mainly because their particles settle at an appreciable rate owing to the force of gravity. They may exist only when the dispersing liquid is in turbulent flow. In systems with very small particles, Brownian movement is sufficiently vigorous to prevent sedimentation.

There is a probability of solid particles moving in a disperse system to collide, and as a consequence, adhesion of colliding particles may result in aggregation. Such a coarsening process is called *coagulation* or *flocculation.* This process produces changes in the properties of the colloid system, and consequently (1) a higher degree of turbulence is necessary to maintain dispersion of the system, and (2) the specific surface area of the system decreases.

There are five major types of destabilization mechanisms, namely, double layer repression and primary coagulation, specific sorption of potential-determining ions, bridging, interaction between particles of dissimilar nature (O'Melia, 1970), and a secondary coagulation.

1. Double Layer Repression (Coagulation by Electrolytes): Not every collision will result in adhesion. As can be seen from Figure 8.5, for large values of interparticle separation, h, the potential energy of interaction parallels the energy of attraction, V_{vw}. The potential gradient dV/dh is positive, indicating a mutual attraction between the particles. At shorter distances, the repulsion component V_{el}, which increases exponentially with decreasing h, can overcompensate V_{vw} and reverse the signs of dV/dh in the direction of repulsion. At the interparticle separation where dV/dh changes from positive to negative, a potential well is obtained, known as a *secondary minimum.* On further re-

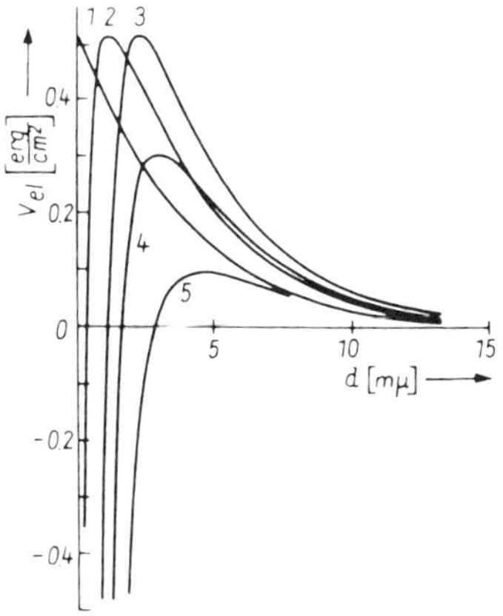

Fig. 8.6. Electrostatic repulsion energy per unit surface for plate-shaped particles of like charge. Ionic strength, 0.01 M. *(1)* $\varphi\delta_1 = \varphi\delta_2 = 46.1$ *(2)* $\varphi\delta_1 = 76.8$ mV, $\varphi\delta_2$ = 46.1 mV; *(3)* $\varphi\delta_1$ = 153.6 mV, $\varphi\delta_2$ = 46.1 mV; *(4)* $\varphi\delta_1$ = 153,6 mV, $\varphi\delta_2$ = 35.8 mV; *(5)* $\varphi\delta_1$ = 153.6 mV, $\varphi\delta_2$ = 20.5 mV. (From Sonntag and K. Strenge, 1972: Used with permission of Keter Publishing House)

duction of the gap, V_{vw} should again predominate, reversing the sign of dV/dh in the direction of attraction. There is thus a repulsion maximum in the function $V = V(h)$, which can be easily found from the conditions $\delta V/\delta h = 0$ (the *primary maximum* in Fig. 8.5).

The primary maximum in the interaction energy curve (Fig. 8.5) is an *energy barrier.* The height of this peak shows the minimum energy required to bring about a successful collision and adhesion. The stability of the disperse system depends on the fraction of particles that have the necessary thermal energy to overcome this barrier. This fraction decreases with increasing height of the energy barrier, and consequently the stability of the disperse system increases with this height. From Figure 8.4 it follows that the addition of electrolytes reduces the energy barrier, and consequently the stability of the disperse system is reduced.

Peptization, which was defined (Ch. 3) as the process of the conversion of unconsolidated particulate rock material into colloid solutions and suspensions due to the interaction with flowing water, can be regarded as the reverse process of coagulation. The interparticle separation in the nondispersed coagulated rock material corresponds to the primary minimum of Figure 8.5. The gap is increased by the mechanical energy of the flowing waters. In the presence of electrolytes the primary maximum is low and the peptized particles will recoagulate, but if the electrolyte is washed away by the flowing waters, this maximum increases and the dispersed particles remain suspended. However, in the complete absence of electrolyte, the primary maximum will be so high that the mechanical energy of the flowing waters will not be sufficient to overcome the energy barrier and coagulated particles will not peptize.

Colloids can be regarded as hydrophobic if their interaction energy curve shows a primary minimum. If such a minimum exists, colloid dispersions are metastable. The stability of such systems depends on the energy barrier and is a question of kinetics and rates of coagulation rather than of thermodynamics. In kinetics of the coagulation process *rapid* and *slow coagulations* are distinguished.

In the case of *rapid coagulation* so much electrolyte has been added that the energy barrier is reduced and is absent or too small to prevent particles from coming into contact (Fig. 8.4, curve 5). Every collision between two particles results in irreversible aggregation, and coagulation is fast. The probability of collision and of coagulation is the same and is proportional to the product of the constant of mutual diffusion and the distance of closest approach. The mutual diffusion is determined by the normal Brownian movement of the particles in undisturbed water and by the eddy diffusion in turbulent systems. The distance of closest approach is determined by the geometric configuration of the particle.

For monodisperse systems of spheres in undisturbed water the rate of rapid coagulation is described by

$$dc/dt = -kc^2 = -8\Pi rDc^2 \tag{8.11}$$

where c is the concentration of the original single, noncoagulated, particles, k is the velocity constant, D is the diffusion coefficient, and r is the radius of the sphere (Smoluchowski, 1916). The diffusion coefficient varies inversely with the radius of the diffusing particle. It may therefore be concluded that in monodisperse systems the rate of coagulation is not dependent on the size of the particle.

In polydisperse systems of spheres having radii r_i and r_j, effective radii $R_{ij} = r_i + r_j$ and effective diffusion coefficients $D_{ij} = D_i + D_j$ must be considered for the calculation of the rates of collision and coagulation. According to Eq. (5.14) (Ch. 5) $D_i = (kT/6\Pi\eta r_i) = (K/r_i)$ and $D_j = (K/r_j)$ where $K = kT/6\Pi\eta$. Eq. (8.11) can be written

$$(dc/dt) = -2\Pi R_{ij} D_{ij} c_i c_j. \qquad (8.11^*)$$

The product $D_{ij} R_{ij}$ can be written as follows:

$$D_{ij} R_{ij} = K\,(r_i^{-1} + r_j^{-1})\,(r_i + r_j) = K\,\{4 + [\sqrt{(r_i/r_j)} - \sqrt{(r_j/r_i)}]^2\}.$$

The term $\sqrt{(r_i/r_j)} - \sqrt{(r_j/r_i)}$ becomes zero in monodisperse systems when $r_i = r_j$ and in this case $D_{ij} R_{ij}$ is equal to $4K$ and Eq. (8.11*) is equal to (8.11). For a polydisperse system $D_{ij} R_{ij} > 4K$. It is evident that the rate of coagulation between particles of different sizes is much greater than the rate of coagulation of particles of the same size (Müller, 1926). From the same calculation it appears that in a polydisperse flocculating system the smaller particles disappear much more quickly than the larger ones.

Sols in which the particles are other than spherical coagulate more rapidly than sols of spheres. As nonspherical particles are subjected to rotational Brownian movement, the collision diameters will be of the order of the larger of the particle diameters. The diffusion constant, however, is a function of the mean diameter, which in the case of rods may be considerably smaller than the length of the rod. Consequently nonspherical particles have a relatively large collision diameter and a large diffusion constant and this combination results in a high probability of collision (Overbeek, 1952).

If the collisions result from Brownian motion of the colloid particles, the coagulation is said to be of a *perikinetic* type. If they result from velocity gradients within the suspended fluid or from differential settling rates of suspended particles the coagulation is of an *orthokinetic* type. Orthokinetic coagulation occurs markedly in natural systems of moderate concentrations of suspended materials that contain some particles large enough to settle at appreciable speed under gravity together with a large number of smaller particles with which they can collide. The large particle sweeps a certain volume clear of small particles on its passage downward. The velocity of coagulation tends to increase as the large particles grow, until practically all the small particles are used up (Marshall, 1964).

In the case of *slow coagulation* enough electrolyte has been added to the system to reduce appreciably the interaction energy barrier but not to eliminate it (Fig. 8.4, curves 3 and 4) and it has a positive value. Since only a fraction of the particles has the energy necessary to overcome the barrier, the coagulation is slow. The rate of coagulation depends on the fraction of collisions that results in aggregation. It increases with the concentration of the electrolytes, which in turn decreases the height of the energy barrier.

At a certain value of electrolyte concentration the interaction energy barrier becomes equal to zero. Coagulation becomes rapid as the concentration increases from this value. This is the *critical electrolyte concentration*, c_{cr}, which is characterized by the fact that V becomes zero at an interparticle distance h, where $\delta V/\delta h = 0$. For this special case summation of Eqs. (8.1) and (8.5) is equal to zero as follows

$$\delta V/\delta h = -64 c_{cr} RT\gamma^2 e^{-\kappa_{cr} h_{cr}} + K/h_{cr}^3 = 0 \tag{8.12}$$

and (8.10) gives

$$V = \frac{64 c_{cr} RT\gamma^2}{\kappa_{cr}} e^{-\kappa_{cr} h_{cr}} - \frac{K}{2h_{cr}^2} = 0 \tag{8.13}$$

where the expression $1/\kappa_{cr}$ corresponds to the compressed thickness of the diffuse double layer at an electrolyte concentration of c_{cr}, and h_{cr} is the distance of the peak of the energy barrier from the particle surface at the same electrolyte concentration. Dividing (8.12) by (8.13) gives

$$h_{cr} = 2/\kappa_{cr} \tag{8.14}$$

From Eqs. (2.80), (8.14), and (8.12) or (8.13), c_{cr} can be calculated. The following relation is obtained for an ion of a charge z with a sign opposite to that of the particle:

$$c_{cr} z^6 = \text{const.}, \tag{8.15}$$

where the constant includes other quantities except z and c_{cr}. From (8.15) it follows that the critical concentrations of ions of various valencies are inversely proportional to the sixth power of their valency. Ratios of critical concentrations for a specific colloid are as follows:

$$\begin{aligned} c_{cr(1)} : c_{cr(2)} : c_{cr(3)} : c_{cr(4)} &= 1 : (1/2)^6 : (1/3)^6 : (1/4)^6 \\ &= 1 : 1.6 \times 10^{-2} : 1.4 \times 10^{-3} : 2.4 \times 10^{-4}. \end{aligned}$$

Ions having a charge of the same sign as that of the particle have almost no effect on the coagulation process and on the critical concentration of the electrolyte. Critical concentrations for the coagulation of positively charged ferric and aluminum hydroxides are listed in Table 8.2. Although the individual values are scattered because of the poor accuracy of the methods available for the determination of c_{cr}, the mean values are in good agreement with Eq. (8.15). The polyvalent anions are the most effective coagulating agents for these hydroxides. Polyvalent cations, on the other hand, such as aluminum and ferric iron, are the most effective coagulating agents for silica and clay minerals. Double layer repression-type coagulation occurs with iron and aluminum only for pH below 3 (O'Melia and Stumm, 1967; Hahn and Stumm, 1968).

2. *Specific Sorption of Potential-determining Ions:* From Eqs. (8.6) and (8.10) it is seen that both γ and the interparticle interaction energy barrier depend on the electric potential at the solid surface. Specific sorbed ions, which cause a lowering of the surface charge and of the potential toward zero, possess a particularly high coagulating power. The lowest stability of the disperse system is attained when the surface charge and potential become equal to zero. At higher electrolyte concentrations, there is further sorption of these potential-determining ions. If enough potential-determining ions are adsorbed, particle charge reversal occurs that may result in sol restabilization as a second zone of dispersion stability is attained (Vrale and Jorden, 1971). Dilute suspensions of

Table 8.2. Critical coagulation concentrations (in mmol/l) for positively charged sols. (After Overbeek, 1952)

Electrolyte	Fe_2O_3 sol	Mean value	Al_2O_3 sol	Mean value
NaCl	9.25		43.5	
KCl	9.0		46.0	
NH_4Cl	–		43.5	
1/2 $BaCl_2$	9.65		–	
KBr	12.5		–	
KI	16.0	11.8	–	52
KNO_3	12.0		60.0	
1/2 $Ba(NO_3)_2$	14.0		–	
HCl	>400		–	
1/2 $Ba(OH)_2$	0.42		–	
K_2SO_4	0.205		0.30	
$MgSO_4$	0.22		–	
$MgCrO_4$	–	0.21	0.95	0.63
$K_2Cr_2O_7$	0.195		0.63	
H_2SO_4	0.5			
$K_3Fe(CN)_6$	–		0.080	0.080
$K_4Fe(CN)_6$	–		0.053	0.053

hydrous oxides and clay minerals are stabilized by the specific sorption of inorganic phosphates. There is a correlation between the amounts of adsorbed phosphates, the zeta potential, and the stability of the suspension (Moriyama, 1976).

The effect of pH on the stability of a suspension is as follows. When the particles all have charges of the same sign, coagulation is difficult. If the acidity of the aqueous suspension changes, the particles will coagulate and settle out as their isoelectric point is reached. The particles will repeptize when they acquire a charge opposite in sign to their previous charge by further change of the pH. (See also pages 215–218).

Specific sorption of amphipathic ions and the formation of adsorbed layers, reduces the interparticle van der Waals attraction because the density of atoms in the adsorbed layer is less than in the boundary of the solid particle. For example, in the interaction of parallel flat plates the van der Waals attraction can be reduced by a factor of 5 (Becher, 1973). This is further reduced if the sorbed layer is incomplete. Organic ions are effective coagulating agents only if they are in concentrations sufficient to complete the coating of the particles. The subject has been discussed in Chapter 3, in connection with peptization.

3. Bridging Action. Polyvalent ions, such as iron and aluminum, which form polymeric ionic species on the solid surface, act as excellent coagulating agents by serving as bridges, in joining two particles together, as well as by their being specifically sorbed to colloid surfaces. Destabilization of disperse systems by a bridging action occurs if the bridging molecules or ions are large and have segments extending out from the colloid surface (Stumm and Morgan, 1962; Packham, 1963).

Very small amounts of organic polyfunctional polymeric material can sometimes bring about the flocculation of lyophobic colloids. The adsorption of different parts of the same molecule on the surfaces of more than one particle, i.e., bridging, is an important factor in the mechanism of this process (Hall, 1974).

When polymeric ions are in concentrations high enough for coating the particles, so that the particle charge is reversed, a new double layer is created and the particles may peptize. Organic polyelectrolytes, which are hydrophilic colloids and are common in nature, have the same effect on the stability of disperse systems. At low concentrations they may act as coagulating agents, forming bridges between several mineral particles. When the polyelectrolyte concentration becomes high, the salt tolerance of the hydrophobic particle is considerably increased, even to the extent that a colloid solution remains stable in concentrated salt solutions. Because of this effect, polyelectrolytes are called *protective colloids.* The repulsion occurring between the coated particles is called "entropic repulsion" and results from the steric hindrance in a layer of adsorbed surface-active molecules, which occurs when the particles approach each other to within a distance at which the adsorbed molecules begin to interfere.

The effects of various types of anionic polymeric surfactants with molecular weights ranging between 900 and 4500 on the stabilities of aqueous suspensions of several metal oxides and clay minerals have been examined by Moriyama (1976). At a high polymer/particle ratio, the polymer serves as a protective colloid and the particles become negatively charged. At a low ratio the polymer acts as a bridge bonding two or more particles into an aggregate. At low concentrations of disperse inorganic material, a complete breakdown of the initial aggregates into independent primary particles is obtained by dispersion, and the stabilities of the suspensions increase with the concentration of the organic polymer. On the other hand, at high concentrations of disperse particles the breakdown of the initial aggregates is incomplete and the organic polymer does not succeed in stabilizing the suspension. In this case the stabilities of the suspensions decrease markedly with the increase in the concentration of the organic polymer. The stability of the dilute suspensions is closely related to the zeta potential and to the adsorption of the organic polymer, whereas the stability of the concentrated suspensions is not related to the zeta potential. (For a discussion of peptization by means of protective colloids, see Ch. 3.)

4. Interaction Between Particles of Dissimilar Nature: Particles carrying charges of opposite signs attract each other at any interparticle separation. Particles carrying charges of the same sign but with unlike potentials attract each other at short distances. However, aqueous suspensions of dissimilar particles may exist if there are considerable osmotic or entropic repulsions. The two particles adhere, at least initially, because of van der Waals attraction forces acting between the atoms in the two surfaces. Van der Waals forces are operative effectively over very small distances. Hence, in order that particles may adhere, the atoms in the surfaces must be brought close enough together for these forces to become operative. If both dissimilar particles had an absolutely smooth (on an atomic scale) planar surface, every interparticle collision of parallel dissimilar planar surfaces would result in coarsening, and all attempts to separate them mechanically would be unsuccessful. However, actual surfaces differ from these ideal surfaces in that they are rough and contaminated. Both of these imperfections contribute to a greatly decreased area of actual contact between the surfaces. Trapped water molecules may penetrate into the surface indentations of the solid particles and may provide a great deal of bridging. At the same time the trapping of water molecules is accompanied by a decrease in entropy because of the decrease of the number of free water molecules, thus decreasing the degree of disorder in the system. If there is a considerable loss in entropy, flocs will not form unless they are chemically very stable, and the suspension is said to be stabilized by *entropic repulsion*.

If the water spreads spontaneously on the solid surface, the fluid can flow more completely into the micro- or submicroscopic pores, displacing gas pockets and other contaminations, and as a result the interfacial area of contact increases even if the thickness of the adsorbed water layer is small. Consequently, the number of interfacial water molecules is small, and a large number of these molecules bridge between the solid surfaces forming bonds with both surfaces at the same time. The floc thus obtained is very stable, and the entropic repulsion is not sufficient to separate the particles and to stabilize the suspension.

Coagulation of dissimilar particles is called mutual coagulation or *heterocoagulation*. The principal factors governing heterocoagulation are the same as those controlling coagulation of identical particles, namely, the electrostatic forces due to the double layer interaction and the van der Waals attraction forces.

5. Secondary Coagulation: A secondary minimum exists in the interaction energy curve (Fig. 8.5). Particles are in a stable configuration when they are separated by a distance equal to the distance of the secondary minimum from the surfaces. A secondary minimum can be found in most systems of disperse minerals. However, for many systems, particularly those involving very small particles, the secondary minimum is extremely shallow. For interactions between relatively large particles, on the other hand, the secondary minimum can represent a deep potential well, which can be expected to lead to coagulation. From the form of the interaction energy curve it is obvious that secondary coagulation is a reversible phenomenon. According to Hogg and Yang (1976) the kinetics of secondary coagulation are less sensitive to electrolyte concentrations than is the case with coagulation as represented by the primary minimum.

2.3 Coagulation in Natural Waters

2.3.1 Coagulation and Sedimentation in Surface Waters

Coagulation of river-derived suspended matter is the dominant mechanism by which sedimentation of suspended matter occurs in estuarine environments (Schubel, 1968; Edzwald, 1972). Sediment particles, especially the fine particles, are not transported by marine water in the form in which they occur in the sediment. Inorganic sediment suspended in the sea as independent individual mineral particles is unstable and occurs as coagulated aggregates with settling rates many times faster than those of the constituent grains (Berthois, 1961; Biddle and Miles, 1972; Sheldon, 1968). According to Feely (1976) near-bottom oceanic nepheloid layers represent regions of maximum sedimentation due to high rates of coagulation.

Samples from waters rich in inorganic sediments, such as tidal inlets, narrow straits, and the bottom waters of muddy areas have characteristic particle-size spectra[1] composed of regular, almost symmetric broad distributions with well-developed modes (Fig. 8.7). The particles forming these distributions consist of unsorted mixtures of flocs and single grains (Kranck, 1973). When samples with regular, normal size spectra are deflocculated and the single grain distributions are measured, the size spectra acquire a broader, flatter distribution with smaller mode size. The regular spectra of coagulated systems differ in both composition and spectral shape from the straight featureless spectra of samples of waters with low particle concentrations of any kind; these latter contain relatively few aggregates and a greater number of single grains relative to the aggregates (Kranck, 1975). These spectra are also quite distinct from irregular wavy spectra characteristic of plankton-rich samples (Sheldon et al., 1972). Aggregates composed largely of organic matter have been observed in the sea (Riley, 1963; Gordon, 1970). Organic matter may be an important factor in the coagulation of inorganic particulates.

Flocculated species in seawater are very stable. Agitation or increased salinity has no effect on the flocculation–deflocculation processes in seawater suspensions (Gripenberg, 1934; Whitehouse et al., 1960). Baylor and Sutcliffe (1963) showed that aggre-

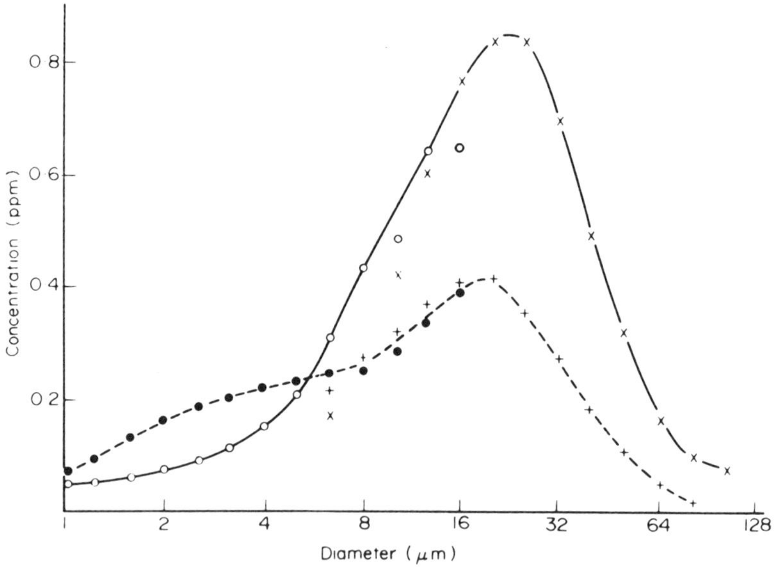

Fig. 8.7. Example of typical grain-size spectra of suspended particulate matter from coastal waters (Peptpeswick Inlet) with high inorganic content. The solid curves show the natural flocculated particle distributions and the broken curves show the deflocculated inorganic grains. (From Kranck, 1975: Used with the permission of Blackwell Scientific Publications Ltd)

1 Size spectrum is the curve of the relationship of concentration versus particle size (particle-size distribution).

gates are obtained in seawater in the presence of bubbles during flotation, whereas Riley (1963) showed that they can also form in the absence of bubbles.

Kranck (1973) proposed the following mechanism for the flocculation of a suspension of single grains into a stable flocculated distribution. In salinities above about 3%, fine particles in a suspension are unstable and flocculate readily on contact. A high degree of particle collision will occur in the initial unflocculated suspension, because gravity and turbulent forces cause a random variation in rate and direction of transport of particles of different sizes. Smaller particles will flocculate most readily (Müller, 1926), and larger grains adhere to flocs composed of many smaller grains by the mechanism of heterocoagulation. As progressively larger flocs are formed, the transport velocities become more uniform and particle collisions less frequent. Eventually, a state is reached in which the settling velocities of the largest flocs equal the velocities of the large independent single grains that have hitherto escaped flocculation, and further flocculation does not occur. (See orthokinetic coagulation, page 345).

When flocs are deposited they lose their identity and become part of the bottom sediment. In an analysis of the sediment one must be careful in directly relating the grain-size distribution of the sediment to the transport mechanisms and dynamic behavior of the particles before deposition.

The stability of colloid solutions of amorphous silica and heterocoagulation of amorphous silica with amorphous alumina in seawater were studied by Willey (1975). Solid amorphous alumina was found to prevent the peptization of silica in seawater or in a 0.6N NaCl solution. In distilled water, alumina inhibited but did not prevent this peptization of silica. In single-component systems that contain only silica, in either NaCl solution or in seawater, total silica concentrations (dissolved plus colloidal) reached 200 mg Si per liter. It appears that under these conditions the colloidal silica is hydrophilic and the electrolyte concentration of seawater is not sufficient to bring about coagulation of the silica. In two-component systems that contain both silica and alumina in contact with these same solutions, the total silica concentrations remain near the solubility values of amorphous silica. The silica adsorbs Al ions and its solubility decreases from 52.7 and 36.5 mg Si to 35.9 and 31.4 mg Si per liter in distilled and seawater, respectively. (See also pages 269–270).

The silica saturation in the seawater systems was exceeded by lowering the temperature from 22° to 4 °C. In the amorphous silica system the dissolved silica in excess of saturation polymerized to become a colloid solution. In the two-component system, however, most of the excess dissolved silica coagulated and precipitated after it had been polymerized, and both the dissolved and the total silica concentrations in solution decreased as the temperature decreased.

Selective flocculation and settling account for the separation of clay minerals in the ocean. From laboratory experiments Whitehouse and Jeffery (1953) suggested that montmorillonite may be deposited near the mouth of a river and illite is carried in suspension to a more distant depositional site.

The flocculation of hydrophilic colloids of iron and aluminum associated with soluble humic substances occurring during the mixing of waters from Scottish rivers with seawaters or with aqueous solutions of the three major sea salts, NaCl, $MgCl_2$, and $CaCl_2$, was studied by Eckert and Sholkovitz (1976). The hydrophilic colloids flocculate after being transformed into a hydrophobic colloid. The flocculation process is rapid. The amounts of flocculated constituents increase with the salinity of seawater. However, when the salinity is above 2%, any increment affects flocculation only slightly. The effectiveness of the salts on flocculation decreases in the order $CaCl_2$, $MgCl_2$, and NaCl. The extent of Al and humate flocculation obtained with $CaCl_2$ is not reached either with $MgCl_2$ or with NaCl even at molarities many times that of $CaCl_2$; consequently, Eckert and Sholkovitz suggest that chemical reactions occurring between Ca and Al humates are involved in the flocculation process, in addition to the double layer repression, which is the chief mechanism for other electrolytes. Only a small fraction of these hydrophilic colloid species are flocculated at Na, Mg, and Ca concentrations significant for seawater. Sholkovitz (1976) demonstrated that there is a close association of Fe, Mn, Al, and P with both river-dissolved humic substances and the seawater-flocculated humates, suggesting the important role of dissolved organic matter in controlling the flocculation behavior of inorganic constituents. (See also page 365).

The kinetics of coagulation of clay minerals in estuaries was studied by Edzwald et al. (1974). The suspended load of estuarine can be placed in two categories, namely, cohesionless coarse particles such as sand and cohesive or fine particles such as silt and clays. The deposition of cohesionless particles in estuaries depends primarily on the hydrodynamics of the systems, since these particles remain as individual particles, regardless of the flow conditions and the solution chemistry. On the other hand, cohesive particles range in size from one micron to several microns, and the effects of the surface and interparticle forces are at least as important as those resulting from hydrodynamics. The rate of coagulation and deposition are determined by particle collisions that are due to fluid motion. Under the conditions of the estuary, an illite suspension coagulates more slowly than kaolinite, which is more stable than montmorillonite. In the estuary they studied, kaolinite was the dominant clay in the upper end of the estuary where salinity is lowest and decreases toward the mouth, where salinity is highest. Illite occurred in minor amounts in the upper end and increased toward the mouth, while montmorillonite is present in minor amounts along the entire length of the estuary.

Heterocoagulation of a marine humic acid with clay minerals was studied by Rashid et al. (1972). Adsorption of humic acids on clays in weak ionic solutions such as fresh water would be minimal due to the extensive double layers surrounding both the clay and humic acid. Adsorption of humic acid on clay minerals in 3.5% NaCl solution would be enhanced due to the repression of the double layer and the increased contribution of attracting van der Waals forces. In the natural environment this process begins when river water carrying clays and humic acids enters the marine estuary. The mechanism of interactions between clays and humic substances will be discussed in more detail in the next section of the present chapter (See also pages 302–305, 310–311).

2.3.2 Filtration of Ground Water During Migration

Heterocoagulation is of great importance in the migration of water through sediment beds, resulting in the filtration of the water from suspended particles. According to Hunter and Alexander (1963) the porous medium may be considered to consist of regions where the liquid flow approximates that in a capillary and regions where the liquid flow is virtually zero. At points between these extremes the shearing rate at any time will vary from zero to a maximum value that depends only on the driving pressure and the structure of the porous medium. By the same token, the shearing force applied to the suspension in the pores will vary from zero up to a maximum value, depending on the shearing rate. Natural waters may contain small quantities of clay and other mineral particles, metal oxides and hydroxides, and complex organic compounds, all in colloid suspension. When such a suspension is passed through a porous medium, the particles diffuse across the streamlines into the regions of low shear due to the concentration gradient initially present. This effect continues until the concentration of the particles in this "dead space" is equal to that in the flowing liquid. Heterocoagulation occurs when the fine particles in the "dead space" approach the surface of the pore; consequently, the mechanical rigidity of the particulate material in the "dead space" may be sufficient to enable it to resist the shearing force of the moving liquid. It will behave as a rigid deposit, and the resulting distortion of the streamlines will cause a reduction in permeability. The preceding authors studied the flow of kaolinite sol in a silica column and demonstrated that the removal of kaolinite particles was improved by an increase of electrolyte concentration and by reversal of the surface charge of silica from negative to positive by the addition of cetyltrimethylammonium ions. Both additives reduce the energy barrier of the function $V = V(h)$ (where V is the potential energy of the adhesion of kaolinite to the silica) and increase the rate of adhesion.

The attachment of particles of polyvinyl chloride and ferric oxide to filter media (sand, glass, and anthracite) as a step in filtration has been examined quantitatively, and equations for the double layer interaction between the adherents have been derived. The higher the energy barrier in the total potential energy curve, the lower is the filtration efficiency. However, there were some cases in which the filtration efficiency was high despite the presence of a rather high potential energy barrier (Gregory, 1969). It seems that the height of the potential barrier is not the only criterion for filtration efficiency.

3. Soil Aggregates

Soil aggregates are formed when "primary" independent, individual particles cluster together into larger separable "secondary" units. Several elementary units together with flocculated forms can be combined in higher order features to form soil aggregates that exhibit definite physical boundaries. Some types of clustering of primary particles in secondary particles are shown in Figure 8.8 (Collins and McGown, 1974). The nature of soils is affected by the characteristics of the aggregates. In the present section, only the types of interactions between the elementary soil particles that comprise aggregates will be considered.

The individual particles in a particular soil aggregate are arranged in a definite manner, being relatively stable in their arrangement, depending on how well they are bound together. If they are coarse, they are not commonly united into aggregates, and the soil structure is called "single-grained". Single-grained structures are found in sandy soils. If the primary particles include clay minerals, aggregation becomes significant and influences the physical behavior of the soil. Clay mineral particles in an aggregate are usually packed together in an oriented manner. The geometry of the aggregate is subjected to change with changing water content due to shrinking and swelling of clay particles. The clay mineral packs in the intergranular space are produced by coagulation from aqueous

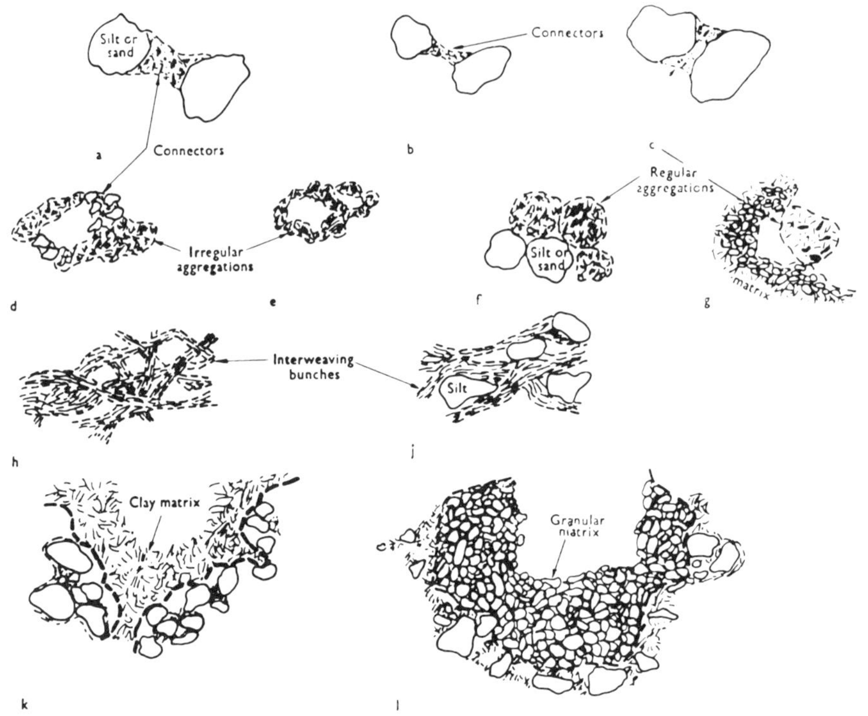

Fig. 8.8 a–l. Schematic representations of particle assemblages in soils and sediments. (From Collins and McGowan, 1974: Used with the permission of the Institution of Civil Engineering, London)

suspension. As will be shown in the next section, a "card-house" structure, in which the clay platelets are arranged in a nonparallel manner, can be obtained in the first stage, if the coagulation occurs from a solution that is poor in electrolyte. Shearing movements, which are the macroscopic results of minute swelling and shrinking strains within the clay packs, will in general lead to orientations of the platelets parallel to the direction of the movements. Dense aggregates are therefore to be expected in soil that has been subjected to repeated cycles of wetting and drying, such as lateritic soils (Marshall, 1962; Mendelovici, 1977).

The structural stability of the aggregates depends on the resistance of the various components to the internal shearing movements (Low, 1968). An aggregate can be regarded as a porous medium in which the tortuosity is derived from the arrangement of the coarse particles in the packing. Because of the tortuosity, small particles in the aggregate are more resistant to shearing than similar particles outside the aggregate. The internal movement of the particles can be further reduced by reducing the void volume of the aggregate (cementation) and by binding between the various individual particles (coagulation and heterocoagulation).

Deposition from soil solutions of amorphous iron and aluminum hydroxides, amorphous silica, calcium and magnesium carbonates in the intergranular space of the aggregate reduces the void volume and leads to stabilization of the aggregate. This process is known as *cementation.* The sesquioxides are adsorbed on the surfaces of clay particles, binding the clayey constituents into coarser particles. Clay particles coated with iron and aluminum hydroxides do not disperse in water unless the sesquioxides are removed (Harward et al., 1962). Soil aggregates become disaggregated after the iron and aluminum have been extracted with organic complexing agents (Giovannini and Sequi, 1976). Carbonates also act as cements in semi-arid soils. In particular cases they can be deposited progressively, especially in an evaporating zone, where nodules or pans of considerable hardness may form.

The presence of crystalline iron and aluminum oxides in the aggregate is associated with some degree of structural stability. There is evidence showing that free aluminum oxides are the more important bonding agents in old, highly weathered and well-oxidized soils (Deshpande et al., 1968) in which the iron oxide is present as discrete particles on clay surfaces, having little effect on the aggregate stability (Follett, 1955; Greenland et al., 1968). Crystalline iron oxides are often associated with a more amorphous phase, "active" iron oxide, which helps to bind individual primary soil particles together (Cornell et al., 1974).

Organic material plays an important role in the formation of soil aggregates (Edwards and Bremner, 1967). The organic soil material is polar and may interact with the inorganic components of the aggregate. Humic substances are amorphous polyfunctional compounds and may bridge between several inorganic particles. Nonpolar groups take part in hydrophobic interactions with surfaces of nonpolar minerals and also provide sites for trapping gases. The following bondings can occur between the inorganic particles and the organic polymers: (1) quartz–organic matter–quartz; (2) quartz– organic matter–clay particle; (3) clay particle–organic matter–clay particle; (4) hydrous oxide–organic matter–clay particle; (5) quartz–organic matter–hydrous oxide; and (6) hydrous oxide–organic matter–hydrous oxide (Emerson, 1959). As a result of these bridges the degree of shrinkage and swelling of the aggregate is limited.

Amphipathic ions may be specifically sorbed by clay minerals and amorphous oxides, leading to flocculation. The mineral coated with the organic ions may hydrophobically interact with the sandy grain.

The small particulate residues left by plants appear to play an important role in stabilizing the structure of the aggregate (Marshall, 1962). Provided that they have not been acted on by microorganisms, their role is similar to that played by the sandy grains, namely, as nuclei for the accumulation of smaller particles. Unlike the inorganic grains, they are not necessarily almost spherical, and thus their steric contribution to the aggregate stability is not equal to that of the sandy aggregate.

3.1 Interaction Between Clay Particles

3.1.1 Modes of Associations of Plate-like Clay Particles

The present model was suggested by van Olphen (1963, 1964) for the interaction between plate-like particles of 2:1-type clay minerals. The interaction may result in three different modes of particle associations; (1) adhesion between flat oxygen planes of two parallel plate-like particles (face-to-face, FF); (2) adhesion between broken-bond surfaces of neighboring particles (edge-to-edge, EE); and (3) adhesion of broken-bond surfaces to a flat oxygen planar surface (edge-to-face, EF). EE and FF types of adhesion result from the overlapping of similar double layers, whereas the EF type results from the overlapping of dissimilar double layers, by the mechanism of heterocoagulation. Since the face and the edge surfaces are negatively and positively charged, respectively, the barrier of the potential energy of EF-type interaction is less than zero, and heterocoagulation is rapid. On the other hand, the formation of EE and FF associations is extremely slow, unless the double layers have been repressed either by the addition of electrolytes or by the removal of water. Moreover, the van der Waals interacting energy will be different for the three types of association, being the highest in the FF association. The resulting interaction energy curve obtained for the FF association has the lowest primary minimum.

From the discussion in the previous section on the frequency of collisions and the rate of *rapid coagulation* of nonspherical particles, it follows that EE coagulation, in which the longest diameter of the plate serves as the collision diameter, is the fastest of the three coagulation types, whereas the FF coagulation, in which the thickness of the plate serves as the collision diameter, is the slowest. Consequently, the three types of associations will not necessarily occur simultaneously or to the same extent when a clay suspension is flocculated.

The FF association leads to thicker and larger flakes, whereas the EE and EF associations will lead to three-dimensional, voluminous "card-house" structures. The various modes of particle association are given in Figure 8.9a–g. Only the EE and the EF types of associations lead to aggregates that can be classed as flocs. The thicker particles, which result from FF association, cannot properly be called "flocs" and the terms oriented aggregates or tactoids are commonly used.

The modes of layer association that predominate in suspensions of smectites containing various amounts of NaCl are summarized here after van Olphen:

1. In an electrolyte-free system, a negative surface charge displays on the faces of the clay platelets due to the diffuse distribution of the exchangeable cations. The

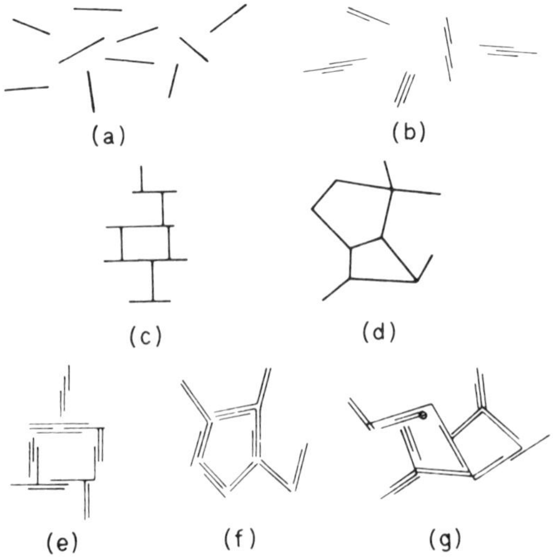

Fig. 8.9. Modes of particle association in clay suspensions **a** dispersed; **b** face-to-face association (tactoids); **c** edge-to-face association; **d** edge-to-edge association; **e** edge-to-face association of tactoids; **f** edge-to-edge association of tactoids; and **g** edge-to-face and edge-to-edge association of tactoids. (From van Olphen, 1963: Used with permission of John Wiley and Sons)
c d: Card house structure; **e f g:** Book house structure

broken-bonds surfaces carry positive charges by the adsorption of potential-determining ions (present as impurities in the system). Since the concentrations of diffuse ions in both double layers are extremely small, the osmotic repulsion that arises from the overlapping of the dissimilar double layers is small and a rapid heterocoagulation results in the formation of an EF mode of particle association.

2. In the presence of trace amounts of NaCl, both double layers on the clay particle are sufficiently well developed, and electrostatic and osmotic repulsions prevent particle association. All modes of coagulation occur at an extremely slow rate and a stable suspension of individual layers is obtained.

3. When the amount of NaCl in the suspension is increased, both double layers are compressed. Once more the subtle balance between edge-to-face attraction and face-to-face repulsion becomes favorable for the formation of EF particle associations. Simultaneously, edge-to-edge association by van der Waals attraction may contribute to the formation of a "card-house" structure.

4. At concentrations of NaCl above the critical coagulation concentration, face-to-face coagulation becomes rapid. Since the FF mode of particle association is thermodynamically more stable than either the EE or EF mode, the number of particles in the "card house" is reduced whereupon "oriented aggregates" and "book-house" structures are formed. Critical coagulation concentrations for montmorillonite are given in Table 8.3. Posner (personal communication, 1977) suggests that the FF coagulation is too slow a process to occur spontaneously, but happens after the application of external forces, such as shearing, or by drying.

3.1.2 Coagulation of Clay Minerals

The results found in the literature for the critical coagulation concentrations of natural clay minerals vary considerably because the natural minerals are coated by amorphous material that affects the surface charge of the particle and also because of the organic and inorganic contaminants that may act as coagulating agents. Laponite is a synthetic magnesium silicate clay very similar in structure and properties to hectorite, but is free from contamination and appears to be a useful model for the study of the critical coagulation concentration (Neumann and Sansom, 1970; Perkins et al., 1974). The particle charge becomes more negative as the pH of an aqueous suspension of laponite increases, which is due to the adsorption of the hydroxide ions on the positive broken-bond surface. As a result the edges become either neutral or slightly negatively charged and the EF mode of association cannot exist. At pH above 12 the concentration of the counterions introduced with the base becomes greater than the corresponding critical coagulation concentration, and oriented aggregates with an FF mode of association are obtained.

Table 8.3. Critical coagulation concentrations (in mmol/l) for montmorillonite sol. (After van Olphen, 1963)

Sol (conc. 1/4%)	NaCl	1/2 CaCl
Na-montmorillonite	12 –16	2.3 –3.3
Ca-montmorillonite	1.0–1.3	0.17–0.23

At pH below 7 the positive charge in the broken-bond surface increases, and flocs with an EF mode of association are obtained.

The dependence of the critical coagulation concentration of various counterions on the pH of the suspension is shown in Figure 8.10. The overall net charge of a laponite particle is negative, and the coagulation data of monovalent and divalent counterions show the expected importance of the ionic charge, namely, that divalent cations are more effective coagulating agents than monovalent cations. The order of cations is irregular and Perkins et al. suggest that these fine differences are due to specific sorption and exchange characteristic of these counterions. The critical coagulation concentration rises with increasing pH.

Kaolinite suspensions flocculate when the pH falls below 4 (Krizek et al., 1975). As one would expect in the electrolyte flocculation of kaolinite, the concentration of the salt required for suspension destabilization decreases with increasing charge of the cations. However, kaolinite appears to be somewhat less sensitive than montmorillonite to the presence of salts (Swartzen-Allen and Matijević, 1974). The amount of face-to-face aggregation of a montmorillonite sol increases, and its colloid stability decreases with increasing charge and unhydrated radius of its counterions (Lahav and Banin, 1968; Fitzsimmons et al., 1970). In kaolinite, the counterions have little or no effect on the degree of face-to-face aggregation (Rebhun and Sperber, 1967). The difference between the flocculation behavior of kaolinite and montmorillonite results from the fact that the kaolinite tactoid is composed of many more unitlayers than the montmorillonite tactoid and consequently, the sum of EF van der Waals plus EF electrostatic attractions becomes approximately equal to the FF van der Waals interaction. The rate of flocculation, which is dependent on the ratios between the longest and shortest axes of a nonspherical particle (page 345), is higher in montmorillonite (Edzwald et al., 1974).

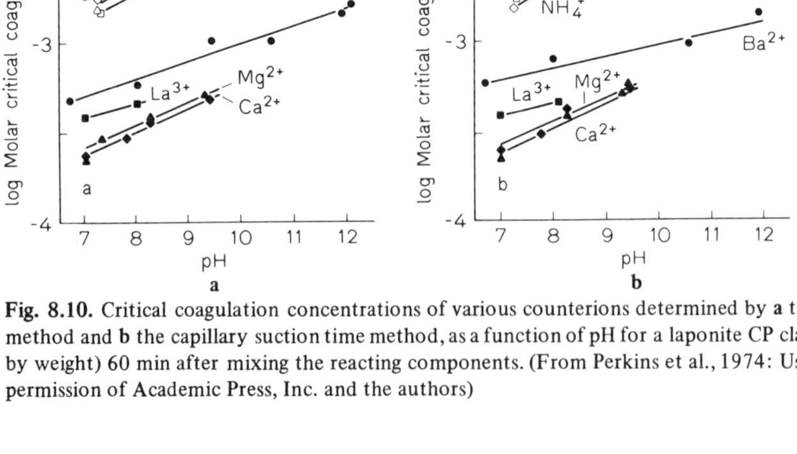

a b

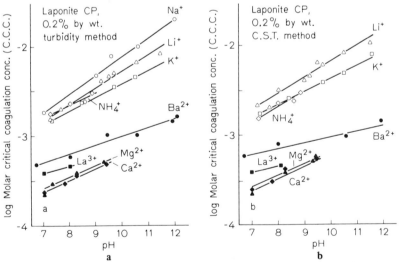

Fig. 8.10. Critical coagulation concentrations of various counterions determined by **a** the turbidity method and **b** the capillary suction time method, as a function of pH for a laponite CP clay sol (0.2% by weight) 60 min after mixing the reacting components. (From Perkins et al., 1974: Used with the permission of Academic Press, Inc. and the authors)

Fitzsimmons et al. (1970) and Greene et al. (1973) differentiate between condensation of elementary silicate unit layers of Ca-montmorillonite by the mechanism of orientation to form tactoids, which they call "quasi-crystals", and FF mode of aggregation. The initial flocculation that occurs on the addition of electrolyte results in the particles approaching one another to distances corresponding to the secondary minimum, with the individual lamellae randomly oriented with respect to one another, with some having FF orientation. Mechanical agitation causes distortion of the flocs, and further collisions between the newly dispersed lamellae results in a more parallel alignment of the initially randomly flocculated platelets. In the third stage, the great attraction of the negative platelets for divalent cations results in their condensation, the distance between parallel layers becoming 1 nm, which corresponds to the primary minimum in the curve of potential energy versus interparticle separation (Fig. 8.5). These quasi-crystals are stable so that condensation into a primary minimum is not readily reversible unless the interlayer cation is exchanged by Na. Sodium ions are less effective than calcium ions in causing collapse of the diffuse double layer on montmorillonite so that on flocculation of Na-montmorillonite with NaCl there is less coalescence between the elementary particles. When Ca-montmorillonite suspension is freshly prepared by treatment of Na-montmorillonite with a Ca-resin, primary platelets are completely dispersed. With time the elementary layers may condense and extend in the *a, b* directions to form large flat particles.

3.1.3 Effect of Iron and Aluminum on the Coagulation of Clay Minerals

Coagulation of expanding 2:1-type minerals by Al or Fe salts results mainly in the FF mode of particle association, followed by the formation of large tactoids. Both Fe and Al form hydroxy complexes in the interlayer spaces of expanding clay minerals, which restrict the hydration and swelling of the minerals (Carstea et al., 1970; Rowell, 1965; Kidder and Reed, 1972). Hydroxy Al interlayers are of widespread occurrence (Rich, 1968; Reneau and Fiskell, 1972), and their effect on structure is important in alkali and nonalkali soils (Deshpande et al., 1968). Hydroxy Fe interlayers are less common, but under conditions of alternating oxidation and reduction where iron hydroxide is being freshly precipitated and the soil pH is below 8.5, which is the point of zero charge of iron hydroxide, their effects are also important in soil aggregate stabilization (El-Rayah and Rowell, 1973). In this mode of association a very high stability is achieved by the bridging effect brought about by the hydrated hydroxy layer (pages 290, 300).

Blackmore (1973) studied coagulation of three clay minerals–kaolinite, illite, and montmorillonite–by the products of ferric iron hydrolysis. Aggregation increases with oxide amount up to a weight ratio of about 0.5, smaller amounts being less effective in aggregating the bentonite than the other clays. The aggregates produced by association with the latter stages of hydrolysis had different properties in both wet and dry forms than those produced by association with earlier stages of the hydrolysis. The former were of powdery consistency when dry and of different color. The suspension resulting from their dispersion was visibly a mixture of clay and oxide, in marked contrast with the dispersion of the more stable material, in which there was homogeneity throughout. The end product of the hydrolysis is a gelatinous hydroxide precipitate, and aggregation of the mineral occurs by the mechanism of heterocoagulation. In the earlier stages of hydrolysis polymeric hydroxy cations of Fe are obtained, and during the aggregation of the clay they form bridges connecting between the particles. Blackmore concludes that, in this case, cementation (or heterocoagulation) is not an effective aggregating process, whereas bridging of the clay particles by the polymeric cations is apparently necessary for the achievement of stable bonding in the soil aggregate. According to El-Swaify and Emerson (1975), upon drying clay minerals and ferric hydroxide particles together, the effective interparticle distances become sufficiently small for H bonds between the clay and the hydroxide to be formed, and heterocoagulation becomes effective. (See also pages 177–180, 301).

Bundy and Murray (1973) studied the effect of aluminum on the aggregation of kaolinite, by surface area and viscosity measurements. Maximum dispersibility of Al-kaolinite occurs at pH 3, when Al has not polymerized. With increasing pH, the increased amount of positively charged aluminum hydroxide precipitated promotes the formation of FF associations. The presence of sulfate ions in a kaolinite-aluminum hydroxide system changes the modes of association. Where relatively high concentrations of sulfate are present in the system, EE and EF structures dominate. It is presumed that the positively charged aluminum hydroxide is bridged to the positively charged edge surface of kaolinite by sulfate anions. With removal of most of the sulfate, the positively charged aluminum hydroxide would tend to aggregate on the negatively charged oxygen plane and thus promote the formation of FF associations.

3.2 Interaction Between Sand Grains

The interaction forces between two quartz grains are very weak for the following reasons: (1) the surface is composed of the "inert" siloxane groups, which do not participate in short-range interactions; (2) the oxygen atoms that comprise the surface boundary layer are packed in a very low density and the van der Waals attraction forces between two surfaces in contact are very weak; and (3) the grain surfaces are not planar and the adhesive area is very small.

The electric potential at the quartz surface is low, and according to Eq. (8.5), the force of repulsion that arises from the overlapping of the diffuse double layer is also low. As a consequence, a dispersion of quartz grains is unstable, and after the particles have been settled, resuspension by means of Brownian movement does not occur. Such a dispersion is therefore known as an "irreversible dispersion".

The attraction and repulsion interaction forces between feldspar grains have not been measured, but it appears that both are somewhat stronger than those between quartz grains because of the higher polarization and cation exchange capacity (Breazeale and Magistad, 1928) of the feldspar surface compared to that of quartz. Still, they are very weak compared to those of clay minerals.

Although there are weak adhesive forces between the "sand" grains and flocculation may occur to some extent, they are commonly known to comprise the "cohesionless" or "nonflocculated" soil fraction, whereas the clay minerals and the hydrous oxides are known to comprise the "cohesive" or "flocculated" soil fraction.

3.2.1 Coagulation of Spherical Particles

The present model is based on the treatment of Barclay et al. (1972). The potential energy curve as a function of distance for interaction between two large spherical particles has essentially the same form as that shown in Figure 8.5. With small particles the secondary minimum is extremely shallow at high electrolyte concentrations and is essentially absent at low electrolyte concentrations. The calculated potential energy curves for various particle radii show that an increase in the height of the primary maximum and an increase in the depth of the secondary minimum occur as the particle size increases at constant surface potential and electrolyte concentration. Thus, with increasing particle size, the formation of flocculated structures in which the particles associate at distances corresponding to that at the secondary minimum becomes possible. This association will not cease at the doublet stage but will continue to triplet and quadruplet formation and so on, so that essentially a nucleation process will occur forming clusters of particles, or flocs, in which the particles are associated at a distance. This type of flocculation does not occur with small particles where the secondary minimum is absent or is very shallow, and Brownian movement is sufficient to disaggregate the floc.

For the interaction of a single particle with a floc the latter will behave approximately like a flat plate of the same material. The secondary minimum for a sphere–flat plate interaction is deeper than for a sphere–sphere interaction, at the same potential, and a firmer association will be formed with the particles moving closer.

Flocs are formed at different sites in the system. Because this association of the primary particles is reversible, the "nucleation" process depends on the concentration. Thus the number of flocs in equilibrium with free primary particles will depend on the depth of the secondary minimum and the number concentration (number of particles per unit volume) of single particles. As the number concentration increases, either due to infiltration of water or due to evaporation, the number of flocs formed will increase, and interaction will occur between flocs as well as between free primary particles and flocs and between free primary particles and other free primary particles. Interaction in a system that contains flocculated structures leads to a more disordered array, and hence the volume fraction of the solid in the floc becomes smaller, the greater the extent of flocculation.

Increasing the concentration of electrolytes causes a decrease in the force of repulsion between the spheres.

The formation of sand aggregates from aqueous soil suspension may occur by the same mechanism. However, the adhesive strength between the sand grains is low and, because of gravity, independent primary particles settle down and the aggregate decomposes. If the adhesive strength of the aggregate increases by interaction, for example, with clay minerals and organic material, the secondary particle will not disaggregate.

3.2.2 Effect of Capillary Liquid on the Force of Adhesion Between Spherical Solid Particles

Let us consider a model system containing two spherical grains, a very small amount of liquid water, and a vapor phase. The four following combination types of solid–liquid and liquid–gas adhesion may occur (Fig. 8.11). In type I only liquid–gas adhesion

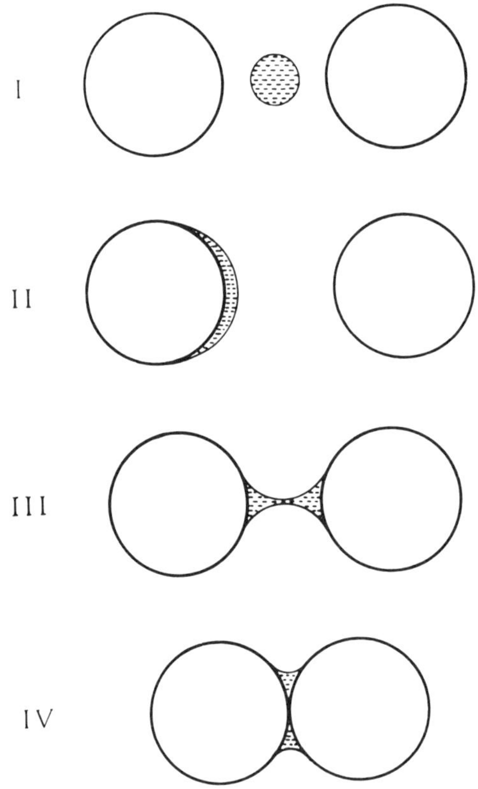

Fig. 8.11, I–IV. Four possible types of adhesion between two mineral grains and a water drop

occurs, the water drop does not touch the grains, and the two following interfaces exist, solid–gas and liquid–gas. In the second type, the water drop is spread on one of the grains. In the third and fourth types of adhesion the water drop bridges both grains and the system becomes similar to a capillary, the liquid possessing a convex surface. In the fourth type the two grains are in contact. Adhesion types II, III, and IV result in three interfaces: solid–gas (SG), liquid–gas (LG), and solid–liquid (SL). Adhesion type I occurs when the solid–liquid interface tension, σ_{SL}, is very high, i.e., when the liquid–liquid cohesion strength is very high and the solid–liquid adhesion strength is very low. Adhesion types II–IV occur when the solid–liquid interface tension becomes low. The type of adhesion depends on the difference $(\sigma_{LG} - \sigma_{SL}) = \Delta\sigma$. The distance between the two grains bridged by adhesion type III decreases with the increase of $\Delta\sigma$. Adhesion type IV is obtained when this value becomes very high. It follows that capillary liquid may bring about adhesion between solid particles. Binding forces caused by liquid bridges between two solid bodies are known as "capillary binding forces". It is obvious that this type of adhesion is not the result of a strong chemical bond formation between the solid particles, but is rather due to the fact that in the presence of small amounts of liquid the stable configuration is that in which the liquid is located in a capillary and the capillary is formed by the bringing of the solid particles into contact.

If one maintains one of the spheres fixed and measures the force required to pull the spheres apart, this force, denoted by F, will be given by

$$F = -\Delta p (\Pi R^2 \sin^2 \Psi) \tag{8.16}$$

where Δp is the capillary pressure due to the curved liquid surface, R is the radius of the spherical particle, and Ψ is a geometric factor that depends on the ratio between the liquid content, the radius of the particle, and the contact angle Θ.

Using elementary capillary theory Gillespie and Settineri (1967) calculated the force of adhesion between spherical solid particles and obtained the relation:

$$F_{\Psi \to 0} = 2\, \Pi R \sigma \cos \Theta. \tag{8.17}$$

The equation shows that the force of adhesion increases with increasing particle size. They also showed that this force decreases with increasing amount of water between the particles. This force increases also with a decreasing contact angle between liquid and solid particle, i.e., with increasing hydration energy of the particle. In the presence of small amounts of water sand grains may become adhered by capillary action. However, when they are coated by amorphous material, the contact angle Θ becomes small and the adhesion is much stronger.

3.3 Reactions of Humic Substances with Mineral Soil Components

Aqueous solutions of humic substances may be either true or colloid solutions of hydrophilic polyelectrolytes. They can be coagulated by different electrolytes. As one would expect, in neutral solutions (pH=7), when the net charge of the colloid is negative, trivalent positive ions are more effective in coagulating humic substances than are divalent ions, and divalent ions are more effective than monovalent ions. Cations of the same valence with the largest ionic radius and the lowest hydration energy, are the most effective coagulating agents, because they can easily move from the bulk solution into the hydrophobic hydration boundary layer (Baver and Hall, 1937). This does not apply to trivalent cations, which tend to polymerize in solution and to be specifically sorbed by the humic substances. Also certain transition metals tend to be specifically sorbed by the humic compounds.

The specific sorption of certain cations by humic substances results from the relative abundance of oxygen-containing functional groups in the organic-soil material. Essentially two types of reactions occur between humic material and metal ions: a major one, involving simultaneously both acidic COOH and phenolic OH groups, and a minor one, in which only less acidic COOH groups participate (Schnitzer, 1969), the latter occurring only at high pHs. If the humic substance contains two or more donor groups that occupy adjacent sites on aromatic rings so that one or more five- or six-membered rings are formed with the metal, the resulting structure is a chelate. This may be illustrated by the interaction of Cu^{2+} with a humic substance as follows

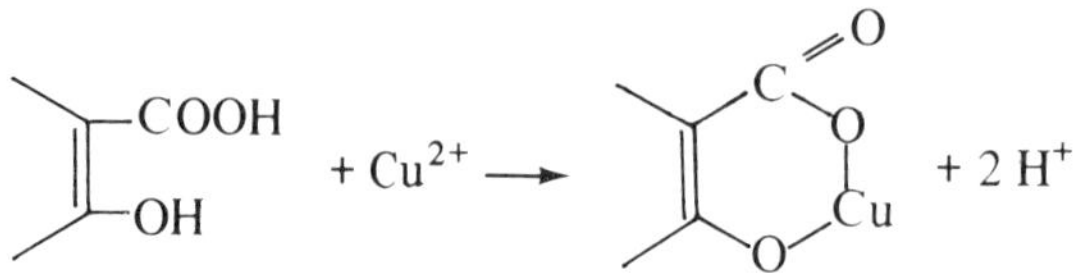

(8.I)

The most widely distributed functional groups in humic substances that have been shown to participate in the metal chelation, in addition to one COOH group, are: a second COOH group (Stevenson, 1976), phenolic OH (Schnitzer and Skinner, 1963), C=O, N=O (DeMumbrum and Jackson, 1956), and NH_2 (Schnitzer and Khan, 1972). The stability of complex formation for divalent cations with humic acid decreases in the following sequence: Cu > Ni > Co > Zn > Cd > Fe > Mn > Mg (Khanna and Stevenson, 1962). The following order of complex stability is obtained for divalent cations with fulvic acid: at pH 3.5, Cu > Fe > Ni > Pb > Co > Ca > Zn > Mn > Mg and at pH 5, Cu > Pb > Fe > Ni > Mn > Co > Ca > Zn > Mg (Schnitzer, 1969). Ferric iron forms the most stable complex with fulvic acid, followed by aluminum. The stability constants increase with increase in pH.

The net negative surface charge of humic substances decreases with decreasing pH of the solution, and in the absence of salts, when the pH becomes less than 3, humic acid solution coagulates. Peptization usually occurs at a somewhat higher pH than coagulation. In the presence of electrolytes the pK_a of the humic acid decreases (Posner, 1964) and the pH of peptization may rise to 4.5–5.0. The specific sorption of cations may reduce the net negative surface charge of humic acid and the critical coagulation concentration of electrolytes. The order of increasing effectiveness of metal ions for coagulating humic acids is Mn < Ni < Zn < Cu < Fe < Al (Khan, 1969). Aluminum and ferric humates become negatively charged only when the pH of the solution rises above 7. At pH between 4 and 7 they coagulate easily. Sulfate is more effective than nitrate and chloride in coagulating these complexes.

Since humic acids dispersed in water dissociate, that is, the carboxylic and phenolic groups become negatively charged, a mutual repulsion of the functional groups occurs and the polyelectrolytes adopt stretched configurations (Ong and Bisque, 1968). Such configurations are highly hydrated. In the presence of high electrolyte concentrations "salting out" manifest itself by a great portion of the cations becoming "anhydrous" or almost so, and attaching themselves to the negatively charged sites on the humic macromolecule, thus causing a reduction in the intramolecular repulsion in the polymer chain and possible coiling of the chain. "Salting out" also expels a portion of the water of hydration that surrounds the polymer, leaving it less hydrated and more capable of coiling. As the electrostatic repulsion and hydration are reduced, van der Waals interaction between the polymers may result in coagulation. The coagulation of hydrophilic colloids results predominately from "salting out" processes rather than from double layer repression and may occur only at very high electrolyte concentrations.

Gels of humic acid behave as reversible, swelling systems with high moisture contents. Below a certain moisture content they cease to swell again with the addition of water. Heating to 105 °C causes certain irreversible changes. After such treatment humic acid can be brought into solution by heating with alkali or by prolonged treatment with it in the cold. The exchangeable cations have considerable effect on the water content of the gel. For example, Ca humate has a higher moisture content at a given vapor pressure than the humic acid samples (Marshall, 1964).

3.3.1 Reactions of Metal Hydroxides and Oxides with Humic Substances

Two types of reactions between metal hydroxides and humic substances have been identified: (1) sorption of the organic substance by the inorganic mineral and (2) dissolution of the hydroxide (in part or totally) through the formation of a stable soluble or sol-forming complex between the humic substance and the metallic cation (Schnitzer and Khan, 1972). The type of reaction depends on the environmental conditions. A concentrated solution of humic acid percolating at a high rate leaches the metallic cations whereas a slowly migrating dilute solution gives rise to adsorption of the acid by the mineral. In the latter case, the organic polyelectrolyte spends a long time near the surface of the hydroxide and more and more functional groups form bonds with the hydroxide surface. On the basis of the mechanism for specific anion sorption (Ch. 6) and entropy considerations Schnitzer and Khan proposed the following structure:

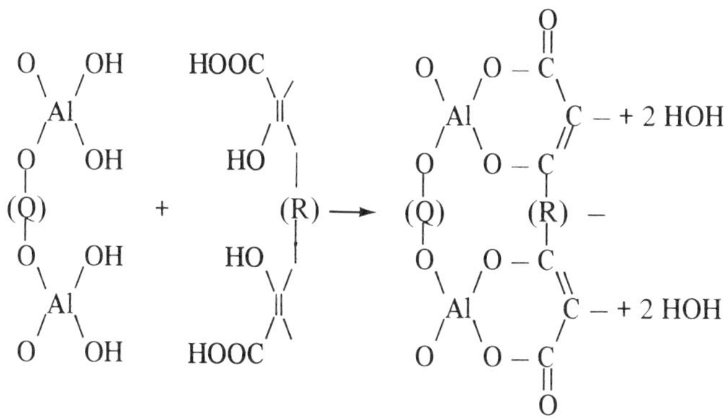

(8.II)

wherein (Q) and (R) are the cores of $Al(OH)_3$ and the humic substance, respectively. The formation of a great number of bonds between the polymeric species and the surface of one hydroxide particle is accompanied by an increase in entropy because of the great number of chemisorbed water molecules being desorbed from the surface, thus increasing the degree of disorder in the system. The humic substance thus adsorbed becomes fixed on the surface.

Freshly precipitated ferric and aluminum hydroxides adsorb humic substances, with aluminum hydroxide being more active in this respect than ferric hydroxide. Coagulation through bridge formation between several hydroxide particles occurs at low concentrations of organic matter or when the inorganic particles are at a short distance from each other. Higher concentrations of humic substances and also the occurrence of a great interparticle separation at the initial stage of the adsorption bring about peptization as the particles achieve a negative charge.

3.3.2 Reactions of Clay Minerals with Humic Substances

Humic substances can be adsorbed by clay minerals, giving rise to organo–clay complexes of varying stabilities. The formation mechanisms of these complexes are complicated, but they can, at least in part, be envisaged from our knowledge of reactions between clays and simple organic molecules. The principal mechanisms that may apply to reactions between clay minerals and humic substances involve (pages 302–312):

1. "Long-range" interaction and heterocoagulation by nonspecific anion exchange sorption onto the broken-bond surfaces or by the slow association of particles that have a charge of the same sign but have different surface potentials (Tan and McCreery, 1975).

2. If the pH of the solution is below 4, which is rare in natural waters, humic substances become positively charged, and long-range heterocoagulation may occur by nonspecific cation-exchange sorption of humic substances onto the flat oxygen plane surfaces (Schnitzer and Kodama, 1969).

3. Specific anion sorption onto the broken-bond surfaces through the "short-range" interaction with an Al atom at the edge of the crystal (Greenland, 1965 a, b).

4. Chelate formation with aluminum or iron located in the interlayer space or at the broken-bond surface, either as exchangeable monomeric cation or more probably, as a polyhydroxy cationic complex (Greenland, 1965 a, b) with a structure similar to that of the chelate in (8.II).

5. Chelate formation with transition metals that have a strong tendency to form coordination compounds, such as copper, located at the interlayer space or at the broken-bond surface, where the 1,3-diketone groups in the humic substances react with the metal to form acetylacetone-type chelates (8.III) (Schnitzer and Kodama, 1972; Kodama and Schnitzer, 1974 b).

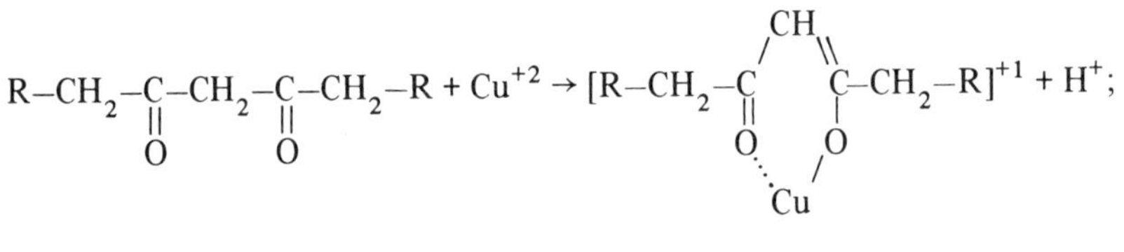

$$R-CH_2-\underset{\underset{O}{\|}}{C}-CH_2-\underset{\underset{O}{\|}}{C}-CH_2-R + Cu^{+2} \rightarrow [R-CH_2-\underset{\underset{O\cdots}{\|}}{C}\underset{\underset{Cu}{/}}{\overset{\overset{CH}{/}}{\underset{}{\diagdown}}}\underset{\underset{O}{/}}{C}-CH_2-R]^{+1} + H^+; \quad (8.III)$$

6. Carboxylic acid groups are linked to exchangeable cations through a "water bridge", in which the bonding between the water molecule and the COOH group is of the short-range hydrogen bond type, in one of the following ways

$$M^{m+}\cdots\underset{|}{\overset{|}{O}}-H\cdots O=C-R \qquad\qquad M^{m+}\cdots\underset{|}{\overset{|}{O}}\cdots HO-\overset{\overset{O}{\|}}{C}-R$$

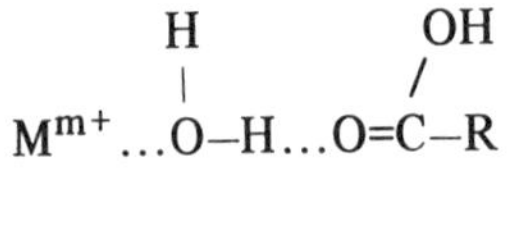

(8.IVa) (8.IVb)

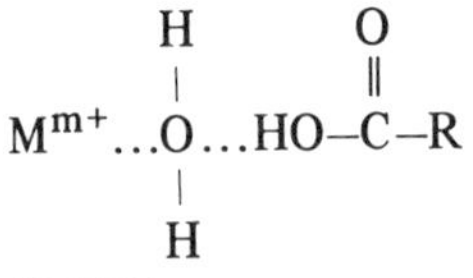

(Yariv, et al., 1966).

7. Phenolic groups are linked to exchangeable cations through a water bridge, in which the bonding between the water molecule and the OH group is again of the short-range hydrogen bond type, in one of the following ways

$$\begin{array}{c} \mathrm{H} \quad\quad \mathrm{H} \\ | \quad\quad\;\; | \\ \mathrm{M^{m+}\ldots O{-}H\ldots O{-}Ph} \end{array} \qquad\qquad \begin{array}{c} \mathrm{H} \\ | \\ \mathrm{M^{m+}\ldots O\ldots HO{-}Ph} \\ | \\ \mathrm{H} \end{array}$$

(8.Va) (8.Vb)

(Fenn and Mortland, 1972; Saltzman and Yariv, 1975).

8. Carbonyl groups, which are abundant in soil humic acid and humin, are linked to exchangeable cations through a water bridge, in which the C=O group acts as a proton accepter, forming a hydrogen bond as follows

$$\begin{array}{l} \mathrm{H} \\ \mathrm{M^{m+}\ldots O{-}H\ldots O{=}CH{-}R} \end{array} \qquad (8.VI)$$

(Parfitt and Mortland, 1968).

9. Short-range interaction of the aromatic rings through π electrons to exchangeable transition metal cations, such as Cu and Ag (Clementz and Mortland, 1972) and to the flat oxygen sheet of the clay mineral (Yariv et al., 1976).

10. Small organic molecules with at least two functional groups, e.g., amino acids, are able to provide a bridge link between clay minerals and humic acid (Scharpenseel, 1970).

Schnitzer and Kodama (1969) and Kodama and Schnitzer (1974a) studied the adsorption of fulvic acid (with a $pK_a = 4.5$) by expanding and nonexpanding clay minerals. The magnitude of adsorption of the acids by the various clay minerals is related to the degree of ionization of the acidic functional groups and depends on the pH of the suspension. At pH < 4 relatively few of the COOH groups are ionized, so that the fulvic acid behaves like an uncharged molecule that can penetrate into the interlayer space. As the pH rises, more and more functional groups ionize, resulting in an increased negative charge while repulsion of negatively charged fulvic acid by negatively charged clay mineral results in less adsorption. The interlamellar adsorption of fulvic acid by montmorillonite at pH 2.5 is rapid. After 1 min of shaking, 68 mg of fulvic acid were adsorbed by 100 mg of Na-montmorillonite. After 18 h the adsorbed amount increased to 83 mg of fulvic acid per 100 mg clay. Small amounts of the acid were desorbed from the clay just by washing with water. The amount of fulvic acid adsorbed by montmorillonites depends on the exchangeable cation and decreases in the following order: Pb > Cu > Na > Zn = Co > Mn > Mg = Ca > Fe > Ni. Of the total fulvic acid adsorbed by montmorillonite, about 57% is in the interlayer spaces of the clay with the remainder on external surfaces. At pH 2.5 the adsorption of fulvic acid by kaolinite (fine fracture), muscovite, and sepiolite is 7.3, 4.9, and 13.1 mg acid per 100 mg clay. With various kaolinites it was found that the adsorption is directly proportional to the surface area of the mineral particles.

Tan (1976) studied adsorption of humic acid with molecular weights ranging from 1000 to > 30,000 and found that only the low molecular fraction (1000–5000) was adsorbed by kaolinite and montmorillonite at pH above 7.

3.3.3 Reactions of Quartz with Humic Substances

The types of associations that are obtained during the interactions between sand minerals and humic substances will be regarded here from our knowledge of reactions between quartz and simple organic molecules. Sorption of organic material by quartz occurs by two types of mechanisms, namely, electrostatic and hydrophobic associations (Fuerstenau and Healy, 1972). Feldspar grains behave similarly.

At pH > 2 the quartz is negative and long-range electrostatic attraction between quartz and humic substances may occur so long as the surface charge of the humic species is positive, that is, at pH < 4.

Di- and polyvalent metallic cations are specifically adsorbed on negatively charged quartz reversing the sign of the charge from negative to positive and leading to the adsorption of the humic substances at pH $\geqslant 4$. The pH at which the sign of the surface charge of the quartz particle changes depends on the cation; it becomes positive at that pH at which the metal ion begins to hydrolyze, and adsorption of humic substances ceases at the pH at which the metal hydroxide begins to precipitate. This is explained as follows. Unhydrolyzed monomeric ions, regardless of their charge, are not adsorbed onto quartz surfaces due to ion hydration. All metal ions, and particularly polyvalent ones, are strongly hydrated in aqueous solutions. Unhydrolyzed ions coordinated by water, as well as single water molecules, are not adsorbed by quartz since the charge of quartz is low (Matijević, 1973). Hydrolyzed metal ions form hydrates of stabilities lower than those of the highly charged unhydrolyzed ions. Obviously, water molecules coordinated with unhydrolyzed ions are a more efficient barrier to interactions between the ions and the solid surfaces than when coordinated to hydrolyzed ions. Activation of quartz by the following cations (2×10^{-4} M, measured by the sorption of sulfonate anions, Fuerstenau and Healy, 1972) occurs at the following pH values: Fe, 2–4; Al, 4–7; Pb, 6–12; Ba, ~7; Mn, 7.5–9; Mg, 10–11.5; Ca, 10.5–13.5.

At lower concentrations of organic material every macromolecule of the humic substance is adsorbed as an individual species, and because there are almost no sites of short-range interactions the adsorption forces are very weak. At higher concentrations several macromolecules associate by means of van der Waals interaction of the hydrophobic moieties of the macromolecules adsorbed in the Stern layer. These associated, adsorbed species may be regarded as *two-dimensional aggregates* and are called "hemimicelles". It is obvious that this type of adsorption is not the result of a strong chemical bond formation between the quartz grain and the organic substances, but rather is due to the fact that the hydrophobic moieties, which form unstable hydrates, are repelled from the aqueous phase into the liquid–solid interface, where the stability of the water structure is less than in liquid water. This type of adsorption, which is called "hydrophobic adsorption", is more important for sand minerals than for clay minerals and hydrous oxides, which have high surface charge densities and hydrophilic hydration structures. Tschapek et al. (1973) showed that the wettability of quartz by water decreased after the quartz grains had been treated with Na-humate at pH 3.5. This organo–quartz complex is an important soil constituent and some organo-quartz grains may form metastable hydrophobic caves for air trapped in the soil aggregate.

3.4 Pore Space in Soils

Pore space is important in determining water migration in soils, soil aggregate stability, and soil aeration. Four groups of soil and soil aggregate pore space can be distinguished: (1) intra-elemental; (2) intra-assemblage; (3) interassemblage; and (4) transassemblage pores (Collins and McGown, 1974). A schematic representation of the occurrence of the various pore space types is given in Figure 8.12. *Intra-elemental* pores occur within the various elementary particle arrangements and include *interparticle pores,* i.e., those occurring between similar primary individual particles and *intergroup pores,* i.e., those occurring between clay tactoids inside a floc. *Intra-assemblage pores* occur within particle assemblages that are obtained between sets of dissimilar elementary particles or between small flocs of a single mineral contained within the aggregate. *Interassemblage* pores occur between secondary units that are combined into higher order assemblages of any level of complexity. A degree of overlap in definition exists between intra-assemblage and interassemblage pores. *Transassemblage pores* are the pores traversing the soil fabric without any specific relationship to the occurrence of the individual microfabric features. The walls of the intra-elemental pores are characterized by being homomineralic. Those built from clay minerals are capillaceous and are extremely aerophobic.

3.5 Stability of Aqueous Soil Dispersions and the Migration of Soil Constituents

Transfer of mobile constituents in soils and sediments is principally effected by soil solution and ground-water currents in voids and channels, which are the result of precipitation, evaporation, and fluctuations of the ground-water table. At the soil surface the composition of the solution may approach that of pure (or rain) water. At shallow

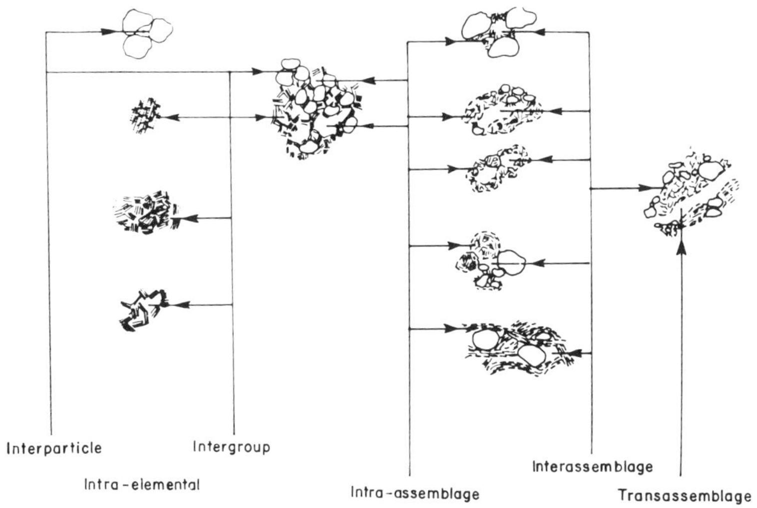

Fig. 8.12. Schematic representation of pore space types in soils. (From Collins and McGowan, 1974: Used with the permission of the Institution of Civil Engineering, London)

depth it becomes enriched with nutrient salts, soluble and hydrophilic colloidal organic compounds, silica, and finely dispersed clay minerals and hydrated oxides of iron and aluminum.

Soil aggregates in water systems can disaggregate either by wetting or under the influence of shear. The heavy particles settle down whereas the clay particles peptize and may form stable suspensions. There is a *critical soil/water ratio* below which mechanically dispersed clay particles can form a stable suspension (D'Hoore, 1970). At soil/water ratios above the critical value a mechanically dispersed soil colloid flocculates and settles down into an immobile form. The higher the critical value, the higher is the potential mobility of the bulk of the natural soil colloid and the greater the possibility that stable suspensions may migrate with the soil solution, moving in saturated or slightly saturated flow at the surface or through the larger soil pores. Such soil/water ratios may be occasionally reached in surface layers and larger soil voids even in dry climates with scant rainfall. In deeper layers coarse and medium soil fractions are usually immobile. Small amounts of very fine clay, after they have been dispersed, remain mobile even at relatively high soil/water ratios, and migrate to lower soil horizons (Fig. 8.13).

The dispersion of the clay fraction and its mobility are enhanced by a low electrolyte concentration in the pore fluid, by monovalent ions, by a high pH (above 7) preventing positive charges on the edges of the clay particles, and by a high water content increasing interparticle distances. A high pH value also brings about lowering of the net positive charges of the hydrous oxides of iron and aluminum and consequently a lower degree of their heterocoagulation with the clay minerals. When the pH of the pore fluid is above the pK of carboxylic acid groups of the humic substances, the interaction between the humic substances and the clay minerals becomes weak, and if the concentration of exchangeable polyvalent ions is small, peptization and mobilization of both the clay minerals and the humic substances may occur.

There are two types of stresses leading to the breakdown (slake) of a dry soil aggregate when the latter is immersed in water, namely, stresses induced by entrapped air (Yoder, 1936) and by swelling (Henin, 1938).

1. When water penetrates into the soil aggregate by capillarity, the entrapped air is compressed. Aggregates are suddenly disrupted when the pressure of the included air is greater than the adhesion between the soil components (Concaret, 1967). This breakdown factor is called "air explosion".

2. Once aggregates are immersed in water, an *osmotic stress* arises between the negatively charged clay particles. This stress increases as the soluble salts initially present in the aggregate diffuse out, and the thickness of the double layers increases. The increase may be sufficient to cause dispersion of the clay, that is to say, peptization.

The water dispersibility of the clay fraction depends on the chemical and mineralogic constitutions of the soil aggregate. In Ca-rich soils exchangeable Na is required to disperse the clay fraction (see Ch. 3) and at least 12% of the total exchange positions in the clay fraction must be occupied by sodium. However, some dispersion of the clay is already evident when 7% of the total exchange position is occupied by Na. Dispersibility is high for smectites and is lower for illite and kaolinite. The presence of carbonates (such as calcite) in the aggregates prevents the clay from dispersing. As the soluble salts diffuse outward the percentage of exchangeable Na present on the as yet undis-

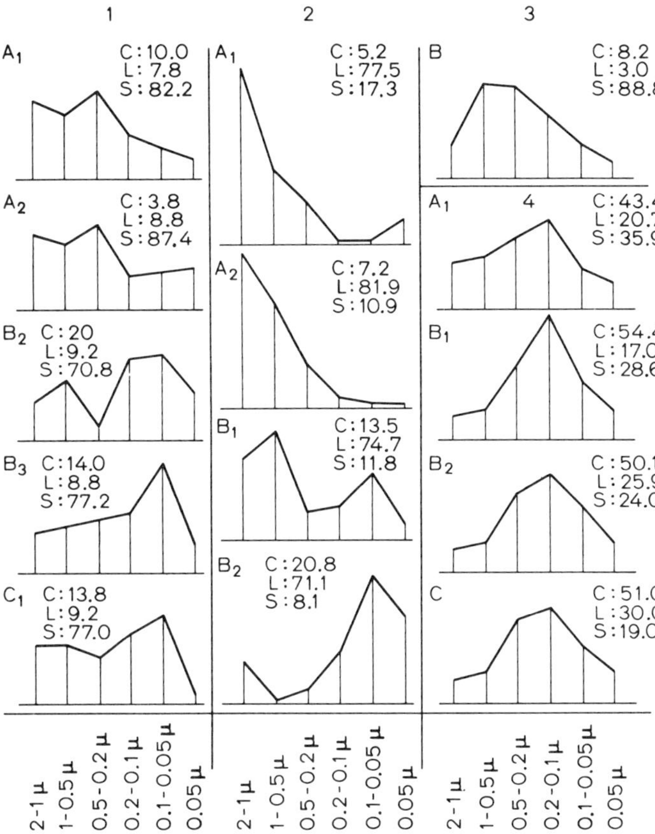

Fig. 8.13. Clay subfraction distribution in the following soils: (*1*) Tropudult, Sao Paolo, Brasil; (*2*) Glossudalf, Meerdaal Forest, Belgium; (*3*) Argillic horizon in bands (lamellae), Meerdaal Forest, Belgium; (*4*) Oxisol over hardpan, Bugesera, Rwanda. *A, B, C* = soil horizons.
C = percentage of clay, *L* = percentage of silt, and *S* = percentage of sand. (From D'Hoore, 1970: Used with the permission of the Belgian Soil Science Society)

persed clay will be gradually reduced by exchange of Na ions for Ca ions derived from the carbonate, making the clay indispersible. The presence of gypsum in the aggregate has the same effect. Aggregates from acid soils, with a median pH value of 5.5, can be dispersed only after thoroughly shaking with great amounts of water, probably because of the high coagulating power of aluminum, which is always present in acid soils. Soluble salts initially present in dry aggregates that can be broken down merely by wetting with water can only reduce the degree of dispersion and the mobility of the clay fraction by double-layer compression (Emerson, 1967).

Gillman (1974) studied the influence of net charge on water-dispersible clay in samples from various depths of a krasnoze profile, under a virgin rain forest from Gregory Falls in North Queensland. It is to be expected that the presence of a net surface charge, whether negative or positive, produces stable clay dispersions (Table 8.4). The

Table 8.4. Data for water-dispersible clay, total clay, zero point of charge, soil pH, organic matter, and clay mineralogy in a krasnozem profile from Gregory Falls, North Queensland, Australia. (After Gillman, 1974)

Depth (cm)	Total clay (%)	Water dispers-ible clay (%)	Organic matter (%)	Point of zero charge	Soil pH[a]	Clay mineralogy K	Go (%)	G
0–10	60	33.7	11.5	–	5.6	10–20	40–50	20–30
10–20	80	32.8	6.8	4.5	5.7	10–20	40–50	20–30
20–30	86	13.8	4.6	–	5.4		n.d.	
30–60	78	0.1	3.1	–	5.5	20–30	40–50	30–40
60–90	n.d.	< 0.1	2.0	–	5.4		n.d.	
90–120	75	0.1	1.3	5.4	5.4	20–30	40–50	30–40
210–240	65	49.2	–	5.8	5.3	20–30	40–50	30–40
450–510	60	< 0.1	–	5.2	5.1	> 80	5–10	5–10

[a] Soil pH in 1 : 5 soil: water at 25 °C.
n.d., not determined. *K*, kaolinite; *Go*, goethite; *G*, gibbsite.

position of the point of zero charge in the upper horizons is lower than would be expected from the dominant oxides, which have points of zero charge at about pH 8–9.2. The point of zero charge approaches higher pH values as the depth increases to the 210–240 cm horizon. Since the clay mineral does not vary over this depth, the effect may be due to the decrease in organic matter content (van Raij and Peech, 1972). Negatively charged organic matter is held at positive sites on the mineral surfaces and a decrease of organic matter content with depth is accompanied by the exposure of increasing numbers of positive sites. The degree of dispersion of the clay fraction depends on the organic matter content. Humic acid, being a polyelectrolyte, may serve either as a peptizer or as a coagulating agent. At upper soil horizons organic matter is high in concentration and it acts as a protective colloid. It coats the positively charged iron and aluminum hydrous oxides, changing the surface charge of the particles to negative, and a large fraction of the clay is dispersed (Fig. 8.14). At shallow depths organic matter concentration is low and it bridges between the particles. Clay aggregates are obtained and the degree of dispersion becomes extremely low. At greater depths (210–240 cm below surface) the content of organic matter is too low to have any appreciable effect on the stability of the clay dispersion. The surface charge of the soil is positive and its charge density is high, giving rise to a stable dispersion. In the 450–510 cm horizon, the clay mineralogy is dominated by kaolinite. The point of zero charge is lower than that of the 210–240 cm horizon, because the net charge of the kaolinite is negative. Furthermore, heterocoagulation between kaolinite and the hydrous oxides of aluminum and iron is evident; consequently, the stability of the clay dispersion is reduced.

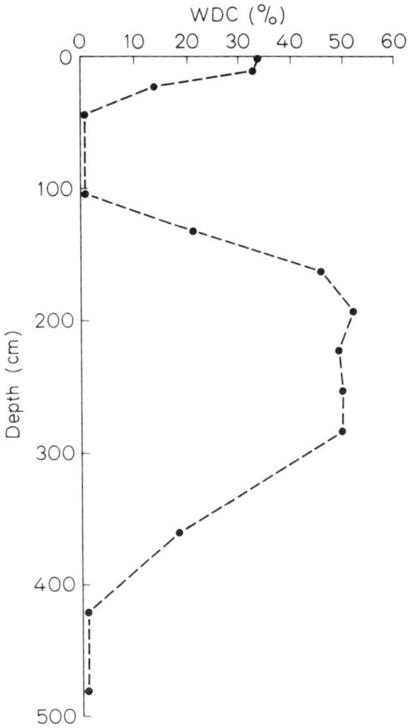

Fig. 8.14. The content of water-dispersible clay (WDC, in percent from total clay) versus depth in a krasnozen soil profile, under virgin rain forest from Gregory Falls (Australia). (From Gillman, 1974: Used with permission of the publisher)

References

Barclay, L., Harrington, A., Ottewill, R. H.: The measurement of forces between particles in disperse systems. Kolloid. Z. Z. Polym. *250,* 655–666 (1972)

Baver, L. D., Hall, N. S.: Colloidal properties of soil organic matter. Univ. Mo. Agr. Expt. Sta. Res. Bull. No. 267 (1937)

Baylor, E. R., Sutcliffe, Jr., W. H.: Dissolved organic matter in sea water as a source of particulate food. Limnol. Oceanogr. *8,* 369–371 (1963)

Becher, P.: Effect of adsorption on the van der Waals interaction of parallel flat plates. J. Colloid Interface Sci. *42,* 645–646 (1973)

Berthois, L.: Observations directe des particules sédimentaires fins dans l'eau. Revue Géogr. Phys. Geol. dyn. IV *1,* 39–42 (1961)

Biddle, P., Miles, J. H.: The nature of contemporary silts in British estuaries. Sediment. Geol. *7,* 23–33 (1972)

Blackmore, A. V.: Aggregation of clay by the products of iron (III) hydrolysis. Aust. J. Soil Res. *11,* 75–82 (1973)

Braezeale, J. F., Magistad, O. C.: Base exchange in orthoclase. Univ. of Ariz. Agr. Expt. Sta. Tech. Bull. No. 24 (1928)

Bundy, W. M., Murray, H. H.: The effect of aluminum on the surface properties of kaolinite. Clays Clay Miner. *21,* 295–302 (1973)

Carstea, B. D., Harward, M. E., Knox, E. G.: Comparison of iron and aluminum hydroxy interlayers in montmorillonite and vermiculite. I. Formation. Soil Sci. Soc. Am. Proc. *34,* 517–521 (1970)

Clementz, D. M., Mortland, M. M.: Interlamellar metal complexes in layer silicates. III. Silver(I)-arene complexes in smectites. Clays Clay Miner. *20,* 181–187 (1972)

Collins, K., McGown, A.: The form and function of microfabric features in a variety of natural soils. Geotechnique *24,* 223–254 (1974)
Concaret, J.: Etude des méchanismes de la destruction des agrégats de terre au contact de solution aqueuses. Ann. Agron. *18,* 65–144 (1967)
Cornell, R. M., Posner, A. M., Quirk, J. P.: The dissolution of synthetic goethites. The early stage. Soil Sci. Soc. Am. Proc. *38,* 377–378 (1974)
De Mumbrum, L. E., Jackson, M. L.: Infrared adsorption evidence on exchange reaction mechanism of copper and zinc with layer silicates and peat. Soil Sci. Soc. Am. Proc. *20,* 334–337 (1956)
Deshpande, T. L., Greenland, D. J., Quirk, J. P.: Changes in soil properties associated with the removal of iron and aluminum oxides. J. Soil Sci., *19,* 108–122 (1968)
D'Hoore, J.: Induced migration of clay and other moderately mobile soil constituents. I. General introduction. II. Natural clay fractionation. Pedologie *20,* 51–61 (1970)
Eckert, J. M., Sholkovitz, E. R.: The flocculation of iron, aluminum and humates from river water by electrolytes. Geochim. Cosmochim. Acta *40,* 847–848 (1976)
Edwards, A. P., Bremner, J. M.: Microaggregates in soils. J. Soil Sci. *18,* 47–63 (1967)
Edzwald, J. K.: Coagulation in estuaries. Sea Grant Publ. UNC-36-72-06. Univ. N. C. (1972)
Edzwald, J. K., Upchurch, J. B., O'Melia, C. R.: Coagulation in estuaries. Environ. Sci. Technol. *8,* 58–63 (1974)
El-Rayah, H. M. E., Rowell, D. L.: The influence of iron and aluminum hydroxides on the swelling of Na-montmorillonite and the permeability of Na-soil. J. Soil Sci. *24,* 137–144 (1973)
El-Swaify, S. A., Emerson, W. W.: Changes in the physical properties of soil clays due to precipitated aluminum and iron hydroxides. I. Swelling and aggregate stability after drying. Soil Sci. Soc. Am. Proc. *39,* 1056–1063 (1975)
Emerson, W. W.: The structure of soil crumbs. J. Soil. Sci. *10,* 235–244 (1959)
Emerson, W. W.: A classification of soil aggregates based on their coherence in water. Aust. J. Soil Res. *5,* 47–57 (1967)
Feely, R. A.: Evidence for aggregate formation in a nepheloid layer and its possible role in the sedimentation of particulate matter. Mar. Geol. *20,* M7–M13 (1976)
Fenn, D. B., Mortland, M. M.: Interlamellar metal complexes on layer silicates. II. Phenol complexes in smectites. Proc. Int. Clay Conf., Madrid. Serratosa, I. M. (ed.) Madrid: Div. Ciencias C. S. I. C., 1973, pp. 591–603
Fitzsimmons, R. F., Posner, A. M., Quirk, J. P.: Electron microscope and kinetic study of the flocculation of calcium montmorillonite. Isr. J. Chem. *8,* 301–314 (1970)
Follett, E. A. C.: The retention of amorphous, colloidal ferric hydroxide by kaolinites. J. Soil Sci. *16,* 334–341 (1965)
Fuchs, N.: The mechanics of aerosols. New York: Pergamon Press, 1964
Fuchs, N., Sutugin, A. G.: Highly dispersed aerosols. Ann Arbor, Mich. Ann Arbor Science 1970
Fuerstenau, D. W., Healy, T. W.: Principles of mineral flotation. In: Adsorptive bubble separation techniques. Lemlich, R. (ed.). New York: Academic Press 1972
Gillespie, T., Settineri, W. J.: The effect of capillary liquid on the force of adhesion between spherical solid particles. J. Colloid Interface Sci. *24,* 199–202 (1967)
Gillman, G. P.: The influence of net charge on water dispersible clay and sorbed sulphate. Aust. J. Soil Res. *12,* 173–176 (1974)
Giovannini, G., Sequi, P.: Iron and aluminum as cementing substances of soil aggregates. I. Acetylacetone in benzene as an extractant of fractions of soil iron and aluminum. J. Soil. Sci. *27,* 140–147 (1976)
Greene, R. S. B., Posner, A. M., Quirk, J. P.: Factors affecting the formation of quasi-crystals of montmorillonite. Soil Sci. Soc. Am. Proc. *37,* 457–460 (1973)
Greenland, D. J.: Interactions between clays and organic compounds in soils. Part I. Mechanisms of interaction between clays and defined compounds. Soils Fert. *28,* 415–425 (1965a)
Greenland, D. J.: Interactions between clays and organic compounds in soils. Part II. Adsorption of soil organic compounds and its effect on soil properties. Soils Fert. *28,* 521–532 (1965b)
Greenland, D. J., Oades, J. M., Sherwin, T. W.: Electron microscope observations of iron oxides in some red soils. J. Soil Sci. *19,* 123–126 (1968)
Gregory, J.: The calculation of Hamaker constants. Adv. Colloid Interface Sci. *2,* 396–417 (1969)

Gripenberg, S.: A study of the sediments of the Northern Baltic and adjoining seas. Fennia *60* (3), 1–231 (1934)
Gordon, D. C.: Microscopic study of organic particles in the North Atlantic ocean. Deep-Sea Res. *17*, 175–185 (1970)
Hahn, H. H., Stumm, W.: Kinetics of coagulation with hydrolyzed Al, the rate limiting step. J. Colloid Interface Sci. *28*, 134–144 (1968)
Hall, D. G.: The role of bridging in colloid flocculation. Colloid Polymer Sci. *252*, 241–243 (1974)
Harward, M. E., Theisen, A. A., Evans, D. D.: Effect of iron removal and dispersion methods on clay mineral identification by X-ray diffraction. Soil Sci. Soc. Am. Proc. *26*, 535–541 (1962)
Henin, S.: A physical chemical study of the stability of soils. Nat. Centre Agron. Res. Paris 1938
Hogg, R., Yang, K. C.: Secondary coagulation. J. Colloid Interface Sci. *56*, 573–576 (1976)
Hunter, R. J., Alexander, A. E.: Surface properties and flow behavior of kaolinite. Part III. Flow of kaolinite sols through a silica column. J. Colloid Interface Sci. *18*, 846–862 (1963)
Khan, S. U.: Interaction between the humic acid fraction of soils and certain metallic cations. Soil Sci. Soc. Am. Proc. *33*, 851–854 (1969)
Khanna, S. S., Stevenson, F. J.: Metallo-organic complexes in soil: I. Potentiometric titration of some soil organic matter isolates in the presence of transition metals. Soil Sci. *93*, 298–305 (1962)
Kidder, G., Reed, L. W.: Swelling characteristics of hydroxy-aluminum interlayered clays. Clays Clay Miner. *20*, 13–20 (1972)
Kodama, H., Schnitzer, M.: Adsorption of fulvic acid by non-expanding clay minerals. Trans 10th Int. Congr. Soil Sci., Moscow. Zyrin, W. G. (ed) Moscow: Nauka, 1974a, vol. 2, pp. 51–56
Kodama, H., Schnitzer, M.: Further investigations on fulvic acid-Cu-montmorillonite interactions. Clays Clay Miner. *22*, 107–110 (1974b)
Kranck, K.: Flocculation of suspended sediment in the sea. Nature (London) *246*, 348–350 (1973)
Kranck, K.: Sediment deposition from flocculated suspensions. Sedimentology *22*, 111–123 (1975)
Krizek, R. J., Edil, T. B., Ozaydin, I. K.: Preparation and identification of clay samples with controlled fabric. Eng. Geol. (Amsterdam) *9*, 13–38 (1975)
Krupp, H.: Particle adhesion. Theory and experiment. Adv. Colloid Interface Sci. *1*, 111–239 (1967)
Lahav, N., Banin, A.: Effects of various treatments on the optical properties of montmorillonite suspensions. Isr. J. Chem. *6*, 285–294 (1968)
Low, P. F.: Mineralogical data requirements in soil physical investigations. In: Mineralogy in soil science and engineering. Soil Sci. Soc. Am. Spec. Publ. *3*, 1–34 (1968)
Marshall, C. E.: The physical chemistry and mineralogy of soils. New York: John Wiley & Sons, 1964
Marshall, T. J.: The nature, development and significance of soil structure. Trans. Joint Meeting Com. IV and V, Intern. Soc. Soil Sci., N. Z. 243–257 (1962)
Matijevic, E.: Colloid stability and complex chemistry. J. Colloid Interface Sci. *43*, 217–245 (1973)
Mendelovici, E.: Formacion y composicion de las lateritas y bauxitas. Centro de Ingeneria y Computacion, Instituto Venezolano de Investigaciones Cientificas, RD-B 76-01 1977
Moriyama, N.: Stabilities of aqueous inorganic pigment suspensions. Colloid Polym. Sci. *254*, 726–735 (1976)
Müller, H.: Die Theorie der Koagulation polydisperser Systeme. Kolloid-Z. *38*, 1–2 (1926)
Neumann, B. S., Sansom, K. G.: The study of gel formation and flocculation in aqueous clay dispersions by optical and rheological methods. Isr. J. Chem. *8*, 315–322 (1970)
Ohshima, H.: Diffuse double layer interaction between two parallel plates with constant surface charge density in an electrolyte solution. I. The interaction between similar plates. Colloid Polym. Sci. *252*, 158–164 (1974)
O'Melia, C. R.: Coagulation in water and waste water treatments. In: Water quality improvement by physical and chemical processes. Gloyna, E. F., Eckenfelda, W. W., Jr. (eds.). Austin: Univ. Texas Press, 1970, pp. 219–236
O'Melia, C. R., Stumm, W.: Aggregation of silica dispersions by iron (III). J. Colloid Interface Sci., *23*, 437–447 (1967)
Ong, H. L., Bisque, R. E.: Coagulation of humic colloids by metal ions. Soil Sci. *106*, 220–224 (1968)
Overbeek, J. Th. G.: Kinetics of flocculation. In: Colloid science. Kruyt, H. R. (ed.). Amsterdam: Elsevier, 1952, Vol. 1, pp. 278–301

Packham, R. F.: The coagulation process, a review of some recent investigations. Proc. Soc. Water Treatm. Exam. *12,* 15–35 (1963)

Parfitt, R. L., Mortland, M. M.: Ketone adsorption on montmorillonite. Soil Sci. Soc. Am. Proc. *32,* 355–363 (1968)

Perkins, R., Brace, R., Matijevic, E.: Colloid and surface properties of clay suspensions. I. Laponite CP. J. Colloid Interface Sci. *48,* 417–426 (1974)

Posner, A. M.: Titration curves of humic acid. 8th Int. Cong. Soil Sci. Bucharest Trans. *2,* 161–174 (1964)

Rashid, M. A., Buckley, D. E., Robertson, K. R.: Interactions of a marine humic acid with clay minerals and natural sediment. Geoderma *8,* 11–27 (1972)

Rebhun, M., Sperber, H.: Optical properties of diluted clay suspensions (letter to the editor). J. Colloid Interface Sci. *24,* 131 (1967)

Reneau, Jr., R. B., Fiskell, J. G. A.: Mineralogical properties of clays from Panama soils. Soil Sci. Soc. Am. Proc. *36,* 501–505 (1972)

Rich, C. I.: Hydroxy interlayers in expansible layer silicates. Clays and Clay Miner. *16,* 15–30 (1968)

Riley, G. A.: Organic aggregates in seawater and the dynamics of their formation and utilization. Limnol. Oceanogr. *8,* 372–381 (1963)

Rowell, D. L.: Influence of positive charge on the inter- and intra-crystalline swelling of oriented aggregates of Na-montmorillonite in NaCl solutions. Soil Sci. *100,* 340–347 (1965)

Saltzman, S., Yariv, S.: Infrared study of the sorption of phenol and p-nitrophenol by montmorillonite. Soil Sci. Soc. Am. Proc. *39,* 474–479 (1975)

Scharpenseel, H. W.: Aufbau und Bindungsform der Ton-huminsäurekomplexe, Teil IV. Fällungsradiometrie mit aminosäure-belegtem Ton. Z. Pflanzenernaehr. Bodenkd. *125,* 111–115 (1970)

Schnitzer, M.: Reactions between fulvic acid, a soil humic compound and inorganic soil constituents. Soil Sci. Soc. Am. Proc. *33,* 76–81 (1969)

Schnitzer, M., Khan, S. U.: Humic substances in the environment. New York: Marcel Dekker 1972

Schnitzer, M., Kodama, H.: Reactions between fulvic acid, a soil humic compound, and montmorillonite. Isr. J. Chem. *7,* 141–147 (1969)

Schnitzer, M., Kodama, H.: Reactions between fulvic acid and Cu-montmorillonite. Clays Clay Miner. *20,* 359–367 (1972)

Schnitzer, M., Skinner, S. I. M.: Organo-metallic interactions in soils: I. Reactions between a number of metal ions and the organic matter of podzol B_h horizon. Soil Sci. *96,* 86–93 (1963)

Schubel, J. R.: Suspended sediment of the Northern Chesapeake Bay. Tech. Rep., Chesapeake Bay Inst. *35,* (Ref. No. 68-2) (1968)

Sheldon, R. W.: Sedimentation in the estuary of the River Crouch, England. Limnol. Oceanogr. *13,* 72–83 (1968)

Sheldon, R. W., Prakash, A., Sutcliffe, W. H.: The size distribution of particles in the ocean. Limnol. Oceanogr. *17,* 327–340 (1972)

Sholkovitz, E. R.: Flocculation of dissolved organic and inorganic matter during the mixing of river water and sea water. Geochim. Cosmochim. Acta *40,* 831–845 (1976)

Smoluchowski, von, M.: Drei Vorträge über Diffusion, Brownsche Molekularbewegung und Koagulation von Kolloidteilchen. Physik. Z. *17,* 557–571 (1916)

Sonntag, H., Strenge, K.: Coagulation and stability of disperse systems. Kondor, R. (trans.) Jerusalem: Israel Program for Scientific Translations 1972

Stevenson, F. J.: Binding of metal ions by humic acids. In: Environmental biochemistry. Vol. 2, Metals transfer and ecological mass balances. Nriagu, J. O. (ed.). Ann Arbor, Michigan: Ann Arbor Sci. Pub. Inc. 1976 pp. 519–540

Stumm, W., Morgan, J. J.: Chemical aspects of coagulation. J. Am. Water Works Assoc. *54,* 971–994 (1962)

Swartzen-Allen, S. L., Matijević, E.: Surface and colloid chemistry of clays. Chem. Rev. *74,* 385–400 (1974)

Tan, K. H.: Complex formation between humic acid and clays as revealed by gel filtration and infrared spectroscopy. Soil Biol. Biochem. *8,* 235–239 (1976)

Tan, K. H., McCreery, R. A.: Humic acid complex formation and intermicellar adsorption by bentonite. Proc. Int. Clay Conf., Mexico. Bailey, S. W. (ed.) Wilmette, IL: Applied Publishing Ltd., 1975, pp. 629–641

Tschapek, M., Pozzo, G., de Bussetti, S. G.: Wettability of humic acid and its salts. Z. Pflanzenernaehr. Bodenkd. *135,* 16–31 (1973)

Usui, S.: Heterocoagulation. Prog. Surf. Membr. Sci. *5,* 223–266 (1972)

van den Tempel, M.: Interaction forces between condensed bodies in contact. Adv. Colloid Interface Sci. *3,* 137–159 (1972)

van Olphen, H.: An introduction to clay colloid chemistry. New York: Interscience Pub. 1963

van Olphen, H.: Internal mutual flocculation in clay suspensions. J. Colloid Interface Sci. *19,* 313–322 (1964)

van Raij, B., Peech, M.: Electrochemical properties of some oxisols and altisols of the tropics. Soil Sci. Soc. Am. Proc. 36, 587–593 (1972)

Verwey, E. J. W., Overbeek, J. Th. G.: Theory of the stability of lyophobic colloids. Amsterdam: Elsevier 1948

Vrale, L., Jorden, R. M.: Rapid mixing in water treatment. J. Am. Water Works Assoc. *63,* 52–58 (1971)

Whitehouse, U. G., Jeffrey, L. M.: Differential settling velocities of kaolinitic montmorillonitic and illitic clays in saline water. Chemistry of marine sedimentation. Texas A. and M. Research Foundation Proj. *34A,* Tech. Rpt. No. 1 (1953)

Whitehouse, U. G., Jeffrey, J. M., Debrecht, J. D.: Differential settling tendencies of clay minerals in saline waters. Clays Clay Miner. *7,* 1–76 (1960)

Willey, J. D.: Silica-alumina interactions in seawater. Mar. Chem. *3,* 241–251 (1975)

Yariv, S., Lahav, N., Lacher, M.: On the mechanism of staining montmorillonite by benzidine. Clays Clay Miner. *24,* 51–52 (1976)

Yariv, S., Russell, J. D., Farmer, V. C.: Infrared study of the adsorption of benzoic acid and nitrobenzene in montmorillonite. Isr. J. Chem. *4,* 201–213 (1966)

Yoder, R. E.: A direct method of aggregate analysis of soils and a study of the physical nature of errosion losses. J. Am. Soc. Agron. *28,* 337–351 (1936)

Chapter 9
Rheology of Colloid Systems

1. Flow Behavior of Suspensions

Consider a unit volume of fluid consisting of an infinite number of layers, the lower surface of the bottom layer of which is fixed. This base is indicated in Figure 9.1 as the "reference plane". A tangential force is applied at the top surface of the unit volume, which results in movement of the top layer in the direction of the applied force. The stacked layers slide one above the other by equal relative amounts to each other. The result is that a velocity gradient $D = (dv/dy)$ is established perpendicular to the plane in which the top layer moves, wherein v is the velocity of the sliding layer and y is the distance of the layer from the "reference plane". This velocity gradient is called *rate of shear* (or simply *shear*), and is measured in reciprocal seconds ($m\ s^{-1} \cdot m^{-1} = s^{-1}$). The applied tangential force is called *shearing stress* (or simply *stress*) and is indicated by τ. The stress is measured in dynes/cm^2 or in newtons/m^2.

The rheologic behavior of material under flow conditions can be described by the relation between the shear and the applied stress, $\tau = f(D)$. In Figure 9.2 various shear versus stress relations are presented graphically. Most liquids obey the simplest linear relation, $\tau = \eta D$, which is represented by a curve passing through the origin. This behavior is termed *Newtonian flow* and the fluid a *Newtonian fluid.* The coefficient η (or cot α in Fig. 9.2A) is a constant that is characteristic for every liquid and is defined as *viscosity*. It is given by the following equation

$$\eta = \tau/D. \tag{9.1}$$

Thus the viscosity is the stress per unit shear per unit distance between the boundaries with the units of g cm^{-1} or (dynes cm^{-2})/s^{-1} = dynes cm^{-2}s. The unit of viscosity is called the *poise.*

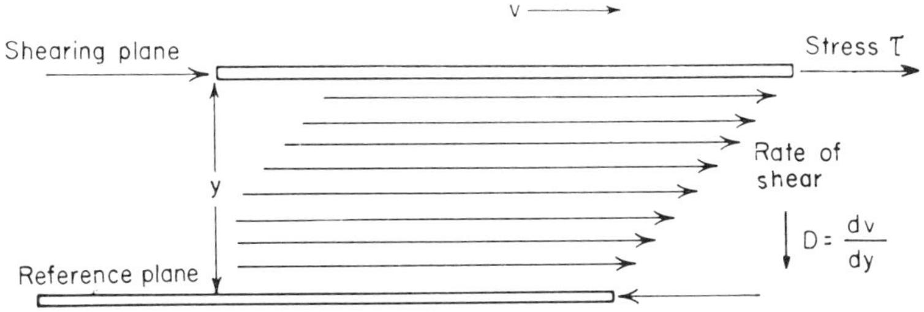

Fig. 9.1. Laminar flow. Definition of shearing stress and rate of shear

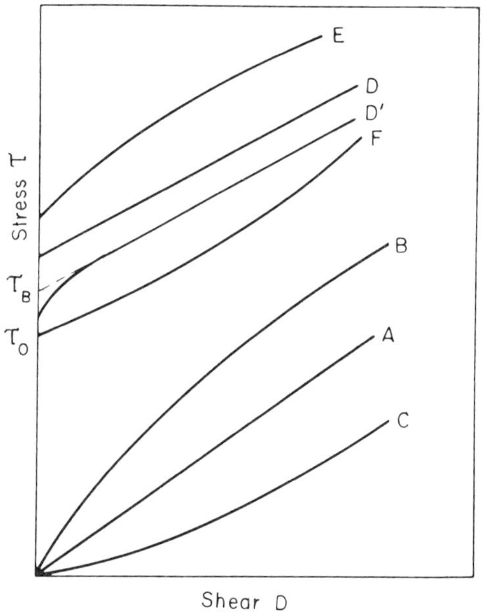

Fig. 9.2. Relations between shearing stress and rate of shear in time-independent systems: *A*, ideal or Newtonian flow; *B*, shear-thinning flow; *C*, shear-thickening flow; *D*, ideal plastic flow; *D'*, Bingham plastic flow; *E*, shear-thinning with a yield value flow, and *F*, shear-thickening with a yield value flow

The reciprocal of viscosity, $1/\eta$, is called *fluidity*. It is sometimes necessary to take into account the density of the fluid, ρ, whereby

$$\nu = \eta/\rho. \tag{9.2}$$

The quantity ν is known as the *kinematic viscosity*. The unit of kinematic viscosity is the *stoke*. Frequently, values of viscosity and kinematic viscosity are reported in units one hundredth of the above-mentioned units. These are the *centipoise* and the *centi-stoke.*

In dealing with solutions it is convenient to define *relative viscosity*

$$\eta_{\text{rel}} = \eta_s/\eta_o, \tag{9.3}$$

where η_s is the solution viscosity and η_o is the solvent viscosity. The *specific viscosity* is defined as follows

$$\eta_{\text{sp}} \equiv \eta_{\text{rel}} - 1 = (\eta_s - \eta_o)/\eta_o = \Delta\eta/\eta_o. \tag{9.4}$$

Reduced viscosity is the viscosity increment per unit concenctration of solute

$$\eta_{\text{sp}}/C \equiv (\eta_s - \eta_o)/\eta_o C,$$

wherein *C*, the solute concentration, is usually given in grams per deciliter. The limit of reduced viscosity given by

$$\lim_{C \to 0} \frac{\eta_{\text{sp}}}{C} \equiv [\eta], \tag{9.5}$$

is known as the *intrinsic viscosity*.

The symbol η is used for both the viscosity in poises and for quantities such as the specific viscosity and the intrinsic viscosity. In the latter usages the symbol is dimensionless and the unit of the viscosity is the reciprocal concentration. The viscosity of water at 20 °C is about 0.01 poise.

According to the Newtonian theory of fluids, friction that exists between adjacent layers moving at unequal velocities gives rise to irreversible conversion of a part of the energy of motion to heat, the dissipation of which results in a velocity gradient, dv/dy. The viscous flow obeys the equation

$$F = \eta\, S(dv/dy) = \eta\, S\, D, \tag{9.6}$$

where F is the total frictional force and S is the contact area of neighboring layers. From Eqs. (9.1) and (9.6) it is obvious that $F/S = \tau$. In other words, the stress has to overcome friction, the amount of which increases with the velocity gradient.

According to the kinetic theory of fluids, the velocity gradient is the result of the random molecular–kinetic movement. The layers moving at higher velocities exchange molecules with layers moving more slowly. The molecules coming from the slower moving layers tend to slow down the layers moving at the higher rates. At the same time, those molecules of a high kinetic energy that enter a layer moving at a slower rate, tend to increase the velocity of this layer.

Colloid systems, in general, display a more complicated flow behavior. They are called *non-Newtonian systems.* Some common examples of non-Newtonian flow behavior are illustrated in Figure 9.2. For any point on these curves there is an *apparent viscosity* given by τ/D, or, alternatively, a *plastic viscosity,* which is also called a *differential viscosity,* given by $d\tau/dD$.

Fluids can be divided into two groups depending on whether or not they exhibit time-dependent rheologic properties. Those fluids whose flow properties are independent of time are conveniently categorized according to two broad characteristics, namely, *yield value* and the shape of the flow curve. Two groups of flow curves, A, B, C and D, E, F are shown in Figure 9.2, denoted respectively, by not having or having a yield value. The second group of curves shows that no flow occurs until the shearing stress reaches a certain value τ_0. This value of the shearing stress at which the system begins to yield is called the *yield stress.* Below this value the fluid behaves as an elastic solid, above it as a liquid. The *ideal* or Newtonian fluid has a yield value equal to zero and a constant ratio of stress to rate of shear. The flow curve is a straight line whose slope is η.

A *shear-thinning fluid* is one in which the ratio τ/D continuously decreases with increase in the rate of shear. A *shear-thickening fluid* has the opposite characteristic, there being an increase of both the apparent and differential viscosities with increasing shear rate. Apparently, systems displaying such a behavior resist deformation the more, the higher the shear rate. This flow behavior is sometimes called *dilatancy.* The classes of suspensions denoted by the curves D, E, F, which have a yield value, also show the three types of dependence on rate of shear.

The same flow characteristics are encountered in systems exhibiting time-dependent properties. In these systems the viscosity measured at a given shear rate will either decrease (*thixotropy*) or increase (*rheopexy*) as shearing continues. Such systems will often exhibit yield values that vary with time and also give a family of time-dependent flow curves. The flow behavior of some common geologic fluidized systems is given in Table 9.1.

Ideal plastic flow (curve D in Fig. 9.2) is an ideal behavior having a yield stress value. Above the yield stress value the rate of shear is proportional to the shearing stress, the differential viscosity, or slope, is constant, and the apparent viscosity decreases with increasing shearing stress. The flow behavior obeys the equation $\tau - \tau_0 = nD$. This linear equation may be written as

$$\tau = \tau_0 + nD, \tag{9.7}$$

indicating that the shearing stress is composed of two components: nD, a shear-dependent viscosity contribution, and τ_0, a constant independent of the rate of shear but dependent on the dissociation of links between particles or molecules of adjoining layers.

Molecules or particles belonging to a single sliding layer are not necessarily interlinked by chemical bonds. A moving molecule or particle pushes those molecules and particles with which it collides, resulting in the movement of the layer in a direction parallel to the direction of the applied force. However, particles in adjoining layers may be linked by various bonds. When particles or molecules in the top layer of a fluid are pushed by the applied tangential force, they attract particles and molecules in the adjoining layer with which they are bonded. These bonded species in the second layer push other particles in the same layer, the latter not necessarily being bonded to species in the top layer. The velocity of the second layer is less than that of the top layer because constituents of this layer may be bonded to those in a third layer. This, in turn, tends to slow the movement of the second layer, and so on.

Table 9.1. Some common suspensions and their flow behavior

Property	Suspension	Class
Time independent	Water	Ideal, Newtonian
	Brine	
	Solutions of biopolymers of high molecular weights (DNA, RNA, and many proteins)	Shear-thinning
	Deflocculated clay slurries	Shear-Thickening
	Some heavy metal phosphate suspensions	
	Starch suspensions	
	Suspensions of glass beads	Plastic
	Certain mineral slurries (with very low interparticle forces, like quartz, feldspar, and fluorspar)	
	Flocculated kaolin slurries	Shear-thinning with yield value
	Monodisperse sand slurries	Shear-thickening with yield value
Time dependent	Phosphate rock slurries	Thixotropic
	Quartz, feldspar, and fluorspar suspended in organic solvents like benzene	
	Bitumens	Rheopectic

Since the movements of adjoining layers are not at the same rate, bonds between linked particles or molecules in adjoining layers are stretched until the links are broken. At the same time particles of adjoining layers become sufficiently close for new links to be formed. Energy is to be dissipated in doing work to dissociate linked particles. This stress, which is equal to τ_0, is dependent on both the strength of the interparticle links and the average time in which the particles of adjoining layers are linked during the shearing process. Links are constantly broken and reestablished in equal numbers and consequently the average time in which these particles are linked during the shearing process is independent of the rate of shear but dependent on the average distance between the particles or molecules and the strengths of the interparticle or intermolecule links.

A slight departure from the flow behavior represented by curve D is shown in curve D$'$. This type of flow, which is common in soil and clay disperse systems, is called *Bingham plastic flow.* The curve shows linear and nonlinear sections. When the straight line portion of the curve is extrapolated to low shear rates, it intersects the shearing-stress ordinate at a point marked τ_B, which is called the *Bingham yield stress.* This yield stress is somewhat higher than the real or true yield stress τ_0. The linear part of the curve is similar to (9.7) and obeys the equation $\tau = \tau_B + nD$. At this stage of the flow the contribution of the thermal dissociation of bonds between particles and molecules in adjoining layers is negligible in proportion to the contribution of shearing.

If thermal dissociation did not occur, the bonds between adjoining layers would not break, and the fluid would not flow at stresses below the Bingham yield stress. In actual fact, at low shear rates, the thermal dissociation of the links becomes important in proportion to the contribution of shearing to bond breaking. With stresses slightly above and below the Bingham yield stress, thermal dissociation may be high enough to enable a flow of the suspension with shearing rates higher than the calculated ones, resulting in the nonlinear region of the flow curve.

2. Rheology of Dispersions in the Hydrosphere

In the present section we shall deal first with dispersions of primary particles, both spherical and ellipsoid, whose concentrations are sufficiently low that the hydrodynamic interaction between particles is negligible. We shall then deal with infinitely dilute dispersions that contain both flocs and independent primary particles. In the last part of this section rheology of concentrated dispersions will be discussed. The following models are useful for describing the effect of suspended minerals and organic macromolecules on the rheologic properties of the hydrosphere. Rheologic properties play a tremendous role in the erosive activity of the hydrosphere and especially in turbidity currents and mud flows.

2.1 Rheology of Dilute Dispersions

2.1.1 Dispersions of Spherical Nonflocculated Particles

The Einstein equation is applicable to aqueous dilute dispersions of rigid, uncharged spherical particles whose dimensions are large compared with molecules of the suspending medium, while still small enough to exhibit significant Brownian movement. It may be written in the equivalent forms (Einstein, 1956):

$$\eta = \eta_o(1 + 2.5\,\varphi), \tag{9.8}$$

$$\eta/\eta_o \equiv \eta_{rel} = 1 + 2.5\,\varphi, \tag{9.9}$$

$$\eta_{sp} \equiv (\eta - \eta_o)/\eta_o = \Delta\eta/\eta_o = 2.5\,\varphi, \tag{9.10}$$

$$[\eta] \equiv (\eta_{rel} - 1)/\varphi \equiv \eta_{sp}/\varphi = 2.5, \tag{9.11}$$

where φ is the volume fraction of spheres and η, η_o, η_{rel}, η_{sp} and $[\eta]$ are the viscosity of the dispersion, the viscosity of the liquid medium, the relative viscosity, the specific viscosity, and the intrinsic viscosity, respectively. The intrinsic viscosity is defined as the "limiting viscosity number" and is given by

$$[\eta] \equiv \lim_{\varphi \to 0} \frac{\eta_{sp}}{\varphi} = \lim_{\varphi \to 0} \frac{\eta/\eta_o - 1}{\varphi}.. \tag{9.12}$$

Here the particle concentration is expressed as volume fraction and $[\eta]$ is a dimensionless quantity. When concentration is expressed in grams per deciliter (g/dl), the intrinsic viscosity will be defined as in Eq. (9.5) and its dimensions are dl/g.

From Eq. (9.11) it follows that the intrinsic viscosity should always be equal to 2.5. Experimentally $[\eta]$ is evaluated from the relationship η_{sp}/φ in the limit as φ approaches zero.

Deviations from the Einstein equation may arise because: (1) the dispersion is not sufficiently dilute and the spheres spend some time as doublets; (2) a non-negligible amount of liquid is adsorbed on the particles; (3) either flocs or stable agglomerates are present, and (4) the liquid penetrates or swells the particles. Provided that the assumptions made in deriving the Einstein equation are satisfied and the flow of the pure liquid is Newtonian, i.e., η_o obeys Eq. (9.1), then the flow of the dispersion of spherical particles is always Newtonian and η also obeys Eq. (9.1), unlike a dispersion of nonspherical particles. Dilute suspensions of sandy grains of quartz, kyanite, and feldspar, which are usually spherical and do not form hydrates or flocs, obey the Einstein equation, whereas clay minerals that have a tendency to hydrate and to flocculate do not obey the Einstein equation, unless they are very dilute.

2.1.2 Dispersions of Ellipsoidal Nonflocculated Particles

In dilute dispersions of ellipsoid or any nonspherical particles, the Einstein equation becomes

$$\eta = \eta_o(1 + n\varphi), \tag{9.13}$$

where $n > 2.5$. In infinite dilution it follows from Eq. (9.12) that n is equal to the intrinsic viscosity. The orientation of a very small nonspherical particle in a quiescent liquid varies randomly with time as a result of Brownian movement. Under shear, however, the particle tends to orient in such a manner that its major axis will be parallel to the direction of the flow. The time-average orientation of the particle is determined by the competition between hydrodynamic forces and Brownian movement. When the rate of shear is sufficiently small, the particles retain their initial random orientation. Assuming this conditions, the intrinsic viscosity was calculated as a function of the axial ratio for both prolate (fibrous minerals) and oblate (platy minerals) ellipsoids. Provided that the axial ratio p (ratio of major to minor axis) is greater than 10, the intrinsic viscosity for prolate ellipsoid is given by (Simha, 1940)

$$[\eta] = \frac{p^2}{15\,(\ln 2p - 3/2)} + \frac{p^2}{5\,(\ln 2p - 1/2)} + \frac{14}{15} \tag{9.14}$$

and for oblate ellipsoids,

$$[\eta] = (16/15)\,(p/\tan^{-1}p). \tag{9.15}$$

With increasing shear rate, the particles progressively become more highly oriented in the flow direction, giving a continuous reduction in the intrinsic viscosity. This can be seen from the calculations of Scheraga (1955). If α is a number related to the rate of shear and inversely related to Θ, the rotary diffusion constant, then $\alpha \equiv D/\Theta$ and $\Theta \equiv [\mathrm{kT}\,(31\,n\,p + 0.57)]/(8\,\pi\eta_o\,a^3)$ where k, T, and a are Boltzmann's constant, absolute temperature, and the length of the major semiaxis, respectively. For prolate ellipsoids when α increases from zero to 60 the intrinsic viscosity decreases from 5.81 to 3.32 when $p = 5$, and from 13.63 to 5.28 when $p = 10$. When p is very large (e.g., 300), the intrinsic viscosity is a little more than four times greater at $\alpha = 0$ than at $\alpha = 60$. From the shear rate dependence of the intrinsic viscosity, the length of a macromolecule or a particle having the shape of the rod can be calculated (Tanford, 1967).

In the complete absence of Brownian movement, i.e., for large particles, the viscosity depends on the initial orientation of the particles. Minimum and maximum values of the intrinsic viscosity were calculated (Goldsmith and Mason, 1967). For $p = 5$ and 10 the maximum values of $[\eta]$ are 3.5 and 5.0, respectively, i.e., the intrinsic viscosity increases with p. It should be noted that these maximum values are essentially equal to values obtained for very small particles in an environment having a very high shear rate.

In the absence of Brownian movement, the energy dissipation reflected by the intrinsic viscosity is determined solely by hydrodynamic effects associated with the perturbation of the flow field; the ellipsoids rotate in the flowing liquid, their long axes remaining oriented in the flow direction for a longer period than in the perpendicular direction. When Brownian movement occurs, it tends to destroy the preferred orientation. Energy has to be dissipated in doing work against the forces of Brownian movement and hence the intrinsic viscosity is increased. The fraction of this dissipated energy decreases with increasing rate of shear. It follows that the flow of a dispersion of nonspherical particles is non-Newtonian, but behaves as a shear thinning fluid. Aqueous solutions of geo-organic polymers of high molecular weight and of many proteins behave in this way.

Very dilute suspensions of clay minerals behave in some respects similarly to dispersions of this type. Shainberg and Otoh (1968) determined the intrinsic viscosity of very dilute suspensions of Na and Ca-montmorillonite from the experimental values of the specific viscosity and volume fraction of the particles. They considered the layer mineral to be an oblate ellipsoid and using Eq. (9.15) they calculated the axial ratio p. Knowing the thickness of one single unit layer, this ratio enabled them to calculate the number of unit layers in a single tactoid. They concluded that the average tactoid of Na-montmorillonite consists of a single unit layer, with almost no sorbed "immobile" water molecules, whereas that of Ca-montmorillonite consists of four or five unit layers with 3 or 4 water layers in the interlayer space, between the aluminosilicate platelets. This water becomes "immobile" and thus behaves as a part of the disperse solid phase. The net volume of the "immobile" water of Ca-montmorillonite suspensions is almost equal to the net volume of the aluminosilicate fraction. Hence, the volume fraction φ being the total solid phase, is greater in the Ca clay system than in the Na clay system, in systems containing the same amounts of dry aluminosilicate. It follows from Eq. (9.13) that the specific viscosity of Ca-montmorillonite, calculated per gram of dry material, is greater than that of Na-montmorillonite.

According to Chen and Schnitzer (1976), viscosity measurements of aqueous solutions of humic and fulvic acids indicate that these soil organic materials form flexible rod-like structures. The pH of the aqueous solutions has little effect on particle thickness, but strongly affects particle length. Humic acid rods at pHs above 7 have approximately the same thickness as those of fulvic acid, but are considerably longer. Humic acid at pH 7 and fulvic acid at pH between 1 and 1.5 behave like uncharged polymers. At higher pH levels, both humic and fulvic acids exhibit strong polyelectrolytic characteristics. For fulvic acid the minimum intrinsic viscosity and the shortest rod-like length is at pH 3.0. As the pH is lowered association of particles occurs so that the viscosity increases. Under these conditions the fulvic acid particles are chargeless and electrostatic repulsion does not occur. As the pH is raised to above 3.0, increased dissociation of the carboxyl and phenolic functional groups takes place, which leads to increased electrostatic repulsion between fulvic acid particles. This is accompanied by the clustering of water molecules around the ionized functional groups, so that the net result is a gradual increase in viscosity as the pH increases.

2.1.3 Dilute Dispersions of Flocs

The rheologic behavior of hydrospheric dilute clay suspensions is best illustrated by a model of three types of species existing simultaneously: (1) kinetically independent primary particles, which are platelets of single unit-layers or tactoids, (2) spherical flocs, and (3) aggregates (Fig. 9.3). With increasing rate of shear, the average size of the

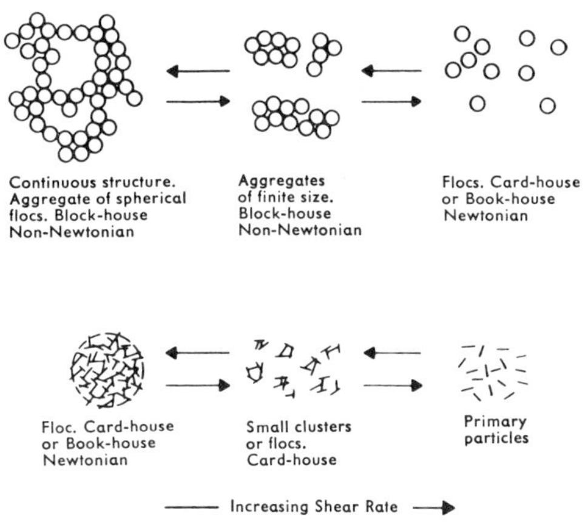

Fig. 9.3. Schematic representation of the disruption of structure of flocculated particles accompanying an increase in flow rate. (From Smith, 1972: Used with permission of the author)

flocs and aggregates decreases until at a sufficiently large shear rate the dispersion contains either dispersed primary particles or small stable flocs, or possibly both types of units. This model is also valuable for the illustration of the behavior of dilute dispersions of humic substances and other polar organic polymers, which tend to flocculate and to maintain water molecules in an "immobile" state. (See also pages 355–357).

Let us consider an infinitely dilute dispersion that contains both spherical flocs and primary particles. If φ_f is the volume fraction of the flocs (particles plus "immobile" liquid) and φ_{pp} is the volume fraction of particles that are kinetically independent and thus do not reside in flocs (silicate unit layers and in the case of smectite and vermiculite tactoids, also the interlayer water) then

$$\varphi_f + \varphi_{pp} = (\varphi + \varphi_{iw}) + \varphi_l, \tag{9.16}$$

where φ is the volume fraction of the dry particulate solid material in the dispersion (excluding "immobile" liquid), φ_{iw} is the volume fraction of water in the interlayer space of smectite or vermiculite tactoid, and φ_l is the volume fraction of "immobile" water trapped in the floc at spaces between the kinetically dependent primary particles. If the primary particles happen to be spherical, then according to the Einstein equation

$$\eta_{sp} = 2.5\,(\varphi_f + \varphi_{pp}). \tag{9.17}$$

In the more general case of clay minerals the primary particles are nonspherical and from Eqs. (9.10), (9.13), and (9.16) the specific viscosity will become

$$\eta_{sp} = 2.5\,\varphi_f + [\eta]_{pp}\,\varphi_{pp}, \tag{9.18}$$

where $[\eta]_{pp}$ is the intrinsic viscosity of the nonflocculated primary particles. Elimination of η_{sp} from Eqs. (9.12) and (9.18) and dividing both sides of the equation by $[\eta]\,\varphi_f$ gives

$$(\varphi_{iw} + \varphi)/\varphi_f = 2.5/[\eta] + [\eta]_{pp}\,\varphi_{pp}/[\eta]\,\varphi_f. \tag{9.19}$$

If the tendency for flocculation is strong, then $\varphi_{pp} \ll \varphi_f$ and $[\eta]_{pp} < [\eta]$. Hence, except in the case of the shear rate being sufficiently high to destroy most of the flocs, Eq. (9.19) reduces to

$$\varphi_{ppf} \equiv (\varphi_{iw} + \varphi)/\varphi_f = 2.5/[\eta], \tag{9.20}$$

where φ_{ppf} is the volume fraction of primary particles in a floc. The fraction of non-interlayer "immobile" water in a floc is thus $1 - \varphi_{ppf}$.

Michaels and Bolger (1962a) measured φ_{ppf} for flocs of kaolinite (where $\varphi_{iw} = 0$), in very dilute aqueous dispersions. The values ranged from 0.015–0.029, depending on the pH and salt content of the dispersion. The corresponding average diameter of the flocs varied from 0.260–0.099 mm. Flocs of other clay minerals and of metal oxides are also known to contain a very low fraction of solid material and a very high fraction of immobile water, and hence the effect of these minerals on the rheologic properties of the hydrosphere is far greater than that to be expected from their concentrations.

2.2 Rheology of Concentrated Dispersions and Muds

2.2.1 Dispersions of Nonflocculated Particles

Concentrated dispersions of uniformly dispersed sand-mineral spheres commonly exhibit Newtonian flow as long as their concentration is below about 20–30% by volume. At higher concentrations they are found to exhibit shear thinning, i.e., the viscosity decreases with increasing shear rate. The value of the viscosity and whether or not the flow is Newtonian depend, among other factors, on the electrolyte content. Newtonian flow can be preserved provided that sufficient electrolyte is present. When the electrolyte content is reduced, the viscosity increases markedly and the flow becomes highly non-Newtonian. Such behavior is explainable in terms of the electric double layer around the particles and the effect of electrolyte content in decreasing both the thickness of the double layer and the fraction of the immobile water. Certain polysaccharides have the same effect on suspensions of sand-minerals. They are adsorbed by these minerals and immobile water molecules are released into the bulk liquid. Newtonian flow is thus preserved even at high concentrations of the minerals.

The viscosity increase associated with the addition of large amounts of the spheres results from: (1) the presence of a high proportion of water having a hydrophobic structure, which causes an increase in the viscosity of the available fluid; and (2) a reduction in the volume/sphere ratio, which increases the probability that one sphere will collide with or slide over another sphere during shearing.

Several equations have been proposed to represent the viscosity of dispersions of spheres. For a monodisperse system, the dependence of viscosity on volume fraction φ is usually represented by a power series expression for the relative viscosity η_{rel} (Sherman, 1970),

$$\eta/\eta_o = 1 + k_1\varphi + k_2\varphi^2 + k_3\varphi^3 + \dots \tag{9.21}$$

The factor k_1 determines the linear concentration range where there is no interaction between particles. The higher order factors k_n allow for the additional effects of simultaneous collisions of n particles at increasing concentrations. The equation derived by Mooney (1951), which is widely used, is:

$$\eta/\eta_o = \exp\,[2.5\,\varphi/(1 - k\varphi)], \tag{9.22}$$

where k, ($k \neq 1$), is a crowding factor. The quantity 2.5 was introduced to give $\eta/\eta_o = 1 + 2.5\,\varphi$, as φ approaches zero. The crowding factor is, in fact, concentration independent. It equals the reciprocal of the volume fraction of spheres at which $\eta = \infty$, that is, $\eta \to \infty$ as $\varphi \to \varphi_{max} \equiv 1/k$.

In practice, the spheres are polydisperse, being composed of n groups, each of uniform size, and the obtained equation is

$$\ln \eta/\eta_o = 2.5 \sum_{i=1}^{n} \frac{\varphi_i}{1 - \Sigma\lambda_{ji}\varphi_i}, \tag{9.23}$$

where λ_{ji} is the crowding factor that accounts for the effect of the jth group of spheres on the ith group. Following Mooney's approach, Brodnyan (1959) derived an equation for the Newtonian viscosity of dispersions of ellipsoid particles.

2.2.2 Concentrated Dispersions of Flocs and Extended Loose Structures

When the primary particles tend to associate, a three-dimensional structure extending throughout the medium may be formed. If such a structure is obtained a stress greater than the yield stress, τ_o, must be applied to initiate flow. Under an increasing shear rate, aggregates of finite size, formed when flow begins, are progressively broken down into flocs, and these are further broken down into smaller flocs or clusters of particles, until, at very high shear rates, only primary particles or highly stable flocs exist. The breakdown of aggregates and flocs is reversible. For each shear rate, there is an equilibrium, or steady-state distribution of particles, flocs, and other structures. This equilibrium distribution, and hence the steady-state viscosity, is shear-dependent. Because a finite time is required to establish a steady-state condition following the sudden application of a constant shear rate, the viscosity is dependent on time as well as on shear-rate. These effects are indicated schematically in Figure 9.4. Under the constant shear

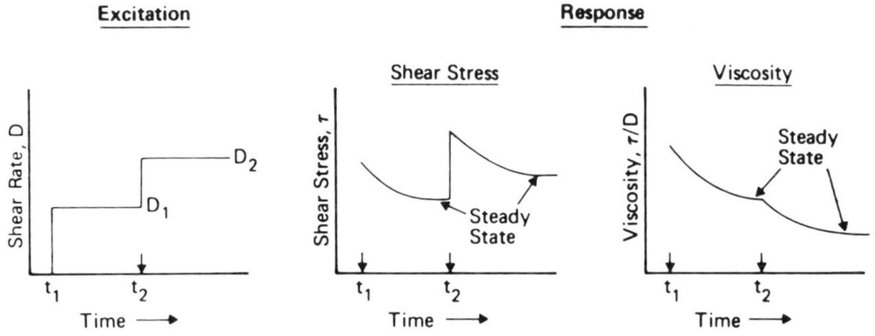

Fig. 9.4. Time-dependent response of a thixotropic dispersion to a stepwise-imposed shear rate. (From Smith, 1972: Used with permission of the author)

rate $D = D_1$, the shear stress and the viscosity, η, decrease to constant values. When the shear rate is now increased to $D = D_2$, the shear stress increases suddenly and decays to a new constant level. The viscosity, however, decreases and eventually attains a constant value that is less than under shear rate D_1. If the material has the characteristics shown in Figure 9.4, the D vs. τ curve obtained when D is increased (an upcurve) will differ from that observed when the shear rate is decreased (a downcurve) (Fig. 9.5). The phenomenon in which upcurve and downcurve do not overlap is known as *hysteresis.* The intercept of the downcurve is considered to be the yield stress, as indicated in the Figure. The downcurve may or may not represent equilibrium conditions. This type of behavior is characteristic for thixotropic materials.

Aqueous dispersions of montmorillonite are thixotropic when the concentration of the mineral is above 7 wt %. According to Gabrysh et al. (1963) from repetition of the cyclic deformation the thixotropic loop becomes small but does not disappear, and after three of four cycles it reaches a constant reproducible shape. After repeated experiments the suspensions are allowed to stand for 30 min and the first subsequent loop obtained appears to be nearly as large as the original first loop. The montmorillonite suspension appears to be of a Bingham plastic flow type. The yield stress becomes smaller with decreasing clay concentration. Organic compounds greatly affect the rheologic behavior of montmorillonite. Gabrysh et al. (1963) showed with increasing concentrations of glycerin in aqueous suspensions of montmorillonite the flow behavior of the system approaches that of a Newtonian fluid. A suspension of montmorillonite in pure glycerin in weight concentrations of up to 30% behaves as a Newtonian fluid.

The flow behavior of montmorillonite was explained using two different dispersion mechanisms. According to Weymann (1965) at finite shear rates the flocs of montmorillonite are spherical and contain immobile liquid. The volume fraction of the flocs and the size of the largest flocs decrease with increasing shear rate until at infinite shear rate all flocs are disrupted. Thus, the decrease in viscosity of the dispersion with increasing shear rate results from the release of immobile liquid accompanying the decrease in floc size.

According to Gabrysh et al. (1963) concentrated montmorillonite suspensions contain two types of disperse particles: (1) disentangled units, whose behavior is considered *Newtonian* and (2) entangled units of *non-Newtonian* character. The Newtonian units, which are small flocs or clusters, have the "card-house" structure. The suspension medium penetrates into the vacant portion resulting in almost spherically shaped units. The non-Newtonian units are those having the "blockhouse" arrangement of

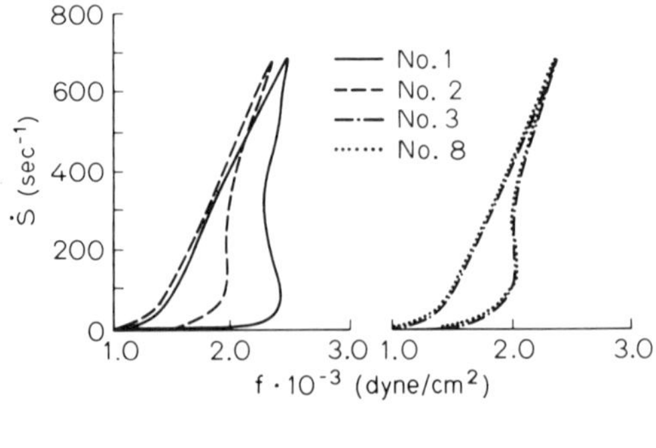

Fig. 9.5. Shear stress-response of a thixotropic dispersion to a shear rate that increases and then decreases with time. Family of curves obtained by successive cycling to reproducibility. 9% montmorillonite suspension at pH = 8.9 and 30 °C, 60-s up–down cycles. (Only cycles 1, 2, 3, and 8 are shown). (From Gabrysh et al., 1963: Used with permission of the American Ceramic Society)

platelets. Since these particles are porous the suspension medium penetrates into the vacant portion giving flexibility to the flowing particles. The non-Newtonian units are transformed to the smaller Newtonian units as a result of shear. If the stress is relieved, the transformed unit tends to return to its original state. The rate of build-up is negligible compared with that of the breakdown of structure. Equilibrium (i.e., type 1 $\rightleftharpoons$ type 2) is maintained between these units at low shear rates, but when the shear rate increases, the breakdown of the non-Newtonian units is faster than the reverse reaction and hysteresis occurs. The concentration of the entangled units increases with the increase in the concentration of the clay, and consequently the departure of the suspension from Newtonian behavior increases. In glycerin suspensions the swelling of montmorillonite is limited and viscosities are lower than in aqueous suspensions. Furthermore, a non-Newtonian unit is obtained only when the clay concentration is above 30 wt % and consequently suspensions at lower concentrations behave like Newtonian fluids. (See Fig. 9.3).

Michaels and Bolger (1962b) distinguished between aggregates and flocs in kaolinite suspensions, in which aggregates are built up from spherical flocs. Under increasing shear rate, the aggregates are progressively reduced in size until at high shear rates only flocs exist. They concluded that flocs of kaolinite are stable at high shear rates and thus are the basic units which at low shear rates cluster into aggregates. The values of φ_{ppf} for flocs in concentrated aqueous dispersions of kaolinite at high shear rates of up to 10,000 s^{-1}, were estimated by Michaels and Bolger to be about 0.1, which means that 90% of the volume of the floc consists of immobile water. They showed that the amount of immobile liquid in aggregates is about four times greater than in flocs.

Brückner (1966) studied the influence of mechanical oscillations of high frequency (ultrasonic) and of low frequency (vibration) on the flow curves of kaolinite suspensions. With both types of oscillation a thixotropic breakdown occurred, but in the case of the ultrasonic oscillation this breakdown was only partial. In the latter case the equilibrium flow curves did not show a Bingham yield point. In the case of the vibration oscillations a total breakdown may occur and the flow curves becomes straight lines, characteristic of Newtonian flow.

Müller-Vonmoos and Jenney (1970) studied the effect of mechanical oscillation on the rheologic behavior of opalinuston, a carbonate-rich clay sediment, commercially known as 'Opalit'. Ultrasonic oscillation causes the breaking of calcite bridges between the flake-shaped silicate minerals, resulting in the disintegration of the aggregates into single grains. This significantly improves the stability of the suspensions and the degree of hydration of the clay fraction and thereby the viscosity greatly increases. The difference in viscosity between the oscillated and the nonoscillated Opalit suspension depends on the ratio between water and the mineral and increases with increasing mineral concentration. It is 9 times higher for 50% suspensions (80 and 9 centipoise) but only 3, 2, and 1.5 for 33%, 25%, and 10% suspensions respectively.

3. Rheology of Sediments of Silicate Minerals

The postdepositional flow of sediments under the lithostatic pressure and the shape of the deformed sediment depend heavily on the rheological properties of the minerals.

"Stress" refers to the average load per unit area of the sediment, and "strain" is the ratio of the average change in a dimension to the original value of that dimension. Depending on the sedimentation process the microfabric of sediments of silicate minerals can be divided into two main groups: (1) sediments obtained from stable suspensions, and (2) sediments obtained from flocculated suspensions (Fig. 9.6).

1. Nonflocculated. Repulsion forces that acted between the particles while they were in suspension prevented the flocculation or aggregation of the solid phase. Sediments of stable suspensions obtain a high degree of compaction in the sediment. When the individual particles of a stable dispersion reach the floor of the sedimentation basin, they are able to slide and roll past each other because of their mutual repulsion, and hence they reach the lowest position in the sedimentation environment. A considerable fraction of the volume of the sediment is occupied by the mass of the particles. With particles of a size in the range of colloids, interparticle porous voids are capillaceous and the sediments may be impermeable (van Olphen, 1963).

2. Flocculated. Attraction forces between the particles in suspension lead to flocculation and aggregation. Sediments of flocculated suspensions are usually much more voluminous than those of stable suspensions of the same concentration. The haphazardly formed voluminous flocs settle as such and pile up at the bottom of the sedimentation basin to form a voluminous sediment with large void spaces in and between the agglomerates. Such sediments are permeable (van Olphen, 1963).

The voluminous sediment of a flocculated suspension behaves more or less like a Bingham system, as does a concentrated flocculated suspension, and strain is proportional to stress. The closely packed sediment of the nonflocculated suspension, on the other hand, behaves like a shear-thinning system, as does a very concentrated suspension

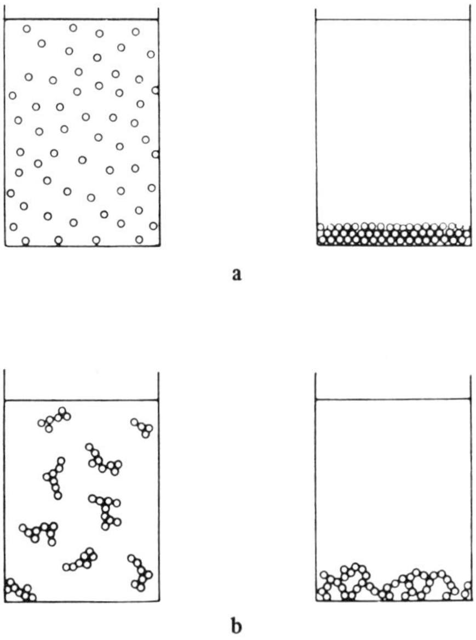

Fig. 9.6a and b. Sedimentation in a peptized and in a flocculated suspension. **a** Peptized suspension–dense, closely packed sediment; **b** flocculated suspension–loose, voluminous sediment. (From van Olphen, 1963: Used with permission of John Wiley & Sons)

of nonflocculated primary particles, and the increment of strain is smaller than that of stress.

The shear-thinning behavior of the nonflocculated sediment can be explained as follows: These particles are easily shifted with respect to each other by a small shearing stress. However, during the displacement of the particles liquid must flow through the interstitial spaces. If the displacement of the particles is slow, the liquid is able to keep pace with the change in the packing, but during a rapid displacement, the capillary flow of the liquid becomes the bottleneck in the shearing motion of the sediment, and a high resistance against shear develops. The displacement of the particles involves a disturbance of the condition of close packing of the sediment, and the interstitial volume increases. Consequently, water is drawn into the interior of the sediment, which outwardly appears dry. The dry appearance of the sand around the foot when stepping on the wet sand of the beach is a familiar example of this phenomenon.

Shear-thinning behavior is shown by systems of more or less spherical particles. Upon the addition of flocculant, such a system may be converted into a Bingham plastic system. The latter is not true for sediments of coarse sand grains, which do not tend to flocculate. A sand system remains a shear-thinning fluid when a flocculating salted seawater replaces fresh water in the interstitials, whereas a clay system is flocculated by seawater. Shear-thinning behavior is seldom observed in clay sediments, probably because they always contain adsorbed organic and inorganic flocculants.

In sediments containing nonspherical particles, rigidity may develop when these particles are entangled haphazardly in the manner of a heap of twigs. By applying a shearing stress the original packing is destroyed by orientation of the particles in a more or less parallel arrangement, giving rise to strain. Such a system will flow only when a shearing stress greater than the yield stress is applied.

The rheologic behavior of clay sediments and soils changes with a change in fluid content. Swelling in the presence of water and compaction under pressure with loss of water are characteristics of these systems. Two ranges of interaction may be distinguished during the swelling of a dry clay. In the first stage of hydration of the dry clay particles, water is adsorbed in successive monolayers on the surfaces and forces the particles apart. Water is also adsorbed onto the interlayer spaces of expanding clay minerals. In the latter case the volume changes accompanying this stage of swelling may be as high as 100% of the original volume of the dry clay. When the hydration takes place on the exterior surfaces of nonexpanding clays, the volume changes become smaller with increasing size of the particles. In the second stage of hydration swelling occurs due to double layer repulsion. This stage is also known as "osmotic swelling" (Ch. 8) and is identified with osmosis arising from concentration differences of dissolved ions in different parts of the system. This stage is accompanied by greater volume changes than in the first stage. The reverse reactions occur during compaction. The *swelling pressure* of a clay paste can be determined by measuring the confining pressure that is required to prevent further uptake of water by the paste, which is in contact with water via a semipermeable membrane. Swelling pressures decrease with increasing water content of the pastes.

Unlike clays, sand sediments hardly swell or shrink when becoming wet or dry, respectively. This is because the adsorbed water layer that covers particles of sand materials is very thin, the surface charge and the concentration of exchangeable ions are extremely low, and the double layer repulsion is negligible. In this dense, closely packed sediment, the position of the grains is controlled by gravity forces and the penetrating water may fill the voids between the particles but no more than that. Consequently, every sand particle is in contact with its near neighbors even when the sediment is in contact with liquid water. Quartz sediments may retain up to only 0.7% water whereas clay sediments hold between 3 and 11.5% water. Heat of hydration of quartz sediments is about 0.147 cal/g of dry material whereas the heat of hydration of clay minerals may be more than 3 cal/g of dry material.

3.1 Wet Sediments of Sand Grains

It is possible to build simple structures from wet sand which may carry their own weight. These structures will not be preserved when the sand dries unless it contains some clay material. The rheologic properties of the sand–water system are the consequence of mutual repulsion between the grains themselves and attraction between them and water molecules (Ch. 8). Liquid water may cause sand grains to adhere by filling intergrain capillaries (capillary attraction forces), giving some stability to structures built from wet sand (Weyl and Ormsby, 1960).

Organic polar molecules may be weakly linked to sand grains. This type of interactions increases the hydrophobic character of the sand grains, and consequently the adsorption of water decreases. Thus, in the presence of organic matter, the ratio between immobile and mobile water and the viscosity of the system is lower than in the absence of organic matter. Sediments of sand minerals that have settled in environments poor in organic materials are able to preserve structural features characteristic of the geologic environment during the sedimentation whereas structures of sand sediments deposited in environments rich in organic materials are highly affected by the pressures occurring during burial.

The relation between stress and strain and the rheologic behavior of sand sediments are sometimes associated with partial dissolution of the minerals. When sand sediments are buried they will gradually come under the influence of pressure. The pressure distribution in the sediment will become highly inhomogeneous and any one of the following processes may occur to reduce the water content and the porosity of the sediment, grain re-orientation, grain breakage, plastic deformation, or dissolution-reprecipitation reactions. From geologic observations it appears that grain re-orientation alone reduces porosity to minimum values of about 25%. The presence of pore water will promote both fracturing and plastic deformation (Swolfs, 1972; Griggs, 1967), and in any sample, even those subjected to extreme stresses caused by tectonics, grain breakage, and plastic deformation are only minor contributors to the degree of compaction and lithification (Lowry, 1956). In many sediments the process of "pressure solution" is responsible for a high degree of compaction and lithification. This process involves dissolution inside the grain-to-grain contacts, accompanied by diffusion through an adsorbed water layer (Weyl, 1959), and after the material at or near the grain contacts has been removed, neighboring grains interpenetrate. The dissolved material will diffuse away and precipitate near places where there is no grain-to-grain contact. According to de Boer (1977) under high pressure solutions can reach fairly high supersaturation. However, the only place where a reasonable degree of supersaturation may occur is in capillary-structured water inside grain contacts. This process is extremely slow because of the low mobility of the capillary-structured water. The "pressure solution" process may speed up when a clay mineral, and especially smectite, is present between the grains of quartz. There is a better diffusion of dissolved silica through water structured between contacting clay-quartz grains than through the immobile water boundary layer on quartz. The rheologic properties of sand sediments depend on the existence of such a pressure solution.

3.2 Argillaceous Sediments

The present treatment is based on that of Rieke and Chilingarian (1974). A change in the moisture content of the clay system results in a material having different mechanical properties. It is possible to identify four stages of hydration in clay sediments; the first stage does not lead to swelling whereas the other stages lead to swelling, the mechanism of which was previously discussed.

(1.) When the clay is dry the anhydrous silicate particles are in contact with each other. The net negative charge of the particles leads to a mutual repulsion of the particles and the solid clay is therefore friable or brittle. Because of the diversity in the charged sites of the surfaces of clay minerals, some weak electrostatic linkage between particles may occur where oppositely charged sites of two different particles interact by Coulombic attraction. In the first stage of the hydration, linkages are not broken. The penetrating water fills the voids between the particles and the volume of the sediment does not change. During this stage the system is regarded as "solid".

(2.) The interparticle linkages are broken while the distances between the particles increase, and the volume of the sediment increases in proportion to the volume of water adsorbed in this stage. However, there are enough interparticle linkages, either Coulombic or via water molecule bridges, to keep the system rigid. During this stage the system is regarded as "semi-solid".

(3.) The water content is high enough to form a discrete phase of water layers sandwiched between the clay particles. The water layers are positively charged due to a higher than average proton density whereas the clay particles are negative. According to Weyl and Ormsby (1960) Coulombic attraction forces come into play, keeping the mixture "sticky". During this stage the system has the characteristics of a "plastic" material.

(4.) The water content is so high that it becomes the dispersing medium and the diffuse double layers at the water–clay interfaces are fully developed. During this stage the system has the properties of a fluid.

This process occurs during fluidization and erosion of argillaceous sediments. The various reaction stages are reversible and the reverse reaction occurs during sedimentation and compaction of argillaceous sediments. On the floor of a sedimentation basin there is a zone where the sediment acts as a thick suspension with the flow properties of a highly viscous fluid. As the water content is reduced through subsequent burial and compaction, the flow becomes that of a paste-like system with a very high yield-stress. On further compaction the clay water content decreases and the flow properties change respectively.

The *shrinkage limit* is defined as the water content at which the argillaceous sediment changes from a "solid" to a "semi-solid" state. The *plastic limit* (*PL*) is the water content at which the transformation between "semi-solid" and "plastic" states occurs. The *liquid limit* (*LL*) of a clay is the water content at which the transformation between "plastic" and "fluid" states occurs. The liquid and plastic limits are two properties of clay systems, known as *Atterberg limits,* which are used to provide a quantitative measure of the degree of plasticity of clays. The *plasticity index* (*PI*) is defined as the difference between the two Atterberg limits:

$$(PI) = (LL) - (PL). \tag{9.24}$$

The liquid limits of clay systems are firmly associated with the ease of their peptization. Clays with lower liquid limits show higher tendencies to peptize. As with peptization, the liquid limit depends on the exchangeable cation and increases with increasing charge of the cation. It decreases with decreasing particle size and increasing specific surface area and amorphous fraction of the system. Organic and inorganic peptizers reduce the liquid limit whereas flocculating agents tend to increase it (pages 196–201).

Both Atterberg limits of soils or sediments reflect the amount and the type of clay minerals present therein and are therefore functions of such properties as total surface area and cation exchange capacity. The natural water content of a system can be compared directly with its Atterberg limits by a ratio defined as the *liquidity index* (*LI*), which is given by

$$(LI) = [w - (PL)]/[(LL) - (PL)], \tag{9.25}$$

where w is the water content of the system, which is usually defined as the ratio of the weight of water to the weight of dry solids, given in percent. Skempton (1970) showed that for a wide variety of normally compacted argillaceous sediments, the liquidity index lies within a narrow range of values, at any given affective pressure, although the corresponding natural water contents of the sediments may vary between very wide limits. If two clay samples from different depths have the same natural water content, it is possible to infer that the greater pressure applied on the deeper clay sample had caused no additional compaction. However, the liquidity index should decrease with depth, and it is therefore concluded that the liquid limit of the lower sample is greater than that of the upper sample.

It is common to evaluate the rheologic behavior of sediments according to their *consistency,* which is the resistance to flow. It is related to the internal friction and the cohesion. The internal friction is the rolling and sliding friction, together with the resistance resulting from the geometry and relative position of the clay particles whereas the cohesion is the result of the clay resistance, which is due to the force of attraction existing between the clay mineral particles. The cohesion is significant mainly during the "plastic" stage, when water layers serve as bridges linking the solid phase particles.

4. Viscosity of Magmas

The viscosity of a magma influences the velocity of the flow of this liquid and its ability to penetrate into cracks. It also influences the velocity of motion of solid bodies and gas bubbles in the fluid. Also the diffusion of molecules, ions, and radicals depends on the viscosity of the magma, and this factor plays an important role during the crystal growth of the various minerals and the separation of gas and liquid bubbles.

Most magmas in the earth are viscous although they exist at high temperatures. Room temperature viscosities of some familiar liquids and high temperature viscosities of some rock and mineral melts are shown in Table 9.2. The viscosity of rock melts is a function of their composition, being in acidic rocks rich in silica up to 10^3-10^4 times that of basic rocks. The presence of divalent metallic cations having a coordination number 6, lowers viscosity. Mg, however, lowers viscosity to a lesser extent than does Fe. Divalent cations with a coordination number higher than 6, such as Ca, and alkali cations (mainly with high Na/K ratios), also lower the viscosity of a magma but to a smaller extent. Increasing concentrations of elements having a coordination number 4, such as Si, Al, and Ti, result in increased viscosity.

Table 9.2. Typical viscosities of some common liquids, rock melts, and magmas. (After Macdonald, 1972; Zavaritskii and Sobolev, 1964; Goranson, 1934; Minakami and Sakuma, 1953)

Liquid	Temperature (°C)	Viscosity (poises)
Water	20	1×10^{-2}
Glycerine	20	1×10
Gelatine	20	$1 \times 10^6 - 1 \times 10^8$
Diopside	1400	3.8
Anorthite	1400	3.8×10
Albite	1400	4×10^4
Orthoclase	1400	1×10^8
Diabase	1400	$6 \times 10 - 9 \times 10$
Basalt	1400	5.5×10
Nepheline basalt	1400	8×10
Olivine basalt	1150	9×10^2
Olivine basalt	1200	5×10^2
Olivine basalt	1300	2×10^2
Olivine basalt	1400	1.2×10^2
Andesine basalt	1150	8×10^4
Andesine basalt	1200	3×10^4
Andesine basalt	1300	2.6×10^2
Andesine basalt	1400	1.4×10^2
Vesuvius lava	1400	2.5×10^2
Obsidian	1400	1.7×10^5
Basalt from Mauna loa, Hawaii	940	1×10^4
(measured in 1950)	1070	4×10^3
Basalt from Oshima, Japan	1038	2.3×10^5
(measured in 1951)	1083	7.1×10^4
	1108	1.8×10^4
	1125	5.6×10^3

The differences in the effects of the elements on the viscosity of the magma can be explained as follows. As was stated in Chapter 1, the silicate melt contains small mobile species of low molecular weights and flexible magmaphilic colloid polymeric species, which may be regarded as immobile. The viscosity of the system increases with increasing volume fraction of the immobile polymeric species. Elements having a coordination number of 4 may substitute Si in the polymeric species whereas elements having other coordination numbers act as structure breakers and may lead to the depolymerization of the immobile magmaphilic polymeric speices. Consequently, the molar fraction of this phase and the viscosity of the system decrease (Shaw, 1972). Aluminum plays an ambivalent role. When added to pure silica liquid it becomes 6-fold coordinated and breaks the polymers of silica, reducing the viscosity of the melt. If aluminum is added together with other cations, it becomes 4-fold coordinated. It thereby acts as a structure maker of polymeric species and thus increases the viscosity of the melt (Bottinga and Weill, 1972). On the basis of the above-mentioned role of certain elements as structure breakers and of other elements as structure makers, Shaw (1972) and Bottinga and Weill (1972) introduced empirical methods to calculate the viscosity of magmas, which are based on their chemical compositions.

The presence of magmaphobic solid fragments (either phenocrysts or fragments from foreign origin) in magma affects its viscosity, and according to Eq. (9.13) the viscosity increases with the concentration of the solid fraction. The surface area and the surface activity of the solid fragments also have an effect on the viscosity, since mobile silica species can be sorbed onto the interface and become immobile.

Most volatile molecules dissolved in the magma lead to depolymerization of immobile species and consequently decrease its viscosity. For example, increasing the water content by one percent reduces the viscosity of a melt by about one order of magnitude. Thus, a pantelleritic rhyolitic magma with 1% water has a viscosity similar to that of a calcalkaline rhyolitic magma with 2% water at the same temperature. In fact, field observations of many peralkaline supersaturated lava flows show structures very similar to calcalkaline rhyolitic lavas, implying that viscosities of such lavas were probably similar to those of calcalkaline varieties (Schmincke, 1974). On the other hand, dissolved volatiles that form stable magmaphobic solvation structures (Ch. 1) give rise to high viscosities.

Gas bubbles increase the viscosity of a magma if they are very abundant, because some of the energy is dissipated on elongating the shapes of the bubbles from spherical into ellipsoidal or almond shaped and thereby increasing their surface area.

Under a high pressure, when the solubility of the volatiles is high, the viscosity of magmas is low. When the pressure is released the solubility of the volatiles decreases, gas bubbles and silicate polymers are formed, and consequently the viscosity of the magma increases. The penetration of highly acidic magmas into very narrow cracks occurs in the presence of volatiles, under a very high pressure, when the viscosity of the melt becomes low. Since the molar fraction of the mobile species in a magma melt depends on the temperature, the viscosity of the magma also depends on the temperature. The higher the temperature of the magma, the lower is its viscosity. For example, lava flows flowing away from the vent lose heat and polymerize, and their viscosity steadily increases.

Granitic magmas are markedly more viscous than basaltic magmas, even under conditions of high water pressure. A granitic magma saturated with water at depths in the crust equivalent to 1 to 2 kbar pressure has a viscosity of 10^6 poises. Under similar conditions a basaltic magma has a viscosity of less than 10^3 poises (Shaw, 1965). Under these conditions intrusions of a basaltic magma up a restricted dyke system would be more than 10^3 times more rapid than intrusions of granitic magmas.

References

Boer, R. B. de: On thermodynamics of pressure solution-interaction between chemical and mechanical forces. Geochim. Cosmochim. Acta *41,* 249–256 (1977)

Bottinga, Y., Weill, D. F.: The viscosity of magmatic silicate liquids: a model for calculation. Am. J. Sci. *272,* 438–475 (1972)

Brodnyan, J. G.: The concentration dependence of the Newtonian viscosity of prolate ellipsoids. Trans. Soc. Rheol. *3,* 61–68 (1959)

Brückner, R.: Der Einfluß mechanischer Schwingungen auf das Fließverhalten von Kaolin-Wasser-Mischungen. Ber. Dtsch. Keram. Ges. *43,* 709–717 (1966)

Chen, Y., Schnitzer, M.: Viscosity measurements on soil humic substances. Soil Sci. Soc. Am. Proc. *40,* 866–872 (1976)

Einstein, A.: Investigations of the theory of the Brownian movement. New York: Dover Pub. 1956

Gabrysh, W. F., Eyring, H., Lin-Sen, P., Gabrysh, A. F.: Rheological factors for bentonite suspensions. J. Am. Ceram. Soc. *46,* 523–529 (1963)

Goldsmith, H. L., Mason, S. G.: The micro rheology of suspensions. In: Rheology. Eirich, F. R. (ed.). New York: Academic Press, 1967, Vol. 4, pp. 85–250.

Goranson, R. W.: A note on the elastic properties of rocks. Wash. Acad. Sci. J. *24,* 419–428 (1934)

Griggs, D.: Hydrolytic weakening of quartz and other silicates. Geophys. J. *14,* 19–31 (1967)

Lowry, W. D.: Factors in loss of porosity by quartzose and sandstone of Virginia. Bull. Am. Assoc. Petrol. Geol. *40,* 489–500 (1956)

Macdonald, G. A.: Volcanoes. Englewood Cliffs, New Jersey: Prentice-Hall, Inc. 1971

Michaels, A. S., Bolger, J. C.: Settling rates and sediment volumes of flocculated kaolin suspensions. Ind. Eng. Chem. Fundam. *1,* 24–33 (1962a)

Michaels, A. S., Bolger, J. C.: The plastic flow behavior of flocculated kaolin suspensions, Ind. Eng. Chem. Fundam. *1,* 153–162 (1962b)

Minakami, T., Sakuma, S.: Report on volcanic activities and volcanological studies concerning them in Japan during 1948–1951. Bull. Volcanol. (ser. 2) *14,* 79–130 (1953)

Mooney, M.: The viscosity of a concentrated suspension of spherical particles. J. Colloid Sci. *6,* 162–170 (1951)

Müller-Vonmoos, M., Jenny, F.: Einfluß der Beschallung auf Körnung. Rheologische Eigenschaften, Sedimentations-Verhalten und Injizierbarkeit wäßriger Opalit-Suspensionen. Schweiz. Mineral. Petrogr. Mitt. *50,* 227–243 (1970)

Rieke III, H. H., Chilingarian, G. V.: Compaction of argillaceous sediments. Amsterdam: Elsevier 1974

Scheraga, H. A.: Non-Newtonian viscosity of solutions of ellipsoidal particles. J. Chem. Phys. *23,* 1526–1532 (1955)

Schmincke, H. U.: Volcanological aspects of peralkaline silicic welded ash-flow tuffs. Bull. Volcanol. *38,* 594–636 (1974)

Shainberg, I., Otoh, H.: Size and shape of montmorillonite particles saturated with Na/Ca ions (inferred from viscosity and optical measurements). Isr. J. Chem. *6,* 251–259 (1968)

Shaw, H. R.: Comments on viscosity, crystal settling and convections in granitic magmas. Am. J. Sci. *263,* 120–152 (1965)

Shaw, H. R.: Viscosities of magmatic silicate liquids: an empirical method of prediction. Am. J. Sci. *272,* 870–893 (1972)

Sherman, P.: Industrial Rheology. London: Academic Press 1970

Simha, R.: The influence of Brownian movements in the viscosity of solutions. J. Phys. Chem. *44,* 25–37 (1940)

Skempton, W. A.: The consolidation of clays by gravitational compaction. Q. J. Geol. Soc. London *125,* 373–411 (1970)

Smith, T. L.: Rheological properties of dispersions of particulate solids in liquid media. J. Paint Technol. *44,* 71–79 (1972)

Swolfs, H. S.: Chemical effects of pore fluids on rock properties. Mem. Am. Assoc. Petrol. Geol. *18,* 224–234 (1972)

Tanford, C.: Physical chemistry of macromolecules. New York: John Wiley & Sons 1967

van Olphen, H.: An introduction to clay colloid chemistry. New York: John Wiley & Sons 1963

Weyl, P. K.: Pressure solution and the force of crystallization–a phenomenological theory. J. Geophys. Res. *64,* 2001–2025 (1959)

Weyl, W. A., Ormsby, W. C.: Atomistic approach to the rheology of sand-water and clay-water mixtures. In: Rheology, theory and applications. Eirich, F. R. (ed.). New York: Academic Press, 1960, Vol. 3, pp. 249–297

Weymann, H. D.: On the viscosity of thixotropic suspensions. Proc. 4th Int. Congr. Rheol. 1963. Lee, E. H. (ed.). 1965, Part 2, pp. 573–591

Zavaritskii, A. N., Sobolev, V. S.: The physicochemical principles of igneous petrology. Kolodny, J., Amoils, R. (trans.). Jerusalem: Israel Program for Scientific Translations 1964

Chapter 10
Colloid Geochemistry of Argillaceous Sediments

The term "argillaceous sediments" as used here includes all the more "solid" and "semisolid" (mud) forms of clay–water systems. The important properties of clay minerals, namely, their very small particle size and their high surface energy, are recognizable in the physical and chemical reactions that take place after the particles have settled out (pages 395–396).

There is a close relationship between the geologic age of an argillaceous sediment and its clay mineralogy. Younger shales and recent argillaceous sediments are relatively enriched in kaolinite, smectite, and mixed-layer clay that has a high proportion of expandable layers, whereas older shales (especially pre-Mesozoic) are depleted in these phases but enriched in illite and chlorite (Weaver, 1967). Smectites are usually absent from moderately old shales whereas kaolinites are usually absent from the older shales (Foscolos and Kodama, 1974). Changes in bulk chemistry of shales accompany the variation of mineralogy with age. Younger sediments are relatively enriched in CaO, MgO, Na_2O, and SiO_2, whereas older sediments may be slightly enriched in K_2O (Garrels and MacKenzie, 1972).

Depth-dependent mineralogic variations in a sequence of an argillaceous sediment may be attributed to mineralogic transformation (diagenesis), but they may also reflect changes through time of the mineralogic composition of the detritus received in the sedimentary basin. For example, Weaver and Beck (1971) reported the abrupt occurrence of chlorite in one of the horizons they studied from a well in the Anadarko basin and stressed that this mineral was from a different formation than their other samples. In one horizon of a well in the Gulf Coast, Perry and Hower (1970) obtained abnormally large amounts of chlorite and discrete illite along with the illite/smectite that was present throughout the entire sequence. This was observed in a part of the section where the depositional environment shifted from the middle and outer neritic to nearshore. At greater depths in the section, where the deposition again occurred farther offshore, the chlorite and discrete illite were present in reduced amounts, and the illite/smectite mixed-layer mineral type was again dominant. The chlorite and discrete illite were determined to be detrital.

The conversion of smectite to a mixed-layer illite/smectite and further to illite is a diagenesis that is attributed to burial metamorphism. The mineralogy of argillaceous sediments has been used to estimate the degree of sediment diagenesis, to reconstruct the geochemical history of the sedimentary basin, and to assess the potential of rocks as sources of petroleum. This alteration process includes (1) dehydration of the mud under the high pressure-temperature conditions of deep burial (Burst, 1969) and (2) compositional changes in the minerals by the interaction with chemical components derived from the decomposition of unstable detrital minerals present in the sediment and in part supplied from other localities (Weaver and Beck, 1971; Heling, 1974).

During the first few hundred meters of burial, pore water is expelled from shales by overburden load. At depths greater than 1300–2000 m, as the smectite component changes to illite, the water released from the interlayers moves into the remaining pore space of the sediment and may flush out inorganic and organic material (Burst, 1969). Perry and Hower (1972) describe a third pulse of water expulsion at still greater depth, related to the appearance of ordered illite/smectite interlayering. Burst's and Perry and Hower's water escape curves are compared in Figure 10.1.

If the expulsion of pore water or of interlayered water is hindered by low permeability and/or poor drainage, the shales cannot reach compaction equilibrium, and the pore water has to support a part of the overburden pressure. In such circumstances the pore pressure becomes abnormally high, resulting in a state described as "compacted disequilibrium". Shales with abnormally large amounts of pore water are called *undercompacted shales.* According to Magara (1975) such an abnormal pressure is responsible for abnormal shale porosity, abnormal low shale density, and abnormal high subsurface temperature. Under these circumstances the temperature-dependent clay mineral diagenesis is most likely to take place at shallower depths than normally.

Magara (1976) examined water expulsion from clastic sediments and concluded that in an interbedded sand-shale sequence expelled water moves horizontally whereas in a continuous shale sequence the expelled water moves vertically. The electrolyte content of the sediment decreases during compaction. The electrolyte concentration

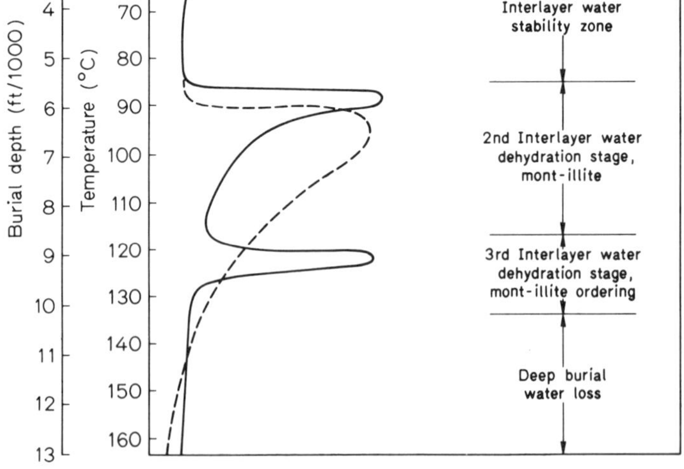

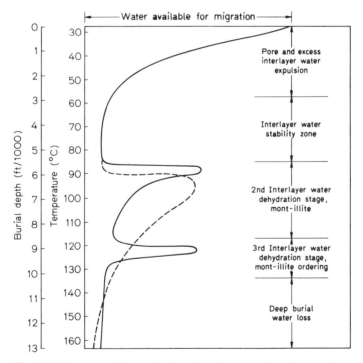

Fig. 10.1. Water expulsion vs. depth of burial curves, *dashed line*, after Burst (1969) and *full line*, after Perry and Hower (1972). (From Johns and Shimoyama, 1972: Used with permission of the publisher)

of water expelled by the overburden pressure is higher than that of the remaining pore water (Bolt, 1961; Apello, 1977; von Engelhardt and Gaide, 1963). This "negative salt adsorption" is caused by the Donnan expulsion of co-ions (Ch. 2) when compaction brings the charged particles closer to each other.

Various types of organic species are found in sediments, which may include aliphatic and aromatic hydrocarbons, nonhydrocarbon geopolymers *bitumen* and *kerogen*. Bitumen is extractable by organic solvents. The term kerogen is applied to the finely dispersed amorphous organic matter of sedimentary rock that is insoluble in organic solvents, mineral acids, and bases. A certain proportion of sedimentary kerogen is composed of coaly particles that can be recognized as structural plant material. Occurrence and distribution of these particles in a sediment are strongly influenced by the environment of deposition. Sedimentary kerogen also contains different proportions of unstructured, diagenetically formed organic particles. Kerogens can be divided into (1) alginite-type material, originating from algal debris, with a relatively high H/C ratio consistent with being more paraffinic; (2) huminite- or vitrinite-type material, originating from humic or plant debris, with a lower H/C ratio typical of a more aromatic structure; and (3) a mixture of both types of material.

The organic material from dead organisms undergoes diagenetic alterations by biogenous and nonbiogenous mechanisms. According to Welte (1973) the pathway of organic matter from biopolymers to geopolymers includes four stages:

1. In the first stage, biopolymers are transformed into degradation products that contain soluble compounds such as free amino acids, sugars, and unsaturated lipids. After the death of a living organism, its biopolymers become unstable and are subjected to mechanical disintegration and chemical hydrolysis. The more stable biopolymers, such as nucleic acids, proteins, and polysaccharides, are attacked by microorganisms and are enzymatically broken down into partly degraded polymers and microbially

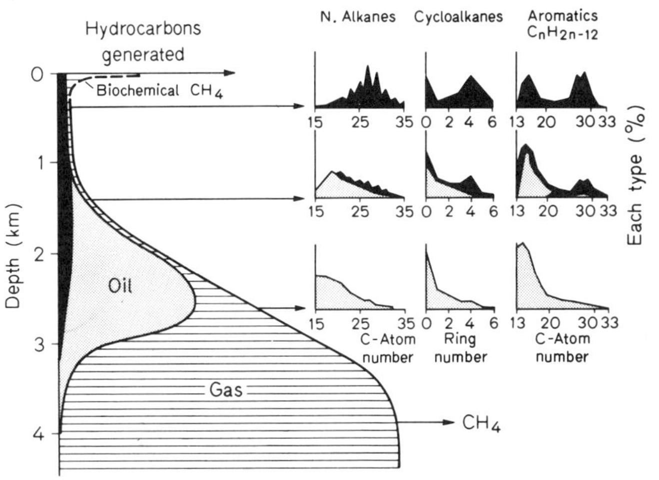

Fig. 10.2. General scheme of hydrocarbon generation with depth. Black-hydrocarbon from living organisms and degradation products of biopolymers. White-Thermal cracking products of geopolymers. (From Tissot et al., 1974: Used with permission of the publisher)

newly formed soluble compounds. At depths, due to elevated temperatures and burial pressures and due to the dehydration and compaction of the sediment, the organic compounds achieve closer contacts. Consequently, the second and third stages of transformations of organic matter are characterized by various types of polymerization.

2. In the second stage the monomers and the soluble compounds are randomly polymerized into the hydrophilic colloids fulvic and humic acids, due to the partial dehydration of the system and to the slight degree of oxidation.

3. During the third stage the degree of random polymerization and that of cross-linking are gradually further increased and the hydrophilic colloids are converted into heterogeneous hydrophobic geopolymers, known as kerogens, which are not soluble in alkalies or in acids, and are also insoluble in organic solvents. This stage is usually reached at depths of a few hundred meters.

4. In the fourth stage organic matter undergoes cracking reactions to yield petroleum hydrocarbons. With increasing diagenesis lower molecular weight hydrocarbons become predominant and with excessive thermal alteration petroleum hydrocarbons are cracked to methane; the residual organic matter tends to form graphite according to the equation

$$C_nH_{2n+2} \rightarrow \frac{n+1}{2} CH_4 + \frac{n-1}{2} C. \quad (10.I)$$

The depth at which these transformations occur can be related to geothermal gradients and, to some degree, to the period of time for which the sediments have been exposed to elevated temperatures. Organic matter transformations and clay mineral diagenesis are parallel, and Foscolos et al. (1976) used the following three parameters to evaluate the degree of diagenesis and oil generation potential of shales: (1) the clay mineralogy of a shale; (2) the chemical properties of the coal fragments; and (3) the organic compounds in the shale.

1. Microstructure of Argillaceous Sediments

Three different "solid" and "semisolid" clay–water systems can be distinguished, namely, jelly, paste, and compacted. The first two are commonly known as "gels".

Normally, clay particles tend to cohere when they contact one another at favorable positions, forming ramifying flocs termed "card-house" or "book-house", in which water molecules become immobile. At the bottom of the sedimentation basin, where the concentration of the flocs becomes high, association between flocs takes place. When this process proceeds further, whereby the whole liquid at the bottom of the sedimentation basin becomes immobile, the flocs lose their substantiality and the sol becomes a jelly in progressive stages. In the jelly state the plate-like aluminosilicate unit layers are sufficiently linked to cause all the water molecules at the bottom of the basin to be enmeshed in the loose framework formed by the joined flocs. It is obvious that for the formation of a jelly two properties are necessary; first, strong adhesion of water molecules to the clay particles and second, strong cohesion of the particles at points of contact. A jelly has the continous "block-house" structure (Fig. 9.3, page 387).

Stresses occurring at the bottom of the basin even if these are small, have great effect on particle orientation in the jelly. A jelly once buried or subjected to loading will have suffered some degree of structural change (Greene-Kelly and Mackney, 1970). As a result of compressive stress the jelly loses water and its volume becomes smaller. A paste is obtained that differs from a jelly in that it bears a much closer relation to the layered structure of the clay mineral. In a paste one may expect an increase in the parallelism of the particles (Figs. 1.5a and c, page 35).

The transformation of the jelly-type gel into the paste-type gel requires the breaking of EE and EF modes of association and the formation of FF modes. The process may proceed via two mechanisms: (1) under the effect of stress the platelets slide with respect to each other from a nonparallel into a parallel position, being in contact during the whole process, and (2) the EE and EF links dissociate and the interparticle separation increases before the platelets achieve a position parallel to the stress. A very high activation energy is required for the transformation to occur via mechanism (1) whereas a much lower activation energy is required for the transformation to occur via mechanism (2), and it is equal to the activation energy of the deflocculation of the system, (see Ch. 8). However, the presence of water is essential for the transformation to proceed via mechanism (2).

According to Diamond (1971) if a clay soil is compacted with water content appreciably lower than the optimum water content, a randomly oriented fabric is obtained. Increasing the water content will reduce the randomness of particle arrangement. If the sediment is compacted at a water content above the optimum, the compacted clay will have a partially oriented fabric. The greater the water content, the more oriented the fabric will be. The larger particles of illite and kaolinite develop a preferred orientation more easily than small particles (Bolt, 1956).

According to Blackmore and Miller (1961) as pressure is increased on a clay, the number of particles in each tactoid also increases. They found that for montmorillonite at 100 kg/m^2 there were about eight platelets to each tactoid. Sloane and Kell (1966) showed that compaction causes kaolinite to form tactoids. During the formation of tactoids additional water must be repelled and the interparticle separation becomes smaller. Meade (1964) found that the larger clay mineral particles appear to form tactoids more rapidly than the smaller clay mineral particles. (See also pages 200, 357).

The geologic phenomena associated with the changes in the microstructure of sediments due to overburden pressure, namely, particle orientation and tactoid formation, are accompanied by water migration, reduction in sediment volume, and porosity and increase in sediment density.

1.1 Microstructure of Fresh Sediments

In this section a fresh clay sediment is considered as one having undergone only a negligible postdepositional compaction by the pressure of superincumbent sediment. The spatial arrangement of the sedimented particles is influenced by the chemical composition of the solution from which they are deposited. Marine clays that have been deposited from an electrolyte-rich solution have a different fabric from clays deposited from fresh water. Sediments formed in electrolyte-rich solutions have a "salt-flocculated" structure, whereas those formed in electrolyte-free solutions are of "nonsalt-flocculated" structure. A third type of sediment structure is that obtained by clay sediments formed in solutions rich in organic substances that act as peptizers, and it is known as a "dispersed" structure (Lambe, 1958).

The various modes of platelet association that predominate in clay suspensions containing various concentrations of NaCl were described in Chapter 8. Sedimentation in a marine environment, where the NaCl concentration is above the critical coagulation concentration and the FF mode of particle association becomes more stable than either the EE or EF mode, results in the individual platelets aggregating together in "books" and the books then forming an open arrangement referred to as "book-house".

In sedimentations in river or lake environments, where the electrolyte concentration is below the critical coagulation concentration of the FF mode of particle association, EE and EF modes are obtained, forming an open arrangement referred to as "card-house". Both card-house and book-house arrangements form voluminous sediments that may be considered as jellies, the first forming larger void spaces.

Sedimentation in an environment rich in peptizers is extremely slow and leads to a high degree of compaction in the sediment (see Ch. 9). The resultant fabric obtained is totally oriented, the platelets being horizontal. With such an arrangement the solid mass reaches the lowest position in the sedimentation environment.

In the considerations that led to the preceding conclusions on the various arrangements of platelets in fresh clay sediments, two factors were neglected: (1) sea currents may deflocculate associations of EE and EF modes, whereas FF modes of association are more stable and may survive these currents, and (2) in deep-sea sediments orthokinetic coagulation is very significant, whereby the settling tactoid picks up single small platelets. As a result, when the tactoid reaches the bottom of the sedimentation basin it is coated by a nonuniform clay layer. The resultant fabric of the sediment is composed of tactoids or "turbostratic groups" having oriented packets of clay flakes in turbulent array. Such an arrangement has been described by Aylmore and Quirk (1960).

Silt particles lead to orthokinetic coagulation of the clay particles in the sea, the silt serving as a focal point for orienting clay particle around it. The resultant structure is known as a "honeycomb", in which the clay particles form flocs with enmeshed silt particles (Casagrande, 1932). (See also Fig. 8.8, page 354).

1.2 Microstructure of Compacted Sediments

Compaction of argillaceous sediments may result in either lithified fissile *shales* or in massive *claystones* and *mudstones*. Hedberg (1936) envisaged four major stages in the progressive compaction of clays: (1) rearrangement of particles, producing a "paste"-packing mode of greater stability and the expulsion of the majority of free water; (2) expulsion of water electrochemically bound to the particles and growth of tactoids; (3) contact of secondary particles with one another, but since they are more irregularly shaped than primary particles, appreciable contact and further reduction in volume is only possible by mechanical deformation of the tactoids; and (4) reduction of voids to below about 10%, which may occur by mineralogic transformation and recrystallization of the clay minerals, i.e., diagenesis.

Shales differ from mudstone in their fissile laminated structure. Fissility is associated with a high degree of preferred orientation, whereas the massive structure of mudstones is associated with the very poor degree of preferred orientation of the clay minerals. With compaction there is an increase in particle orientation, and it has been thought by many investigators that preferred orientation and hence fissility must increase with burial depth of the sediment. Baldwin (1971) gives a 350-m depth as the point of the beginning of the fissility in a clay, the first stage of shale formation. According to von Engelhardt and Gaida (1963) clay sediments form shales at depths between 100 and 3000 m.

There is no obvious relation between fissile structure and overburden depth (Odom, 1967). White (1961), for example, showed that some marine argillaceous rocks from a depth of one kilometer were less fissile than those found nearer the surface. Gipson (1965) found that planes of fissility tend to occur along zones of organic material. Odom (1967) recognized that samples with a high organic content generally show the highest degree of clay particle orientation and fissile structure. White (1961) associated a dispersibility of the system, and hence fissility in the argillaceous rocks, simply with a low electrolyte content. The concentration of the latter would be insufficient to flocculate the clay and would result in single independent particles being produced and taking up a parallel orientation. According to Meade (1964, 1966), in addition to the importance of electrolyte concentration, the types of cations present in the system and their coagulating power have some effect on the fissility of shales. The effect of biologic activity on the sediment structure was suggested by Gipson (1966), according to whom the action of benthonic organisms may disorganize some sediment layers and not others, thus resulting in juxtaposed strata of preferred and random clay mineral orientation. According to Moon (1972), for a certain clay concentration at or very near the depositional interface, particle clusters take up the tactoid arrangement in flocculated and in dispersed systems. The orientation of the tactoids under compaction will be greatly influenced by electrolyte concentration; a dispersed solution will be conducive to a structure with a high degree of preferred orientation of tactoids, because of the complete absence of interactions between the tactoids. In a flocculated medium this may not be so, and compaction will not necessarily give rise to preferred orientation of the tactoids but rather result in the incorporation of more primary particles in each tactoid.

It is likely that a fissile shale is obtained when a paste is dehydrated, whereas a massive mudstone results from the dehydration of a jelly. The degree of preferred orientation of clay minerals in the compacted sediments will depend on the transformation reaction of the jelly into a paste during the occurrence of the "semisolid" state of the sediment.

The conversion of a jelly into a paste is a time and liquid water-consuming process. Two possibilities can occur: (1) the water content of the gel is sufficiently high for the conversion of the jelly into a paste, and does not diminish before the process has been completed and (2) the water content of the gel is sufficient for the conversion to occur only up to a certain point, but then becomes small, resulting in only partial conversion of the jelly into a paste. In the first case compacted sediments will have a very good degree of preferred orientation, similar to that of peptized sediments. In the second case the compacted sediments will have a lower degree of preferred orientation. Jellies obtained from stable flocs, e.g., from electrolyte solutions of concentrations above the critical coagulation concentration, are only very slowly converted into pastes. If the dehydration of the gel by compaction is very fast, it may result in massive mudstone.

Ideal peptized clay systems do not form jellies. As the primary particles settle out, the initial interparticle separation is not less than that of a secondary coagulation, and a small overburden pressure is sufficient to convert the nonrigid layer lying in the bottom of the sedimentation basin into a paste. Dehydration of these peptized clays by compaction always results in lithified shales.

2. Aging and Diagenetic Alteration of Smectities in Argillaceous Sediments

When the particle of the clay mineral reaches the bottom of the sedimentation basin it enters a new environment and may become unstable. The particle may undergo changes while aging in contact with the solutions of the sediment; the chemical composition of the particle may change and finally a new equilibrium is attained. If the particle is very small, it will have a high solubility and may disappear. At the same time larger particles, which are less soluble, grow at the expense of the smaller ones. This effect is known as "coarsening" and unless the liquid in the sediment is in a highly turbulent flow, particles of a larger average size are formed as a result of "aging" and the total surface area of the solid phase decreases.

When pore water contacts the particle some ions go into solution until saturation is reached. Ions from a small particle are more likely to go into solution than ions from a coarse particle. The solution in the vicinity of the coarse particle is less concentrated than that near the small particle, and ions diffuse from the concentrated to the more dilute environment. Furthermore, a solution that is saturated only with respect to small particles may become supersaturated with respect to coarse particles. However, supersaturation is not sufficient for the generation of a new crystallization nucleus, and the dissolved ions condense on the coarse crystal by the orientation mechanism (see Ch. 3). The solution, which at this stage is in equilibrium with the coarse particle, becomes undersaturated with respect to the small particle. Again ions from this small particle will go into solution and later will precipitate onto the coarse particle, and so on, until the small particle has completely dissolved, whereas the coarse particle has increased in coarseness. This type of process is known as "recrystallization".

Since recrystallization occurs in an environment that differs from the environment where the original mineral was formed, one would expect that the products obtained from recrystallization are not necessarily the same as the original reactants. Changes are expected to occur in the mineralogic as well as in the chemical composition. One would expect that those minerals which are stable under a high pressure and temperature will be formed at the cost of the original, less stable minerals. Compounds from various origins are gathered in the sediment and they all contribute to establishment of chemical equilibrium.

Other chemical reactions occurring during the aging of the particles in the buried sediments are ion exchange, dehydration, and sorption–desorption of organic compounds.

Clay minerals that are obtained from the weathering of rock-forming silicates and that are stable in soils and in the hydrosphere, are not necessarily stable in the buried sediments. The elapsed time represented by the top few meters of unlithified sediment, i.e., the gel fraction, comprises a few million years. Temperatures rarely exceed 30 °C, and at such low temperatures, recrystallization of one clay mineral into another, although favored thermodynamically, may not take place during early diagenesis because it requires a high activation energy. Higher temperatures and longer times are needed, and the evidence for chemical diagenesis is to be found in deep buried sediments and ancient sedimentary rocks. Mineralogic changes reported by various investigators for metamorphic sequences in buried argillaceous sediments all show progressive conversion with increasing burial depth of smectite into illite or into a mixed-layer illite/smectite with a high proportion of illite layers.

The conversion of smectite to an illite/smectite must involve chemical changes in phases other than smectites present in the shale. The chemical equation that describes the conversion reaction is

$$[(Al_{1.5}Mg_{0.5})(Si_{4.0})O_{10}(OH)_2]Na_{0.5} + 0.8\,K^+ + 0.3\,Al^{3+} + 1.2\,OH^-$$
$$\rightarrow [(Al_{1.5}Mg_{0.5})(Si_{3.7}Al_{0.3})O_{10}(OH)_2]K_{0.8} + 0.5\,Na^+ + 0.3\,H_4SiO_4 \quad (10.II)$$

Potassium is fixed in the interlayer space of illite. It may be supplied either by the decomposition of unstable potassium-bearing minerals or through pore water from sources outside the rock. The diagenesis also involves an increase in the negative charge of silicate unit layers either by the substitution of divalent for trivalent cations in the octahedral sheet, or, more likely, by the substitution of aluminum for silicon in the tetrahedral sheet (Towe, 1962). Such reactions require a source other than smectite to supply the cations involved. The specific mineralogic and chemical reactions that occur during the burial metamorphism of the sediment depend on the original composition of the mud and of the pore water.

Hower et al. (1976) studied the mechanism of burial metamorphism of a shale cutting from a well in Oligocene-Miocene sediment of the Gulf Coast of the United States to a depth of 5500 m. Such a sediment seems to be representative for this type of process. They observed the following depth-dependent mineralogic variations:

1. Calcite decreases in abundance with depth, disappearing from progressively coarser particle-size fractions.

2. Illite/smectite is the most abundant mineral in the shale. Its composition changes from a phase having less than 20% illite-like unit layers to one having 80% illite-like unit layers over the stratigraphic interval from 2000 to 3700 m, after which the percentage of illite-like layers remains constant. In the fine-size fraction of a given sample, the percentage of illite-like layers is inversely proportional to particle size.

3. Potassium feldspar is present in the shallow shale samples, decreasing in amount between 2500 and 3700 m, and below which it is absent.

4. Chlorite is absent in all samples found at depths shallower than 2500 m. It increases in concentration between about 2500 and 4300 m, below which its concentration remains constant.

5. Kaolinite content abruptly decreases below 3400 m.

6. Mica is present in all samples but is more abundant in the coarser fractions of the shale from the shallow depths.

Except for the $CaCO_3$ content, the original chemical composition of the shale in this stratigraphic section did not vary greatly throughout its depositional history. The effect of burial metamorphism was accompanied by the removal of CaO and CO_2 and by the loss of a small amount of sodium during the reduction in cation exchange capacity of the illite/smectite. Hower et al. found the < 100 nm fraction to be almost pure illite/smectite and to have a number of significant depth-dependent chemical variations. In this fraction Ca and Na, present as exchange ions, decrease with depth in accordance with the reduced cation exchange capacity of the clay when it is converted to illite. Mg, Fe, and Si, present in the smectite crystal samples from shallow depths, decrease with increasing depth. The loss in silicon is in accordance with (10.II) and characteristic for conversion of smectites into illites with the removal of tetrahedral silicon. The loss in magnesium and iron may indicate that there is a change in the chemical composition of the octahedral sheet in addition to the changes occurring in the tetrahedral sheet. Hower et al. suggest that the magnesium and iron lost from the fine fraction of illite/smectite probably take part in chlorite formation. The chlorite makes its appearance with increasing burial depth, being a part of the coarser fraction of the rock (see also Velde and Byström-Brusewitz, 1972).

As one would expect from (10.II), aluminum and potassium concentrations in the clay fraction increase with depth, the latter increasing strikingly from 2.0 to 5.5%, calculated as K_2O, which is parallel to the increase of the illite-like layers. On the basis of depth-dependent mineralogic changes, Hower et al. suggest that the source of potassium and aluminum required for the conversion of smectite to illite layers is potassium feldspar. This mineral is concentrated in the coarse-size fraction, which loses potassium and aluminum to the clay-size fraction.

According to Hower et al. the silicon that is released during the decomposition of feldspar forms quartz, as does Si released from the tetrahedral layers of the clay. This gives rise to increasing quartz content in the fine clay fraction with increasing depth of the shale. Quartz was not detected in the very fine fraction (< 100 nm). The mechanism of quartz formation in the sediment during burial is not known. Some of the released silica is transferred to the solute fraction of ground water (see Sect. 2.4, Ch. 6).

The main factor controlling the extent of diagenesis appears to be temperature, although other variables such as the length of time the shale has been buried may also be important. All diagenetic changes take place up to a depth of 3700 m. No significant changes take place between 3700 m and the bottom of the Gulf Coast (5776 m). The mineralogic changes that occur take place in different formations and are not associated with formation boundaries or age. After the illite/smectite has reached a content of about 80% of illite-like layers and potassium feldspar has disappeared, the assemblage remains constant over the temperature range of 95° to 175 °C. There is some evidence that at 200 °C the mineral assemblage in the shale may undergo a different type of recrystallization, approaching the lower greenschist-facies assemblage for pelitic rocks.

The temperature interval over which recrystallization of this type took place in the Beltian argillite has been determined to be between 200 ° and 300 °C (Eslinger and Savin, 1973).

Hower et al. came to the conclusion that the chemical and mineralogic evidence indicates that the smectite-to-illite diagenesis proceeds by the redistribution of potassium and aluminum present in the rock, rather than by their introduction from some outside source. The diagenesis may occur as long as the sources for these elements are present in the system. Weaver and Beck (1971) suggest that after complete decomposition of potassium feldspar, the potassium necessary for the conversion of smectite to illite is introduced into the shale from great depths.

According to Heling (1974) the rate of the illitization process depends not only on temperature but also on the potassium concentration in the pore solutions. Illitization in the shaly marls of the Hydrobienschichten in Germany occurred at a high rate. These argillaceous sediments are intercalated by numerous thin limestone, calcarenite, and lumachelle horizons. These porous intercalations give the Hydrobienschichten a high permeability which is preserved even under appreciable overburden. Apparently the good mobility of the pore solutions due to the high permeability contributes to a better potassium supply in this formation causing a greater reaction rate in the illitization process. Here, smectite is no longer present in samples that reached temperatures above 40 °C. In highly bituminous clayey sediments, where the mobility of pore water is poor, the alterations of smectite should be slower than those of nonbituminous clays.

Several mechanisms have been proposed to describe the diagenesis process and the conversion of smectite minerals into illite. Most of the mechanisms proposed presuppose that the aluminosilicate layers are preserved during the diagenetic process. Burst (1969) attributed the loss of expandibility to irreversible loss of interlayer water. However, the temperatures at which diagenesis occurs in nature are much lower than those in which irreversible dehydration occurs in the laboratory. According to Dunoyer de Segonzac (1970) the discrepancy between the experimental data and those from field observations is due to the fact that the reaction rate of the process is slow and equilibrium is not reached during the short time of the laboratory tests. Thus sufficient geologic time is necessary to complete the reaction at low potassium concentration and low temperatures.

According to Heling (1974) the alteration process from smectite to illite includes the incorporation of potassium ions, which are initially adsorbed in exchange positions and later, at higher temperatures and during dehydration, are fixed at interlayer positions.

Powers (1959) suggested that on deep burial magnesium migrates into the octahedral layer, where it replaces aluminum, which in turn displaces tetrahedral silicon, thus causing an overall increase in layer charge. Pollard (1971) proposed that in the first stage aluminum enters into the interlayer space of the smectite (by ion exchange mechanism) and from there into the hexagonal holes of the oxygen plane. Subsequently, under conditions of low grade metamorphism, temperatures become sufficiently high to permit "switching" of Si–O bonds to Al–O bonds, accompanied by rotation of the tetrahedra, geometric adjustment of the octahedral sheets, and migration of protons from OH groups to apical oxygen. The layers are regarded as activated complexes requiring a critical concentration of Al for the rotation to become irreversible.

Heller-Kallai (1975) suggested a different type of mechanism for illitization. She showed that heating montmorillonite with alkali halides causes changes in the structure of the clay layers. On the basis of their X-ray powder diffraction patterns, infrared spectra, cation exchange capacity, and chemical composition they could be regarded as illite/smectite interstratifications. She concluded that irreversible deprotonation of some of the network OH groups may occur by solid state thermal diffusion, in the presence of proton acceptors at the temperatures encountered in the early stages of metamorphism. Feldspars may serve as proton acceptors during this process. Structural formulas of minerals that may be regarded as illite/smectite, prepared from montmorillonite by thermal deprotonation in the presence of proton acceptors (KBr), are given in Table 10.1. They were calculated by the classical method of Ross and Hendricks (1945) based on 44 anionic charges and by a method based on the assumption that the composition of the layers remains unchanged except for some deprotonation. The conventional methods give chemical formulas for the heated montmorillonites that are very similar to those widely known in the literature for illites, with greater Al for Si substitution, which increases with the duration of the KBr heat treatment. This apparent increase is greater for Al-saturated samples but occurs also if there is no increase in the total Al content. She concluded that deprotonation, rather than increased Al for Si substitution, may be responsible for the fixation of K and for illitization in some natural occurrences.

It seems plausible that in general the diagenetic process and the conversion of smectite into illite occurs by recrystallization. This reaction may proceed as long as the sediment contains sufficient water for the sedimented particle to depolymerize and dissolve and then migrate to the site where the new phase is formed. Both the tetrahedral and octahedral sheets are subjected to this process (see Ch. 7 on the solubility of clay minerals). The new clay phase is affected by the chemical composition of the pore solution and may become poorer in certain elements as compared with the sedi-

Table 10.1. Structural formulae of Mg-montmorillonite (Wyoming bentonite) untreated and heated with KBr at 520 °C. (From Heller-Kallai, 1975)

Time of heating, h	Structural formula calculated on assumption of deprotonation	Structural formula calculated on basis of 44 anionic charges
0	$[Si_{7.69}Al_{0.31}]\ [Al_{3.07}Fe^{3+}_{0.36}$	
	$Fe^{2+}_{0.05}Mg_{0.51}]\ O_{20}(OH)_4Mg_{0.45}$	
1	$[Si_{7.69}Al_{0.31}]\ [Al_{3.07}Fe^{3+}_{0.40}$	$[Si_{7.63}Al_{0.37}]\ [Al_{2.95}Fe^{3+}_{0.40}$
	$Mg_{0.51}]\ O_{20.70}(OH)_{3.30}Mg_{0.34}K_{0.90}$	$Mg_{0.76}]\ O_{20}(OH)_4K_{0.84}$
72	$[Si_{7.69}Al_{0.31}]\ [Al_{3.07}Fe^{3+}_{0.40}$	$[Si_{7.40}Al_{0.60}]\ [Al_{2.67}Fe^{3+}_{0.38}$
	$Mg_{0.51}]\ O_{22.35}(OH)_{1.65}Mg_{0.42}K_{2.39}$	$Mg_{0.80}]\ O_{20}(OH)_4K_{2.10}$

mented clay mineral. The pore solutions are never highly supersaturated with respect to illite and the new phase is condensed by the mechanism of orientation.

As a result of recrystallization of the clay fraction, both the crystallinity of the illite mineral particles and the average particle size increase with the depth of burial and time of diagenesis, whereas the surface area and the cation exchange capacity of the clay fraction decrease. After long periods of diagenesis at elevated temperatures kaolinite disappears from the sediment, being converted into illite (Foscolos and Kodama, 1974), probably by recrystallization, whereby a great part of the aluminum that was originally located in octahedral sheets now appears in tetrahedral sheets.

If the pore solution does not contain sufficient K and Al, recrystallization may result only in coarsening of illite crystals and not in new phase illitization. Hower et al. (1976) observed, for instance, that at depths below 3700 m, where potassium feldspar was no longer present, illitization did not continue with increased burial, but small particles increased in size.

Hiltabrand et al. (1973) performed experiments on the diagenesis of argillaceous sediments in artificial seawater at temperatures of 200 °C and 100 °C. They found that the soluble silica concentration stabilized near the saturation values for quartz. Chlorite and illite were formed during diagenesis in the sediment. Small particles of montmorillonite (< 1000 nm fraction) were less stable than the coarser ones (1000–2000 nm fraction) and disappeared during diagenesis, but no change was observed with the coarser fraction. Kaolinite and feldspar in the original sediment were essentially destroyed during the experimental process. Quartz remained unchanged in the sediment. The content of amorphous material increased. Relative percentage of illite increased at high sediment/fluid ratios approaching those of natural sediments when the temperature was 100 °C, but barely increased at 200 °C, when the stability of the illite is lessened (Eslinger and Savin, 1973) and the amorphous phase becomes more abundant. Since the diagenetic changes do not occur if the sediments are kept at 100 °C in the absence of seawater, the preceding experiments support the suggestion that the diagenetic alteration of the smectites occurs by the mechanism of recrystallization.

3. Surface Chemistry of Solutes Flow Through Argillaceous Sediments

There are two major types of migration fluxes of soluble and colloid species in sediments—that on the solid particles and that in interstitial water. The behavior of chemical species in sediments during migration is determined by the following processes: (1) chemical interactions between the solid phase and pore solutions; (2) diffusional transport in pore water and on solid surfaces; and (3) advective transport by the moving waters and by suspended particles in the pore waters. According to Lerman (1977), temperature and porosity gradients have only a small effect on the distribution of the soluble species in interstitial waters. He also showed that molecular diffusion in interstitial waters and the chemical reactions of dissolution, precipitation, and adsorption are the main processes controlling the geochemistry of such ions as calcium, magnesium, and ammonium and of silica.

3.1 Nature of Pores in Shales

There is little information available on the size of grains and pores in compacted shale. Montmorillonite is present predominantly as extremely fine grains. As it is converted to mixed-layer clays and illite, it becomes more coarsely grained. However, such data are not useful for calculating porosity of the sediment, since quartz is a major component of most shales and its particles may on the average be larger than the clay particles, giving rise to disturbed compacted structures, similar to a "honeycomb" microstructure, with some larger pore sizes.

Cahen et al. (1965) found the diameter of pores in Sahara shales to be about 5 nm. Slabaugh and Stump (1964) showed that in a sample of recent sediments from the Pacific Ocean nearly all the pores had diameters less than 5 nm. The porosity of Pennsylvanian shales studied by Dickey (1975) ranged from 4–13%. Most of the flakes of clay minerals are nearly parallel, but some wrap around spherical grains of quartz. The pores seem to be long and flat and their dimensions are indefinite, apparently as much as 5000 nm in length and 100 nm in width. Permeability to fluid flow across the bedding is obviously very small. Porosity decreases with depth and with the age of the sediment at any given depth.

In fissile shales, where the degree of preferred orientation is very high, we may differentiate between "interlayer space" pores with negatively charged surfaces and pores limited by the broken-bond surfaces, the charge of which is determined by the specific sorption of ions. In acid media, from which Al ions are specifically sorbed, the pore has a positive charge and in alkaline or in organic-rich media, it acquires a negative charge. In more massive sediments a floc may be considered as the unit that builds the walls of some of the pores. The most likely shape of such a floc is a sphere, within which the platelets form the card-house or book-house structures and the liquid is immobile. The net charge of such a sphere is negative. During compaction this floc shrinks, but it preserves its spherical shape and the surface charge of the pores is negative. Highly compacted massive sediments contain, in addition to pores made from flocs, "interlayer space" pores and "broken-bond" pores.

Unless the walls of the pore are coated by organic hydrophobic matter, water normally wets the surfaces of the pore. According to Low (1959) the hydration boundary layer may extend as far as 10 nm from the surfaces. The degree of "immobilization" of porous water is greater near the pore walls and less in the center of the pore (Miller and Low, 1963). The hydration boundary layer may lose its structure and liquefy if a certain threshold pressure gradient is exceeded. Increasing temperature and pressure with depth of burial tends to reduce the amounts of immobile water, but the proportion of immobile pore water may increase with depth of burial. The *effective porosity* for solute flow is determined by that fraction of the bulk volume of the rock occupied by mobile-liquid water. If the pores of the shale have a diameter less than 10 nm, the water and solutes cannot flow unless the threshold pressure gradient is exceeded. *Permeability* is the capacity of the porous medium to transmit water liquid and is therefore proportional to the effective porosity.

3.2 Effect of Migrating Water on the Migration of Solutes

Migrating water in argillaceous sediments serves as the chief carrier for inorganic and organic solutes. The higher the solubility of the solute, the higher will be the rate of migration. The adsorbing clay tends to lower the rate of migration. The chemical reaction occurring during this process may be formulated for uncharged solutes as follows:

$$\begin{aligned} &\mathrm{Clay}^1 \cdot \mathrm{Sol}_{(S)} + (x+y)\,H_2O_{(L)} \underset{u_1'}{\overset{u_1}{\rightleftharpoons}} \mathrm{Clay}^1 \cdot (H_2O)_{y(S)} + \mathrm{Sol} \cdot (H_2O)_{x(L)} \\ &\mathrm{Clay}^2 \cdot (H_2O)_{z(S)} + \mathrm{Sol} \cdot (H_2O)_{x(L)} \underset{u_2}{\overset{u_2'}{\rightleftharpoons}} \mathrm{Clay}^2 \cdot \mathrm{Sol}_{(S)} + (x+z)\,H_2O_{(L)}, \end{aligned} \qquad (10.III)$$

where Sol denotes an uncharged solute, (L) and (S) denote liquid and solid phases respectively, Clay^1 denotes a solid phase located below Clay^2, and u_1, u_1', u_2 and u_2' are the reaction rates.

If the solute is charged, its migration is associated with ion exchange reactions and hydrolysis as follows:

$$\begin{aligned} &\mathrm{Clay}^1 \cdot A_{(S)} + B_{(L)} \underset{u_1'}{\overset{u_1}{\rightleftharpoons}} \mathrm{Clay}^1 \cdot B_{(S)} + A_{(L)} \\ &\mathrm{Clay}^2 \cdot C_{(S)} + A_{(L)} \underset{u_2}{\overset{u_2'}{\rightleftharpoons}} \mathrm{Clay}^2 \cdot A_{(S)} + C_{(L)}, \end{aligned} \qquad (10.IV)$$

and

$$\begin{aligned} &\mathrm{Clay}^1 \cdot A_{(S)} + H_2O_{(L)} \underset{u_1'}{\overset{u_1}{\rightleftharpoons}} \mathrm{Clay}^1 \cdot H_{(S)} + A \cdot OH_{(L)} \\ &\mathrm{Clay}^2 \cdot H_{(S)} + A \cdot OH_{(L)} \underset{u_2}{\overset{u_2'}{\rightleftharpoons}} \mathrm{Clay}^2 \cdot A_{(S)} + H_2O_{(L)}, \end{aligned} \qquad (10.V)$$

where A, B, and C are cationic species and H is a proton.

With increase of u_1 and u_2 and decrease of u_1' and u_2' the rate of the migration of Sol in (10.III) and of A in (10.IV) and (10.V) increases. These values are characteristic for the various soluble species. It follows that the sediment may restrict and even prevent the passage of some components while allowing relatively unrestricted flow of others. The process is known as *ultrafiltration* and the sediment behaves as a *semipermeable membrane.*

The rate of migration of solutes also depends on the properties of the sediment. With decreasing size of pore diameter, there is a higher probability for collisions and interactions between solutes and walls of the pores. The types of minerals and their surface charge densities influence the process of ultrafiltration. Migration of ions may be restricted in the presence of minerals on which they may become fixed.

As stated previously the restriction of the flow resulted from specific short-range interactions between solutes and the walls of the pore. When the pores are fine so that the double layers from adjacent walls overlap, Donnan exclusions of co-ions from the porous medium results in the lowering of their concentrations in the pore solution compared to their concentrations in the external solution. Electroneutrality should be preserved during all the stages of the ultrafiltration, and the migration of a salt through a membrane requires that both components–the cation and the anion–enter and leave the pore together. Since the penetration of the co-ion into the pore is restricted, the penetration of counterions, in quantities higher than those necessary to neutralize the charge of the pore walls, is also restricted. The system will slow down the migration of salts.

3.2.1 Ultrafiltration of Salt Solutions

This mechanism has been employed to explain the formation of oilfield brines (Hitchon et al., 1971) and geothermal brines (Berry, 1969). Soils also behave as semipermeable membranes.

Mokady and Low (1968) observed that the flux of H_2O in clay membranes is about 500 times greater than the flux of NaCl. Hanshaw and Coplen (1973) studied sodium ion exclusion by compacted clay membranes and showed that the exclusion depends upon excess negative charge of the clay mineral. The highly charged clay pores retard the passage of ionic species more effectively than the weakly charged clay pores. Ion exclusion becomes more effective with decreasing porosity of the membrane. Ultrafiltration is highly effective when the solutions are very dilute, becoming less effective with increase in concentration. Hanshaw and Coplen attribute this to the fact that when the concentration of the input solution is high and the diffuse double layer at the walls of the pore is compressed, the concentration of anions admitted into the clay pores is also high, and therefore the ultrafiltrate concentration approaches the residual solution concentration.

Kharaka et al. (1973, 1976) showed that the efficiency of a given membrane increases with increasing compaction pressure but decreases slightly at higher temperatures for solutions of the same ionic concentrations. The retardation sequence for monovalent cations at room temperature is $Li < Na < K < Rb < Cs$ and for divalent cations, $Mg < Ca < Sr < Ba$. At high flow rates, monovalent cations (except for Li) are generally retarded with respect to divalent cations. With decreasing flow rates the membrane efficiency for monovalent cations decreases and the opposite occurs with divalent cations. At field flow rate Ca and other divalent cations are retarded in comparison with Na and other monovalent cations. Field data actually show Ca hyperfiltration as related to Na.

3.2.2 Migration of Organic Polar Molecules

The reaction rates, u_1 and u_2, of Eq. (10.III) depend on the availability of water molecules to be exchanged for all functional groups in the organic molecule that form bonds with various sites in the clay, either as electron donors or acceptors. These rates depend on the site of the functional group, namely, whether it is at the edge of or within the molecule. The probability of exchanging all functional groups for water molecules decreases with increasing number of functional groups, aromatic character, and molecular weight. Migration of macromolecules such as humic substances and kerogen, which are found in sediments and are rich in polar groups and aromatic rings, is very slight, despite their hydrophilic character.

Amphipathic molecules, such as fatty acids, may join migrating water in the sediment. The sorption energy increases with increasing surface concentration of the organic material (see Sect. 2.1.1, Ch. 7). It follows that u_1'/u_1 and u_2'/u_2 of Eq. (10.III) increase with increasing interface concentration of the organic material. Migrating water will carry these organic compounds from pores where their surface density is low to pores where their surface density is high. This allows the complete saturation of a pore with organic ions or polar molecules. Such a pore is called "organopore". In reality the number of these pores out of the total number of pores in a source rock is extremely small. The distribution of aqueous soluble organic material in a clay rock will not be homogeneous. This material tends to be concentrated along the lenghts of some of the pores. At a certain stage during compaction of the rock, the formation of a continuous network of organopores is possible (Yariv, 1976).

The life span of an organopore depends on the supply of suitable organic material. When this ceases, the amphipathic molecules may be desorbed from the clay surface and join the migrating water. The chemical stability of these organic compounds, which tend to decompose, giving hydrocarbons, also determines the life span of the organopore.

3.2.3 Migration of Nonpolar Molecules (Primary Migration of Petroleum)

The aqueous solubility of hydrocarbons is very low and decreases with increasing chain lengths. The solubility of *n*-octane and heavier paraffins is less than 1 mg/l water. The naphthenes are more soluble, and the aromatics even more, but never exceed 0.2% (McAuliffe, 1969). Increasing the temperature increases the solubility of hydrocarbons in water. *n*-Octane, for example, has a solubility of 0.431 mg/l at 25 °C, 1.12 at 99 °C, and 11.8 mg at 149 °C (Price, 1973). The presence of salt in the water greatly decreases the solubility of hydrocarbons. This effect, which is known as the "salting out" effect, is the result of the breakage of the water hydrophobic hydration structures due to the formation of stable hydrates by the ionic components of the salt (Frank and Wen, 1957).

The migration of petroleum-like paraffins as an aqueous solute takes place only to a very small extent. On the basis of initial water content of a sediment and the quantity of generated petroleum, Dickey (1975) estimated that petroleum migration via aqueous dissolution is possible only with an improbable petroleum concentration of 12000 or more mg hydrocarbons per liter water. Hydrocarbons are generated in buried sediments on the surfaces of clays (Shimoyama and Johns, 1971). Isolated hydrocarbons are extracted by upwardly migrating water. As a result of their diffusive flow and due to migration of soluble hydrocarbons near newly formed hydrocarbons, regions supersaturated with hydrocarbons are obtained in the porous water, giving rise to their separation as droplets. Surface tension of petroleum-like paraffins is lower than that of water. According to Ross (1973), in the absence of specific interactions or electric effects at the surface of the pore, the more stable liquid-in-liquid films are those formed by the liquid having the lower surface tension, which surrounds and separates droplets of the liquid of greater surface tension. That is, a system containing water drops in paraffin is more stable than one containing paraffin drops in water. When two droplets of dispersed paraffins are near each other, molecules of the intervening water are simultaneously pulled by the homogeneous water–water attraction forces emanating from the water medium. Since the heterogeneous parrafin–water attraction is weaker than the homogeneous water–water attraction, the intervening liquid will retract and the two droplets of paraffin will coalesce (pages 111–112).

The driving force for migration of drops of hydrocarbon through a porous rock is $\Delta P = P_2 - P_1$ where P_1 and P_2 are the hydrostatic liquid pressures at two horizontal parallel planes, the distance between which is d. The rate of flow of the hydrocarbons across the rock is

$$\bar{U} = (\mu_2 - \mu_1)/\bar{r} = \bar{v}\Delta P/\bar{r}, \tag{10.1}$$

where $\bar{v}$ is the molar volume of the hydrocarbon, μ is the appropriate chemical potential, and $\bar{r}$ is the resistance to flow. The major resistance to flow is offered by the rock, and $\bar{r} = d/D_{\text{eff}}$ where D_{eff} is a modified diffusion coefficient for the hydrocarbons in the rock analogous to the Fick diffusion coefficient. D_{eff} relates flow rate to chemical potential gradient or to pressure gradient, just as the Fick diffusion coefficient relates flow rate to concentration gradient (see also Eqs. 2.26 and 5.1).

D_{eff} is conventionally related to a constant D_o, which may be considered to be a modified form of the bulk diffusivity in a nondisturbed environment by

$$D_{\text{eff}} = D_o\, \Theta K_p\, K_r/\tau, \tag{10.2}$$

where Θ is the pore volume fraction, τ is the tortuosity, K_p is the equilibrium partition coefficient (that is, the ratio of concentration inside the pore to concentration outside the pore at equilibrium), and K_r is the fractional reduction in diffusivity within the pore that results from the diameters of the droplets and those of the pores being of comparable size.

The concept of an equilibrium partition coefficient was originally introduced to describe the geometric exclusion effect of rigid molecules. It is also applicable to the geometry of drops. A chemical interaction may occur between the surface of the pore and molecules in the bulk of the drop or at the thin film surrounding the drop, leading to a change in the size of the drop. Consequently, K_p is also dependent on the chemical properties of the pore surface. If hydrocarbon drops enter into the field of polar surfaces of a nonorganic pore, then the films that surround these drops will become unstable and drops will coalesce. Since their size increases, there will be more geometric disturbances and $\bar{U}$ will decrease. If there is a strong chemical interaction between the paraffin molecules and the surface of the pore, the water in the surrounding film will be repelled when a hydrocarbon drop approaches this surface. The drop of paraffin will disintegrate and in this case, the values of D_o, K_r, and K_p will depend on the size of the individual hydrocarbon molecules and not on the size of the whole drop. Consequently $\bar{U}$ will greatly increase. This in fact occurs in the organopores. When such pores are saturated with hydrocarbons, water can be present only in the form of droplets surrounded by hydrocarbons. The migration rate of water in such pores is very low since D_o, K_r, and K_p will depend on the size of the water drops and not on the size of the water molecule.

It may be concluded that pores with organophilic surfaces are hydrocarbon permeable and water impermeable while pores with hydrophilic surfaces are the reverse.

4. Diagenesis of Organic Matter and Oil Generation in Argillaceous Sediments

Animal and vegetable organic matter accumulated in argillaceous sediments or fine-grained carbonate rocks is the principal source material for petroleum in general. Analyses of recent sediments have shown that the disseminated organic detritus is largely in the form of insoluble kerogen, together with smaller amounts of soluble nitrogen, sulfur, and oxygen compounds. Most ancient rocks, on the other hand, contain appreciable quantities of low-molecular-weight hydrocarbons.

Progressive heating of sedimentary rocks in the absence of oxygen causes a sequential two-stage release of organic compounds (Claypool and Reed, 1976). Material released in the 30–400 °C heating interval consists of volatile compounds present as such in the rock. This material can be extracted from the rock by organic nonpolar solvents. It is composed mainly of hydrocarbons present in the rock together with some other low-molecular-weight organic compounds. This group of materials is not present in recent sediments (Barker, 1974a). At higher temperatures (400–800 °C), nonvolatile organic matter, such as kerogen, thermally decomposes to yield volatile compounds containing both saturated and aromatic hydrocarbons (McIver, 1967). The residual organic matter forms graphite, mineral charcoal, and other substances incapable of yielding hydrocarbons.

The temperature at which the nonvolatile organic matter thermally decomposes depends on the geologic history and maturation of the rock and on the type of organic matter. Vitrinites separated from coals of various maturation ranks give a maximum yield of volatile compounds at temperatures ranging from 422 °C at 68.9% C to 660 °C at 92.7% C. Lignites decompose at lower temperatures than vitrinites whereas anthracites decompose at higher temperatures.

Bordenave et al. (1970) distinguished between continental and marine organic matter in sediments. The first group gave products rich in aromatics whereas the latter produced more paraffinic hydrocarbons during low-temperature pyrolysis. A sediment deposited in a marine environment incorporates mainly plankton from which the predominant decomposition products are various proteins, carbohydrates, and lipids. Deposition in nonmarine environments is associated with detritus of higher plants from which the predominant decomposition products are made up of lignins, cellulose, and skleroproteins. The original marine organic matter has a higher H/C ratio and is more paraffinic whereas the original nonmarine organic matter has a lower H/C ratio and is more aromatic.

The group of organisms of greatest importance as the source of petroleum precursors is phytoplankton. Coalification, on the other hand, occurs mainly when original plant material becomes buried. The conditions favorable for preservation of source material and petroleum precursors are (1) a reducing environment; (2) the absence of destructive organisms; and (3) active deposition of fine-grained sediments (Hedberg, 1964). The original matter from which petroleum was generated consisted of fundamental organic compounds such as proteins, fats, waxes, and their decomposition compounds, and also humus and the like. Clay minerals collect organic material of the different types by adsorption, either on the land or in the oceans, carrying this material into the sediment. Powers (1967) proposed that the organic fraction of an argillaceous bed must have been deposited originally as an organic complex of smectite-type minerals, since this mineral group has the highest adsorption capacity for organic material.

Stevens et al. (1956) pointed out the extreme difference in composition of the normal C_{22}–C_{34} hydrocarbons from recent sediments on the one hand and of those from crude oils and ancient sediments on the other hand. They observed that in recent sediments normal paraffin molecules with an odd number of carbon atoms strongly predominate over those with even numbers, whereas in several ancient sediments and in crude oils there are approximately equal concentrations of molecules with odd and even numbers of carbon (Fig. 10.2).

The explanation for the preponderance of odd-carbon number hydrocarbons in recent sediments is that normal paraffins synthesized by many living organisms contain in the main only normal hydrocarbons with odd carbon numbers. Because of their biogenous origin, the normal hydrocarbons from recent sediments, which are a mixture derived from many species of living organisms, also contain mostly odd-carbon hydrocarbons. The fact that the normal hydrocarbons from many ancient sediments and from crude oils are so very different from recent-sediment hydrocarbons and exhibit smooth carbon-number distribution curves is due primarily to the diagenesis of the organic material and the oil generation process occurring in the sediment (Philippi, 1965, 1975).

Fatty acids are abundant in organisms and are structurally similar to hydrocarbons. Decarboxylation of fatty acids results in the formation of hydrocarbons. Most of the free fatty acids of organisms disappear under reducing conditions, soon after the deposition of the incorporating muds. Only about 0.1% of the organic matter in recent sediments consists of fatty acids, and it was suggested that they are transformed to kerogen, the insoluble organic fraction of the sediment (Abelson, 1967). Terpenoids and alcohols, which are also abundant among plants and animals and are also petroleum precursors (Mair, 1964), may also be transformed to kerogen at a certain stage of the diagenesis.

The formation of oil in argillaceous sediments seems to be associated with the transition of the oil precursors from the organic to the inorganic fraction of the sediment. Larger molecules, being more organiphilic than smaller ones, are sorbed by the kerogen to a greater extent than the latter, both chemically and physically. At high temperatures, thermal desorption and thermal decomposition of the precursors, which may be catalyzed by the inorganic fraction (Henderson et al., 1968), result in the formation of petroleum.

Thermal alteration of organic matter in sediments results from temperature rise due to burial. The general mechanism suggests thermal cracking of organic compounds as an essential part of petroleum formation in addition to hydrogen transfer and polymerization reactions (Philippi, 1975). From thermodynamic considerations one would expect that the final products of the thermal diagenetic process would be short-chain hydrocarbons (methane) and graphite. However, activation energies are very high and the generation process commonly is incomplete. The generation of hydrocarbons is the result of the combined influence of temperature, catalytic effect of the mineral fraction, and time. In horizontal space, overly brief diagenesis is the reason for the absence of commercially usable hydrocarbons in the youngest rocks, whereas over-extensive diagenesis rather than geologic age limits the hydrocarbon content of ancient rocks (Landes, 1967).

The product of diagenesis lie within a wide range of solid and liquid petroleums and a range of gaseous hydrocarbons on the one hand and coal and matured carbon-rich kerogen on the other hand. Natural gas varies greatly in composition. Unprocessed gases contain 60–80% methane (CH_4), 5–9% ethane (C_2H_6), 3–18% propane (C_3H_8), and 2–14% higher hydrocarbons, mainly butane (C_4H_{10}). These low-molecular-weight aliphatic hydrocarbons have low boiling points. They are soluble in water and mobile to a higher extent than hydrocarbons with longer chains. Boiling and melting points and organophilic properties of the hydrocarbons increase with increasing chain length.

Over a certain time and depth span (more than about 1000 m), the hydrocarbon content of an argillaceous sediment remains low. Then, beyond a certain threshold, new hydrocarbons are generated, first gradually, then rapidly, by thermal breakdown of the insoluble kerogen. These hydrocarbons have no characteristic structure distribution pattern, and the original "biogenic hydrocarbons" are diluted progressively by the newly formed species. This is the principal stage of oil formation. According to Cordell (1972) the shallowest depth for significant oil formation should be at 1500 m, at which depth the temperature is about 65 °C. At depths greater than 3000 m the temperature is likely to be very high and the decomposition reactions of the organic matter are consequently very fast. Cracking of carbon–carbon bonds increases and light gaseous hydrocarbons are generated from kerogen and from preexisting high-molecular-weight hydrocarbons as well, the latter also forming coal (Eq. (10.I), Fig. 10.2).

Harrison (1976) showed that fatty acids are released from the kerogen matrix at 160 °C, at rates exceeding their degradation rate. These liberated acids could migrate in the argillaceous sediment, wherein final degradation, mainly decarboxylation, could occur. The presence of varying amounts of fatty acids in oil reservoir waters has been shown by Cooper and Bray (1963), and Willey et al. (1975), who were unable to detect fatty acids in fresh-water samples. Harrison suggested that the general scheme of fatty acid diagenesis includes: (1) early decarboxylation of free and lightly bound fatty acids; (2) cleavage of bound fatty acids from the kerogen matrix, and (3) final decarboxylation of the acids that were released from the kerogen at lower temperatures according to

$$R - CH_2 - COOH \rightarrow R - CH_3 + CO_2. \qquad (10.VI)$$

The aqueous soluble organic compounds, such as fatty acids, which are present in the sediment from the time of sedimentation, or after they have been released from the kerogen, join the migrating water and may be sorbed by clay minerals. It is likely that they will be concentrated on the surfaces of only some of the pores (Yariv, 1976). Pores may thereby become permeable to hydrocarbons (organopores). The formation of continuous networks of organopores in an oil source bed is likely. The time required to obtain organopores depends on the supply and rate of migration of water-soluble polar organic molecules. Since these molecules are strongly bound primarily to the kerogen, their extraction requires large amounts of water. Very dilute aqueous solutions are obtained and a large amount of water will pass through the rock before adsorption by the pore is high enough to render it organophilic. This process, together with the decomposition of the organic material and the formation of water-soluble ions or polar molecules, requires long periods.

Hydrocarbons generated at lower depths migrate through these organopores in the direction of the pressure gradient. This is known as "primary migration" of oil, and it is the process of expulsion of petroleum from the source rock to the permeable rock. Migrating water is the driving force for this process, the mechanism of which was discussed in the previous section (pages 416–418).

With further sedimentation above the source rock, the temperature of the bed having the organopores becomes sufficiently high for the fatty acids and other polar molecules, which are sorbed on the clay minerals and form the walls of the organopores, to decompose to hydrocarbons. The lowest edge of the organopore decomposes first and the newly formed hydrocarbons rise through the remainder of the organopore.

Crude oils are seldom found reservoired in shales, and more than half of the world's known recoverable reserves are in sandstones, even though these sands are deposited in environments that are not conducive to the preservation of organic matter. Hydrocarbons are generated in shales that are named "source rocks" and from there they migrate by the mechanism of primary migration into the coarser rocks. The coarse rock is permeable and the hydrocarbons continue to migrate in this rock until they reach the reservoir. The migration that takes place in the coarser permeable rocks is known as the *secondary migration.*

Rocks that generate hydrocarbons are "potential source rocks" and can only be classed as "source rocks" after commercially exploitable quantities of petroleum have migrated into reservoirs. Large volumes of undercompacted, abnormally pressured shale are important to the primary migration process because they can supply the large volumes of water to the two stages of primary migration: (1) raising the petroleum precursors that are necessary for the formation of the organopores after their desorption from the kerogen and (2) raising the hydrocarbons by the hydrostatic pressure developed, after their thermal generation (Evans et al., 1975).

The hydrocarbons that are found in shales were generated from the adsorbed petroleum precursors that were isolated from the kerogen during the diagenesis process and were not able to migrate from the shale because of not reaching any organopore.

4.1 Petroleum Origin Related to Kerogen

Kerogen is the most abundant reservoir of organic carbon on earth, and it has been estimated that there are 3×10^{15} tons of kerogen, compared with 5×10^{12} tons of coal in the earth's crust (Forsman, 1963). The precise structure of this organic material is unclear. Kerogens from various samples have different structures depending on the types of organic material from which they were generated and on the variations in the environmental conditions they experienced during and after their formation. The general structure usually proposed for kerogen is similar to that of coal but kerogen is richer in aliphatic groups. It is made up of cyclic polycondensed nuclei, bearing alkyl chains and functional groups and linked by heteroatomic bonds, containing in particular oxygen.

Kerogen is formed by condensation or polymerization of microbially altered products originating from carbohydrates, lipids, proteins, and other organic materials

(McIver, 1967). Shimoyama and Ponnamperuma (1975) showed that organic soluble material, such as lipids and fatty acids, is converted into insoluble humic substances during the early stages of diagenesis of the sediments. Philp and Calvin (1976) showed that kerogen-like material can be isolated from recently deposited algal mats. On oxidative degradation this material gives rise to products similar to those obtained by degradation of ancient kerogens known to be of algal origin and they concluded that kerogen-like precursors or protokerogens are formed very rapidly on a geologic time scale and are found in recent sediments rich in organic matter.

The following mechanisms, which may occur simultaneously, have been suggested for the formation of kerogen: (1) intracellular chlorophyll becomes grafted onto cellular macromolecules, and with increasing maturation, finally gives rise to kerogen (Oehler et al., 1974); (2) insoluble residues, which are cellular macromolecules, are the basic building blocks for kerogen, and with the passage of time additional soluble lipid material can join this basic framework by irreversible chemical reactions, and (3) condensation of the polymeric debris gives rise to the kerogen matrix (Philp and Calvin, 1976).

The burial of sediments at depths which result in a temperature and pressure increase induces progressive transformation of kerogen. Experiments have shown that the initial thermal decomposition product of kerogen is bitumen, which is the organic fraction that is extractable with organic solvents and richer in aliphatic nonpolar components than kerogen, and is subsequently decomposed to oil or tar. The bitumen components consist of polymeric nitrogen–sulfur–oxygen compounds and asphaltic hydrocarbons (Forsman, 1963). They represent intermediate products between kerogen and hydrocarbons (Louis and Tissot, 1967).

As the burial period and the temperature increase, heteroatomic bonds are broken successively, starting with some labile carbonyl and carboxyl groups. Oxygen is removed as CO_2 and H_2O. As the temperature continues to increase, hydrocarbons, particularly of the aliphatic type, are freed, This is the principal stage of oil formation. In the final stage, breaking of C–C bonds becomes very important and cracking generates gas from both kerogen and preexisting petroleum. During the transformation the carbon concentration increases in the residual kerogen, and it becomes more condensed (Tissot et al., 1974).

The successive steps of increasing thermal evolution can be described by an "evolution path" (Fig. 10.3) in which the atomic ratio H/C is plotted against the atomic ratio O/C. Progressive evolution results first in a partial elimination of oxygen, marked by a decrease of the O/C ratio, due to the progressive elimination of the C=O groups. In the following stage, elimination of hydrogen, especially in the form of hydrocarbons, causes the H/C ratio to decrease (Tissot et al., 1974).

Kerogens of marine origin, being rich in aliphatic structures, generate abundant amounts of oil. They originally have a high H/C ratio and a relatively low O/C ratio. Kerogens with a predominantly aromatic structure accumulate in nonmarine environments. They have a high original oxygen content, with an O/C ratio that may reach 0.2 or 0.3 in some samples, and a relatively low H/C ratio. They generate abundant CO_2 and HOH and produce methane and coal at great depths, but little or no oil (Tissot et al., 1974). Many of the known kerogens contain both aliphatic and aromatic structures.

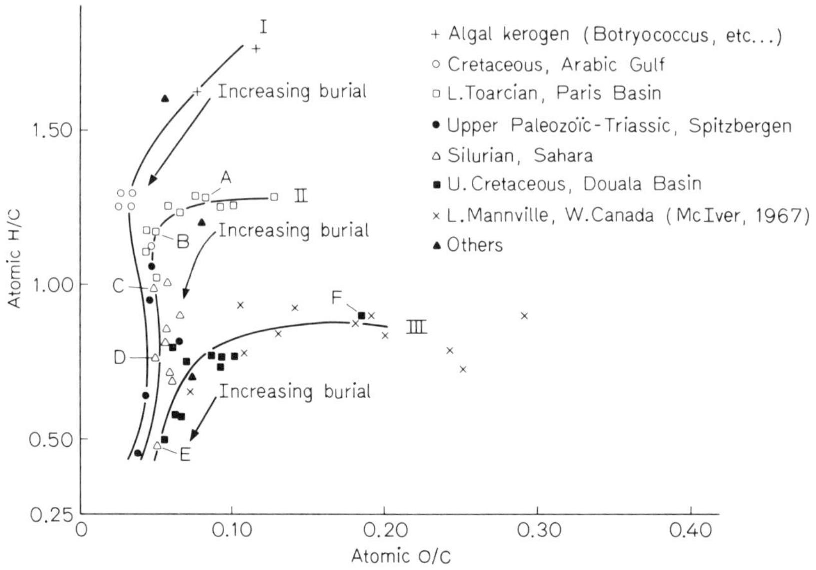

Fig. 10.3. Examples of kerogen evolution paths. Path I includes algal kerogen and excellent source rocks. Path II includes good source rocks. Path III corresponds to less oil-productive organic matter, but may include gas source rocks. Evolution of kerogen composition with depth is marked by an arrow along each particular path. (From Tissot et al., 1974: Used with permission of the publisher)

4.2 Generation of Hydrocarbons

Reactions of several types, which may occur simultaneously during the diagenesis of organic material, are responsible for the formation of hydrocarbons in sediments. It appears that clay minerals have a great catalytic effect on the reactions.

At and just below the depositional surface, the organic matter of argillaceous sediments commonly undergoes vigorous attack by bacteria. In the upper few centimeters the main action is generally caused by aerobic bacteria, but on deeper burial these are replaced by anaerobic ones (Hedberg, 1974). Boon et al. (1975) suggest that algal fatty acids are rapidly resynthesized into bacterial fatty acids before their incorporation into sedimentary organic matter. This therefore has a great effect on the molecular weight spectrum of hydrocarbons developed during deep burial. The molecular weight spectrum of deeply buried fatty acids corresponds to that of bacterial fatty acids rather than to that of algal acids.

Almon and Johns (1975) showed that the 2:1 clay minerals, such as illite, talc, and smectites, promote the decarboxylation of fatty acids to alkanes containing one carbon atom less than the parent acid in addition to generating both shorter and longer alkane chains as well as branched alkanes and other products. The 1:1 clay minerals promote the reaction in much the same manner. The presence of water has a profound effect on the distribution of the isomers. In the presence of water the branching ratio (branched alkanes/normal alkanes) is of the order of 0.1 while in anhydrous system the branching ratio is 4.5.

The catalytic effect of smectites on the thermal behavoir of various alcohols was studied by Galwey (1972). The catalytic processes were largely independent of the particular organic alcohol. The qualitative compositions of the product mixtures were very similar, though the yields varied according to the reactions that took place. In addition to thermal dehydroxylation, cracking of the carbon chain and isomerization also occurred. The thermal reactions yielded mixtures of alkanes in which 2- and 3-methylpentane and 2- and 3-methylhexane were major products but *n*-alkanes were also formed. A surface reaction between ethanol and *n*-dodecanol resulted in the formation of *iso*-hexane, a complex surface process that involves cracking, polymerization, skeletal isomerization, and hydrogen transfer processes. The compositions of the product mixtures obtained from reactions of the hydrocarbons were similar to those obtained from the alcohols.

Large amounts of gaseous hydrocarbons are formed from organic matter during early diagenesis, probably by biochemical processes. These hydrocarbons are derived more or less directly from the organisms incorporated in the sediment. Some are incorporated in the sediment without any change, such as *n*-alkanes from higher plants or from algae, others such as steroids or pentacyclic triterpenoids derive from biogenic molecules through an early transformation that does not affect the hydrocarbon skeleton, but only entails the loss of a functional group. There is no supporting evidence for the catalytic conversion of these gas molecules to oil. Cordell (1974) stated that the countless small gas accumulations, and even a few of the largest, originate from organic matter during relatively early diagenetic transformations. Liquid long-chain hydrocarbons and potential precursors, such as soaps and fatty acids, are relatively immobile at shallow depths, since they have a strong affinity for, and are sorbed by kerogen.

Concentrations of aliphatic saturated and aromatic hydrocarbons from total organic carbon, up to a depth of 4000 m in the Douala Basin, near the Niger River (Cameroon), was studied by Albrecht et al. (1976). The original organic matter is generally the same in composition and content all along the sequence of sedimentation (Durand and Espitalié, 1976), and the observed changes are to be attributed to the effect of burial. The evolution of the hydrocarbons according to depth is shown in Figure 10.4. The analyses give the amounts of hydrocarbons that have not yet migrated from the rock. Three distinctive zones can be observed:

1. At depths less than 1200 m the concentration of hydrocarbons is very low (1–3 mg/g of organic carbon). This represents a very small portion of the total organic matter in that horizon and according to Philippi (1965) this is characteristic for recent or immature sediments.

2. Between 1200 and 2200 m the concentration of hydrocarbons increases rapidly to a maximum of 50 mg/g of organic carbon. The bulk of these alkanes is formed by loss of functional groups from polar molecules and by genesis from insoluble kerogen. At this depth the temperature at which this hydrocarbon generation step is discerned is 65–70 °C. In the much younger Los Angeles Basin, this step is discerned at a greater depth with a temperature of 115 °C (Philippi, 1965), and in the much older formation of the Paris Basin (Louis and Tissot, 1967), at a lesser depth with 60 °C.

3. Below 2200 m the concentration of hydrocarbons decreases rapidly, and it reaches very low values below 3000 m. The evolution is mainly due to cracking of the long chains leading to light hydrocarbons and gas. Aromatic hydrocarbons display a

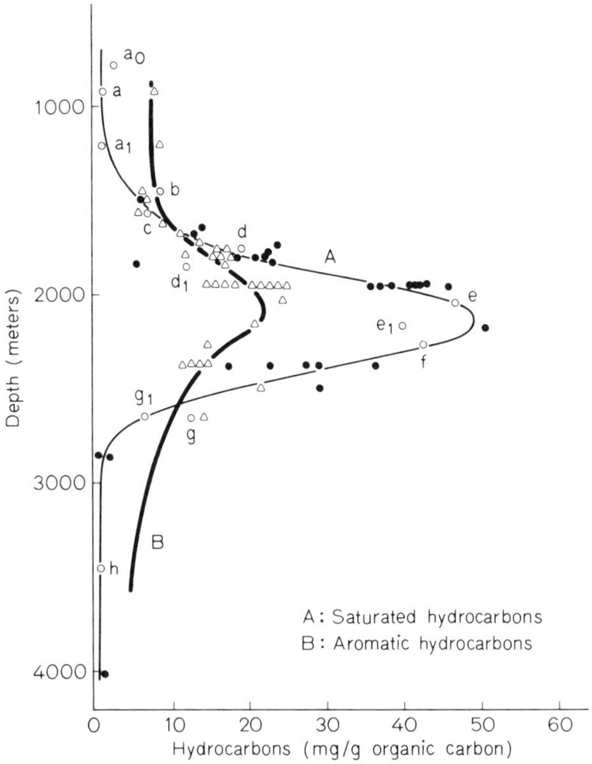

Fig. 10.4. Evolution of hydrocarbons with burial in Logbaba series, Douala Basin (Cameroon). Curve A, saturated hydrocarbons, curve B, aromatic hydrocarbons. (From Albrecht et al., 1976: Used with permission of the publisher)

similar distribution all along the series, with a predominance of purely aromatic structures with short side chains. They are generated during maturation to a lesser degree than the aliphatic hydrocarbons, but they disappear less rapidly with depth, due to their greater stability. This sequence is characteristic for hydrocarbon generation in detrital sediments.

References

Abelson, P, H.: Conversion of biochemicals to kerogen and *n*-paraffins. In: Researches in geochemistry. New York: John Wiley & Sons, 1967, Vol. 2, pp. 63–86

Albrecht, P., Vandenbroucke, M., Mandengué, M.: Geochemical studies on the organic matter from the Douala Basin (Cameroon)–I. Evolution of the extractable organic matter and the formation of petroleum. Geochim. Cosmochim. Acta *40,* 791–799 (1976)

Almon, W. R., Johns, W. D.: Petroleum forming reactions: Clay catalyzed fatty acid decarboxylation. Proc. Int. Clay Conf., Mexico, Bailey, S. W. (ed.). Wilmette, IL.: Applied Publishing Ltd., 1975, pp. 399–409

Appelo, C. A. J.: Chemistry of water expelled from compacting clay layers: A model based on Donnan equilibrium. Chem. Geol. *19,* 91–98 (1977)

Aylmore, L. A. G., Quirk, J. P.: Domain or turbostratic structure of clays. Nature (London) *187*, 1046–1048 (1960)
Baldwin, B.: Ways of deciphering compacted sediments. J. Sediment. Petrol. *41*, 293–301 (1971)
Barker, C.: Pyrolysis techniques for source-rock evaluation. Am. Assoc. Petrol. Geol. Bull. *58*, 2349–2361 (1974a)
Barker, C.: Programmed-temperature pyrolysis of vitrinites of various rank. Fuel *53*, 176–177 (1974b)
Berry, F. A. F.: Relative factors influencing membrane filtration effects in geologic environments. Chem. Geol. *4*, 295–301 (1969)
Blackmore, A. V., Miller, R. D.: Tactoid size and osmotic swelling in calcium montmorillonite. Soil Sci. Soc. Am. Proc. *25*, 169–173 (1961)
Bolt, G. H.: Physico-chemical analysis of the compressibility of pure clays. Geotechnique *6*, 86–93 (1956)
Bolt, G. H.: The pressure filtrate of colloidal suspensions. I. Theoretical considerations. Kolloid-Z. *175*, 33–39 (1961)
Boon, J. J., de Leeuw, J. W., Schenck, P. A.: Organic geochemistry of Walvis Bay diatomaceous ooze–I. Occurrence and significance of the fatty acids. Geochim. Cosmochim. Acta *39*, 1559–1565 (1975)
Bordenave, M., Combaz, A., Giraud, A.: Influence of the origin of organic matter and of its degree of evolution on the pyrolysis products of kerogen. In: Adv. Organic Geochem. 3rd Int. Cong. Proc. 389–405 (1970)
Burst, J. F.: Diagenesis of Gulf Coast clayey sediments and its possible relation to petroleum migration. Am. Assoc. Petrol. Geol. Bull. *53*, 73–93 (1969)
Cahen, R. M., Marechal, J. E., della Faille, M. P., Fripiat, J. J.: Pore size distribution by a rapid continuous flow method. Anal. Chem. *37*, 133–137 (1965)
Casagrande, A.: The structure of clays and its importance in foundation engineering. J. Boston Soc. Civ. Eng. *19*, 168–221 (1932)
Claypool, G. E., Reed, P. R.: Thermal analysis technique for source-rock evaluation: Quantitative estimate of organic richness and effects of lithologic variation. Am. Assoc. Petrol Geol. Bull. *60*, 608–612 (1976)
Cooper, J. E., Bray, E. E.: A postulate role of fatty acids in petroleum formation. Geochim. Cosmochim. Acta *27*, 1113–1127 (1963)
Cordell, R. J.: Depths of oil origin and primary migration: a review and critique. Am. Assoc. Petrol. Geol. Bull. *56*, 2029–2067 (1972)
Cordell, R. J.: Depths of oil origin and primary migration: A geologist's discussion–Reply. Am. Assoc. Petrol. Geol. Bull. *58*, 1857–1861 (1974)
Diamond, S.: Microstructure and pore structure of impact-compacted clays. Clays Clay Miner. *19*, 239–250 (1971)
Dickey, P. A.: Possible primary migration of oil from source rock in oil phase. Am. Assoc. Petrol. Geol. Bull. *59*, 337–345 (1975)
Dunoyer de Segonzac, G.: The transformation of clay minerals during diagenesis and low grade metamorphism: a review. Sedimentology *15*, 281–346 (1970)
Durand, B., Espitalié, J.: Geochemical studies on the organic matter from the Douala Basin (Cameroon)–II. Evolution of kerogen. Geochim. Cosmochim. Acta *40*, 801–808 (1976)
Engelhardt, W. von, Gaida, K. H.: Concentration changes of pore solutions during the compaction of clay sediments. J. Sediment. Petrol. *33*, 919–930 (1963)
Eslinger, E. V., Savin, S. M.: Oxygen isotopic geothermometry of the burial metamorphic rocks of The Precambrian Belt Supergroup, Glacier National Park, Montana. Geol. Soc. Am. Bull. *84*, 2549–2560 (1973)
Evans, C. R., McIvor, D. K., Magara, K.: Organic matter, compaction history and hydrocarbon occurrence, Mackenzie Delta, Canada. Proc. 9th World Petrol. Cong., Tokyo, Panel Dis. *3*, 149–157 (1975)
Forsman, J. P.: Geochemistry of kerogen. In: Organic geochemistry. Breger, I. A. (ed.). New York: Macmillan Co., 1963, Ch. 5, pp. 148–182
Foscolos, A. E., Kodama, H.: Diagenesis of clay minerals from lower cretaceous shales of Northeastern British Columbia. Clays Clay Miner. *22*, 319–335 (1974)

Foscolos, A. E., Powell, T. G., Gunther, P. R.: The use of clay minerals and inorganic and organic geochemical indicators for evaluating the degree of diagenesis and oil generating potential of shales. Geochim. Cosmochim. Acta *40,* 953–966 (1976)

Frank, H. S., Wen, W, Y.: Ion-solvent interaction. Structural aspects of ion-solvent interaction in aqueous solutions: Suggested picture of water structure. Discuss. Faraday Soc. *24,* 133–140 (1957)

Galwey, A. K.: The rate of hydrocarbon desorption from mineral surface and the contribution of heterogeneous catalytic-type processes to petroleum genesis. Geochim. Cosmochim. Acta *36,* 1115–1130 (1972)

Garrels, R. M., Mackenzie, F. T.: Sedimentary cycling in relation to the history of the continents and oceans. In: The nature of the solid earth. Robertson, E. C. (ed.). New York: McGraw-Hill 1972, pp. 93–121

Gipson, M.: Application of the electron microscope to the study of fissility in shales. J. Sediment. Petrol. *39,* 90–105 (1965)

Gipson, M.: A study of the relation of depth, porosity and clay mineral orientation in Pennsylvania shales. J. Sediment. Petrol. *36,* 888–903 (1966)

Greene-Kelly, R., Mackney, D.: Preferred orientation of clay in soils, the effect of drying and wetting. In: Micromorphological techniques and applications. Osmond, D. A., Bullock, P. (eds.). Application. Soil Surv. Techn. Monogr. *2,* 43–52 (1970)

Hanshaw, B. B., Coplen, T. B.: Ultrafiltration by a compacted clay membrane–II. Sodium ion exclusion at various ionic strengths. Geochim. Cosmochim. Acta *37,* 2311–2327 (1973)

Harrison, W. E.: Thermally induced diagenesis of aliphatic fatty acids in Holocene sediment. Am. Assoc. Petrol. Geol. Bull. *60,* 452–455 (1976)

Hedberg, H. D.: Gravitational compaction of clays and shales. Am. J. Sci. *31,* 241–287 (1936)

Hedberg, H. D.: Geological aspects of origin of petroleum. Am. Assoc. Petrol. Geol. Bull. *48,* 1755 –1803 (1964)

Hedberg, H. D.: Significance of high-wax oils with respect to genesis of petroleum. Am. Assoc. Petrol. Geol. Bull. *52,* 736–750 (1968)

Hedberg, H. D.: Relation of methane generation to undercompacted shales, shale diapirs and mud volcanoes. Am. Assoc. Petrol. Geol. Bull. *58,* 661–673 (1974)

Heling, D.: Diagenetic alteration of smectite in argillaceous sediments of Rhinegraben (SW Germany). Sedimentology *21,* 463–472 (1974)

Heller-Kalai, L.: Montmorillonite-alkali halide interaction: A possible mechanism for illitization. Clays Clay Miner. *23,* 462–467 (1975)

Henderson, W., Englinton, G., Simmonds, P., Lovelock, J. E.: Thermal alteration as a contributory process to the genesis of petroleum. Nature (London) *219,* 1012–1016 (1968)

Hiltabrand, R. R., Ferrell, R. E., Billings, G. K.: Experimental diagenesis of Gulf Coast argillaceous sediments. Am. Assoc. Petrol. Geol. Bull. *57,* 338–348 (1973)

Hitchon, B., Billings, G. K., Klovan, J. E.: Geochemistry and origin of formation waters in the Western Canada sedimentary basin. III. Factors controlling chemical composition. Geochim. Cosmochim. Acta *35,* 567–598 (1971)

Hower, J., Eslinger, E. V., Hower, M. E., Perry, E. A.: Mechanism of burial metamorphism of argillaceous sediment. I. Mineralogical and chemical evidence. Geol. Soc. Am. Bull. *87,* 725–737 (1976)

Johns, W. D., Shimoyama, A.: Clay minerals and petroleum forming reactions during burial and diagenesis. Am. Assoc. Petrol. Geol. Bull. *56,* 2160–2167 (1972)

Kharaka, Y. K., Berry, F. A. F.: Simultaneous flow of water and solutes through geological membranes–I. Experimental investigation. Geochim. Cosmochim. Acta *37,* 2577–2603 (1973)

Kharaka, Y. K., Smalley, W. C.: Flow of water and solutes through compacted clays. Am. Assoc. Petrol. Geol. Bull. *60,* 973–980 (1976)

Lambe, T. W.: The structure of compacted clays. Proc. Am. Soc. Civ. Eng. *91.SM4,* 85–106 (1958)

Landes, K. K.: Eometamorphism and oil and gas in time and space. Am. Assoc. Petrol. Geol. Bull. *51,* 828–841 (1967)

Lerman, A.: Migration processes and chemical reactions in interstitial waters. In: The sea: Ideas and observations on progress in the study of the seas. Goldberg, E. D. (ed.). New York: John Wiley & Sons, 1977, pp. 695–738

Louis, M., Tissot, B.: Influence de la température et de la pression sur la formation des hydrocarbons dans les argiles à kérogene. Proc. 7th World Petrol. Cong. Mexico. Barking, England: Elsevier, 1968, vol. 2 pp. 47–60

Low, P. F.: Viscosity of water in clay systems. Clays Clay Miner. *8,* 170–182 (1959)

Magara, K.: Reevaluation of montmorillonite dehydration as cause of abnormal pressure and hydrocarbon migration. Am. Assoc. Petrol. Geol. Bull. *59,* 292–303 (1975)

Magara, K.: Water expulsion from clastic sediments during compaction, directions and volumes. Am. Assoc. Petrol. Geol. Bull. *60,* 543–553 (1976)

Mair, B. J.: Terpenoids, fatty acids and alcohols as source materials for petroleum hydrocarbons. Geochim. Cosmochim. Acta *28,* 1303–1321 (1964)

McAuliffe, C.: Determination of dissolved hydrocarbons in surface brines. Chem. Geol. *4,* 225–234 (1969)

McIver, R. D.: Composition of kerogen–clue to its role in the origin of petroleum. Proc. 7th World Petrol. Cong. Mexico. Barking, England: Elsevier, 1968, vol. 2 pp. 25–36

Meade, R. H.: Removal of water and rearrangement of particles during the compaction of clayey sediments–review. U. S. Geol. Surv. Prof. Paper, *497–B, B*1–*B*23 (1964)

Meade, R. H.: Factors influencing the early stages of compaction of clays and sands–review. J. Sediment. Petrol. *36,* 1085–1101 (1966)

Miller, R. J., Low, P. F.: Threshold gradient for water flow in clay systems. Soil Sci. Soc. Am. Proc. *27,* 605–609 (1963)

Mokady, R. S., Low, P. F.: Simultaneous transport of salt and water through clays. I. Transport mechanisms. Soil Sci. *105,* 112–131 (1968)

Moon, C. F.: The microstructure of clay sediments. Earth Sci. Rev. *8,* 303–321 (1972)

Odom, I. E.: Clay fabric and its relation to structural properties in Mid-Continent Pennsylvanian sediments. J. Sediment. Petrol. *37,* 610–623 (1967)

Oehler, J. H., Aizenshtat, Z., Schopf, J. W.: Thermal alteration of blue-green algae and blue-green algal chlorophyl. Am. Assoc. Petrol. Geol. Bull. *58,* 124–132 (1974)

Perry, E. A., Hower, J.: Burial diagenesis in Gulf Coast pelitic sediments. Clays Clay Miner. *18,* 165–177 (1970)

Perry, E. A., Hower, J.: Late stage dehydration in deeply buried pelitic sediments. Am. Assoc. Petrol. Geol. Bull. *56,* 2013–2021 (1972)

Philippi, G. T.: On the depth, time and mechanism of petroleum generation. Geochim. Cosmochim. Acta *29,* 1021–1049 (1965)

Philippi, G. T.: The deep subsurface temperature controlled origin of the gaseous and gasoline-range hydrocarbons of petroleum. Geochim. Cosmochim. Acta *39,* 1353–1375 (1975)

Philp, R. P., Calvin, M.: Possible origin for insoluble organic (kerogen) debris in sediments from insoluble cell-wall materials of algae and bacteria. Nature (London) *262,* 134–136 (1976)

Pollard, C. O.: Semidisplacive mechanism for diagenetic alteration of montmorillonite layers to illite layers. In: Clay water diagenesis during burial: How mud becomes gneiss. Weaver, C. E., Beck, K. C. (ed.). Geol. Soc. Am. Spec. Paper, No. *134,* 1971, pp. 79–93

Powers, M. C.: Adjustment of clays to chemical change and the concept of equivalent level. Clays Clay Miner. *6,* 327–341 (1959)

Powers, M. C.: Fluids release mechanisms in compacting marine mudrocks and their importance in oil exploration. Am. Assoc. Petrol. Geol. Bull. *51,* 1240–1254 (1967)

Price, L. C.: The solubility of hydrocarbons and petroleum in water as applied to the primary migration of petroleum. Ph. D. Dissert., Univ. Calif., Riverside, 1973

Ross, S.: Adhesion versus cohesion in liquid–liquid and solid–liquid dispersions. J. Colloid. Interface Sci. *42,* 52–61 (1973)

Ross, C. S., Hendricks, S. B.: Minerals of the montmorillonite group. U. S. Geol. Surv. Prov. Paper, pp. 205–208 (1945)

Shimoyama, A., Johns, W. D.: Catalytic conversion of fatty acids to petroleum-like paraffins and their maturation. Nature (London) *232,* 140–144 (1971)

Shimoyama, A., Ponnamperuma, C.: Organic material of recent Chesapeake Bay sediments. Geochem. J. *9,* 85–95 (1975)

Slabaugh, W. H., Stump, A. D.: Surface areas and porosity of marine sediments. J. Geophys. Res. *69*, 4773–4778 (1964)

Sloane, R. L., Kell, T. R.: Fabric of mechanically compacted kaolin. Clays Clay Miner. *14*, 289–296 (1966)

Stevens, N. P., Bray, E. E., Evans, E. D.: Hydrocarbons in sediments of Gulf of Mexico. Am. Assoc. Petrol. Geol. Bull. *40*, 975–983 (1956)

Tissot, B., Durand, B. Espitalié, J., Combaz, A.: Influence of nature and diagenesis of organic matter in formation of petroleum. Am. Assoc. Petrol. Geol. Bull. *58*, 499–506 (1974)

Towe, K. M.: Clay mineral diagenesis as a possible source of silica cement in sedimentary rocks. J. Sediment. Petrol. *32*, 26–28 (1962)

Velde, B. D., Byström-Brusewitz, A. M.: The transformation of natural clay minerals at elevated pressures and temperatures. Geol. Foeren. Stockholm Foerh. *94*, 449–458 (1972)

Weaver, C. E.: Potassium, illite and the ocean. Geochim. Cosmochim. Acta *31*, 2181–2196 (1967)

Weaver, C. E., Beck, K. C.: Clay-water diagenesis during burial: How mud becomes gneiss. Geol. Soc. Am. Spec. Paper, *134*, 1–79 (1971)

Welte, D.: Recent advances in organic geochemistry of humic substances and kerogen, a review. In: Adv. Geochem. Proc. 6th. Int. Meeting Organic Geochem., Rueil-Malmaison, 1973, pp. 3–13

White, W. A.: Colloid phenomena in the sedimentation of argillaceous rocks. J. Sediment. Petrol. *31*, 560–570 (1961)

Willey, L. M., Kharaka, Y. K., Presser, T. S., Rapp, J. B., Barnes, I.: Short chain aliphatic acid anions in oil field waters and their contribution to the measured alkalinity. Geochim. Cosmochim. Acta *39*, 1707–1711 (1975)

Yariv, S.: Organophilic pores as proposed primary migration media for hydrocarbons in argillaceous rocks. Clay Sci. *5*, 19–29 (1976)

Author Index

Page numbers in *italics* refer to the references.

Mineral Index

(including group names)

Subject Index

Handbook of Geochemistry

Executive Editor: K.H. Wedepohl
Editorial Board: C.W. Correns, D.M. Shaw, K.K. Turekian, J. Zemann

Volume I
1969. 60 figures. XV, 442 pages
ISBN 3-540-06577-6

Volume II Part 1
1969. 172 figures. X, 586 pages
ISBN 3-540-06578-4

Volume II Part 2
1970. 105 figures. IV, 667 pages
ISBN 3-540-04840-5

Volume II Part 3
1972. 142 figures. IV, 845 pages
ISBN 3-540-05125-2

Volume II Part 4
1974. 103 figures. VI, 898 pages
ISBN 3-540-06879-1

Volume II Part 5
1978. Approx. 200 figures.
Approx. 1500 pages
ISBN 3-540-09022-3

The *Handbook of Geochemistry* offers a critical selection of important facts about the distribution of the chemical elements and their isotopes in the earth and the cosmos. Specialist authors have made this selection from the flood of information which resulted from improved investigative methods.

As this book clearly shows, geochemistry and cosmochemistry are intimately linked with a number of other disciplines.

The loose-leaf system enables the contributions to be published in random order, regardless of their position in the book, and revisions to be made as desired.

"Both editors and authors are to be congratulated on collecting and organizing so much relevant geochemical data from world sources into a single comprehensive text. With the bulk of material still to appear, a final judgment on the handbook is impossible, but if the present standard is maintained and revisions are incorporated with the passage of time, then an authoritative source book of geochemical information and reference will be available for many years to come."
Nature

"...Enough of the work is now available to permit some real assessment of its ultimate worth and, although the treatment of individual elements naturally varies both in thoroughness and in emphasis, it is clear that on completion we shall have a standard reference work of inestimable value. The standard format of the layout for each chapter is logical and fits the treatment of most elements well: it is easy to turn quickly to the right place to find any particular piece of information ..."
Mineralogical Magazine

Springer-Verlag
Berlin
Heidelberg
New York

G. Millot

Geology of Clays

Weathering, Sedimentology, Geochemistry,

Translators: W.R. Farrand, H. Paquet

1970. 85 figures, 2 plates in color, 15 tables.
XVI, 429 pages
Co-Production with Masson et Cie, Paris
ISBN 3-540-05203-8

"The book begins with a brief review of the composition and structure of the various clay minerals; this review is followed by a statement of the values and dangers of nomenclature which emphasizes that the various clay minerals generally do not in reality fit into narrow compartments. The author emphasizes this point which is often missed by enthusiasts for a rigid classification and nomenclature of the clay minerals.... is essentially a lucid consideration of the origin of clay minerals in weathering progresses and their distribution in sediments deposited in various environments.... presents the best up-to-date summary in any language or the present concepts of weathering processes and of the chemistry of the solution and precipitation of silica and silicates under ambient conditions. ...Everyone interested in the distribution and origin of the clay minerals will find this an essential volume. The author provides excellent summaries at the end of each chapter and often within the chapters, following the consideration of specific matters...."

Ceramic Abstracts

"...Very valuable throughout the book are the author's numerous concluding parts summarizing opinions and presenting his general view on the topics just discussed. With its fundamental basis on geology the book supplements in an excellent way the normal textbooks on clay mineralogy, and it is a real treasure chest for all workers interested in clays, geologists as well as pedologists.... "

Chemical Geology

Soil Components

Volume 1: **Organic Components**
Editor: J.E. Gieseking
1975. 188 figures. IX, 534 pages
ISBN 3-540-06861-9

"This book is a comprehensive discussion and review of soil organic matter. The topics dealt with include humic substances, saccharides, lipids, organic nitrogen, phosphorus and sulfur, micromorphology, and the general nature of humus in different soil groups. Each chapter is authored by one or several international scholars who are recognized for work in a specialized area of soil organic matter research. John Gieseking has provided the editorial effort to make the book a coordinated presentation. The individual chapters, some originally in a foreign language, are all similarly organized and written in a clear, easily understood English style. Each chapter includes many good graphical illustrations and an appropriate amount of data is presented in well-organized tables to supplement the text of the chapter.... is certainly not out of date, however, and will be found highly useful to serious students of soil organic matter."

Soil Science Society of American Proceedings

Volume 2: **Inorganic Components**
Editor: J.E. Gieseking
1975. 212 figures. XI, 684 pages
ISBN 3-540-06862-7

"... the discussions in this book are reviews in the best sense of this word; that is, they are critical examinations and scholarly evaluations of the subjects covered.... This book is indeed a comprehensive reference text which will prove valuable to soil chemists, agronomists, mineralogists, geologists, and ceramists interested in inorganic soil components."

Geoderma

Springer-Verlag
Berlin
Heidelberg
New York